Principles of Veterinary Animal Physiology

This textbook for advanced graduate and postgraduate veterinary students, introduces animal behaviour, offering insights into its origins, cognitive aspects, communication, environmental influences, biological mechanisms, complex behaviours, adaptive strategies, and practical applications. The initial chapters present fundamental principles underpinning animal behaviour, elucidating the roles of evolution, genetics, and ecology. Subsequent chapters introduce the role of natural selection, habitat selection, and pheromones, alongside exploring conflicts, predator-prey dynamics, and the impact of domestication on behaviour. The book further delves into topics such as habitat selection, foraging strategies, predator-prey dynamics, and the effects of domestication on behaviour. It unravels the mysteries of animal defences, altruism, social dominance, territoriality, and the finely tuned art of food and habitat selection. Additionally, it covers the biological mechanisms governing behaviour, unearthing the roles played by neuroendocrinology, biological clocks, and genetics. Towards the end, the textbook examines the practical relevance of behavioural insights in veterinary science.

Key Features:

- Offers a comprehensive exploration of animal behaviour, encompassing a wide range of topics, from evolutionary principles to intricate behavioural patterns.
- Provides understanding of the origins of animal behaviour, including the roles of evolution, genetics, and ecology.
- Highlights the practical applications of behavioural insights in veterinary science and related fields.
- Delves into specialized areas of animal behaviour, such as innate behaviours, animal memory, pheromones, and cooperation.
- Explores how animals adapt to their environments, covering topics like foraging behaviour, predator detection, and self-defensive behaviours.
- Addresses the impact of environmental factors, hormones, and drugs on behaviour, offering a holistic view of animal responses to their surroundings.

Principles of Veterinary Animal Physiology

Edited by
Tanmoy Rana

CRC Press
Taylor & Francis Group
Boca Raton London New York

CRC Press is an imprint of the
Taylor & Francis Group, an **informa** business

Designed cover image: Shutterstock

First edition published 2026
by CRC Press
2385 NW Executive Center Drive, Suite 320, Boca Raton FL 33431

and by CRC Press
4 Park Square, Milton Park, Abingdon, Oxon, OX14 4RN

CRC Press is an imprint of Taylor & Francis Group, LLC

ISBN: 9781032731278 (hbk)
ISBN: 9781032731285 (pbk)
ISBN: 9781003426851 (ebk)

DOI: 10.1201/9781003426851

Typeset in Times
by Deanta Global Publishing Services, Chennai, India

Contents

Preface

This textbook depicts principles of animal physiology with morphological information in relation to cells, tissues, and organs involved in the physiological system. It describes the physiological purview of all systems for the better understanding of the students and is a valuable learning resource with an interrelationship with veterinary medicine. The book emphasizes cellular physiology to body system function to homeostasis and immune function. It also contains an introductory chapter presenting an overview of animal physiology in this regard and elaborately describes cellular interaction with various functions of tissues and organs. The physiological processes with clinical correlations are nicely represented in the book. The book is very useful for the veterinary students, researchers, academicians and veterinary nurses for the better understanding about the physiology in the mechanism of veterinary disease, clinical diagnosis and treatment of animals. The contributors read many textbooks, reference books, and websites to compile knowledge incorporated in this book. To the best of our knowledge, the book will help readers to assimilate all of the information. Suggestions of ways to improve this textbook in subsequent editions are always welcome.

Tanmoy Rana
Department of Veterinary Clinical Complex
West Bengal University of Animal & Fishery Sciences
Kolkata, India

Acknowledgement

I express my gratitude to our Hon'ble Vice Chancellor, West Bengal University of Animal & Fishery Sciences, Kolkata, India, for providing me the opportunity to edit the book. Immense gratitude is due to the contributing authors who completed their chapter within the scheduled time. I also convey my gratitude to all physiology experts across the world for acquiring knowledge in this regard. I would like to express my sincere regards to all professionals of the Taylor and Francis group who actively provide me the opportunity to edit the book. I would also like to thank my family for providing great support and time to finalize the book with great love and support that sustains my existence.

Tanmoy Rana
Department of Veterinary Clinical Complex
West Bengal University of Animal & Fishery Sciences
Kolkata, India

About the Editor

Dr. Tanmoy Rana is an accomplished veterinary professional and academician currently serving as Assistant Professor at the esteemed West Bengal University of Animal & Fishery Sciences (WBUAFS) in Kolkata, India. Previously, he has obtained his Bachelor's degree in Veterinary Science and Animal Husbandry (BV.Sc & A.H.) and a Master's degree (M.V.Sc.) in Veterinary Clinical Sciences from WBUAFS. He has pursued a Ph.D. from the University of Calcutta, focusing on veterinary clinical studies. Before joining academia, Dr. Rana contributed to the welfare of animals and communities as a dedicated Veterinary Officer in the ARD department of the Government of West Bengal, India. In addition to his research endeavours, Dr. Rana shares his knowledge through teaching. He has conducted numerous lectures on arsenicosis at the Indian Veterinary Research Institute's Eastern Regional Station, making a significant impact on future veterinarians. His dedication to science and academia has earned him numerous awards, including merit fellowships for academic excellence and national scholarships during his higher secondary and secondary education. He has published several research articles in journals of high repute and has contributed several chapters and edited numerous books. He is editorial board member of several international journals and organized international and national conferences.

1 General Introduction to Animal Physiology

B. Deepika[1] *and Tanmoy Rana*[2]

[1]Department of Veterinary Physiology and Biochemistry, Veterinary College and Research Institute, Theni, Tamil Nadu Veterinary and Animal Sciences University, Tamil Nadu, India
[2]Department of Veterinary Clinical Complex, West Bengal
University of Animal & Fishery Sciences, Kolkata, India

1.1 INTRODUCTION

Introduction to Animal Physiology provides students with an in-depth knowledge and understanding of the fundamental functions of the various systems of the body. It illustrates the physiological processes across the animal kingdom. The book covers a wide range of topics, including cell structure and functions and the physiology of all vital systems of the body. The way that the various parts of living things adapt to keep their internal environments consistent is always fascinating and unexpected. This process is called homeostasis and is the fundamental concept of physiology. Physiology is the branch of biology that examines the different roles that living things play in the world. This field of study encompasses a wide range of activities, from cellular processes to the interactions between organ systems that support the efficient operation of extremely intricate biological processes. Similar to cell biology, within physiology, other subfields are more well-recognized such as biochemistry, biophysics, functional biology, and endocrinology. Additionally, scientists watch and study how certain bodily systems – such as the respiratory, circulatory, and neurological systems – cooperate and function separately to sustain life. On the other side, ecological physiology shows how animals evolved or developed particular biological processes to deal with a given environment. One example is the characteristic of dark skin, which shields those who live in tropical areas from the sun's harmful rays. Cell biology, often known as cellular physiology, is the study of the composition and operations of individual cells. Since physiological processes are regulated by signaling molecules and reactions cascade down at the cellular, biochemical, and molecular levels, it is impossible to study animal physiology in isolation. The organism's physiology is influenced by behavior as well. The study of the functional similarities and differences across living organisms is known as physiology. In a way, studying physiology is studying life.

1.2 PRINCIPLES OF PHYSIOLOGY

1.2.1 Homeostasis

Homeostasis is the basic principle of animal physiology. Physiologists use the term homeostasis to mean the maintenance of nearly constant conditions in the internal environment. To maintain homeostasis, the body must continually monitor its internal conditions. From body temperature to blood pressure to levels of certain nutrients, every physiological state has a certain threshold. A set point is the physiological value around which the normal range fluctuates. A normal range is a restricted set of values that is optimally healthy and stable. Although not all species regulate all physiological variables to the same level, all homeostatic processes maintain the internal environment. Animals may conform to some physiological variables, with the internal variable being the same as the exterior variable. The important role of homeostasis in animals is the regulation of extracellular environment features that differ from the exterior environment to generate an optimal internal environment in which the cells function.

The preservation of equilibrium around a certain value of an element of the body or its cells is referred to as a set point, and it is the aim of homeostasis. The body's systems often try to return to the established point, even though there are occasional variations from it. A receptor detects a change in the internal or external environment, which is referred to as a stimulus. The system's response is to modify its operations to bring the value back closer to the set point. For example, if the body temperature rises too high, modifications are made to cool the body. If blood glucose levels rise after a meal, adjustments are made to lower them and deliver the nutrition into the tissues that require it, or to store it for future use.

To maintain a stable internal environment for the body and cells, an animal must adapt when its surroundings change. A feedback system includes a receptor that detects

DOI: 10.1201/9781003426851-1

changes in the surroundings. The receptor picks up the stimuli, which could be temperature, glucose, or calcium levels. The brain, which is frequently the control center, receives information from the receptor and uses it to send appropriate signals to an effector organ. The effector organ can then produce an appropriate change, either up or down, based on the information the receptor sends.

1.2.2 Feedback Loops

Animals must: (1) detect external conditions; (2) initiate compensating reactions as needed; and (3) keep vital parts insulated against unfavorable changes to preserve homeostasis. To maintain homeostasis, animals typically use a reflex control mechanism.

A stimulus is provided by a change in the internal or external environment. The stimulus is subsequently followed by a response. Animals use antagonistic controls to fine-tune physiological reactions.

Animals maintain their body temperature by controlling the generation and dissipation of heat. Numerous antagonistic controls are mediated by hormones. The antagonistic regulators of glucose levels are glucagon and insulin.

1.2.2.1 Negative Feedback Loops

A negative feedback loop occurs when the stimulus's strength is decreased by the reaction sending a signal back to it. For instance, the food that enters the stomach causes it to expand when animals eat. A negative feedback loop that lowers appetite is triggered by the shift in stomach volume and early-digestion products. This loop operates through the brain. A set point is a desired physiological state that is protected in many physiological systems via feedback loops.

The body temperature of animals has a fixed point that is roughly 37°C. Animals' bodies sweat to regulate body temperature when it rises, and they shiver to return to their set point when their body temperature drops. While the average animal body temperature is approximately 37°C, each animal's precise body temperature set point differs and fluctuates during the day.

A negative feedback system consists of three basic components: a sensor, also called a receptor, a component of a feedback system that monitors a physiological value and reports it to the control center, which is the component in a feedback system that compares the value to the normal range and activates an effector if the value deviates too much from the set point. Negative feedback is a mechanism that reverses a deviation from the set point, therefore maintaining body parameters within their normal range. The maintenance of homeostasis by negative feedback occurs throughout the body at all times, so comprehension of negative feedback is essential to understanding human physiology. In a feedback system, an effector is the part that initiates a change to reverse the circumstance and return the value to the usual range.

A stimulus must drive a physiological parameter (i.e., go beyond homeostasis) beyond its typical range to initiate the system. A particular sensor "hears" this stimulus. For instance, in the regulation of blood glucose, the pancreas's unique endocrine cells identify excess glucose, or the stimulus, in the blood. In response to the elevated blood glucose level, these pancreatic beta cells release the hormone insulin into the bloodstream. Insulin instructs the liver, adipocytes, and skeletal muscle fibers to absorb extra glucose and eliminate it from the bloodstream. Insulin release ceases when pancreatic alpha cells detect a decline in glucose concentration in the bloodstream, which is a negative feedback loop. By doing this, blood sugar levels are kept from falling below the usual range.

Similar feedback mechanisms govern temperature in humans, whereby heat gain or loss is encouraged. The brain's "heat-loss center" is a group of brain cells that is stimulated when the temperature regulation center receives information from the sensors showing that the body temperature is higher than normal. Three main outcomes result from this stimulation:

- The skin's blood vessels start to widen, allowing more blood to come from the body's core to the skin's surface, radiating heat into the surrounding area.
- Sweat glands are activated to produce more sweat when blood flow to the skin increases. Heat is released into the surrounding air when sweat evaporates from the skin's surface. An animal's depth of respiration increases, and it may breathe through its mouth rather than its nasal passages. As a result, the lungs lose more heat.

On the other hand, blood flow to the skin is decreased when the brain's heat-gain center is activated by cold, and blood returning from the limbs is redirected into a system of deep veins. This configuration limits heat loss and traps heat closer to the body's center. Severe heat loss causes the brain to send more random messages to the skeletal muscles, which makes them contract and shiver. Shaking releases heat through the contraction of muscles while depleting ATP. Thyroid hormone is released by the thyroid gland in response to a brain signal. This hormone raises heat production and metabolic activity in all of the body's cells. Additionally, the brain triggers the adrenal glands to release the hormone known as adrenaline, or epinephrine, which breaks down glycogen into glucose, which the body may utilize as fuel. The conversion of glucose from glycogen also raises metabolism and produces more heat.

The body's water content is essential for healthy operation. Without conscious effort on the part of the individual, the body maintains extremely strict control over hydration levels.

1.2.2.2 Positive Feedback Loops

Positive feedback loops regulate a few physiological systems. Positive feedback loops increase the number of changes in the regulated variable as opposed to negative feedback, which minimizes those changes. For instance, the stomach's muscles are generally controlled to contract and relax predictably to gently combine food. On the other hand, strong contractions caused by a positive feedback loop are activated upon detection of a toxin, pushing food back up the esophageal tube and causing vomiting.

Positive feedback loop pathways start gently but quickly gain intensity. To prevent an action from getting out of hand, a positive feedback loop needs a signal that enables the animal to halt the process when it's time.

Rather than undoing a physiological state change, positive feedback amplifies it. More change and a further departure from the typical range are experienced by the system when there is a divergence from it. The body only responds positively when there is a clear end goal in sight. Positive feedback loops that are typical but only active when necessary include parturition and the body's reaction to blood loss.

A circumstance when maintaining the current physical state is not desired is the birth of a child at full term. To evacuate the baby at the end of the pregnancy, the mother's body must undergo enormous changes. Furthermore, once parturition has started, it must proceed quickly to its end to save both the mother's and the child's lives. Positive feedback systems are responsible for the intense physical exertion involved in labor and delivery.

The fetus is forced towards the cervix, the lowest point of the uterus, by the stimulus of the first labor contractions. Stretch-sensitive nerve cells (the sensors) in the cervix track the amount of stretching. The pituitary gland at the base of the brain releases the hormone oxytocin into the bloodstream as a result of the messages these nerve cells transmit to the brain. Stronger contractions of the uterine smooth muscles, or effectors, triggered by oxytocin push the baby into the delivery canal. The cervix stretches considerably more as a result. Only when the baby is born does the cycle of stretching, oxytocin release, and progressively stronger contractions come to an end. At this moment, the cervix stops expanding, preventing the release of oxytocin.

Reversing severe bodily harm is the subject of a second illustration of positive feedback. Excessive bleeding is the most immediate danger after a penetrating wound. Reduced blood pressure and blood perfusion – the amount of blood that reaches the brain and other important organs – are the results of having less blood moving. Essential organs will shut down, and the person will die if perfusion is drastically decreased. In response to this potential disaster, the body releases chemicals into the damaged blood vessel wall, which triggers the onset of the clotting process. More clotting substances are released as the clotting process progresses. This quickens the clotting and closing off of the injured area.

Because clotting proteins are strictly regulated in their availability, clotting is confined to a small region. This sequence of events is adaptive and life-saving.

1.2.2.3 Thermoregulation

Animals can be categorized into two groups: those whose body temperature remains constant despite changes in the surrounding temperature and those whose body temperature is environment-dependent, meaning it fluctuates with changes in the surrounding temperature. Ectotherms are animals without an internal temperature regulation system. These species' body temperatures are typically comparable to those of their surroundings, while some may take actions to maintain their body temperatures somewhat above or below that of their surroundings. This can involve taking a daytime nap in the sun on a chilly day or digging a hole beneath the surface on a hot day. Although the phrase "cold-blooded" may not apply to an animal in the desert with a very warm body temperature, the ectotherms have been described as such.

An endotherm is an animal that maintains a fixed body temperature despite changes in its surroundings. Because these creatures produce internal heat that keeps their cellular processes running smoothly even in cold environments, they can maintain an activity level that an ectothermic species cannot.

In many different ways, animals can dissipate or store heat. Endothermic animals have insulation of some kind. They might be bald, plump, or have feathers. Thick-furred or feathered animals produce an air cushion between their skin and internal organs. Even though they swim and live in a subfreezing climate, polar bears and seals can maintain a steady, warm body temperature. When it rolls up to sleep in chilly weather, the Arctic fox, for instance, uses its fluffy tail as additional insulation. By shivering, or convulsively contracting their muscles, mammals can generate more heat within their bodies. A person's arrector pili muscles can also contract when they get chilly, which causes individual hairs to stand up. The insulating power of the hair is increased as a result. The reaction that is still present in humans results in "goosebumps" rather than the desired outcome on our comparatively hairless bodies. Animals also insulate with layers of fat. The capacity to retain heat will be hampered by a substantial reduction in body fat.

Circulatory systems are used by ectotherms and endotherms to help regulate body temperature. Vasodilation, or the widening of arteries to the skin due to the relaxation of their smooth muscles, increases the amount of blood and heat that reaches the skin's surface, allowing the body to cool through radiation and evaporative heat loss. By driving blood towards the core and important organs, vasoconstriction – the narrowing of blood vessels to the skin by the contraction of their smooth muscles – reduces blood flow in peripheral blood vessels and preserves heat. Certain animals' circulatory systems have evolved to allow them to move heat from arteries to veins that are next to one another, warming blood that is returning to the heart. This keeps the heart from being cooled by the icy venous blood; it's known as a countercurrent heat exchange. This countercurrent adaptation is shared by bony fish, hummingbirds, sharks, and dolphins.

Certain ectothermic animals employ behavioral adjustments to control their body temperature. To avoid overheating during the warmest portion of the day in the desert, they merely look for cooler spots. On a chilly desert night, the same animals might scale rocks in the evening to gain some heat before heading into their burrows.

The neurological system regulates body temperature. In the sophisticated animal brain, the hypothalamus is the primary location for temperature regulation functions. The hypothalamus regulates body temperature through reflexes that result in shivering or sweating, vasodilation or constriction, and other physiological responses. The hypothalamus, which regulates the sympathetic nervous system, controls the reactions that cause temperature variations and bring the body back to the set point. In certain cases, the set point might be changed. Pyrogens are substances that are created during an infection and travel to the hypothalamus, where they raise the temperature. This permits the body's temperature to rise to a new homeostatic equilibrium point, resulting in what is usually known as fever. The increase in body heat reduces the body's optimal conditions for bacterial development and enhances cell activity, allowing it to better fight the illness.

1.3 HISTORY AND BACKGROUND

The history of physiology can be traced back to the ancient civilizations of India and Egypt, where it was first studied by Hippocrates (referred to as the "father of medicine") in the year 420 BC.

Claudius Galenus (130–200 AD) was the first to conduct experimentation on the systems of the body. Hence, he is referred to as the founder of experimental physiology.

It was Jean Fernel (1497–1558), a French physician, who first introduced the term "physiology," from an Ancient Greek word meaning "study of nature, origins."

Harvey was the first to explain how blood moves from the heart to the brain and other body systems, as well as the circulatory system. An Anatomical Dissertation Upon the Movement of the Heart and Blood in Animals, written by William Harvey, was published in 1628 and marked a significant advance in the history of physiological science.

The arrival of Matthias Schleiden and Theodor Schwann's cell theory in 1838, which proposed that the body is composed of minuscule individual cells, marked a paradigm shift in thinking.

From here, the study of physiology developed and advanced swiftly:

Joseph Lister (1858) discovered life-saving antiseptics while first researching coagulation and inflammation following an injury.

Dogs' conditioned physiological responses were studied by Ivan Pavlov in 1891.

The discovery of how capillaries regulate blood flow earned August Krogh the Nobel Prize in 1920.

The ionic mechanism responsible for nerve impulse transmission was identified in 1952 by Andrew Huxley and Alan Hodgkin.

Physiologists did not distinguish between animal and medical physiology until the twentieth century. The majority of physiological tests on animals were conducted with the purpose of better understanding the human body in both health and disease. However, throughout the twentieth century, biologists grew interested in applying newly developed physiological knowledge to animals living in a variety of settings, as well as in attempting to understand the nature of physiological diversity.

Per Scholander (1905–1980) was a pioneering and prominent comparative physiologist. Scholander investigated a wide range of physiological reactions, including the mechanisms involved in diving vertebrates, warm-blooded animals' adaptations to cold surroundings, and how fish fill their swim bladders (air-filled organs used for buoyancy). In the study program detailed in the opening essay of this chapter, Scholander also organized the influential Alpha Helix excursions.

C. Ladd Prosser's (1907–2002) accomplishments include the discovery of so-called central pattern generators. Many rhythmic behaviors, including breathing and walking, are coordinated by these groupings of neurons. Prosser also established the connection between conduction speed and muscle diameter. As part of the Manhattan Project, he investigated the effects of radiation on animal life during World War II.

Knut Schmidt-Nielsen (1915–2007) dedicated his professional life to learning how animals survive in challenging and peculiar habitats. He demonstrated that the camel's nose has a countercurrent exchanger that enables it to recover moisture from exhaled air, leading to an over 60% reduction in water loss compared to other mammals in his seminal early study on the adaptations of the camel to desert living.

The study of an organism's interactions with its surroundings is known as ecological physiology, and it was founded by George Bartholomew (1923–2006). To evaluate the evolutionary significance of modifications or adaptations in animals to their environments, Bartholomew merged the study of animal behavior, ecology, and physiology. He stressed the significance of variety in physiology and recognized the individual as the primary unit of natural selection.

The topic of adaptational biochemistry was founded by George Somero (1941–) and Peter Hochachka (1937–2002). They have expanded our understanding of how animals adapt to harsh environments at the subcellular level by applying the ideas and methods of biochemistry to questions of comparative physiology. This has given rise to new insights into the biochemical mechanisms that enable animals to live in a variety of habitats, including the deep sea, the Antarctic oceans, high mountain peaks, and tropical rain forests.

The importance of physiology was acknowledged by the 1901 Nobel Prize in Physiology or Medicine. Emil von Behring's ground-breaking research on the treatment of tetanus and diphtheria earned him the first Nobel Prize in Physiology or Medicine. He injected healthy animals with weakened strains of the germs that caused these ailments, and the animals' immune systems rendered the bacterial poisons inert. He then administered this blood serum to the sick animals. Treatment of the infection's symptoms prevented the sick animals from dying of their illnesses. In the end, this was applied to humans, saving many lives.

1.4 DEFINITIONS AND DIVISIONS OF PHYSIOLOGY

Physiology is defined as the study of the normal functioning of organs and organ systems, i.e., the study of homeostasis and biological functioning of an organism. Human and animal physiology are part of the same discipline and may be covered as one coherent whole. Sometimes the two disciplines are separated, though. Within these sub-disciplines, the field can be further divided into other branches, including pathophysiology, which examines the physiology of disease, and system physiology, which evaluates body systems such as the cardiovascular system. Each branch of physiology itself involves lots of research areas.

1.4.1 Cell and Molecular Physiology

Cell and molecular physiology deals with the physiological mechanisms that occur at the cellular level and these have important consequences for the higher levels of organization. Cell and molecular physiologists study molecular genetics, signal transduction, metabolic biochemistry, and membrane biophysics.

Physiology is the study of how animals work and how they solve the challenges of surviving in the natural environment. Though we often think of animal physiology as a study of organs, systems, and whole animals, it is important to recognize that the reasons for many of these features can be traced back to underlying rules of chemistry, biochemistry, and cell biology. Many of the properties of organs and systems emerge from the regulation of cellular processes, such as energy production, membrane transport, cellular anatomy, and gene expression. While the physiology of an animal is much more than the sum of these molecular and cellular processes, an awareness of how cells work is vital to understanding complex physiological processes

1.4.2 Systemic Physiology

Systemic Physiology includes each organ system of the body and its functions. It is the detailed study of individual organ systems like the Respiratory System, Nervous System, Renal System, Digestive System, and Reproductive System. Physiologists will master a particular system of their interest; for example Neurosciologists will study the nervous system. The major systems of animal physiology include

1.4.2.1 The Integumentary System

The skin and its derivatives are parts of the integumentary system. These consist of many types of glands, hair, and nails. The system is responsible for protecting the body, controlling body temperature, eliminating waste, synthesizing vitamin D3 with solar radiation, and responding to different stimuli that are interpreted as pressure, pain, and temperature.

1.4.2.2 The Skeletal System

The term skeleton is derived from the Greek word skeleton, which means "dried up."Bones, joints, cartilage, and ligaments are all part of the skeletal system. The joints allow the body to move and provide flexibility. However, from a structural standpoint, the human skeletal system is made up of two forms of supportive connective tissue: bone and cartilage. Skeletal system functions include the following:

1. It serves as the internal structure that supports and anchors all soft organs.
2. Bones protect soft body organs.
3. Movement: Skeletal muscles associated with the skeletal system act as levers to move the body and its parts.
4. Fat storage: Fat is stored in the cavities of bones. Bone is a mineral storage in and of itself. Calcium and phosphorus are two of the most important.
5. Blood cell production happens in the marrow cavities of specific bones

1.4.2.3 The Muscular System

Muscle tissue encompasses all contractile tissues in the body, including skeletal, cardiac, and smooth muscle. The skeletal muscle system, on the other hand, refers to the skeletal muscle tissue and connective tissues that comprise particular muscle organs such as the biceps brachii muscle. Cardiac muscle tissue is found in the heart and is thus a part of the cardiovascular system. The intestines' smooth muscle tissue is part of the digestive system, while the urine bladder's smooth muscle tissue is part of the urinary system, and so on.

1.4.2.4 The Nervous System

Every bodily system depends on the others to function. All of them depend on one another and function as a single entity to enable the body to function normally. The nervous system and the endocrine system are the two main communication networks that allow the body's billions of cells to be controlled. Though they accomplish it in different ways, both systems transfer information from one area of the body to another. The nervous system uses nerve impulses to quickly transfer information from one part of the body to another. The endocrine system uses substances produced into bloodstream by ductless glands and circulated from glands to other regions of the body to relay information more slowly. The main coordinating body is the nervous system.

The nervous system's function is to react to these internal and external changes, also referred to as stimuli, to help the body adjust to new circumstances. Conditions both within and outside the body are always changing. The nervous system maintains the equilibrium between an individual and their surroundings as well as the internal harmony of the person through the nerve impulses it sends to different organs. The brain and spinal cord serve as switching centers, while the nerve trunks serve as cables that transmit messages to and from these centers. This has led to comparisons of the nervous system to a telephone exchange.

1.4.2.5 The Endocrine System

The two primary coordinating and governing systems in the body are the neurological system and the endocrine system. These two systems differ in several ways. For instance, the endocrine system has more extensive, slower-acting, and longer-lasting effects than the neurological system, which is primarily activated by chemical and electromagnetic inputs. Additionally, the endocrine system affects more general functions like growth, metabolism, and reproduction. Despite these distinctions, there exists a close relationship between the two systems.

1.4.2.6 Cardiovascular System

The circulatory system is the body's transport system that transports food, oxygen, water, and other needs to tissue cells while also transporting waste products away. It is divided into three sections: 1. The blood, which is the fluid that transports materials to and from the tissue. 2. The heart, which is the driving force behind blood circulation. 3. The blood vessels, which carry blood to and through tissues before returning to the heart.

1.4.2.7 Respiratory System

Oxygen is a basic necessity for all bodily cell functions and growth. It is also required for the body to obtain energy from meals. The respiratory system's primary functions are to remove carbon dioxide, a gaseous waste product, from each tissue cell and to deliver oxygen to each one. The inhalation and exhalation of air is referred to as breathing or ventilation. The mixture of gases that makes up air includes nitrogen, carbon dioxide, oxygen, and other gases. The pressure of these gases changes with altitude above sea level. The first phase, known as external expiration, only occurs in the lungs, where blood is filled with oxygen from the surrounding air and is subsequently evacuated with carbon dioxide. Gas exchanges occur between the blood and the body's cells during the second phase, known as internal respiration. Oxygen leaves the blood and enters the cells at the same time that carbon dioxide exits the cells and enters the blood. Air enters the lungs through the complex network of passages and compartments that make up the respiratory system. The pharynx, which is shared by the respiratory and digestive systems, the voice box, or larynx, the windpipe, or trachea, the nasal cavities, and the lungs themselves, with their conducting tubes and air sacs, are examples of these areas. You could see the entire system as an airway connecting the blood and the atmosphere.

1.4.2.8 Digestive System

Every cell in the body requires a steady supply of nutrients to provide energy and building blocks for the production of bodily components. Food, as we consume it, is too big to penetrate the cells. It must first be broken down into small enough particles to pass past the cell membrane. This is referred to as digestion. Food must be delivered to the cells in every region of the body via circulation after digestion. Absorption is the process by which food enters the circulation. The digestive system's two primary activities are digestion and absorption.

1.4.2.9 The Urinary System

The body's excretory system, commonly known as the urinary system, is responsible for removing waste materials from the blood and expelling them from the body. The two kidneys, which remove waste from the blood, balance bodily fluids, and produce urine, make up the urinary system. A tube called the ureter transports urine from the kidneys to the bladder. The urinary bladder: the two ureters deliver urine to this reservoir, which it then stores. The tube known as the urethra transports pee from the bladder to the exterior of the body for excretion.

1.4.2.10 The Reproductive System

Replication is the reproductive system's function. Sexual reproduction creates new individuals and allows genetic features to be handed down from both parents to their offspring. Producing and delivering sperm to the female reproductive system is the male's reproductive responsibility. However, a female's duty in reproduction is to create an ova and carry the developing embryo. The growth and operation of the reproductive organ as well as sexual behaviour and urges are significantly influenced by sex hormones.

1.4.3 Clinical Physiology

Clinical physiology is a discipline of physiology that is concerned with functional aspects of a disease's pathogenesis. Clinical physiology is a diagnostic study where samples are collected from the animals and interpreted. Clinical physiology refers to the in vitro study of disease and accessing the functional changes of an organ and an organ system.

1.4.4 Environmental Physiology

Environmental physiology is a branch of physiology about the relationship between an organism and its environment. It emphasizes the physiological adaptions of an animal to its environment. Several environmental factors including temperature, humidity, altitude, weather, rainfall, radiation, etc. will affect an animal's physiological processes. Environmental pollution and global warming often affect the homeostasis of animals. The study of environmental physiology is inevitable in this modern era facing a lot of environmental challenges.

1.4.5 Comparative Physiology

Comparative Physiology deals with the comparison of physiological processes of different animals of the same species and also between different species. It is the *science* concerned with the differences in the *vital* physiological *processes* in different species of *organisms*, particularly relating to the *adaptation* to the *specific* needs of the species, to illuminating the evolutionary relationships among different species, or to establishing other interspecific generalizations and relationships.

Another area of physiology that sheds light on the functional mechanisms of many living things is comparative physiology. The discipline itself encompasses a wide range of scientific disciplines, including environmental studies, archaeology, and evolution. Therefore, comparative physiology emerged as a result of a necessity and a curiosity. Research on animals that resembled people in many ways was very desirable since doctors needed to find new ways to treat human patients. In the end, the usefulness of comparisons developed into a general scientific interest in the similarities and differences between different animal species. Numerous functions in organisms are studied and contrasted by comparative physiologists. Every aspect of an organism's anatomy has a purpose, and almost all living things have similar basic needs, such as the ability to breathe, digest food, regulate their body temperature, and sustain their hearts. By examining the mechanisms underlying these requirements, such as blood circulation and cell-based exchanges, researchers can amass a plethora of comparable data. Only until a scientist comprehends how each organism's physical makeup enables it to perform the functions necessary for daily existence can accurate comparisons be made. Therefore, the physiological aspect of comparative physiology could include anything from examining how an organism moves through the use of its limbs or other appendages to researching how it breathes.

Ecophysiology, or the study of the interaction between organisms and their surroundings, is a significant component of comparative physiology. Diverse species may respond extremely differently to the same physical environment. For example, a fish's chances of surviving in a desert are far lower than in its native water habitat. On the other hand, because of their anatomical structure, a land-dwelling lizard accustomed to colder weather would not be able to survive in an aquatic environment. Therefore, in terms of comparative physiology, ecophysiology and its research of adaptation-related topics can provide a deeper understanding of all animal groups.

One area of comparative physiology that has gotten more attention throughout time is the use of phylogenic comparative techniques. These methods are used by scientists to investigate potential evolutionary links between various living animals and to chronicle any noteworthy changes that a particular animal group may have undergone since its beginning. Researchers may investigate the physical similarities between different creatures or how different organisms have evolved comparable functional elements, such as lungs or gills for breathing.

As a result, the investigation may find shared ancestors among various species and establish an evolutionary link. Examination of fossil remnants and other archaeological material may also aid comparative physiologists in understanding how an animal group has altered and adapted from ancient times to the present.

1.4.6 Pathophysiology

Pathophysiology or physiopathology is a branch of physiology and is an intersection of pathology and physiology, concerning disordered *physiological processes* that cause, result from, or are otherwise associated with a *disease* or *injury*. Pathophysiology generally deals with functional changes happening when the animal is diseased. Pathophysiology is inevitable to diagnose and cure the disease.

1.4.7 Developmental Physiology

Developmental physiology deals with how structures and functions change as animals grow through the various stages of life from the embryo to adult, to senescence and death.

1.4.8 Evolutionary Physiology

Evolutionary physiology deals with explaining how specific physiological traits arise within lineages throughout multiple generations. The evolutionary physiologist is interested in studying the origin of variations within the populations of a single species, or the closely related groups of animals.

1.5 THE ASSOCIATION OF PHYSIOLOGY WITH OTHER SCIENCES

Physiology, as a discipline, exhibits correlations with other sciences such as physics and chemistry, which in turn facilitate the exploration of novel ideas and innovations. The physiological processes of respiration, fluid balance, and osmoregulation are inherently physical phenomena.

The realm of biology makes extensive use of a variety of instruments such as X-ray machines, microscopes, and ultrasound detectors, all of which find their applications in physics. Physiology, on the other hand, encompasses numerous subjects including biochemistry, biophysics, general physiology, and molecular biology. It holds a crucial position among the functional sciences that are closely associated with the field of medicine. Tackling the unresolved challenges in physiology will require the technical and research expertise of individuals from various fields such as physics, chemistry, mathematics, engineering, and so forth.

The field of comparative biochemistry has laid the groundwork for understanding the physiology of microorganisms beyond just the parasitic forms that affect human health. This has led to a greater appreciation for the essential roles that microorganisms play in the world. Physiological research seeks to integrate the diverse functions of cells, tissues, and organs within an organism. By using both analytical and integrative methods, new challenges are identified and addressed, ultimately leading to practical applications in medicine and a deeper understanding of human and animal biology.

1.6 THE ROLE OF PHYSIOLOGY IN SCIENCE AND MEDICINE

The field of animal physiology provides valuable insights into the functioning of animals. Its practical applications are vast and it aids in the comprehension of health and disease. The study of animal physiology as a model for human physiology is also beneficial. Additionally, the effects of pollutants and heavy metals on the physiological system can be studied to understand their harmful impact and take necessary actions for mitigation. Physiology is a crucial discipline in biology as it integrates various levels of organization in the animal body. By studying physiology, students can gain a comprehensive understanding of how organs and organ systems work, including the nervous and circulatory systems, as well as control mechanisms such as hormones and enzymes.

BIBLIOGRAPHY

Berry, A. K., & K. Berry (2008) *A Textbook of Animal Physiology*, Emkay publications.

Campbell, A. M., & C. J. Paradise (2016) *Animal Physiology*, Momentum Press.

Hall, J. E. (2015) *Guyton and Hall Textbook of Medical Physiology*, 13th Edition, Relx India Pvt. Ltd.

Hill, R. W., G. A. Wyse, & M. Anderson (2016) *Animal Physiology*, 4th Edition, Sinauer.

Lister J. (1858) “On the early stages of inflammation”. Philosophical Transactions of the Royal Society of London. 148. *Royal Society*: 645–702.

Moyes, C. D., & P. M. Schulte (2016) *Principles of Animal Physiology*, 2nd Edition, Pearson Education India.

Randall, D., W. Burggren, & K. French (2002) *Eckert Animal Physiology*, W. H. Freeman and Company.

Rastogi, S. L. (1997) *Essentials of Animal Physiology*, New Age International Publishers, India.

Reznikova, Z. (2007) *Animal Intelligence: From Individual to Cognition*, Cambridge University Press.

Schmidt-Neilson, K (2002) *Animal Physiology: Adaptation and Environment*, Cambridge University Press.

Sherwood, L., H. Klandrof, & P. Yancy (2012) *Animal Physiology: From Genes to Organisms*, Cengage Learning.

Singh, H., Ltika Singh, & M. Yadav (2018) *Fundamentals of Medical Physiology*, Elsevier, Relx India Pvt. Ltd. India

Verma, P. S., B. S. Tyagi, & V. K. Agarwal (2000) *Animal Physiology*, S. Chand & Co. India.

2 Cellular Structure and Functions

Rachana Sharma[1], Poornima Gumasta[2], Priya[1] and Pallavi Khajuria[1]
[1]College of Veterinary Science, Rampura Phul, GADVASU-Ludhiana
[2]College of Veterinary Science, Kishanganj, BASU-Patna

2.1 INTRODUCTION

All organisms consist of cells, which are the fundamental building blocks of life. A single cell possesses all the necessary characteristics to sustain life, making it the smallest unit with these traits. Cells serve as both the structural and functional components of living organisms.

2.1.1 Structure of the Cell

Cells are highly ordered chemical systems and schematically depicted in Figure 2.1 The fundamental components of a cell are the plasma membrane, cytoplasm, and nucleus. The cell membrane restricts and bounds the cell. The cytoplasm, which is the ground substance that creates the background in which the various elements are suspended; and the nucleus is surrounded by a nuclear membrane and separated by the cytoplasm. Thus, the structure of the cell is analyzed under three headings:

1. Cell membrane
2. Cytoplasm
3. Nucleus

2.2 CELL MEMBRANE

The cell membrane or "plasma membrane" was given by Nageli and Cramer. It is the outermost boundary of an animal cell. The cell membrane envelops the cell body like a protective sheath. The membrane separates the intracellular fluid (ICF), which is the fluid inside the cell, from the extracellular fluid (ECF), which is the fluid outside the cell.

The cell membrane is a semipermeable membrane; thus, certain compounds can be freely exchanged between the ICF and ECF. The thickness of the cell membrane ranges from 75 to 111Å.

The Cell membrane is mainly composed of three types of substances:

1. Proteins (55%)
2. Lipids (40%)
3. Carbohydrates (5%)

The cell membrane is called a **unit membrane** or a three-layered membrane. Three layers of the cell membrane are visible under the electron microscope: two electron-dense layers and a middle layer that is electron-lucent. One electron-dense layer is on each side of the central layer (electron-lucent). The core layer is a lipid layer, and two protein layers are on either side of the lipid layer. The cell membrane contains some carbohydrate moieties also.

The phospholipids, which are the primary lipids, are organized in a bilayer structure. Within the membrane, these lipids are positioned with their polar heads facing the outer sides and their hydrophobic tails facing the inner region. The predominant lipid constituents of the membrane are primarily phosphoglycerides. Protein layers provide protection to the central lipid layer. The glycoprotein compounds are mainly present in protein layers. Proteins and the glucose (carbohydrate) molecules found in cell membranes combine to form glycoproteins, also known as proteoglycans. Glycolipids are created when certain carbohydrates are joined to lipid molecules. The glycocalyx is a thin, loose layer of carbohydrates that covers the whole surface of the cell membrane.

1. Cell membrane serves to safeguard the cytoplasm and cell organelles.
2. Cell membrane acts as a semipermeable membrane (selective permeable), permitting certain substances to cross while blocking others.
3. Nutrients are absorbed by the cell through the cell membrane, serving the absorptive function.
4. Waste products and metabolites are eliminated from the cell by passing through the cell membrane.
5. The cell membrane facilitates the exchange of gases, allowing oxygen to enter the cell from the blood while carbon dioxide exits the cell and enters the blood.
6. The cell membrane is accountable for preserving the shape and size of the cell.

2.3 CYTOPLASM

The cell's cytoplasm is the jelly-like substance that contains 80% water. It contains a clear liquid portion called the cytosol and various particles of different shapes and sizes. The nature of these particles includes proteins, carbohydrates, lipids, or electrolytes. The cytoplasm also contains numerous organelles with distinct structure and function.

DOI: 10.1201/9781003426851-2

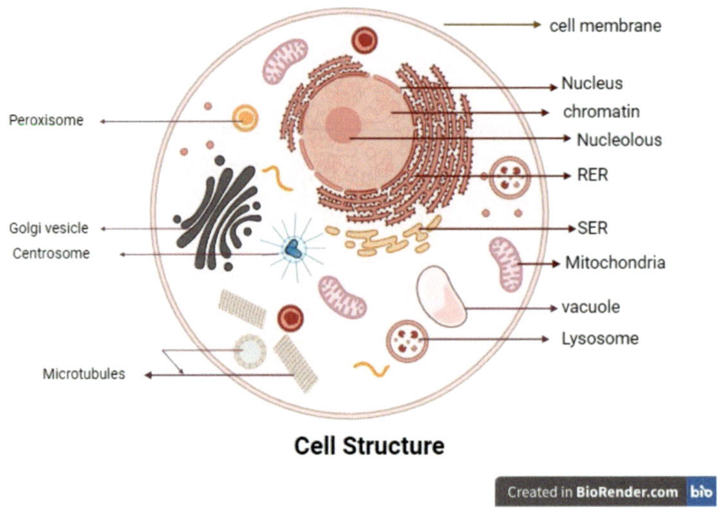

FIGURE 2.1 Cell structure

Two zones in the cytoplasm:

1. Ectoplasm: The portion of the cytoplasm that is located just below the cell membrane.
2. Endoplasm: The inner part of the cytoplasm present between the ectoplasm and the nucleus.

Numerous cellular structures immersed in the cytoplasm are called cytoplasmic organelles. Cytoplasmic organelles are considered small organs of the cell. Some organelles are bound by limiting membranes (endoplasmic reticulum, Golgi apparatus, lysosome, peroxisome, centrosome, secretory vesicles, mitochondria, and nucleus) while others do not have limiting membranes (ribosomes and microtubules). Each organelle has a definite structure and functions

2.4 ENDOPLASMIC RETICULUM

Endoplasmic reticulum is a single-membrane-bound organelle. The endoplasmic reticulum is a network of highly branched tubular and microsomal vesicular structures that are interconnected. One end is connected to the nuclear membrane whereas the other end connects to the plasma membrane. It is often termed the 'system of membranes.' It is covered by a limiting membrane formed by bilayered lipids and proteins. The endoplasmic matrix is a fluid material found in the lumen of the endoplasmic reticulum. The diameter of the lumen is between 400 to 700Å. The endoplasmic reticulum maintains the link between the nucleus and cell membrane.

2.4.1 Types of Endoplasmic Reticulum

There are two types of endoplasmic reticulum: rough endoplasmic reticulum (RER) and smooth endoplasmic reticulum (SER). Both the types are interconnected and continuous with each other. The rough endoplasmic reticulum transitions to smooth endoplasmic reticulum and vice versa depending upon the activities of the cells,

This roughness is why it is referred to as the granular endoplasmic reticulum. Additionally,

Rough endoplasmic reticulum or RER	Smooth endoplasmic reticulum or SER
Ribophorins (glycoproteins) aid in the attachment of ribosomes to their surface, resulting in a rough appearance.	In the absence of ribosomes, they exhibit a smooth surface.
This roughness is why it is referred to as the granular endoplasmic reticulum.	The smooth endoplasmic reticulum displays a sleek appearance and is alternatively known as agranular reticulum.
It plays a crucial role in the active synthesis of proteins and secretion processes, such as in the goblet cell of the liver.	SER is responsible for producing lipids, visual pigments, and steroids, as well as storing them. Additionally, it plays a role in detoxifying harmful drugs.
It is extensively connected and continuous with the outer membrane of the nucleus.	Smooth muscles contain a specialized structure of SER known as the sarcoplasmic reticulum, which plays a crucial role in muscle contraction.
Eg: Adipose tissue, Muscle	Eg. Liver, Pancreas

2.4.1.2 Functions

1. The endoplasmic reticulum (ER) functions as the cytoskeleton, offering structural support and maintaining the shape of the cell.
2. The RER supplies precursor molecules of lysosomal enzymes to the Golgi apparatus.
3. The smooth endoplasmic reticulum is involved in the production of lipids and steroid hormones.
4. RER are actively participating in protein synthesis and secretion.

- Intracellular exchange – The endoplasmic reticulum (ER) forms an intracellular conducting system. Material transport within the cytoplasm can take place through the ER.
- Detoxification – The Smooth ER is primarily responsible for the detoxification of drugs, pollutants, and steroids.

2.5 GOLGI APPARATUS

The Golgi body was discovered by Camilo Golgi (1889) who observed densely stained reticular structures near the nucleus. The Golgi body is surrounded by cytoplasm that contains fewer or possibly no other organelles. This region is referred to as the Golgi ground substance or the zone of exclusion. The Golgi complex, Golgi apparatus, or Golgi body is a membrane-bound organelle, involved in the processing of proteins. The Golgi body is present in all the cells except red blood cells. The Golgi apparatus is composed of a series of 5 to 8 flattened membranous sacs known as cisternae.

2.5.1 Functions

The primary role of the Golgi body is to **secrete macromolecules**, which involves exporting them.

The Process of Secretion in Three Steps:

(i) Golgi body receives the materials from ER through its cis-face.
(ii) Materials undergo chemical modifications within the Golgi body, such as glycosylation of proteins and glycosidation of lipids, resulting in the production of glycoproteins and glycolipids respectively.
(iii) After undergoing chemical alteration, substances are enclosed within vesicles. These vesicles are separated from the trans face of the Golgi body and released (material) outside the cell through a process known as reverse pinocytosis.
  - The acrosome of sperm is formed during the process of spermatogenesis by Golgi body.
  - It plays a crucial role in the synthesis and storage of the yolk and vitelline membrane of the egg.
  - It enhances mitochondrial activity to increase ATP production.
  - The Golgi body plays a role in facilitating the secretion of hormones by endocrine glands.

2.6 LYSOSOME

It is a round, sac-like structure bounded by a single unit membrane composed of lipoprotein. Lysosomes are present in the cytoplasm as membrane-bound vesicular organelles. The lysosomes are produced by the Golgi apparatus. These are found in abundance in phagocytic cells and secreted cells like WBC, kupffer's cells, pancreatic cells, liver cells, etc. Structure. Lysosome shows polymorphism, i.e., different morphological and physiological stages. It is filled with a granular matrix. It is rich in all types of hydrolytic enzymes called hydrolases.

2.6.1 Functions

- All enzymes (acid hydrolases) present in the lysosome necessitate an acidic pH in order to carry out their functions, thus earning the name acid hydrolases.
- These acid hydrolases function efficiently at a pH level of 5. The lysosome's membrane possesses an active H+ pump mechanism, which creates an acidic pH inside the lysosome's lumen.
- Acid hydrolases have the capacity to break down various forms of macromolecules through hydrolysis.

2.7 PEROXISOMES

Peroxisomes or microbodies are the membrane limited vesicles similar to the lysosomes. Peroxisomes are isolated from the endoplasmic reticulum rather than the Golgi apparatus unlike lysosomes. Peroxisomes contain several enzymes like urate oxidase, amino acid oxidase, peroxidase which catalyse the oxidation of substrates forming H2O2 (Hydrogen Peroxide). Catalase is the fastest acting enzyme known after carbonic anhydrase and it is a marker enzyme of peroxisomes. Peroxisomes are also involved in β–oxidation of fatty acids.

2.7.1 Functions

- Fatty acids are broken down through a process known as beta-oxidation, which serves as the primary function of peroxisomes.
- Detoxification involves the breakdown of harmful substances like hydrogen peroxide and other metabolic byproducts. The liver, being the primary organ responsible for detoxification, contains a significant amount of peroxisomes within its cells. When poisons or alcohol enter the cell, hydrogen peroxide is formed. In the presence of hydrogen peroxide, the peroxisomes rupture and release oxidative enzymes. These enzymes effectively eliminate hydrogen peroxide and the enzymes required for its production.
- The peroxisome is the primary location for oxygen consumption within cells.
- Enhance the production of glucose from fats through gluconeogenesis. Degrade purine to uric acid.
- Contribute to formation of myelin.
- Contribute to the synthesis of bile acids.

2.8 CENTROSOME (CENTRIOLES)

Centrioles play a crucial role in the movement of chromosomes during cell division. The cytoplasm that surrounds centrioles is referred to as the centrosphere. Together, centrioles and centrosphere are known as the centrosome. Each centriole is encircled by a pericentriolar mass called massules, satellite, or MTG. Centrioles are elongated, membraneless structures that display a cartwheel-like formation. Basal bodies are also considered a type of centriole. A centriole consists of 9 evenly spaced peripheral fibrils of tubulin arranged in a 9 + 0 pattern. Each peripheral fibril is a triplet. The hub, located in the central part of the proximal region of the centriole, is connected to the protein-composed radial spokes. These radial spokes link the hub to the tubules of the peripheral triplets.

2.8.1 Functions

1. Centrioles in animal cells have a crucial function in initiating cell division by organizing spindle fibers between the two poles of the cell.
2. The positioning of centrioles during cell division determines the plane of division.

2.9 SECRETORY VESICLES

Secretory vesicles are the membrane bound organelles and contain the secretory substances. These vesicles are formed by the endoplasmic reticulum. Secretory vesicles are present all around the cytoplasm. These vesicles are burst as necessary and released the secretory substances into the cytoplasm.

2.10 MITOCHONDRIA

Mitochondria is known as power house of cell. It is the second largest organelle in animal cells and is semiautonomous organelle. Mitochondria are present in immature RBC. It occur in all cells except RBC (Figure 2.2).

Mitochondria are membrane-bound cytoplasmic organelles responsible for energy production. The quantity of mitochondria in each cell can vary based on the cell's physiological functions. Additionally, there is a noticeable range of diversity in terms of shape and size. Typically, they are sausage-shaped or cylindrical, having a diameter of 0.2-1.0μm (average 0.5μm) and length 1.0-4.1μm. Mitochondria are covered by double unit membrane. The outer membrane is smooth, permeable, and encloses the contents of the mitochondrion. The outer membrane of mitochondria contains various enzymes such as acetyl-CoA synthetase and glycerol phosphate acetyltransferase. The inner membrane projects into the central space in the form of finger-like projections called cristae. "Cristae" is a name given by Palade. The cristae expand the surface area, creating a generous amount of space for metabolic activities. They are selectively permeable and contains all enzymes of the electron transport chain. Cristae contain various enzymes and other protein molecules which are involved in respiration and the synthesis of adenosine triphosphate (ATP). Elementary particles or oxysomes or F0-F1–particles are present on the inner membrane of mitochondria. They are differentiated into three parts: the base (round part of oxysome), stalk (joins head with base), and head (embedded in the membrane). The headpiece (F1) behaves as an ATP synthase complex and thus synthesizes ATP by the virtue of proton gradient development. The inner compartment is filled with a dense homogeneous substance called the matrix. The matrix also possesses a single circular DNA molecule, a few RNA molecules, ribosomes (70S), and the components required for the synthesis of proteins.

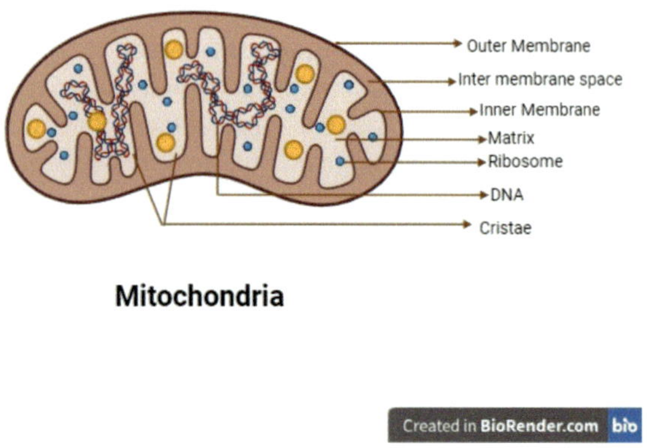

FIGURE 2.2 Mitochondria

2.10.1 Functions

- Mitochondria serves as the location for cellular respiration due to their possession of all necessary enzymes for the process.
- Energy (ATP) is produced within mitochondria as they are the site of cellular respiration.
- Mitochondria play a role in yolk formation during ovum development (vitellogenesis).
- Mitochondria in the formation of the middle piece of sperm during sperm maturation. Additionally, mitochondria aid in the elongation of fatty acids and are involved in cytoplasmic inheritance.

2.11 RIBOSOMES

The Ribosome was discovered by Robinson and Brown (1953) in plant cells of bean roots and by Palade in animal cells. Ribosomes are the smallest dot-like cell organelles with a diameter of 15 nm. They are also known as the "organelle within an organelle" and the "protein factory of the cell." Ribosomes are granular and small dot-like structures. Ribosomes are the smallest cell organelles found in all types of cells (except mature mammalian RBCs). Ribosomes are concerned with protein synthesis in the cell. Ribosomes are scattered in the cytoplasm and attached to the endoplasmic reticulum and outer nuclear membrane.

They are found in large numbers in protein-synthesizing cells. Ribosomes are composed of 35% proteins (ribonucleoprotein) and 65% ribonucleic acid (rRNA). Ribosomal RNA (rRNA) is the RNA found in ribosomes. Each ribosome is composed of two subunits, i.e., larger and smaller subunits. Magnesium ion is essential for binding the ribosome subunits.

2.11.1 Function

The main function of a ribosome is protein synthesis, thus called the protein factory or cell engine.

2.12 CYTOSKELETON

The cytoskeleton is a complex network of filamentous protein structures that can be found within the cytoplasm. It gives support to the cell and determines its shape. It is an extensive network of structures of varying sizes. It is also necessary for the movements and the response of the cell to external stimuli.

The cytoskeleton consists of three types of protein component:

1. Microtubule
2. Intermediate filaments
3. Microfilaments

2.12.1 Functions of Cytoskeleton

Structural support, movement, and maintain the cell morphology.

Microtubules are the tubular structures of the cytoskeleton. These organelles without a limiting membrane are organized in different bundles. The diameter of each tubule is between 20 to 30 nm. The microtubule length varies and it may be 1000 times higher than the thickness. The microtubules are structurally formed by bundles of globular protein called tubulin. Tubulin has two subunits, namely α and β subunit.

2.12.1.1 Function of Microtubule

They form the cytoskeleton and are responsible for maintaining the shape of the cell and providing structural strength to the cell. The centrosphere surrounding the centriole emits these structures, which then assemble into spindle fibers. They constitute the cytoskeleton of cilia and flagella. Additionally, they play a crucial role in facilitating the movement of chromosomes during cell division.

Intermediate filaments are structures that form a network around the nucleus and extend to the cell's periphery. Intermediate Filaments are near about 10 nm diameter. The intermediate filaments are made up of fibrous proteins.

2.12.1.2 Functions of Intermediate Filaments

They help to maintain the structure and shape of the cell. These filaments also link to the nearby cells through desmosomes.

Microfilaments are long and fine structures that resemble threads. The diameter of microtubules is near about 3 to 6 nm. Actin and myosin are non-tubular contractile proteins that make up these filaments. Microfilaments are present throughout the cytoplasm (ectoplasm and endoplasm both). The endoplasmic microfilament contains both actin and myosin protein whereas ectoplasmic microfilament contains only actin protein.

2.12.1.3 Functions of Microfilaments

Provides structural strength of the cells and resistance to the cell against the pulling force. They are responsible for movements of cells such as contraction, gliding and cytokinesis.

2.13 NUCLEUS

Robert Brown discovered the cell nucleus in 1831. The nucleus is the most noticeable and the largest cellular organelle of the cell. The diameter of the nucleus is about 10 μ to 22 μ and occupies approximately 10% of the total volume of the cell. Nucleus is present in all the cells of the body except the red blood cells. The nucleus is required for cell division. Most of the cells are uninucleated (one nucleus) cells but some types of cells like skeletal muscle are multinucleated (more than one nucleus). Generally, the nucleus is spherical in shape and located in the center of the cell (Figure 2.3).

The nucleus is a double membrane-bound dense protoplasmic body that contains many components (Major components of the nucleus are nucleoplasm, chromatin, and nucleolus) and controls all cellular metabolism and encloses the genetic information of the cell. Electron microscopy has

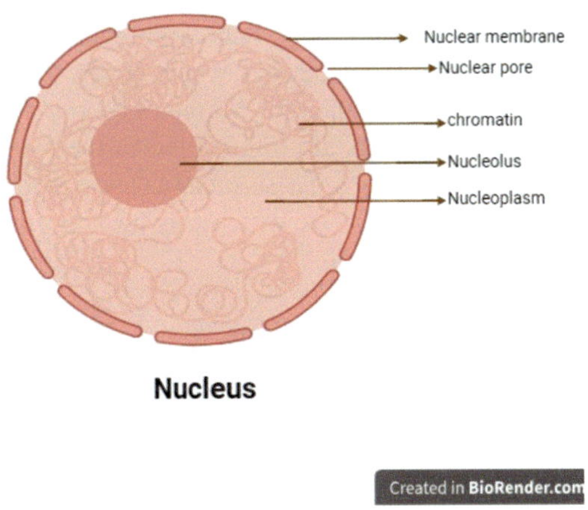

FIGURE 2.3 Nucleus

revealed that the nuclear envelope or **nuclear membrane** is double-layered and porous in nature. The nuclear membrane acts as a barrier between the materials present inside (nucleoplasm) and outside (cytoplasm) of the nucleus. The nuclear membrane's outer layer remains connected to the endoplasmic reticulum, which also carries ribosomes on its surface. The nuclear membrane is perforated by minute nuclear pores, which are guarded by protein molecules. The nuclear pore plays an important role in exchanging materials between the nucleoplasm and cytoplasm. The inner nuclear membrane is lined by the nuclear lamina, which is formed by filaments of lamin protein.

Nucleoplasm is a very viscous fluid that makes up the ground substance of the nucleus. Chromatin and nucleolus are embedded in the nucleoplasm. It also contains a dense fibrillar network of proteins known as the nuclear matrix and many substances such as nucleotides, ATPs, proteins and enzymes of RNA and DNA polymerases, endonucleases, minerals, cations (Ca++, Mg++) etc.. The nuclear matrix creates the structural framework for organizing chromatin. The soluble liquid part of the nucleoplasm is called nuclear hyaloplasm.

Chromatin is the intranuclear thread-like fine material that is mostly made up of DNA molecules. Histone is a specialized basic protein that helps pack the DNA molecules tightly.

DNA is a double helix that encircles the central core of eight histone molecules and forms the fundamental packing unit of chromatin called a nucleosome. A chromatin fiber is created when nucleosomes are firmly packed together with the aid of a histone molecule. The chromatin condenses to form the chromosome just prior to cell division.

Nucleolus is a naked or membrane-less small, round granular structure of the nucleus, which is generally associated with chromatin (chromosome) at a specific area called NOR (Nucleolar Organizer Region). The nucleolus is made up of RNA and some proteins that are similar to ribosomes. The RNA is made up of five different pairs of chromosomes and stored in the nucleolus. Subsequently, it condenses to create the ribosome subunits. Ribosome synthesis is the primary activity of the nucleolus, which is why it is called the ribosome factory of the cell. All the ribosomal subunits formed in the nucleolus are transported to the cytoplasm through the nuclear pores. These subunits unite in the cytoplasm to become ribosomes, which play an important role in the synthesis of proteins.

2.13.1 Chromosomes

The term "chromosome" was first proposed in 1889 by Waldeyer. The chromosome is the rod-shaped nuclear structure that contains a full blueprint of all the inherited characteristics of the species. Chromosomes are best studied at the metaphase stage. This is because the size of the chromosome is reduced during the metaphase because the chromatin threads are very thin. Each chromosome is made up of a single DNA molecule arranged around histone molecules, each of which contains many genes. Generally, chromosomes are not visible under the microscope. All the dividing cells in the body, except reproductive cells, contain 23 pairs of chromosomes (humans). Each pair is made up of one inherited chromosome from the mother and one inherited from the father. The cells with 23 pairs of chromosomes are known as diploid cells, and the cells with 23 single chromosomes are known as haploid cells. The reproductive cells, called gametes or sex cells (haploid), contain only 23 single chromosomes.

2.13.2 Functions

Control of all cellular functions such as metabolism, synthesis of proteins, growth, reproduction (cell division), RNA synthesis, ribosome formation, genetic instruction to the cytoplasm for protein synthesis (mRNA), control of cell division by genes, storage of hereditary information (genetic information), and transformation of that information from one generation to next.

2.14 DEOXYRIBONUCLEIC ACID

Deoxyribonucleic acid (DNA) is a nucleic acid that transmits the genetic information to an organism's progeny. DNA is the chemical basis of inherited characters. It includes instructions on how to make proteins in ribosomes. A gene is a molecule inside a DNA molecule. DNA is found in the nucleus (chromosome) and mitochondria of the cell. The nucleus of a cell contains the DNA responsible for the synthesis of RNA. RNA regulates the synthesis of proteins through ribosomes. Mitochondrial DNA is called non-chromosomal DNA.

2.15 STRUCTURE OF DNA

DNA is a complex double-stranded nucleic acid composed of two polynucleotide chains, which are twisted with each other into a double helix structure. It is formed by phosphoric acid, deoxyribose, and four different types of bases. Each DNA molecule consists of two polynucleotide chains, which are twisted around one another in the form of a double helix. The phosphate and deoxyribose molecules come together to create the two chains that make up the DNA molecule. Additionally, organic bases connect the two DNA strands to each other (Figure 2.4).

The DNA molecule chain is composed of numerous nucleotides. Each nucleotide is made by:

1. Deoxyribose – sugar
2. Phosphate
3. One of the following organic (nitrogenous) bases: Purines {Adenine (A) and Guanine (G)} and Pyrimidines {Thymine (T) and Cytosine (C)}.

The arrangement of DNA strands involves specific pairing of bases, where each strand is bound by particular pairs.

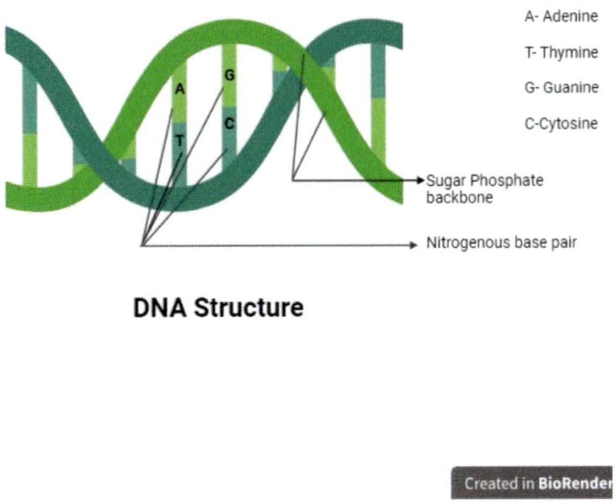

FIGURE 2.4 Nucleus

One strand's adenine specifically binds with the opposite strand's thymine, while the cytosine of one strand binds with the guanine of the other strand.

Chromosomes are composed of DNA, which contains the genetic information. This genetic information, stored in DNA, is referred to as the genome. Genes are the individual units into which a DNA molecule is divided.

BIBLIOGRAPHY

Cunningham, J.G., and Klein, B.G. (2007) *Textbook of Veterinary Physiology*, 4th Ed. Saunders Elsevier, St. Louis, Mo

Frandson, R.D., Wilke, W.L., and Fails, A.D. (2003) *Anatomy and Physiology of Farm Animals*, 6th Ed. Lippincott, Williams and Wilkins, Philadelphia. 450P. [introduction to physiology of farm animals].

Golgi C. (1889). "Sul ciclo evolutivo dei parassiti malarici nella febbre terzana : diagnosi differenziale tra i parassiti endoglobulari malarici della terzana e quelli della quartana" [On the cycle of development of malarial parasites in tertian fever: differential diagnosis between the intracellular parasites of tertian and quartant fever]. *Archivio per le Scienza Mediche*. 13: 173–196.

Reece, W.O. (1997) *Physiology of Domestic Animals*. Williams and Wilkins, Baltimore. 464P [A concise presentation of introductory systems physiology in animals].

Robinson, E., and Brown, R. (1953) Cytoplasmic particles in bean root cells. *Nature* 171, 313.

Sembulingam, K. and Sembulingam, P. (2012) *Essentials of Medical Physiology*, 6th Ed. Jaypee Brothers Medical Publishers (P) Ltd., New Delhi.

Swenson, M., and Reece, W. (2004) *Duke's Physiology of Domestic Animals*, 12th Ed. Cornell University Press [A thorough and comprehensive presentation of domestic animal physiology for advance students]

3 Developmental Physiology

Archana Mahapatra[1], Prasanta Kumar K. Mishra[2], Satish Kumar Pathak[3] and Manish Kumar[4]*

[1,3]Department of Veterinary Anatomy, Faculty of Veterinary and Animal Sciences, Institute of Agricultural Sciences, Rajiv Gandhi South Campus, Barkachha, Banaras Hindu University, Uttar Pradesh, India
[2]Department of Veterinary Biochemistry, C.V.Sc. & A.H., Odisha University of Agriculture & Technology, Bhubaneswar-751003, Odisha, India
[4]Department of Veterinary Physiology and Biochemistry, Faculty of Veterinary and Animal Sciences, Institute of Agricultural Sciences, Rajiv Gandhi South Campus, Barkachha, Banaras Hindu University, Uttar Pradesh, India

3.1 INTRODUCTION

Developmental physiology explains how intricately the biological processes are regulated and integrated in an animal's life from its existence (from the zygote) till its death. The developmental journey of an animal extends beyond birth, encompassing aging and eventual death as natural processes. Throughout life, cells continue to divide and differentiate, organs undergo growth and maturation until puberty, and numerous tissues undergo ongoing developmental changes. The development of an animal is classified into prenatal and postnatal development. The process of prenatal development, classified as growth and development from fertilization until birth within the mother's womb, is separated into three periods: Prenatal development unfolds in three distinct periods: firstly, the germinal period, which marks the zygote/ovum stage from conception until the blastocyst attaches to the endometrium; secondly, the embryonic period, characterized by the formation of major organ systems; and thirdly, the fetal period, primarily dedicated to the growth and development of the conceptus (Green and Winters, 1945; Dean and Grizzle, 2011). Multiple biological and environmental factors contribute to the development in an animal. Within this context, our discourse delved into the intricate nuances of an animal's developmental physiology during the prenatal phase. Embryogenesis emerges as a pivotal stage, characterized by a sophisticated process wherein the embryo undergoes transformation from a fertilized oocyte to a discernible form. This intricate journey involves a series of coordinated events, including cell division, differentiation, and morphogenesis, ultimately leading to the formation of the complex structures and organs of the developing organism.

3.2 EARLY PRENATAL DEVELOPMENT

Individuals mark their phenomenal existence on the planet from a fertilized oocyte. This fertilized oocyte is the product of fertilization. Fertilization is a complex multistep process where the sperm (male gamete) meets an ovum (female gamete) and forms a zygote. The regulated coordination of the organ systems in an animal's body is well assisted by different organs, tissues, and cells of considerable diversity that are derived from a single cell lineage, namely the fertilized oocyte. The intrauterine development is initiated at fertilization with the formation of a zygote. Zygote develops into a 2-cell embryo followed by the first post-fertilization mitosis. Within the cytoplasm of the oocyte, the nucleus of the sperm undergoes enlargement to form the male pronucleus (Lee et al., 2003).

3.3 FACTORS RESPONSIBLE FOR EARLY PRENATAL DEVELOPMENT

Early prenatal development is a complex process influenced by various factors including genetics, maternal health, nutrition, placental function, hormones, maternal stress and emotional well-being, and uterine environment. Overall, prenatal development is a complex and tightly regulated process influenced by a multitude of factors, with interactions between genetic, environmental, and maternal factors playing critical roles. During the initial stages of development, the zygote is profoundly influenced by the oocyte, with the oocytic cytoplasm assuming a pivotal role in regulating germinal development. This influence is evident in various crucial processes, including the formation of the pronucleus, the reprogramming of the genome, the determination of cell division patterns, as well as transcription and RNA metabolism (Wu and Dean, 2020).

Embryonic development is governed by a precisely orchestrated program of gene expression, both temporally and spatially defined. Events of differentiation are directly or indirectly controlled by the cytoplasmic determinants (eg.Vgl and bicoid) through activation or inactivation of regulatory genes (Georgadaki et al., 2016).

DOI: 10.1201/9781003426851-3

The early embryonic developmental control is shifted from the mother to the zygote by embryonic genomic activation mechanism which regulates further embryogenesis. This is known as maternal to embryonic transition (Telford et al., 1990; Kanka, 2003; Walser and Lipshitz, 2011). Another crucial step in developmental process is the cell fate choice in which the cells differentiate into three lineages that is epiblast which form the embryo proper, primitive endoderm (also known as hypoblast) and trophoectoderm that together form the extraembryonic tissues (Mu Jian, 2022).

The endometrium also holds significance in facilitating pregnancy establishment in domestic animals, as the embryo is implanted to the internal endometrial surface throughout the pregnancy. Endometrium and the conceptus communicate effectively to establish maternal reorganization of pregnancy and uterine receptivity for the implantation by hormonal regulation, immune modulation, and intricate molecular communications (Forde et al., 2011; Sanchez et al., 2019).

3.4 PRENATAL DEVELOPMENTAL CHANGES AT THE CELLULAR LEVEL

Cleavage denotes a sequence of swift mitotic divisions taking place in the early embryo following fertilization. The process of cleavage in the early stages of embryonic development involves a series of mitotic divisions of the zygote, leading to the formation of a multicellular embryo. The onset of the first cleavage involves the emergence of a cleavage furrow, manifesting as a ring encircling the zygote. This furrow aligns parallel to the polar axis (extending from the animal pole to the vegetal pole) and perpendicular to the longitudinal axis of the mitotic spindle. Comprising actin and myosin proteins, the contractile ring responsible for the cleavage furrow formation resides in the cortical region of the zygote. Myosin controls the actin filaments, enabling the contraction necessary for the furrow to deepen and eventually separate the zygote into two blastomeres. During the second cleavage, alignment parallel to the polar axis, akin to the first cleavage, is observed. As the second cleavage concludes, the embryo comprises four blastomeres arranged from the animal pole to the vegetal pole. Cleavage continues as a fractionating process, with subsequent divisions further increasing the number of cells. In the developmental sequence, the third cleavage conventionally proceeds perpendicular to the preceding two, thereby culminating in the formation of an eight-cell stage. Continued divisions lead to the formation of a 16-32 cell stage, culminating in the creation of a compact mass of cells referred to as the morula (Abbott and Ducibella, 2001). During the latter phase of the 8-cell stage cleavage, a phenomenon known as cell compaction ensues, facilitating the establishment of cell junctional complexes and imparting polarization to the cell. Consequently, the cell assumes a distinct structural orientation, with a defined contact area adjacent to neighboring cells, termed the basolateral domain, and an unencumbered outer surface known as the apical domain. Notably, the orchestration of compaction is intricately regulated by extracellular calcium ions (Ca2+), acting through epithelial cadherin (E-cad) (Dietmar et al., 1987). The apical domain accumulates apical polarity proteins such as the Par3-Par6-aPKC complex, along with cytoskeletal components, while the basolateral domain harbors proteins like Parl and Lgl. Apical accumulation of actomycin which is facilitated by PKC activation along with proteins Par3-Par6-aPKC complex establishes polarization of the embryo (Lin et al., 2000; Vinot et al., 2005; Suzuki and Ohno, 2006). Polarization is a critical process that dictates the future differentiation and organization of the embryo's cells. Post-polarization, blastomeres can undergo symmetric and two types of asymmetric divisions. Asymmetric division produces 1. Two polarized daughters 2. Following cleavage, one daughter cell becomes polarized while the other remains non-polarized. Non-polarized cells, situated within the blastomere's interior, undergo differentiation into the Inner Cell Mass (ICM). The ICM possesses pluripotency, giving rise to the embryo proper and certain extraembryonic tissues. Polarized cells, positioned on the blastomere's periphery, differentiate into the Trophectoderm (TE), which constitutes the outer layer of the blastocyst. The TE contributes to placental formation and the development of supportive tissues.

The blastomere maintains a balance between ICM and TE cells by regulating the types of divisions. This balance is essential for proper embryonic development, ensuring that there are enough cells to form both the embryo and its supporting structures (Mu Jian, 2022).

By the 32-cell stage, the outer cells of the embryo become fully committed to the trophectoderm (TE) lineage. These TE cells play a vital role in forming the placenta and other supportive tissues crucial for fetal development. Na^+/K^+ pumps situated on the basolateral (inner) domain of TE cells actively transport sodium ions (Na^+) out of the cells and potassium ions (K^+) into the cells. Additionally, Na^+/H^+ cotransporters located on the apical (outer) domain of TE cells facilitate the movement of Na^+ ions into the cells and protons (H^+) out of the cells. The activity of these pumps and co-transporters heightens the osmotic pressure within the intercellular spaces. This higher osmotic pressure draws water into these spaces, leading to fluid accumulation. The pressurized fluid causes the cell-cell contacts within the TE to fracture, creating hundreds of micrometer-sized lumens (small fluid-filled cavities) (Dumortier et al., 2019). The intercellular spaces merge; giving rise to a singular, larger fluid-filled cavity termed the blastocyst cavity or blastocoel. As fluid accumulates within the blastocoel, the inner cell mass (ICM) remains attached to a specific region of the trophoblast. Subsequent to compaction and cavitation, the initial differentiation of blastomeres yields two distinct cell types: the trophectoderm and inner cell mass. In pigs, the cavitation process initiates at the 16-cell stage of

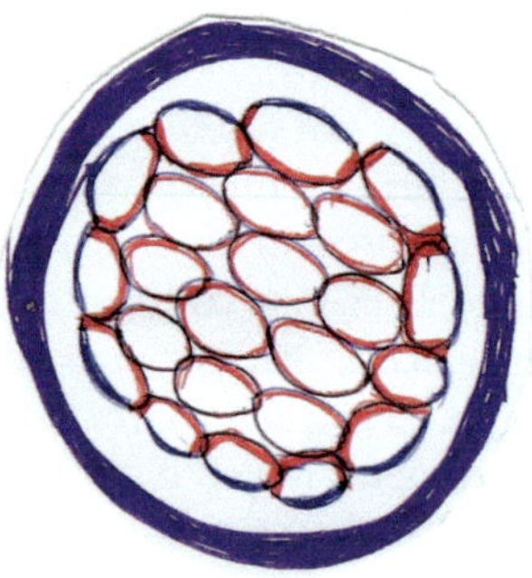

FIGURE 3.1 Events of early embryonic development

the morula. However, in humans, cattle, and sheep, cavitation commences later, at the 64-cell stage of the morula (McGeady et al., 2017) (Figure 3.1).

During early embryonic development, the first major cell fate decision occurs with the formation of the blastocyst. In cattle and sheep, blastocyst formation typically occurs around the 7th day post-fertilization. In pigs, this process usually occurs around the 5th day post-fertilization. In humans, it generally takes place by the 5th or 6th day post fertilization (Yoshinaga, 2018). After blastocyst formation, the trophectoderm contributes to placental formation, while the inner cell mass (ICM) generates the embryo proper and certain extraembryonic tissues. Immediately post-blastocyst formation, it expands and hatches from the zona pellucida. During this hatching process, cells within the ICM face a second fate decision. This decision involves the differentiation of the ICM into two distinct cell lineages: the Epiblast (Epi), which is pluripotent and gives rise to the embryo itself, including all body tissues and organs, and the primitive endoderm, which contributes to extraembryonic tissues, particularly the yolk sac. The yolk sac provides vital support and nutrients to the developing embryo. Specific molecular markers and signaling pathways guide the differentiation of these two cell populations within the ICM (Figure 3.2). This second fate decision is critical for proper embryonic development and the establishment of foundational structures necessary for subsequent growth and differentiation (Mu Jian, 2022).

Growth and expansion of the blastocyst during the pre-implantation period is species-specific. The epiblast, along with the underlying primitive endoderm, forms the embryonic disc and is reshaped. In horses, dogs, cats, and rabbits, the blastocyst appears round to oval, whereas in cattle, sheep, and pigs, it expands like a filament or thread. At 9 days gestation, the porcine blastocyst appears to be a sphere of 10 mm in diameter, which lengthens up to 100 centimeters by the 13th days of gestation. The elongation of the conceptus marks a crucial milestone in early pregnancy across these species, commencing on days 11–12 in pigs, days 12–13 in sheep, and days 13–14 in cattle. This process coincides with pregnancy recognition and implantation (Bazer et al., 2011). **Interferon-Tau (IFN-τ)**, a pivotal signaling molecule produced by the bovine embryo, communicates with the maternal endometrium to inhibit luteolysis and sustain pregnancy. Following implantation, the developing embryo exits the uterine lumen and adheres to the uterine mucosa for the duration of gestation. This timing is crucial for synchronizing the development of the blastocyst with the receptive state of the endometrium, guaranteeing successful attachment and subsequent embryo development.

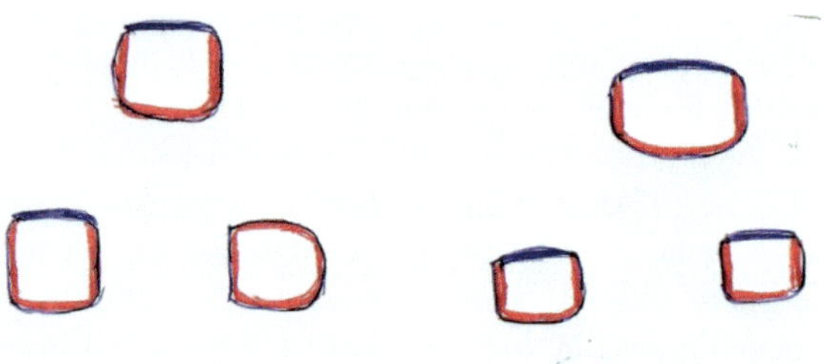

FIGURE 3.2 Cross section of early embryo showcasing polarization polarization

Implantation in sheep typically occurs around 15–16 days after fertilization. In cattle, it generally happens around 18-20 days post-fertilization, while in pigs, it occurs approximately 14–16 days after fertilization (Spencer et al., 2004; Anna Piliszek and Zofia E. Madeja, 2018).

Following blastulation, the embryonic disc is established, and the embryo is now all set for gastrulation. The blastula is reorganized into a three-layered structure comprising the ectoderm, mesoderm, and endoderm. Each of these germ layers will further differentiate into specific tissues and organs, laying the groundwork for the complex anatomy and functionality of the organism. Gastrulation sets the stage for organ formation by establishing the basic body plan and positioning the germ layers. Cells receive signals during gastrulation that guide their future differentiation into specific cell types and tissues. This process plays a vital role in establishing the anterior-posterior, dorsal-ventral, and left-right axes of the embryo, essential for the correct spatial arrangement of tissues and organs. Gastrulation initiates with the emergence of a primitive streak on the surface of the blastula, indicating the future posterior end of the embryo. Cells originating from the outer layer (epiblast) migrate inward through the primitive streak. These migrating epiblast cells differentiate into three primary germ layers: ectoderm, mesoderm, and endoderm. The establishment of the notochord sets the cranial-caudal axis, which is fundamental for the spatial organization of the developing embryo. The ectoderm gives rise to the epidermis of the skin and neural tissue, while the endoderm forms the lining of the gastrointestinal and respiratory tracts. From the mesodermal layer, the urogenital, circulatory, and supportive muscular and skeletal systems emerge (McGeady et al., 2017).

The completion of gastrulation marks the transition to further stages of development, such as the first embryonic organ system: neurulation (formation of the neural tube)

and organogenesis (formation of organs) (McGeady et al., 2017).

In cattle, during the first two weeks of the pre-embryonic stage, the zygote is formed, and the process of implantation starts. The embryonic period commences with trophoectodermal adhesion to the endometrial epithelium. By its conclusion, indicated by the initiation of fetal bone mineralization, all organ systems are structured in their elementary form. Subsequently, the embryo transitions to the fetal period, spanning from the ninth week of gestation until birth canal expulsion (Valadão et al., 2018).

In sheep prenatal development, the blastocyst implants approximately 15 days after fertilization. The embryonic period extends until around the 34th day of intrauterine life, followed by the onset of the fetal period around the 35th day after conception, which continues until birth (Green and Winter, 1945; Valasi et al., 2017).

In the prenatal development, the period of the ovum spans from Days 2 to Day 17, followed by the embryo period from Days 19 to Day 35. Subsequently, the fetal period of the fetus extends from Day 35 until birth (Phemister RD, 1974).

In the prenatal development of the horse, implantation occurs around Day 16, and a series of morphological changes, embryogenesis, and organogenesis follows the embryonic period. The fetal period extends after Days 40 to birth (Ginther, 1983a, 1983b; Enders and Liu, 1991; Sharp, 2000; Allen and Wilsher, 2009; Franciolli et al., 2011).

3.4.1 Molecular and Biochemical Events During the Development of Animals

The cellular identity is attributed to the differential expression of genes which is regulated by various factors. These expressions may determine the fate of the cell through processes like proliferation, interactions with neighboring cells, creating a special group of cells and migration of cells to a definite locus (Vestweber et al., 1987). Cellular environment along with chemical signaling plays a pivotal role in such a process. The fate of the cells is also somewhat regulated by the memory clock ticking inside it. Genes like PURA, which codes protein like Purα, are responsible for the formation of Purkinje cell layer and synapse formation in mice. Apart from that it is important for many other aspects of cellular development. This gene is ubiquitous and binds to DNA fragments containing the -GGCGGA- site present proximal to myelin basic protein (MBP) region. Any abnormality in such protein/homologous proteins may lead to defective neuronal function (Khalili et al., 2003). This indicates the importance of certain generalized factors contributing to the development in animals. Though the genetic complexity in animals is prominent, they share a significant homology between more than half of the genes. The genes which are specific to a particular organism, though are present, contribute less towards the developmental procedure. Also, regulatory genes like Nr2f, ptch, SalI, Smad6, Sp5, Tbx2/3 are universally present and conserved across different species (Frith and Ni, 2023). Some homeobox genes like Gsx, Msx, Pitx are also conserved which bind to a conserved -CCAAT- box. Broadly, two classes of proteins are responsible for development of animals

(i) Protein spanning the membrane and ion channels
(ii) Protein that play a regulatory role.

The former is having a major role in development. The signals to create differentiation between animals are largely controlled by the regular DNA which are mostly non-coding and constitute a major portion of the genome of the animals. More than 2500 enhancers have been identified recently by Wu et al., 2022., with the help of self-transcribing active regulatory region sequence (STARR sequence) in pigs. ZPA regulatory sequence (ZRS) is an established enhancer. It is a functional element gene which is responsible for limb specific (LMBR1) sonic hedgehog (SHH) signaling molecules. Deletion of ZRS results in reduction of limbs in mice. Further, mutation affecting ZRS gene may lead to pre-axial polydactyly and triphalangeal thumb. Different species with body plans of similar kind may also have similar sets of regulatory genes.

Exposing embryos to various ionizing radiation or by transposon mediated nonspecific mutation may help in elucidating the function of a gene. The initial masses of cells of early life may be un-committed or committed or determined as per their differential potency.

Positional value of a cell shows the position specific expression of determinants. A regulatory protein known as T-box protein (Tbx transcription factor) is actively involved in this. Tbx regulates such behavior in formation of limb and heart. The members of T- box gene family bind to the sequence -TCACACCT- of DNA. TBX4 and TBX5 forms the hind and fore limb in case of chickens. Homeobox genes like Hox gene starts the signaling involving Wnt (Wingless and Int-1) signaling pathway and fibroblast growth factor (FGF)signals in formation of limbs. TBX1 and TBX2 are members of this group of protein which control rather development of testis and formation of atrioventricular canal respectively.

Proximate /inductive interaction can also change the fate of the equivalence group of cells which happens due to any local stimuli or inducers. But the ultimate outcome is also dependent upon the interplay between inducers and other regulatory proteins. Competence refers to the capacity to react to a particular inductive signal. The inductive interaction comprises two key components. The first is the inducer, or the tissue generating the signal that alters the behavior of the target tissue. The second component is the responder, which receives the signal. In rat eye tissue formation, Pax6 serves as a competence factor. Interestingly, few inductive events between different species can bring back some lost events. For example, molar mesenchyme from mouse can induce teeth in the jaw of a bird. In contrary to inducers, morphogens have long range and graded induction effect. Graded and non-uniform distribution of morphogens across the tissue forms the basis of pattern formation or morphogenesis. Morphogens like Hedgehog (Hg) and bone morphogenic protein (BMP) are predominant in animals. Not only proteins but non proteinaceous vitamin-A analogue can also function as a morphogen. In some cases, inhibitors of inducers play vital role in deciding the developmental fate rather than the inducers itself (Mehelen et al., 2005). "Chordin" an inhibitor of BMP /TGFβ, navigates the cells towards formation of neurons. As discussed earlier, an internal clock is also set to govern the number of times the cell should undergo a division. Sequential induction generates a diversified population in any organs. It starts with a small population and then propagate to a nearby cell population with production of various kinds of inducers (Lang et al., 2021).

Mammalian embryo needs a very less amount food due to presence of placenta which provides nourishment to it. The fertilized egg transforms into a compact form of morula with 16 cells. These cells are arranged compactly and the tight junction between the cells zips them off and acts as a barrier from the external surroundings. Adherens junctions (AJ) play important role in cell-cell cohesion (Khalili et al., 2003). Cadherins are a group of junctional proteins that initiate this process. Cadherins are classified into E (type -I) and VE (type -II) groups as per their tissue specific expressions. Further, desmosomal cadherins can be classified into desmogleins and desmocollins. Embryo compaction begins at 8 cell stage. It starts from the cell contact points and migrates forward in such a way that the neighboring cells come in the closest contact possible. Two underlying mechanisms may play role in this (i) enhanced formation of filopodia and (ii) surface contractibility mediated by actomyosin. Calcium also plays important role in compaction as it was evident that depriving cells of calcium leads to inhibition of compaction. The polarity regulators like Par3 and Par6 gets distributed to different poles side by side to the compaction. Ezrin (cytovillin), Par6 and Par3 are normally present at the apical domain and JAM-1, Na/K ATPase, E- cadherin are located in the lateral region. Intermediate filaments are joined by desmosomes. It has some additional benefits to the adherens junction like higher tensile strength, Ca^{++} independent adhesive state and plasticity to stretch higher order magnitude of the length spanning. This desmosome assembles initially at 32 cell stage. During gastrulation, junctional proteins like plakoglobin and plakophilin acts as bridging protein for different other proteins involved in such interactions. Further, these junctions are supported by tight junctions or "Zonula Occludens(ZO)". These tight junctions provide space for paracellular diffusion and regulate the passage of different ions and solutes. They also contribute to the polarity by compartmentalizing the plasma membrane to the apical and basolateral parts. The tight junction is made up of junctional adhesion molecules claudin and occluding. ZO (1,2,3) acts as scaffolding proteins (Canse et al., 2023). The epithelium and mesenchyme interact to generate branching structures like lungs. This is otherwise known as "branching morphogenesis". This mostly needs modulation of "tyrosine kinases" mediated signaling. The ligands associated are well expressed in the mesenchyme and cognate receptors are expressed in the epithelium (Lang et al., 2021). Signaling facilitated by glial cell line-derived neurotrophic factor (GDNF) regulates branching morphogenesis in the kidney, while fibroblast growth factor (FGF-10) mediates this process in the lungs (Lopez et al., 2023).

3.4.2 Maternal to Zygote Transition

It is the transition point where the control passes from the mother to the zygote (Li et al., 2013). It starts with eliminating the proteins and mRNA received from mother and synthesizing own protein and RNA. Micro RNA like miR-202-3p is responsible for elimination of these maternal proteins. The cell division increases cells number but potency of the cell decreases which renders the emergence of germ layers. Eomesodermine or T-box brain protein 2 (EOMES) are a group of transcription factors which are essential for development of eomesoderm in vertebrates. A deliberate deletion of "eomes" in mice may result in lethality due to neurodegeneration. RNA binding proteins like Bicaudal-C (Biccl) are also involved in MZT. Biccl regulates the left right patterning in case of mouse and anterior posterior polarity in lower vertebrates like *Xenopus*. Various growth factors also play critical role in development process. Insulin like growth factor (IGF) plays a crucial role in production of embryo. These are proteinaceous factors with significant homology to insulin. Expression of IGFs is well reported in bovine endometrium, pig blastocysts and mouse blastocysts. The role of leukaemia inhibitory factor (LIF) is quite prominent in mice during implantation. TGFα and FGF interact with receptor tyrosine kinases. In mice and rats, its importance in embryo formation has been well documented (Heyner et al., 1993).It can be said that the molecular and biochemical trigger for cellular development is widely regulated through various growth factors, transcription factors and stimulants/ inducers secreted temporally and spatially.

Synchrotron X-ray imaging has revolutionized the study of biological structures and functions. Opening new avenues for research in various biological fields, from molecular biology to comparative anatomy, is the ability to provide detailed, three-dimensional images. (Westneat et al., 2008)

REFERENCES

Abbott, Allison L., and Tom Ducibella. "Calcium and the control of mammalian cortical granule exocytosis." *Frontiers in Bioscience* 6 (2001): D792–D806.

Allen, W. R., and S. Wilsher. "A review of implantation and early placentation in the mare." *Placenta* 30, no. 12 (2009): 1005–1015.

Canse, Ceren, Ecem Yildirim, and Aylin Yaba. "Overview of junctional complexes during mammalian early embryonic development." *Frontiers in Endocrinology* 14 (2023): 1150017.

Dean, R. S., & Grizzle, R. (2011). "Prenatal development." In S. Goldstein & J. A. Naglieri (Eds.), *Encyclopedia of Child Behavior and Development*. Boston, MA: Springer. https://doi.org/10.1007/978-0-387-79061-9_2222.

Dumortier, Julien G., Mathieu Le Verge-Serandour, Anna Francesca Tortorelli, Annette Mielke, Ludmilla De Plater, Hervé Turlier, and Jean-Léon Maître. "Hydraulic fracturing and active coarsening position the lumen of the mouse blastocyst." *Science* 365, no. 6452 (2019): 465–468.

Enders, A. C., and I. K. Liu. "Lodgement of the equine blastocyst in the uterus from fixation through endometrial cup formation." *Journal of Reproduction and Fertility*. Supplement 44 (1991): 427–438.

Forde, N., F. Carter, T. E. Spencer, F. W. Bazer, Olivier Sandra, N. Mansouri-Attia, L. A. Okumu et al. "Conceptus-induced changes in the endometrial transcriptome: how soon does the cow know she is pregnant?." *Biology of Reproduction* 85, no. 1 (2011): 144–156.

Franciolli, André Luis Rezende, Bruna Mascaro Cordeiro, Erika Toledo da Fonseca, Marcio Nogueira Rodrigues, Carlos Alberto Palmeira Sarmento, Carlos Eduardo Ambrosio, Ana Flavia de Carvalho, Maria Angélica Miglino, and Luciano Andrade Silva. "Characteristics of the equine embryo and fetus from days 15 to 107 of pregnancy." *Theriogenology* 76, no. 5 (2011): 819–832.

Frith, Martin C., and Shengliang Ni. "DNA conserved in diverse animals since the Precambrian controls genes for embryonic development." *Molecular Biology and Evolution* 40, no. 12 (2023): msad275.

Georgadaki, Katerina, Nikolas Khoury, Demetrios A. Spandidos, and Vasilis Zoumpourlis. "The molecular basis of fertilization." *International Journal of Molecular Medicine* 38, no. 4 (2016): 979–986.

Ginther, O. J. "Fixation and orientation of the early equine conceptus." *Theriogenology* 19, no. 4 (1983a): 613–623.

Ginther, O. J. "Mobility of the early equine conceptus." *Theriogenology* 19, no. 4 (1983b): 603–611.

Green, Willard Wynn, and Laurence Merriam Winters. *Prenatal Development of the Sheep*, Volumes 167–177. University of Minnesota, Agricultural experiment station, 1945.

Heyner, S., N. Shah, R. M. Smith, A. J. Watson, and G. A. Schultz. "The role of growth factors in embryo production." *Theriogenology* 39, no. 1 (1993): 151–161.

Kanka, J. "Gene expression and chromatin structure in the preimplantation embryo." *Theriogenology* 59, no. 1 (2003): 3–19.

Khalili, Kamel, Luis Del Valle, Vandhana Muralidharan, William J. Gault, Nune Darbinian, Jessica Otte, Ellen Meier et al. "Purα is essential for postnatal brain development and developmentally coupled cellular proliferation as revealed by genetic inactivation in the mouse." *Molecular and Cellular Biology* 23, no. 19 (2003): 6857–6875.

Lang, Christine, Lisa Conrad, and Dagmar Iber. "Organ-specific branching morphogenesis." *Frontiers in Cell and Developmental Biology* 9 (2021): 671402.

Lee, Jang-Won, X. Cindy Tian, and Xiangzhong Yang. "Failure of male pronucleus formation is the major cause of lack of fertilization and embryo development in pig oocytes subjected to intracytoplasmic sperm injection." *Biology of Reproduction* 68, no. 4 (2003): 1341–1347.

Li, Lei, Xukun Lu, and Jurrien Dean. "The maternal to zygotic transition in mammals." *Molecular Aspects of Medicine* 34, no. 5 (2013): 919–938.

Lin, Dan, Amelia S. Edwards, James P. Fawcett, Geraldine Mbamalu, John D. Scott, and Tony Pawson. "A mammalian PAR-3–PAR-6 complex implicated in Cdc42/Rac1 and aPKC signalling and cell polarity." *Nature Cell Biology* 2, no. 8 (2000): 540–547.

López, S. L., Cirio, M. C. and S. Garagna. "Editorial: Early animal development: From fertilization to gastrulation." *Frontiers in Cell and Developmental Biology* 11 (2023): 1146778.

McGeady, T. A., P. J. Quinn, E. S. Fitzpatrick, M. T. Ryan, D. Kilroy, and P. Lonergan. Veterinary Embryology. 2nd ed. Hoboken, NJ: John Wiley & Sons, 2017.

Mehlen, Patrick, Fréderic Mille, and Chantal Thibert. "Morphogens and cell survival during development." *Journal of Neurobiology* 64, no. 4 (2005): 357–366.

Mu, Jian, Zhou Zhou, Qing Sang, and Lei Wang. "The physiological and pathological mechanisms of early embryonic development." *Fundamental Research* 2, no. 6 (2022): 859–872.

Phemister, Robert D. "Nonneurogenic reproductive failure in the bitch." *Veterinary Clinics of North America* (1974): 573–586.

Piliszek, Anna, and Zofia E. Madeja. "Pre-implantation development of domestic animals." *Current Topics in Developmental Biology* 128 (2018): 267–294.

Sanchez, José M., Constantine A. Simintiras, and Patrick Lonergan. "Aspects of embryo-maternal communication in establishment of pregnancy in cattle." *Animal Reproduction* 16 (2019): 376–385.

Sharp, D. C. "The early foetal life of the equine conceptus." *Animal Reproduction Science* 60 (2000): 679–689.

Spencer, Thomas E., Greg A. Johnson, Fuller W. Bazer, and Robert C. Burghardt. "Implantation mechanisms: insights from the sheep." *Reproduction* 128, no. 6 (2004): 657–668.

Suzuki, Atsushi, and Shigeo Ohno. "The PAR-aPKC system: lessons in polarity." *Journal of Cell Science* 119, no. 6 (2006): 979–987.

Telford, Nancy A., Andrew J. Watson, and Gilbert A. Schultz. "Transition from maternal to embryonic control in early mammalian development: a comparison of several species." *Molecular Reproduction and Development* 26, no. 1 (1990): 90–100.

Valadão, Loide, Helena Moreira da Silva, and Fernando Moreira da Silva. "Bovine embryonic development to implantation." In *Embryology-Theory and Practice* B. Wu and H. L. Feng (eds.). London: IntechOpen, 2018.

Valasi, I., M. S. Barbagianni, K. S. Ioannidi, N. G. C. Vasileiou, G. C. Fthenakis, and A. Pourlis. "Developmental anatomy of sheep embryos, as assessed by means of ultrasonographic evaluation." *Small Ruminant Research* 152 (2017): 56–73.

Vestweber, Dietmar, Achim Gossler, Klaus Boller, and Rolf Kemler. "Expression and distribution of cell adhesion molecule uvomorulin in mouse preimplantation embryos." *Developmental Biology* 124, no. 2 (1987): 451–456.

Vinot, Stéphanie, Tran Le, Shigeo Ohno, Tony Pawson, Bernard Maro, and Sophie Louvet-Vallée. "Asymmetric distribution of PAR proteins in the mouse embryo begins at the 8-cell stage during compaction." *Developmental Biology* 282, no. 2 (2005): 307–319.

Walser, Claudia B., and Howard D. Lipshitz. "Transcript clearance during the maternal-to-zygotic transition." *Current Opinion in Genetics & Development* 21, no. 4 (2011): 431–443.

Westneat, Mark W., John J. Socha, and Wah-Keat Lee. "Advances in biological structure, function, and physiology using synchrotron X-ray imaging." *Annual Review of Physiology* 70 (2008): 119–142.

Wu, Di, and Jurrien Dean. "Maternal factors regulating pre-implantation development in mice." *Current Topics in Developmental Biology* 140 (2020): 317–340.

Wu, Yinqiao, Yuedong Zhang, Hang Liu, Yun Gao, Yuyan Liu, Ling Chen, Lu Liu et al. "Genome-wide identification of functional enhancers and their potential roles in pig breeding." *Journal of Animal Science and Biotechnology* 13, no. 1 (2022): 75.

Yoshinaga, Koji. "A historical review of blastocyst implantation research." *Biology of Reproduction* 99, no. 1 (2018): 175–195.

4 Physiology of the Cardiovascular System

Abhishek Kalundia
Cornerstone Pet Clinic Hyderabad - 89, India

4.1 INTRODUCTION

The cardiovascular system, also known as the circulatory system, is a complex network of organs and vessels responsible for circulating blood throughout the body. It consists of the heart, blood vessels (arteries, veins, and capillaries), and blood. The cardiovascular system plays a crucial role in maintaining homeostasis by delivering oxygen, nutrients, hormones, and other essential substances to cells and tissues, while removing metabolic waste products and carbon dioxide.

The importance of cardiovascular function includes

1. Oxygen Delivery: The cardiovascular system transports oxygen-rich blood from the lungs to various tissues and organs, ensuring adequate oxygen supply for cellular respiration and energy production.
2. Nutrient Transport: It carries nutrients absorbed from the digestive system to cells, providing essential building blocks for growth, repair, and maintenance of tissues.
3. Waste Removal: Metabolic waste products, such as carbon dioxide and urea, are transported by the cardiovascular system to organs like the lungs and kidneys for elimination from the body.
4. Hormone Distribution: Hormones produced by endocrine glands are transported through the bloodstream to target tissues, regulating various physiological processes such as metabolism, growth, and reproduction.
5. Thermoregulation: Blood circulation helps regulate body temperature by distributing heat produced by metabolic processes and facilitating heat exchange with the environment through vasodilation and vasoconstriction.
6. Immune Response: White blood cells, antibodies, and other components of the immune system travel through the bloodstream to detect and eliminate pathogens and foreign invaders, contributing to the body's defense against infections.
7. Blood Pressure Regulation: The cardiovascular system regulates blood pressure to ensure adequate tissue perfusion while maintaining vascular integrity and preventing complications such as hypertension and hypotension.

Overall, the cardiovascular system is essential for sustaining life by maintaining the proper distribution of oxygen, nutrients, and waste products throughout the body and supporting various physiological functions. Dysfunction of the cardiovascular system can lead to serious health consequences, including cardiovascular diseases such as heart attacks, strokes, and heart failure, underscoring the critical importance of its proper functioning.

4.2 ANATOMY OF THE CARDIOVASCULAR SYSTEM

The anatomy of the cardiovascular system encompasses the structure and organization of its main components: the heart, blood vessels, and blood.

4.2.1 Heart

A. Structure:
- The heart is a muscular organ located in the thoracic cavity, slightly to the left of the midline.
- It is divided into four chambers: two atria (singular: atrium) and two ventricles.
- The heart is enclosed in a double-layered sac called the pericardium, which provides protection and anchorage.

B. Chambers and Valves:
- Atria: The two upper chambers of the heart receive blood from the veins and pump it into the ventricles.
- Ventricles: The two lower chambers of the heart receive blood from the atria and pump it out to the arteries.

DOI: 10.1201/9781003426851-4

- Valves: Valves within the heart ensure unidirectional blood flow:
  - Atrioventricular valves (AV valves): The tricuspid valve (right side) and mitral valve (left side) separate the atria from the ventricles.
  - Semilunar valves: The pulmonary valve (right side) and aortic valve (left side) separate the ventricles from the pulmonary artery and aorta, respectively (Figure 4.1).

C. Coronary Circulation:

- The heart has its own blood supply through the coronary arteries, which branch off the aorta.
- Coronary veins drain deoxygenated blood from the myocardium into the right atrium (Figure 4.2).

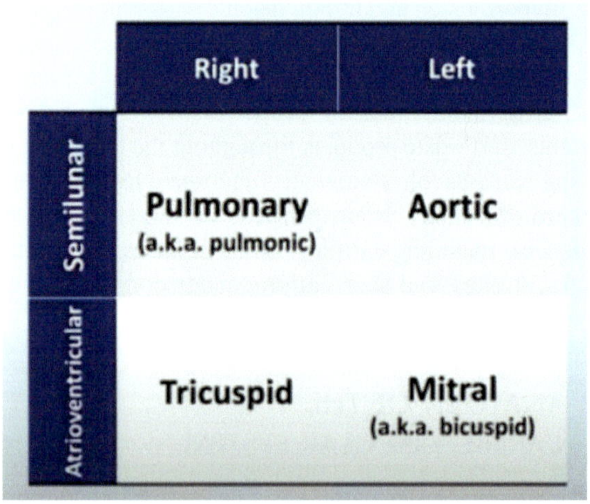

FIGURE 4.1 Pulmonary and Tricuspid valve

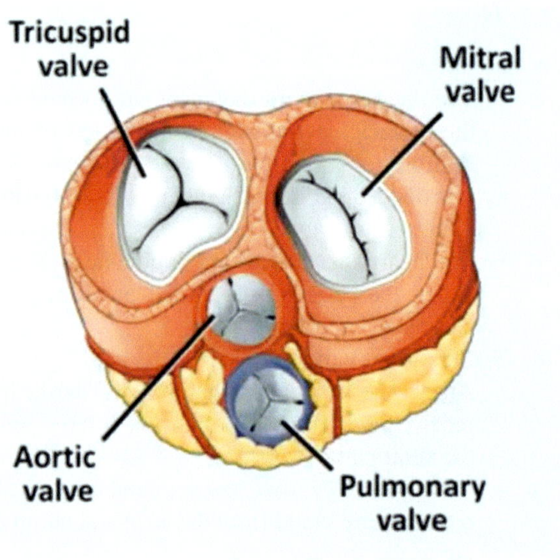

FIGURE 4.2 Coronary circulation

4.2.2 Blood Vessels

A. Arteries:

- Arteries carry oxygenated blood away from the heart to various tissues and organs.
- They have thick, muscular walls to withstand high pressure and maintain blood flow (Table 4.1).

B. Veins:

- Veins carry deoxygenated blood from tissues back to the heart.
- They have thinner walls compared to arteries and contain valves to prevent backflow.

C. Capillaries:

- Capillaries are tiny, thin-walled vessels where gas exchange and nutrient/waste exchange occur between blood and tissues.

TABLE 4.1
Blood Vessels and their Function

Arteries	All blood from the heart leaves into the aorta. From the aorta, they move into the large and medium sized arteries which are thick walled containing a significant amount of elastic tissues and smooth muscles. This allows them to handle the high pressure.
Arterioles	Arteries divide into smaller arterioles which are the sight of greatest resistance to the blood flow through the circulation.
Capillaries	Arterioles further divide into the further microscopic network of vessels called the capillaries. Which are sometimes as small as lines by a single layer of endothelial cells. Capillaries are the place where the actual exchange of fas between the blood vessels and the target peripheral tissue occurs. Although the capillaries are tiny, their numbers cross in billions in the body. Such thattheor net cross sectional area is much larger than any other level of the blood vessels. Capillaries are the place where the blood flows very slowly through the lumen. Hence allowing sufficient time for the gas, nutrients and waste exchange.
Venules	From the capillaries, the blood travels into the venules
Veins	From venules the blood flows into the veins. There is comparatively less elastic tissue in the veins than in arteries. Hence the veins have a larger capacity of distensibility. This allows the veins to act as the reservoir for the blood volume.
Lymphatics	In addition to the blood vessels, there are also conduits. These are responsible to carry the interstitial fluids back to the main circulatory system. Interstitial fluids are the extravascular fluid which surrounds the cells. They sit in the connective tissue. These lymphatic vessels merge together and feed into a conduit called the thoracic duct. Which further joins the large veins of the thoracic cavity.

- They form extensive networks throughout the body, allowing for efficient exchange of substances.

4.3 BLOOD

4.3.1 Components of Blood

1. Plasma: Liquid portion of blood containing water, electrolytes, proteins, hormones, and waste products.
2. Formed Elements:
   a. Erythrocytes (Red Blood Cells): Carry oxygen to tissues and remove carbon dioxide.
   b. Leukocytes (White Blood Cells): Play a role in the immune response.
   c. Platelets: Cell fragments involved in blood clotting and hemostasis.

4.3.2 Functions of Blood

1. Oxygen Transport: Hemoglobin in red blood cells binds and carries oxygen from the lungs to tissues.
2. Nutrient Transport: Glucose, amino acids, lipids, and vitamins are transported to cells for energy and metabolism.
3. Waste Removal: Carbon dioxide and metabolic waste products are transported to organs for elimination.
4. Hormone Transport: Hormones secreted by endocrine glands are carried in the bloodstream to target tissues.
5. Immune Response: White blood cells and antibodies defend against pathogens and foreign substances.
6. Thermoregulation: Blood helps regulate body temperature by distributing heat.

Understanding the anatomy of the cardiovascular system is essential for comprehending its function and the pathophysiology of various cardiovascular diseases.

4.4 CARDIAC CYCLE

The cardiac cycle consists of several phases that describe the events occurring during each heartbeat, while the electrical activity of the heart involves the generation and propagation of electrical impulses that coordinate the contraction and relaxation of cardiac muscle cells.

4.4.1 Phases of the Cardiac Cycle

A. Diastole:
- Early Diastole:
  - Period of ventricular relaxation and filling.
  - Blood flows passively from the atria into the ventricles through the open atrioventricular valves.
- Late Diastole:
  - Atrial contraction (atrial systole) occurs, completing ventricular filling.
  - The atria contract, pushing the remaining blood into the ventricles.

B. Systole:
- Isovolumetric Contraction:
  - Ventricular contraction begins, causing pressure to rise.
  - Atrioventricular valves close (S1 heart sound) as ventricular pressure exceeds atrial pressure, preventing backflow into the atria.
- Ventricular Ejection:
  - Ventricular pressure exceeds arterial pressure, leading to the opening of the semilunar valves.
  - Blood is ejected from the ventricles into the pulmonary artery and aorta.
- Isovolumetric Relaxation:
  - Ventricular relaxation begins, causing ventricular pressure to fall.
  - Semilunar valves close (S2 heart sound) as arterial pressure exceeds ventricular pressure, preventing backflow into the ventricles.

4.4.2 Electrical Activity of the Heart

A. Sinoatrial (SA) Node:
- The SA node is the pacemaker of the heart, located in the right atrium near the opening of the superior vena cava.
- It generates rhythmic electrical impulses that initiate each heartbeat.
- The impulses spread through the atria, causing atrial contraction.

B. Atrioventricular (AV) Node:
- The AV node is located at the junction of the atria and ventricles.

- It delays the transmission of electrical impulses to allow for complete atrial contraction before ventricular contraction.
- The AV node conducts the impulses to the bundle of His.

C. Bundle of His and Purkinje Fibers:
- The bundle of His is a bundle of specialized cardiac muscle fibers that conduct the impulses from the AV node.
- It branches into the left and right bundle branches, which extend through the interventricular septum.
- The Purkinje fibers spread throughout the ventricles, transmitting the impulses rapidly and causing coordinated ventricular contraction.

D. Electrocardiogram (ECG):
- An ECG records the electrical activity of the heart, providing valuable information about its rhythm and function.
- P waves represent atrial depolarization, QRS complex represents ventricular depolarization, and T waves represent ventricular repolarization.

Understanding the phases of the cardiac cycle and the electrical activity of the heart is essential for assessing cardiac function and diagnosing various cardiac conditions.

4.5 BLOOD PRESSURE REGULATION

Blood pressure regulation in dogs and cats involves complex mechanisms that ensure adequate tissue perfusion while maintaining vascular integrity. Several factors influence blood pressure, including neural, hormonal, and local regulatory mechanisms.

4.5.1 Mechanisms of Blood Pressure Regulation

A. Autonomic Nervous System Control:
- Sympathetic Nervous System (SNS):
  - Activation of the SNS leads to vasoconstriction of arterioles, increasing peripheral resistance and raising blood pressure.
  - Norepinephrine released from sympathetic nerve endings stimulates alpha-adrenergic receptors on vascular smooth muscle cells, causing constriction.
- Parasympathetic Nervous System (PNS):
  - The PNS primarily influences heart rate rather than peripheral resistance.
  - Acetylcholine released from parasympathetic nerve endings binds to muscarinic receptors on the heart, slowing heart rate (negative chronotropic effect).

B. Renin-Angiotensin-Aldosterone System (RAAS):
- Renin is released from the kidneys in response to decreased blood flow or decreased sodium levels.
- Renin converts angiotensinogen to angiotensin I, which is then converted to angiotensin II by angiotensin-converting enzyme (ACE) primarily in the lungs.
- Angiotensin II causes vasoconstriction of arterioles, leading to increased blood pressure.
- Angiotensin II also stimulates the release of aldosterone from the adrenal glands, promoting sodium and water retention, which increases blood volume and blood pressure.

C. Baroreceptor Reflex:
- Baroreceptors are stretch receptors located in the walls of blood vessels, particularly in the carotid sinuses and aortic arch.
- They sense changes in blood pressure and send signals to the brainstem (medulla oblongata) to adjust heart rate and vascular tone accordingly.
- Increased blood pressure leads to increased baroreceptor firing, resulting in decreased sympathetic activity and increased parasympathetic activity, causing vasodilation and decreased heart rate to lower blood pressure.
- Conversely, decreased blood pressure leads to decreased baroreceptor firing, resulting in increased sympathetic activity and decreased parasympathetic activity, causing vasoconstriction and increased heart rate to raise blood pressure.

4.5.2 Factors Affecting Blood Pressure

A. Cardiac Output:
- Cardiac output is the product of heart rate and stroke volume and represents the volume of blood ejected by the heart per minute.
- Changes in heart rate or stroke volume can significantly impact blood pressure.

B. Peripheral Resistance:
- Peripheral resistance refers to the resistance encountered by blood flow in the systemic circulation, primarily in the arterioles.
- Factors such as vascular tone, blood viscosity, and vessel diameter influence peripheral resistance.

C. Blood Volume:
- Blood volume affects blood pressure by altering venous return to the heart and cardiac output.

- Factors influencing blood volume include fluid intake, fluid loss (e.g., through urine and sweat), and hormonal regulation of fluid balance.

In dogs and cats, similar mechanisms and factors regulate blood pressure as in other mammals. However, individual differences in breed, size, age, and health status may influence the specific response to these regulatory mechanisms. Understanding these mechanisms and factors is essential for diagnosing and managing conditions such as hypertension and hypotension in veterinary medicine.

4.6 CARDIAC OUTPUT AND STROKE VOLUME

4.6.1 Definition and Calculation

Cardiac output (CO) is the volume of blood ejected by the heart per minute and is determined by the product of heart rate (HR) and stroke volume (SV). Hormonal regulation plays a crucial role in modulating cardiac output and stroke volume in dogs and cats, ensuring adequate tissue perfusion and oxygen delivery. Several hormones, including catecholamines, renin-angiotensin-aldosterone system (RAAS) components, and atrial natriuretic peptide (ANP), are involved in regulating cardiac function.

4.6.2 Regulation of Cardiac Output

A. **Frank-Starling Mechanism:**
- Definition:
  The Frank-Starling mechanism and sympathetic and parasympathetic regulation play key roles in modulating cardiac output in small animals, ensuring efficient adaptation to changing physiological demands and maintaining cardiovascular homeostasis.
  - The Frank-Starling mechanism describes the relationship between preload (end-diastolic volume) and stroke volume.
  - It states that within physiological limits, an increase in preload leads to an increase in stroke volume, resulting in enhanced cardiac output.
- Mechanism:
  - Ventricular Filling:
    - During diastole, the ventricles fill with blood from the atria.
    - Increased venous return, due to factors such as increased blood volume or venoconstriction, leads to increased ventricular filling and preload.
  - Stretch of Cardiac Muscle:
    - Increased ventricular filling stretches the cardiac muscle fibers, leading to increased sarcomere length.
    - This stretch enhances the force of contraction (contractility) of the cardiac muscle fibers, resulting in increased stroke volume.
  - Optimal Length:
    - The Frank-Starling mechanism operates within physiological limits, as excessively high preload can lead to reduced efficiency and impaired cardiac function.

B. **Sympathetic and Parasympathetic Regulation of Cardiac Output:**
- Sympathetic Regulation:
  - Activation:
    - Sympathetic nervous system (SNS) activation occurs in response to stress, exercise, or increased metabolic demand.
    - Release of catecholamines (epinephrine and norepinephrine) from the adrenal medulla and sympathetic nerve endings stimulates β1-adrenergic receptors on cardiac muscle cells.
  - Effects:
    - Increased heart rate (positive chronotropic effect) enhances the rate of ventricular filling and ejection, leading to increased cardiac output.
    - Increased contractility (positive inotropic effect) enhances stroke volume and cardiac output.
    - Arteriolar vasoconstriction increases peripheral resistance, maintaining blood pressure and improving tissue perfusion.
  - Importance:
    - Sympathetic activation allows for rapid adjustment of cardiac output in response to acute physiological stressors or increased demand, such as during exercise or fight-or-flight responses.
- Parasympathetic Regulation:
  - Activation:
    - Parasympathetic nervous system (PNS) activation occurs via the vagus nerve (cranial nerve X).

    - Release of acetylcholine from parasympathetic nerve endings stimulates muscarinic receptors on cardiac muscle cells.
  - Effects:
    - Decreased heart rate (negative chronotropic effect) slows ventricular filling and ejection, reducing cardiac output.
    - Minimal direct effect on contractility but may indirectly influence cardiac function by altering heart rate and ventricular filling time.
  - Importance:
    - Parasympathetic activity predominates during rest and relaxation, helping to maintain baseline cardiac output and conserve energy.

In small animals, including dogs and cats, the Frank-Starling mechanism and sympathetic and parasympathetic regulation contribute to the dynamic control of cardiac output, allowing for efficient adaptation to physiological demands and maintaining cardiovascular homeostasis under various conditions. Understanding these regulatory mechanisms is essential for diagnosing and managing cardiovascular disorders in veterinary medicine.

C. **Hormonal Regulation of Cardiac Output and Stroke Volume:**

- Catecholamines Atrial Natriuretic Peptide (ANP) (Epinephrine and Norepinephrine):
  - Release: Catecholamines are released from the adrenal medulla in response to stress, physical activity, and sympathetic nervous system stimulation.
  - Mechanism:
    - Catecholamines bind to adrenergic receptors on cardiac muscle cells and vascular smooth muscle cells.
    - Stimulation of β1-adrenergic receptors on the heart increases heart rate and contractility, leading to increased stroke volume and cardiac output.
    - Constriction of arterioles via α-adrenergic receptors increases peripheral resistance, further elevating blood pressure and cardiac workload.
- Renin-Angiotensin-Aldosterone System (RAAS):
  - Renin Release: Renin is released from the kidneys in response to decreased blood flow, decreased sodium levels, or sympathetic stimulation.
  - Angiotensin II Formation:
    - Renin converts angiotensinogen to angiotensin I, which is then converted to angiotensin II by angiotensin-converting enzyme (ACE), primarily in the lungs.
  - Mechanism:
    - Angiotensin II causes vasoconstriction of arterioles, increasing systemic vascular resistance and blood pressure.
    - It also stimulates the release of aldosterone from the adrenal glands, promoting sodium and water retention in the kidneys, increasing blood volume, preload, and stroke volume.
    - Increased preload enhances ventricular filling, leading to increased stroke volume and cardiac output.
- Atrial Natriuretic Peptide (ANP):
  - Release: ANP is secreted by atrial myocytes in response to increased atrial pressure and stretch.
  - Mechanism:
    - ANP acts to counterbalance the effects of the RAAS and sympathetic nervous system by promoting natriuresis (excretion of sodium) and diuresis (excretion of water) in the kidneys.
    - This leads to decreased blood volume and preload, reducing ventricular filling pressure and stroke volume.
    - ANP also causes vasodilation of arterioles, decreasing systemic vascular resistance and afterload, which can further improve cardiac output.

In dogs and cats, hormonal regulation of cardiac output and stroke volume involves a delicate balance between sympathetic activation, RAAS activity, and hormonal counter-regulatory mechanisms such as ANP release. Dysregulation of these hormonal pathways can contribute to cardiovascular disorders such as hypertension, heart failure, and volume overload. Understanding these mechanisms is essential for managing cardiac conditions and optimizing treatment strategies in veterinary medicine.

4.6.3 Blood Flow Regulation

Blood flow regulation in animals like dogs and cats involves both local and systemic control mechanisms, ensuring adequate tissue perfusion and oxygen delivery while maintaining overall cardiovascular homeostasis. These regulatory mechanisms operate at various levels, including within individual tissues and organs (local control) and throughout the entire cardiovascular system (systemic control).

I. **Local Control Mechanisms:**

A. Autoregulation:

- Definition: Autoregulation refers to the ability of tissues and organs to adjust their blood flow in response to changes in metabolic demand or perfusion pressure.
- Mechanism:
  - Local factors such as tissue oxygenation, carbon dioxide levels, pH, adenosine, and potassium ions regulate vascular tone.
  - Vasodilation occurs in response to metabolic vasodilators (e.g., adenosine) or decreased oxygen tension (hypoxia), increasing blood flow to metabolically active tissues.
  - Vasoconstriction occurs in response to metabolic vasoconstrictors (e.g., potassium ions) or increased oxygen tension (hyperoxia), reducing blood flow to less metabolically active tissues.
- Importance: Autoregulation helps match blood flow to metabolic demand, ensuring adequate oxygen and nutrient delivery while preventing tissue ischemia or hypoxia.

B. Metabolic Regulation:

- Definition: Metabolic regulation involves the local release of vasodilator substances in response to tissue metabolic activity.
- Mechanism:
  - During increased metabolic activity (e.g., exercise, inflammation), tissues release vasodilator substances such as adenosine, prostaglandins, nitric oxide, and potassium ions.
  - These vasodilators act on arterioles to promote vasodilation, increasing blood flow to meet increased metabolic demands.
- Importance: Metabolic regulation ensures that tissues receive adequate blood flow and oxygen delivery to support their metabolic needs during physiological or pathological conditions.

C. Endothelial Regulation:

- Definition: Endothelial cells lining the inner surface of blood vessels play a crucial role in regulating vascular tone and blood flow.
- Mechanism:
  - Endothelial cells release vasoactive substances, including nitric oxide (NO), prostacyclin, and endothelin-1.
  - Nitric oxide and prostacyclin promote vasodilation by relaxing vascular smooth muscle cells, while endothelin-1 induces vasoconstriction.
- Importance: Endothelial regulation helps maintain vascular homeostasis by balancing vasodilator and vasoconstrictor influences, contributing to blood flow regulation and vascular integrity.

II. **Systemic Control Mechanisms:**

A. Neural Regulation:

- Sympathetic Nervous System (SNS):
  - Sympathetic activation leads to widespread vasoconstriction via α-adrenergic receptors, increasing systemic vascular resistance and blood pressure.
  - Selective vasodilation occurs in certain vascular beds (e.g., skeletal muscle, heart) in response to specific physiological demands (e.g., exercise).
- Parasympathetic Nervous System (PNS):
  - Parasympathetic activity has minimal direct effect on vascular tone but may influence blood flow indirectly by altering cardiac output and metabolic activity.

B. Hormonal Regulation:

- Renin-Angiotensin-Aldosterone System (RAAS):
  - Angiotensin II induces vasoconstriction of arterioles, increasing systemic vascular resistance and blood pressure.
  - Aldosterone promotes sodium and water retention, increasing blood volume and preload, which can affect blood flow and cardiac output.
- Other Hormones:
  - Catecholamines (epinephrine and norepinephrine) from the adrenal medulla can induce vasoconstriction or vasodilation depending on the receptor subtype and target tissue.
  - Vasopressin (antidiuretic hormone) promotes vasoconstriction and water retention, increasing blood pressure.

In dogs and cats, the local and systemic control mechanisms of blood flow regulation ensure optimal tissue perfusion and oxygen delivery while maintaining cardiovascular homeostasis under various physiological and pathological conditions. Understanding these regulatory mechanisms is essential for diagnosing and managing cardiovascular disorders in veterinary medicine.

4.7 BLOOD COMPOSITION AND FUNCTION

Certainly! Blood composition in animals, including dogs and cats, consists of plasma and formed elements, each playing crucial roles in various physiological functions. Here's a detailed explanation of blood composition and their functions:

4.7.1 Components of the Blood

A. Plasma:
- Plasma is the liquid portion of blood, making up about 55% of total blood volume in mammals.
- It is a complex mixture of water, electrolytes, proteins, hormones, nutrients, waste products, and gases.
- Plasma serves as a medium for transporting substances throughout the body and maintaining osmotic balance.

B. Water:
- Function:
  - Provides a solvent for transporting dissolved substances, such as gases, nutrients, and waste products.
  - Helps maintain blood volume and blood pressure.
- Importance:
  - Water is essential for maintaining hydration and facilitating metabolic processes in cells and tissues.

C. Electrolytes:
- Types:
  - Sodium (Na+), potassium (K+), chloride (Cl-), calcium (Ca2+), magnesium (Mg2+), bicarbonate (HCO3-), and phosphate (PO4^3-).
- Functions:
  - Maintain osmotic balance, acid-base balance, and electrical neutrality.
  - Essential for nerve conduction, muscle contraction, and enzyme activity.
- Importance:
  - Electrolyte balance is critical for normal physiological function and cellular homeostasis.

D. Proteins:
- Albumin:
  - Function:
  - Maintains oncotic pressure, helping to regulate fluid distribution between blood and tissues.
  - Binds and transports various substances, including hormones, fatty acids, and drugs.
  - Importance:
  - Albumin plays a key role in maintaining blood volume, pressure, and oncotic pressure.
- Globulins:
  - Function:
  - Include immunoglobulins (antibodies), which play a vital role in the immune response by recognizing and neutralizing pathogens.
  - Transport various substances, such as lipids, hormones, and vitamins.
  - Importance:
  - Globulins contribute to immune function, lipid transport, and hormone distribution.
- Fibrinogen:
  - Function:
  - Essential for blood clotting and hemostasis.
  - Converted to fibrin during clot formation, providing the framework for platelet aggregation and clot stabilization.
  - Importance:
  - Fibrinogen plays a crucial role in preventing excessive bleeding and promoting wound healing.

E. Formed Elements:
- Formed elements are cellular components suspended in plasma, including erythrocytes (red blood cells), leukocytes (white blood cells), and platelets (thrombocytes).
- They are produced in the bone marrow through hematopoiesis and are involved in various physiological functions, including oxygen transport, immune defense, and hemostasis.

F. Erythrocytes (Red Blood Cells):
- Function:
  - Transport oxygen from the lungs to tissues and carbon dioxide from tissues to the lungs.

- Contain hemoglobin, a protein that binds oxygen and carbon dioxide.
- Importance:
  - Erythrocytes play a critical role in oxygen delivery to tissues and maintaining acid-base balance through carbon dioxide removal.
- Regulation:
  - Erythropoiesis is regulated by erythropoietin, a hormone produced by the kidneys in response to hypoxia or decreased oxygen delivery.

G. Leukocytes (White Blood Cells):
- Function:
  - Play a crucial role in the immune response by identifying and neutralizing pathogens, foreign substances, and abnormal cells.
  - Include various types such as neutrophils, lymphocytes, monocytes, eosinophils, and basophils, each with specific functions in immunity.
- Importance:
  - Leukocytes defend the body against infections, regulate inflammation, and contribute to tissue repair and remodeling.
- Regulation:
  - Leukopoiesis is regulated by various cytokines, growth factors, and hormones, depending on the specific type of leukocyte and the physiological state of the organism.

H. Platelets (Thrombocytes):
- Function:
  - Essential for blood clotting and hemostasis.
  - Form platelet plugs at sites of vascular injury and contribute to clot formation and stabilization through release of clotting factors and fibrinogen.
- Importance:
  - Platelets prevent excessive bleeding and promote wound healing by forming and maintaining blood clots.
- Regulation:
  - Platelet production (thrombopoiesis) is regulated by thrombopoietin, a hormone produced by the liver and kidneys in response to thrombocytopenia or increased platelet destruction.

4.7.2 Functions of Blood

Blood and its components in animals serve several crucial functions that are essential for maintaining homeostasis and supporting overall health. Here's an overview of the functions of blood and its components in ten different ways:

A. Oxygen Transport:
- Erythrocytes (red blood cells) contain hemoglobin, a protein that binds and transports oxygen from the lungs to tissues throughout the body.
- Oxygenated blood delivers oxygen to metabolically active tissues, where it is utilized in cellular respiration to produce energy (ATP).

B. Carbon Dioxide Transport:
- Erythrocytes also play a role in transporting carbon dioxide, a waste product of cellular metabolism, from tissues to the lungs for elimination.
- Carbon dioxide is transported in the blood in three forms: dissolved in plasma, bound to hemoglobin, and as bicarbonate ions ($HCO3^-$) formed by the carbonic anhydrase enzyme reaction.

C. Nutrient Transport:
- Blood transports nutrients absorbed from the digestive tract, such as glucose, amino acids, fatty acids, vitamins, and minerals, to tissues and organs for energy production, growth, and repair.

D. Waste Product Removal:
- Blood carries waste products, including carbon dioxide, urea, uric acid, and bilirubin, to organs of elimination (lungs, kidneys, liver) for excretion from the body.

E. Hormone Transport:
- Blood transports hormones produced by endocrine glands to target tissues and organs, where they regulate various physiological processes, including metabolism, growth, reproduction, and stress response.

F. Immune Function:
- Leukocytes (white blood cells) play a crucial role in the immune response, defending the body against pathogens, foreign substances, and abnormal cells.
- Lymphocytes, monocytes, neutrophils, eosinophils, and basophils are involved in different aspects of innate and adaptive immunity, including phagocytosis, antibody production, and cytokine release.

G. Hemostasis and Blood Coagulation:
- Platelets (thrombocytes) and clotting factors are involved in hemostasis, the process of preventing excessive bleeding and maintaining vascular integrity following injury or trauma.

- Platelets form a temporary plug at the site of vascular injury, while clotting factors promote the formation of a fibrin clot to stabilize the hemostatic plug.

H. Temperature Regulation:
- Blood helps regulate body temperature by redistributing heat throughout the body, especially from core organs to peripheral tissues, and by facilitating heat exchange with the environment through vasodilation and vasoconstriction.

I. pH Regulation:
- Blood buffers help maintain the pH of body fluids within a narrow range, ensuring proper cellular function and enzymatic activity.
- Bicarbonate ions (HCO3-) and proteins act as important buffers to regulate acid-base balance in the body.

J. Fluid Balance:
- Blood helps maintain fluid balance in the body by distributing water and electrolytes between tissues and organs and by regulating blood volume and osmolarity through mechanisms such as renal excretion and fluid shifts.

In summary, blood composition in animals consists of plasma and formed elements, each with specific functions essential for maintaining physiological homeostasis and supporting various metabolic processes. Understanding blood composition and its functions is crucial for diagnosing and managing hematological disorders and maintaining overall health in veterinary medicine. Blood and its components play essential roles in transporting nutrients, gasses, hormones, and waste products; regulating immune function, hemostasis, and temperature; and maintaining overall physiological homeostasis in animals.

4.8 HEMOSTASIS AND BLOOD COAGULATION

Hemostasis is the physiological process by which the body prevents excessive bleeding and maintains vascular integrity following injury or trauma. It involves a complex series of events collectively known as blood coagulation or clotting. Here's a detailed explanation of hemostasis and blood coagulation in animals, including dogs and cats:

4.8.1 Hemostasis Process

A. Vascular Spasm:
- Definition:
  - Vascular spasm is the initial response to vascular injury, involving constriction of blood vessels to reduce blood flow and minimize bleeding.
- Mechanism:
  - Contraction of smooth muscle in the vessel wall narrows the lumen, reducing blood flow to the injured area.
- Importance:
  - Vascular spasm helps limit blood loss immediately following injury, providing time for other hemostatic mechanisms to take effect.

B. Platelet Plug Formation:
- Definition:
  - Platelet plug formation is the second stage of hemostasis, involving the accumulation and activation of platelets at the site of vascular injury to form a temporary plug.
- Mechanism:
  - Exposure of collagen and other adhesive proteins in the damaged vessel wall triggers platelet adhesion and activation.
  - Activated platelets release granules containing various substances, including ADP, thromboxane A2, and serotonin, which recruit and activate additional platelets.
  - Platelets undergo shape change, aggregate together, and adhere to the damaged endothelium, forming a platelet plug.
- Importance:
  - Platelet plug formation helps seal small vascular injuries and stabilize the clotting process, preventing further blood loss.

C. Blood Coagulation cascade (Clotting):
- Definition:
  - Blood coagulation is the final stage of hemostasis, involving the formation of a fibrin clot to reinforce the platelet plug and stabilize the injured vessel.
- Mechanism:
  - The coagulation cascade is a series of enzymatic reactions that culminate in the conversion of fibrinogen to fibrin, the insoluble protein meshwork that forms the clot.
  - The intrinsic pathway is activated by contact with negatively charged surfaces (e.g., exposed collagen) within the vessel wall.
  - The extrinsic pathway is initiated by tissue factor released from damaged tissues outside the blood vessel.

- Both pathways converge on the common pathway, leading to the activation of thrombin, which catalyzes the conversion of fibrinogen to fibrin.
- Fibrin strands polymerize and cross-link to form a stable clot, reinforcing the platelet plug and sealing the vascular injury.

- Importance:
  - Blood coagulation provides long-term stabilization of the hemostatic plug, preventing dislodgement and ensuring effective wound healing.

4.8.2 Regulation of Hemostasis

Blood coagulation regulation in animals involves multiple pathways, including the intrinsic and extrinsic pathways, as well as anticoagulation mechanisms and fibrinolysis. Here's how each component works:

A. Intrinsic Pathway:
- The intrinsic pathway is initiated by factors present within the bloodstream and is activated by contact with negatively charged surfaces, such as exposed collagen in damaged blood vessels.
- Key components of the intrinsic pathway include:
- Factor XII (Hageman factor): Factor XII is activated by contact with collagen or other surfaces and initiates the intrinsic pathway.
- Factor XI: Activated factor XII converts factor XI to factor XIa, which in turn activates factor IX.
- Factor IX: Activated factor IX forms a complex with factor VIIIa and calcium ions (Ca2+), leading to the activation of factor X.
- The intrinsic pathway converges with the extrinsic pathway at the activation of factor X, leading to the common pathway and the formation of thrombin.

B. Extrinsic Pathway:
- The extrinsic pathway is initiated by tissue factor (also known as factor III), which is released from damaged tissues outside the bloodstream.
- Key components of the extrinsic pathway include:
- Tissue Factor: Tissue factor binds to and activates factor VII, forming a tissue factor-factor VIIa complex.
- Factor VIIa: Activated factor VIIa, in the presence of tissue factor, activates factor X directly.
- The extrinsic pathway converges with the intrinsic pathway at the activation of factor X, leading to the common pathway and the formation of thrombin.

C. Common Pathway:
- The common pathway is the final stage of the coagulation cascade, where the intrinsic and extrinsic pathways converge to produce thrombin.
- Thrombin catalyzes the conversion of fibrinogen to fibrin, which forms the insoluble meshwork of the blood clot.

D. Anticoagulation Mechanisms:
- Anticoagulation mechanisms maintain hemostatic balance by inhibiting coagulation factors and preventing excessive clot formation.
- Key anticoagulant factors include:
- Antithrombin III: Antithrombin III inhibits the activity of thrombin and other coagulation factors, including factors IXa, Xa, XIa, and XIIa.
- Protein C and Protein S: Protein C, activated by thrombin, inactivates factors Va and VIIIa, limiting further thrombin generation and clot propagation.
- Tissue Factor Pathway Inhibitor (TFPI): TFPI inhibits the extrinsic pathway by blocking the activity of tissue factor and factor VIIa.

E. Fibrinolysis:
- Fibrinolysis is the process of clot dissolution, which prevents excessive or inappropriate clot formation and promotes tissue repair.
- Plasminogen is converted to plasmin by tissue plasminogen activator (tPA) or urokinase, leading to fibrin breakdown and clot lysis.
- Plasmin degrades fibrin into soluble fibrin degradation products (FDPs), which are cleared from the circulation.

Overall, blood coagulation regulation in animals involves a complex interplay of intrinsic and extrinsic pathways, anticoagulant mechanisms, and fibrinolysis to maintain hemostatic balance and prevent excessive bleeding or thrombosis. Dysregulation of these processes can lead to hemorrhagic or thrombotic disorders, highlighting the importance of maintaining coagulation homeostasis for overall cardiovascular health in animals.

4.9 CARDIOVASCULAR DISEASES

Cardiovascular diseases in animals, including dogs and cats, encompass a wide range of conditions affecting the heart and blood vessels. These diseases can be congenital (present at birth) or acquired (developed over time) and may result from genetic factors, aging, lifestyle factors, or underlying medical conditions. Here are some common cardiovascular diseases in dogs and cats:

4.9.1 Congenital Heart Diseases

A. Patent Ductus Arteriosus (PDA):
- In PDA, a fetal blood vessel called the ductus arteriosus fails to close after birth, leading to abnormal blood flow between the aorta and pulmonary artery.
- This condition can result in volume overload of the heart chambers, pulmonary hypertension, and heart failure.

B. Ventricular Septal Defect (VSD):
- VSD is a hole in the septum (wall) between the ventricles, allowing blood to flow from the left ventricle to the right ventricle.
- It can lead to increased pulmonary blood flow, pulmonary hypertension, and right-sided heart enlargement.

C. Atrial Septal Defect (ASD):
- ASD is a hole in the septum between the atria, allowing blood to flow from the left atrium to the right atrium.
- It can cause volume overload of the right heart chambers and pulmonary hypertension.

4.9.2 Acquired Heart Diseases

A. Degenerative Valve Disease (Endocardiosis):
- Degenerative valve disease is a common acquired heart condition in older small breed dogs, particularly Cavalier King Charles Spaniels.
- It involves degeneration and thickening of the heart valves, particularly the mitral valve, leading to mitral regurgitation and volume overload of the left atrium and ventricle.

B. Dilated Cardiomyopathy (DCM):
- DCM is a primary myocardial disease characterized by dilation and impaired contractility of the heart chambers.
- It can lead to systolic dysfunction, decreased cardiac output, and congestive heart failure.
- DCM is more common in certain large and giant dog breeds, such as Doberman Pinschers and Boxers.

C. Hypertrophic Cardiomyopathy (HCM):
- HCM is characterized by hypertrophy (thickening) of the myocardium, particularly the left ventricle, without dilation of the chambers.
- It can lead to diastolic dysfunction, impaired relaxation, and restrictive filling of the ventricles.
- HCM is the most common cardiac disease in cats, particularly certain breeds such as Maine Coons and Ragdolls.

D. Chronic Valvular Disease (Chronic Mitral Valve Disease):
- Chronic valvular disease involves degeneration and thickening of the heart valves, particularly the mitral valve, leading to mitral regurgitation.
- It can result in volume overload of the left atrium and ventricle, pulmonary hypertension, and congestive heart failure.

E. Heartworm Disease:
- Heartworm disease is caused by infection with the parasitic worm Dirofilaria immitis, transmitted by mosquitoes.
- The worms reside in the heart and pulmonary arteries, causing inflammation, obstruction of blood flow, and right-sided heart failure.

F. Arrhythmias:
- Arrhythmias are abnormal heart rhythms that can result from various underlying conditions, such as myocardial disease, electrolyte imbalances, or conduction abnormalities.
- They can manifest as tachycardia (fast heart rate), bradycardia (slow heart rate), or irregular heart rhythms.

G. Myocarditis:
- Myocarditis is inflammation of the myocardium (heart muscle), often caused by infectious agents, toxins, or autoimmune disorders.
- It can lead to myocardial dysfunction, arrhythmias, and heart failure.

These are just some of the cardiovascular diseases that can affect dogs and cats. Early detection, appropriate treatment, and management are essential for improving outcomes and maintaining the quality of life for affected animals. Regular veterinary check-ups and screenings can help identify and manage cardiovascular conditions in pets.

4.10 CLINICAL ASSESSMENT OF CARDIOVASCULAR FUNCTION

The clinical assessment of cardiovascular function in animals, including dogs and cats, involves a combination of physical examination and diagnostic tests to evaluate heart structure, function, and overall cardiovascular health. Here's an overview of the clinical assessment methods used in veterinary medicine:

4.10.1 Physical Examination

A. Inspection:
- Observation of general appearance, body condition, and demeanor.
- Assessment of respiratory rate, effort, and pattern.
- Evaluation of mucous membrane color (e.g., pale, cyanotic), capillary refill time, and presence of edema.

B. Palpation:
- Palpation of peripheral pulses (e.g., femoral, dorsal pedal).
- Assessment of heart rate, rhythm, and strength of cardiac contractions.
- Detection of abnormalities such as murmurs, thrills, or palpable masses.

C. Auscultation:
- Auscultation of heart sounds (S1, S2) using a stethoscope.
- Identification and characterization of additional heart sounds (e.g., S3, S4) or murmurs.
- Assessment of lung sounds (e.g., crackles, wheezes) and presence of murmurs transmitted to the lungs.

4.10.2 Diagnostic Tests

A. Electrocardiography (ECG):
- ECG records the electrical activity of the heart, providing information about heart rate, rhythm, and conduction abnormalities.
- Common ECG abnormalities include arrhythmias (e.g., sinus arrhythmia, atrial fibrillation), conduction delays, and chamber enlargement.

B. Thoracic Radiography (X-rays):
- Thoracic radiographs are used to evaluate heart size, shape, and position, as well as pulmonary vasculature and lung parenchyma.
- Findings may include cardiomegaly (enlarged heart), pulmonary edema, pleural effusion, or evidence of congenital heart disease.

C. Echocardiography (Cardiac Ultrasound):
- Echocardiography allows for non-invasive assessment of cardiac structure and function, including chamber dimensions, wall thickness, and valve morphology.
- Doppler imaging can evaluate blood flow velocity, gradient across valves, and presence of regurgitation or stenosis.
- Echocardiography is essential for diagnosing and monitoring congenital heart disease, acquired heart disease, and evaluating response to treatment.

D. Blood Pressure Measurement:
- Blood pressure measurement is important for assessing systemic perfusion and monitoring response to therapy.
- Techniques include Doppler ultrasonography, oscillometric devices, or direct arterial catheterization.
- Hypertension is a common finding in animals with chronic kidney disease, hyperthyroidism, or primary hypertension.

E. Blood Tests:
- Complete Blood Count (CBC) and Serum Biochemistry:
  - CBC can detect anemia, leukocytosis, or thrombocytopenia, while serum biochemistry assesses renal and hepatic function, electrolytes, and biomarkers of heart disease.
- NT-proBNP (N-terminal pro-B-type natriuretic peptide):
  - NT-proBNP is a biomarker released by the heart in response to myocardial stretch and stress and is used as a screening tool for heart disease and monitoring response to treatment.

F. Holter Monitoring:
- Holter monitoring involves continuous ECG recording over a 24-hour period, allowing for detection and characterization of arrhythmias or rhythm disturbances during normal daily activities.

G. Exercise Tolerance Testing:
- Exercise tolerance testing assesses cardiovascular function under controlled conditions, evaluating exercise capacity, heart rate response, and presence of exercise-induced arrhythmias.

H. Cardiac Catheterization and Angiography:
- Invasive procedures used for precise assessment of cardiac anatomy, hemodynamics, and

TABLE 4.2
Various Cardiovascular Diseases

Sl. No.	Component of the cardiovascular diseases	Associated diseases
1	Valves (Endocardium)	Valvular heart disease / Endocarditis
2	Myocardium	Heart failure / Myocarditis
3	Pericardium	Pericardial effusion / Pericarditis
4	The conduction system	Arrhythmia
5	Coronary artery	Coronary artery disease / heart attacks
6	Arteries & arterioles	Peripheral artery disease
7	Veins & venules	Venous insufficiency
8	Capillaries	Peripheral leahage syndrome / Sepsis
9	Lymphatics	Obstructive lymphadenopathy

coronary artery blood flow. They are typically reserved for specialized centers and complex cases.

The combination of physical examination findings and diagnostic test results allows veterinarians to diagnose cardiovascular diseases, assess disease severity, and develop appropriate treatment plans tailored to the individual needs of the patient. Regular monitoring and follow-up evaluations are essential for managing chronic cardiovascular conditions and optimizing patient outcomes (Table 4.2).

4.11 CONCLUSION

The cardiovascular system in animals is a complex and vital network responsible for transporting oxygen, nutrients, hormones, and waste products throughout the body. Understanding the anatomy, function, regulation, and clinical considerations of the cardiovascular system is essential for veterinarians to diagnose and manage cardiovascular diseases effectively.

Anatomy: The cardiovascular system consists of the heart, blood vessels, and blood. The heart pumps blood through a network of arteries, veins, and capillaries, ensuring efficient delivery of oxygenated blood to tissues and organs.

Function: The primary function of the cardiovascular system is to maintain tissue perfusion and oxygen delivery, supporting metabolic processes and cellular function. The heart generates pressure to propel blood through the vasculature, while blood vessels regulate blood flow and pressure.

Regulation: Cardiovascular function is regulated by neural, hormonal, and local mechanisms. Autonomic nervous system control, hormonal pathways such as the renin-angiotensin-aldosterone system, and local factors like autoregulation and endothelial regulation play key roles in modulating cardiac output, vascular tone, and blood pressure.

Clinical Considerations: Clinical assessment of cardiovascular function in animals involves a combination of physical examination and diagnostic tests, including auscultation, electrocardiography, echocardiography, and blood tests. Common cardiovascular diseases in animals include congenital heart diseases, acquired heart diseases such as degenerative valve disease and cardiomyopathies, and other conditions like heartworm disease and arrhythmias.

Consent of Publication: None Declare

Conflict of Interest: None Declare

Acknowledgement: None Declare

BIBLIOGRAPHY

Conway, A. M., et al. (2018). "The effect of hypoxia on cardiovascular system: A review." *Journal of Biology and Medicine*, 91(3), 229–235.

Dukes, H. H., and M. J. Swenson, eds. (2004). *Dukes' Veterinary Physiology*. 13th ed. Ames: Wiley-Blackwell.

Gan, L. M., et al. (2019). "The role of oxidative stress in cardiovascular diseases." *European Journal of Pharmacology*, 835, 229–236.

Hall, John E. (2021). *Guyton and Hall Textbook of Medical Physiology*. 14th ed. Philadelphia: Elsevier.

Johnson, P. C., et al. (2018). "The role of endothelial cells in cardiovascular system." *Circulation Research*, 123(3), 213–220.

Klabunde, R. E. (2011). *Cardiovascular Physiology Concepts*. 2nd ed. Philadelphia: Lippincott Williams & Wilkins.

Kuroda, H., et al. (2020). "The role of autonomic nervous system in cardiovascular regulation." *Journal of Physiology and Pharmacology*, 71(2), 129–133.

Lukaszyk, D., et al. (2017). "The role of sympathetic nervous system in cardiovascular system." *American Journal of Physiology-Heart and Circulatory Physiology*, 313(6), H1009–H1023.

McLean, M., et al. (2019). "The role of endothelial dysfunction in cardiovascular system." *American Journal of Cardiology*, 124(5), 849–855.

Michalak, G., et al. (2019). "The role of endothelin-1 in cardiovascular system." *Advances in Clinical and Experimental Medicine*, 28(12), 1695–1701.

Moe, G. W. (2017). "The role of natriuretic peptides in the cardiovascular system." *Journal of Cardiac Failure*, 23(8), 630–635.

Mohrman, David E., and Lois Jane Heller. (2017). *Cardiovascular Physiology*. 9th ed. New York: McGraw-Hill Education.

Seghaye, M. C. (2016). "The role of inflammation in cardiovascular system." *Frontiers in Physiology*, 7, 399.

Sherwood, L. (2012). *Animal Physiology: From Genes to Organisms*. 2nd ed. Belmont, CA: Cengage Learning.

West, J. B. (2016). "The physiological challenges of the cardiovascular system during gravity transitions: A review." *Acta Astronautica*, 123, 10–14.

5 Physiology of the Lymphatic and Immune System

Pankaj Kumar Patel[1], *Virendra Kumar Singh*[2] *and Amrita Vasava*[3]
[1]College of Veterinary Science & A.H., Sardarkrushinagar, Kamdhenu University,Gujarat-India
[2]College of Veterinary Science & A.H., Navsari, Kamdhenu University, Gujarat-India
[3]College of Veterinary Science & A.H., Anand, Kamdhenu University, Gujarat-India

5.1 INTRODUCTION

5.1.1 Lymphatic System

The lymphatic system is a complex network that helps maintain fluid balance, fights infections, and prevents diseases. It begins with the formation of interstitial fluid derived from capillary filtration and diffusion, resembling plasma but with lower protein content. This fluid forms a gel-like matrix in the tissue spaces, which, under normal conditions, contains minimal free fluid. Lymphatic capillaries, the starting points of the lymphatic system, are fine vessels that arise from these interstitial spaces. They are equipped with one-way valves formed by overlapping endothelial cells, ensuring the unidirectional flow of lymph. Lymph capillaries collect about one-tenth of the fluid that leaks from the arterial end of blood capillaries, returning it to the bloodstream and bypassing the venous capillaries.

Lymph flow is driven by low interstitial fluid pressure and can be significantly increased by factors such as muscle contractions, body movements, and changes in nearby arterial pressure. The composition of lymph is primarily water, with a small percentage of solids, including proteins, primarily albumin, and other molecules absorbed from the intercellular space. Lymph also contains white blood cells, particularly lymphocytes, which are crucial for the immune response.

Lymph nodes filter harmful substances and contain immune cells that fight infections throughout the body.

The glymphatic system, a component of the brain's waste clearance pathway, is closely related to the lymphatic system. It plays a vital role in brain health by removing waste products from the brain into the cerebrospinal fluid.

The lymphatic system plays an important role in diseases, as it can be implicated in ailments such as lymphoma, which is a form of cancer that affects lymphocytes and organs within the lymphatic system. It also plays a part in other diseases by maintaining fluid balance and supporting the immune system.

5.1.2 Formation of Interstitial Fluid

The interstitial fluid obtained from capillary filtration and diffusion is similar to plasma but has lower protein concentrations due to capillary pores. The fluid is primarily located in the small spaces between proteoglycan filaments, forming a tissue gel. This tissue gel comprises a blend of proteoglycan filaments and trapped fluid. It possesses gel-like characteristics due to the high concentration of proteoglycan filaments that restrict the free flow of fluid. Due to thermal motion, the interstitial fluid generally moves through the tissue gel by diffusion, molecule by molecule. Though most of the fluid remains confined within the gel, small streams of free fluid and vesicles might be present. This free fluid can flow freely since it lacks proteoglycan molecules. In healthy animal tissues, free fluid is scarce, usually less than 1%. But, when tissues develop inflammation and edema, these small pockets and streams of free fluid level increase significantly. In such situations, it's possible for as much as 50 percent or more of the fluid that causes edema to become free-flowing and separate from the proteoglycan filaments.

5.1.3 Lymphatic Capillaries

Lymphatic capillaries are the inception points of the lymphatic system, comprising a fine network of vessels that emerge from interstitial spaces. These capillaries are characterised by their blind-ended structures, termed capillary bulbs, which are fitted with valves that ensure lymph flows in only one direction. Muscle fibres encircle these bulbs, facilitating the propulsion of lymph through the vessels by contraction. The walls of lymphatic capillaries are formed by endothelial cells tethered to the surrounding connective tissue via anchoring filaments. These cells overlap at their edges, forming valves that permit the entry of interstitial fluid into the capillaries while preventing its exit, as backflow prompts the valves to seal shut.

Lymphatic capillaries merge to form progressively larger vessels by converging multiple tributaries. A fraction of the fluid that seeps out from the arterial end of blood capillaries, roughly one-tenth, is collected by these lymphatic capillaries and is subsequently returned to the bloodstream, bypassing the venous capillaries.

These capillaries, along with the larger lymph vessels, can actively pump lymph. When tissue swelling occurs due

DOI: 10.1201/9781003426851-5

to an accumulation of fluid, the anchoring filaments exert a pull on the capillary walls, enabling fluid to enter through the endothelial junctions. Conversely, when tissue compression takes place, the internal pressure of the capillary increases, leading the endothelial cell edges to act as valves and push the lymph into the next segment of the lymphatic system, preventing retrograde flow. Additionally, lymphatic capillary endothelial cells contain contractile actomyosin filaments, which are thought to contribute to the rhythmic contractions that facilitate lymph movement, akin to the contractions observed in smaller blood vessels and larger lymphatic vessels.

Particular body structures, including the outermost layers of the skin, the central nervous system, the cornea, bones, and the alveoli of lungs, lack lymphatic vessels.

5.1.4 Lymph Flow

The lymphatic system functions under significantly low interstitial fluid pressure. The pressure inside the capillaries is usually around 1 mm HG during resting conditions. However, lymph flow can increase by more than twenty times when this pressure rises to atmospheric levels. This increase in lymph flow is facilitated by various factors such as elevated capillary hydrostatic pressure, increased plasma colloid osmotic pressure, increased interstitial fluid colloid osmotic pressure, and increased capillary permeability. All these factors contribute to an increase in interstitial fluid volume and pressure.

As lymph vessel segments fill with fluid, they stretch. This stretching triggers the smooth muscle in the vessel wall to contract, pushing the fluid through the next valve into the following lymphatic segment. This process repeats along the lymph vessels until the fluid drains into the bloodstream.

There is a considerable variation in the flow of lymph in different animals. This variation depends on the animal's age, size, and lactation stage. For instance, in cows, lymph flow from the udder ranges from 13 to 45 ml/h and is characterised by pulsatile episodes with monophasic and multiphasic patterns. In calves, when fed milk, lymph flow from the thoracic or intestinal ducts averages 695 ml/hr and 429 ml/hr, respectively. However, this flow increases to 899 ml/hr and 517 ml/hr when they are fed hay and grain. Interestingly, the lymph flow increases after whole milk feeding, but there is no consistent diurnal pattern in lymph flow associated with feed intake in calves fed hay and grain. In dogs, the volume of lymph in a dog's kidney is about 1% of the blood volume in the cortical peritubular capillaries. Additionally, adult dogs have three times lower lymph flow compared to puppies.

The lymphatic system plays a crucial role in transporting large molecules like proteins that tissues, in any other way, cannot absorb. Lymphatic flow is facilitated by the natural contractions of the lymph vessels and external pressures that intermittently compress these vessels. The following are the external factors that contribute to this process in order of significance: muscle contractions of the muscles surrounding the lymph vessels, body movements, alteration of pressure in nearby arteries, and external pressure applied to the body. During physical activity, the lymphatic system's pumping action is greatly enhanced, potentially increasing lymph flow by 10 to 30 times. In contrast, the movement of lymph tends to be much slower when the body is at rest.

5.1.5 Composition of Lymph

Lymph is a transparent and colourless fluid consisting of 96% water and 4% solids. At its most basic level, lymph is just interstitial fluid that contains an additional 20g/L of protein, primarily albumin. Due to the large size of its molecules, not much fibrinogen can enter the lymphatics. However, even with minimal fibrinogen and without any platelets, lymph can still clot under certain conditions. The average protein concentration in lymph is 4-5 g/100 ml in the thoracic and intestinal ducts, regardless of whether the diet is milk-fed or grain-fed.

Additionally, lymph accumulates other proteins in the intercellular space. These are not like the plasma proteins, which are intentionally secreted into the bloodstream for some purpose. These are products of lytic extracellular matrix processing, tissue remodeling, cellular metabolic activities and the products of cell death. The vast majority of these proteins will undergo phagocytosis by tissue macrophages and antigen-presenting cells long before they reach the lymph nodes, let alone the venous circulation, which is why they never appear in the blood. The proteins of lymph have a lot of regional variation.

Lipid in lymph averaged 0.79 g- 3.14/100 ml in the thoracic and intestinal ducts. Calves that were fed milk consumed 2 to 3 times more lipid compared to those that were fed hay and grain. However, there was not much variation in lipid output throughout the day in the lymph of calves fed either diet. Lymph is modified by passage through the lymph node, whereas data concerning possible changes in the collecting lymphatics are less conclusive.

5.1.6 Lymph Nodes

Lymph nodes, also known as lymphatic nodes or lymph glands, are small glandular structures that are a part of the lymphatic system. A dense connective tissue capsule encloses them and consists of three layers: the cortex, paracortex, and medulla (Figure 5.1).

The cortex of a lymph node comprises primary and secondary lymphoid follicles. Initially, the primary follicle takes shape; later, when antigens infiltrate the body and make their way to the lymph nodes, the cells in the germinal centre region of the primary follicle begin to proliferate. After this multiplication, the primary follicles become secondary follicles. The cortex also contains B lymphocytes, typically grouped into the primary follicles and macrophages.

The paracortex between the cortex and medulla, is rich in T lymphocytes. The medulla, the innermost layer of the

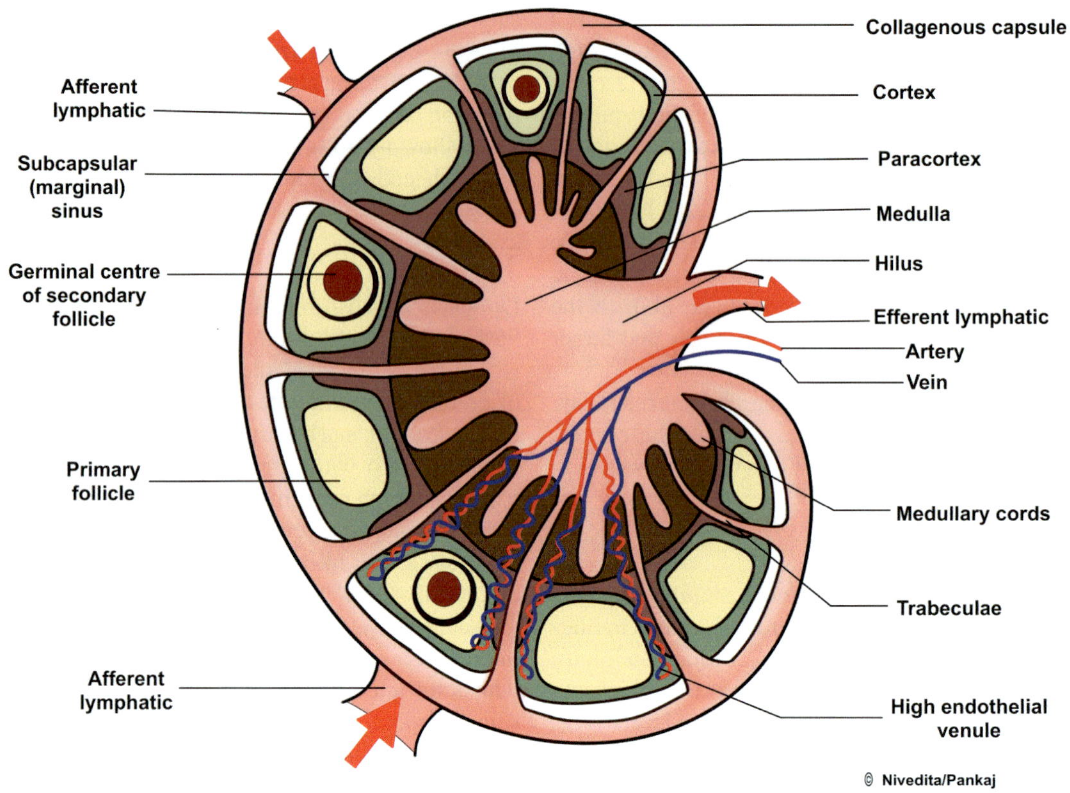

FIGURE 5.1 The structure of the lymph node is arranged in layers of capsule, cortex and medulla

lymph node, contains B and T lymphocytes, macrophages, and large blood vessels. The blood vessels of the lymph node pass through the medulla.

Lymph nodes get lymph through one or two afferent vessels, which divide into small channels. The lymph moves through these vessels and channels, reaching the cortex, and then circulates through the lymph node's cortex, paracortex, and medulla. From the medulla, the lymph exits the node through one or two efferent vessels.

Lymph nodes act as filters, with their primary role being to eliminate bacteria and harmful substances from the lymphatic fluid. As lymph travels through the lymph nodes, it undergoes a filtration process where water and electrolytes are separated, and proteins and lipids are kept within the lymph. Bacteria and other toxic substances are destroyed by the macrophages of the lymph nodes, making them important defence barriers. During an infection or other processes in a specific body region, the lymph nodes' activities in that region increase, causing the lymph nodes to swell.

5.1.7 Role of the Glymphatic System in the Brain

Iliff and colleagues have coined the term glymphatic system, which refers to a perivascular pathway that spans the whole brain and has a vital function in removing interstitial solutes and waste materials from the brain and transporting them into the cerebrospinal fluid (CSF). This system is facilitated by aquaporin-4 (AQP4) channels, which are densely expressed in astrocyte end-foot processes. The deletion of these water channels was found to impair the clearance of extracellular solutes, including amyloid-β (Aβ), a protein associated with Alzheimer's disease.

The glymphatic system serves multiple purposes in neurophysiology. It is essential for delivering nutrients, mainly glucose, the circulation and distribution of apolipoprotein E isoforms produced by the choroid plexus, and astrocytic paracrine signalling with lipid molecules. However, its fundamental role is clearing extracellular metabolites and waste products from the brain parenchyma into the CSF.

The glymphatic system's function is crucial during sleep, where it aids in the clearance of interstitial lactate that accumulates during regular daily brain activity. Disruptions to the glymphatic pathway, such as AQP4 knockout, CSF volume reduction, or alteration of posture, can lead to increased brain lactate levels and reduced cervical lymph node lactate levels.

Impairment of the glymphatic system has been linked to various central nervous system diseases. The influx of glymphatic CSF is reduced in the brain of elderly individuals and in a transgenic mouse model of Alzheimer's disease. As a result, the clearance of Aβ is significantly impaired in both cases. Furthermore, glymphatic impairment is a prominent feature of cerebrovascular disease, including both subarachnoid haemorrhage and multiple microinfarctions. After a traumatic brain injury, reduced clearance of interstitial solutes in the cortex was linked to neurological deficits in both motor and cognitive function.

5.1.8 Lymphatic System Disease

Lymphoma encompasses a group of cancers that arise from the abnormal growth of lymphocytes. Lymphocytes affect organs such as the spleen, bone marrow, and lymph nodes. While most lymphoma cancers occur in these organs, any part of the body can be affected by the disease. Multicentric lymphoma, Alimentary lymphoma, Mediastinal lymphoma, and Extranodal lymphoma are the four most prevalent types of lymphoma in dogs.

Mammary tumours can change the way lymphatic fluid flows, leading to the creation of new drainage paths and the involvement of a large number of lymph nodes. Therefore, understanding the lymphatic drainage of the affected mammary glands is crucial for the surgeon to perform the appropriate surgery and determine the prognosis after the operation. Laverdia-CA1 (Verdinexor tablets) is a medication used to treat lymphoma, which is a type of cancer affecting the lymph nodes and lymphatic system in dogs.

The lymphatic system is essential in maintaining the balance of interstitial fluid protein concentration, volume, and pressure. It acts as a safety valve, returning excess proteins and fluids from the tissue spaces back to the bloodstream, thereby controlling protein concentration. Additionally, it helps absorb fats and fat-soluble vitamins from the digestive system. It supports the immune response by producing lymphocytes and antibodies to combat harmful substances in the interstitial fluid, volume, and pressure.

5.2 IMMUNITY

Immunity is an animal's ability to resist harmful organisms and toxins, which is crucial for maintaining health and preventing diseases. There are two primary forms of immunity, namely innate and acquired. The immune system is a multifaceted structure comprising organs, cells, and proteins that recognise and counteract foreign pathogens such as viruses, bacteria, and toxins. The different parts of the immune system consist of white blood cells, antibodies, and lymphoid organs. The innate immune system reacts quickly and without specificity to a variety of pathogens. While, the adaptive immune system responds specifically to pathogens and retains a memory of the encounter for future protection.

5.2.1 Innate Immunity

Innate immunity, or nonspecific immunity, is the body's first line of defence against pathogens. It is a fundamental immune system component that provides immediate but non-specific defence against infection. The innate immune system has the ability to respond quickly to a wide variety of pathogens, unlike the adaptive immune system, which is more specific and requires time to mount a response.

The innate immune system comprises various physical barriers, cellular components, and chemical factors. The body has natural defences like skin and mucous membranes that protect against the entry of harmful pathogens through respiratory, gastrointestinal, and urogenital systems. These barriers are often supplemented by chemical defences, such as the acidic environment of the stomach and the lysosomal found in bodily fluids like saliva, tears, and reproductive tract pH. Within the respiratory tract, several mechanisms can reduce the antigen load, including the turbulence of air that occurs during inhalation. When microorganisms are carried by this turbulence as well as ciliary action of the respiratory tract, they are pushed onto the walls of the respiratory tract, lined with sticky mucus and bactericidal lysozymes, further reducing the antigen load. The innate ability of urine and milk to flush out infectious antigens is crucial in preventing infectious conditions. Failure to eliminate urine or milk can result in the development of infections. The gastrointestinal tract has a protective layer of symbiotic bacteria that stops harmful bacteria from sticking to its target cells. This mechanism is similar to the protection provided by the skin. On a cellular level, the innate immune system includes various types of leukocytes, such as macrophages, neutrophils, and natural killer cells, which can engulf and destroy pathogens through phagocytosis. Neutrophils release web-like structures called neutrophil extracellular traps (NETs) during processes like NETosis (Figure 5.2).

These NETs are composed of DNA strands associated with histones and are decorated with various proteins, including bactericidal proteins from granules and cytoplasm. When exposed to bacterial lipopolysaccharides (LPS) or proinflammatory cytokines, such as tumour necrosis factor-alpha (TNF-α) and IL-8, neutrophils become activated and initiate the process of NET formation. When tissues are damaged, chemical messengers and chemokines attract neutrophils to the site of infection. This attraction leads to the expression of adhesion molecules on vascular endothelial cells. Neutrophils bind to these cellular adhesion molecules through specific receptors, which trigger them to leave the circulation by crossing capillary walls (diapedesis) and entering the tissues.

In addition to these components, the innate immune system also includes a range of soluble factors, such as the complement system, a group of proteins that can be activated to destroy pathogens, and various antimicrobial peptides and proteins, such as lysozyme, that can directly kill or inhibit the growth of microbes.

Despite its broad range of defences, the innate immune system is not infallible. Pathogens can sometimes evade these defences and establish an infection. However, the innate immune system plays a crucial role in limiting the spread of infection and buying time for the adaptive immune system to mount a more specific and potent response.

5.2.2 Antigens and Their Characteristics

Antigens have the ability to activate lymphocytes, which can cause the body to initiate an immune response. Antibody-binding sites on an antigen are called antigenic determinants or antigenic epitopes. Antigens can be

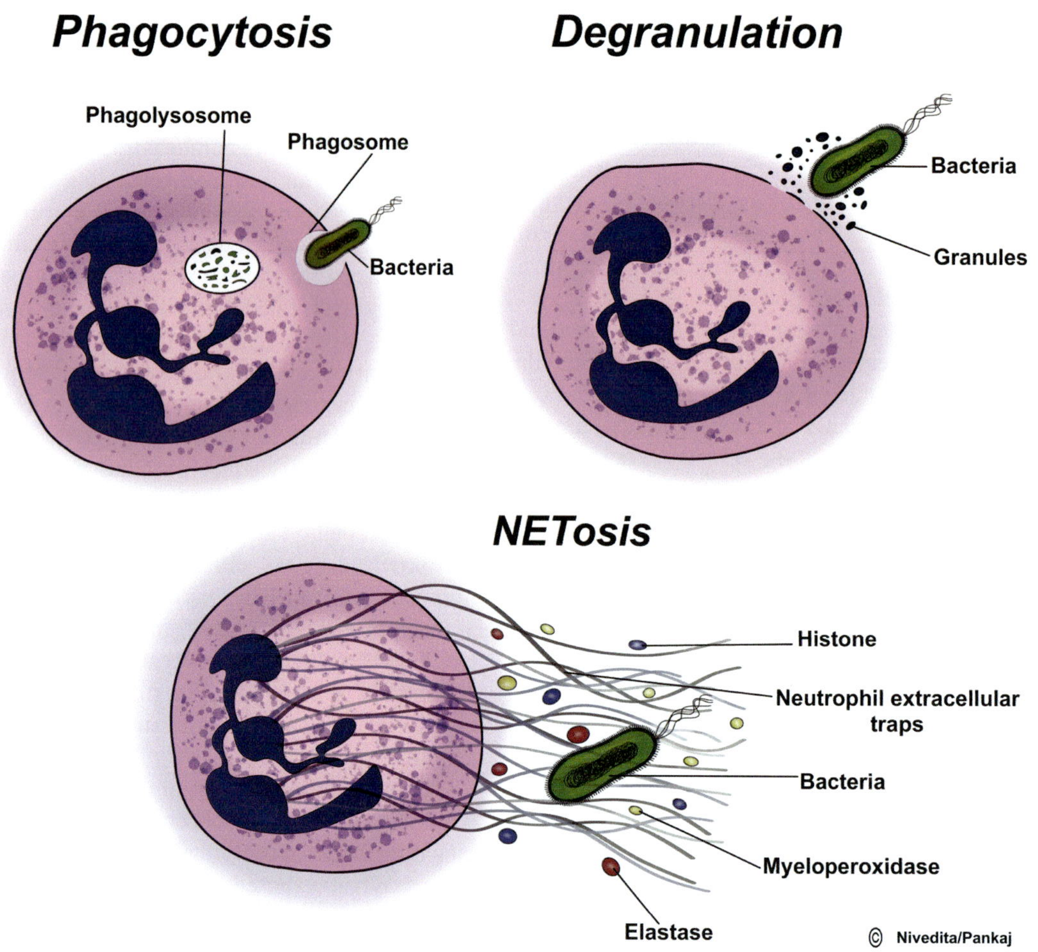

FIGURE 5.2 Different actions of neutrophils to destroy the pathogen

broadly categorised into infectious (microbial) and noninfectious antigens. Infectious antigens include components derived from bacteria, viruses, protozoa, and helminths. In contrast, noninfectious antigens encompass a wide range of substances, including those derived from "self," food, plants, dust, insect and animal venoms, and synthetic and cell surface proteins.

The level of immune response triggered by an antigen is known as antigenicity or immunogenicity. Understanding the features of antigens that can stimulate a strong or weak immune response is important. This knowledge is critical in understanding the body's ability to fight against invading antigens and developing effective vaccine preparations.

Antigens have different features determining their effectiveness: 1. Biochemical structure: More complex molecules can elicit more robust immunogenic responses than simpler molecules. For example, proteins with small sizes but complex structures have better immunogenicity than simple carbohydrates or lipids.2. If an antigen is bigger, it will get processed more effectively by antigen-presenting cells.3. Flexibility and degradability: If the antigen is less flexible, it can generate varied immunogenic responses. However, strong immunogenic responses can be generated with better stability and more degradability. Foreign antigens are more potent than self or host antigens. Moreover, the immune system's response to antigens can be affected by genetics, biomolecules that regulate and modulate immune responses, as well as the amount and method of antigen exposure.

5.2.3 Acquired Immunity

Acquired immunity develops over time in response to specific antigens. This type of immunity is characterised by its ability to learn, adapt, and remember, tailoring its response to each specific antigen encountered. Acquired immunity is mediated by specialised immune cells, including B lymphocytes that produce antibodies in a process known as humoral immunity and T lymphocytes responsible for cell-mediated immunity. These immune responses are initiated in the lymphoid tissues of the body. Acquired immunity can be acquired through various means, such as exposure to an infection, vaccination, or transmitting antibodies from another animal. Upon encountering a specific antigen, the immune system creates a memory that enables a faster and more effective response upon future exposure to the same antigen.

5.2.3.1 Lymphocytes and Their Role in Immunity

Despite their similar appearance under a microscope, they are divided into two primary groups: T lymphocytes and

B lymphocytes. These cells, originating from multipotent haematopoietic stem cells in the embryo, play distinct roles in the immune system. T lymphocytes, or T cells, are involved in cell-mediated immunity, while B lymphocytes, or B cells, are key players in humoral immunity, producing antibodies. The formation of lymphocytes and the development of cell-mediated and humoral immunity are depicted in Figure 5.3. After preprocessing, both T and B cells migrate to lymphoid tissue throughout the body.

A. T cells and cell mediate immunity: T cells derive their name from the thymus gland, which is the location where they travel to and undergo processing. This preprocessing involves rapid division and the development of diverse reactivity against specific antigens. The thymus gland ensures that T cells do not react against the body's proteins or antigens, preventing autoimmunity. This preprocessing primarily occurs shortly before and a few months after birth. During their processing, they are transformed into four categories:

1. Cytotoxic T cells, or CD8+ T cells, are responsible for directly killing infected host cells and tumour cells. Upon activation, cytotoxic T cells induce apoptosis in the target cells, thereby eliminating the source of infection or abnormal cell growth.
2. Helper T cells, or CD4+ T cells, are a heterogeneous collection of cells that can transform into different subsets, including Th1, Th2, Th17, and Tfh. Each subset has distinct effector functions and plays a role in coordinating immune responses. For example, Th1 cells activate macrophages and promote inflammation, while Th2 cells are important for activating B cells and promoting antibody production. Th17 cells are associated with the recruitment of neutrophils and the defence against extracellular pathogens. Tfh cells are essential for the development of long-lived antibody responses
3. Suppressor T cells are a specific type of T cell that has an essential function in preserving the immune system's balance and avoiding autoimmune responses. They suppress other immune cells' activation and effector functions, preventing excessive immune responses and tissue damage.
4. Memory T cells are a specialised subset of T lymphocytes that provide long-term immunity against previously encountered pathogens. Their ability to recognise and eliminate specific pathogens upon re-exposure is essential for preventing the recurrence of infectious diseases.

B. B lymphocytes undergo maturation in two stages: first, in the liver during mid-foetal life, and then in the bone marrow during late foetal life and after birth. The name "B cells" originates from the bursa of Fabricius, an organ in birds where these cells were first discovered. B cells differ from T cells in two main ways: they actively secrete antibodies that react against antigens and exhibit even greater

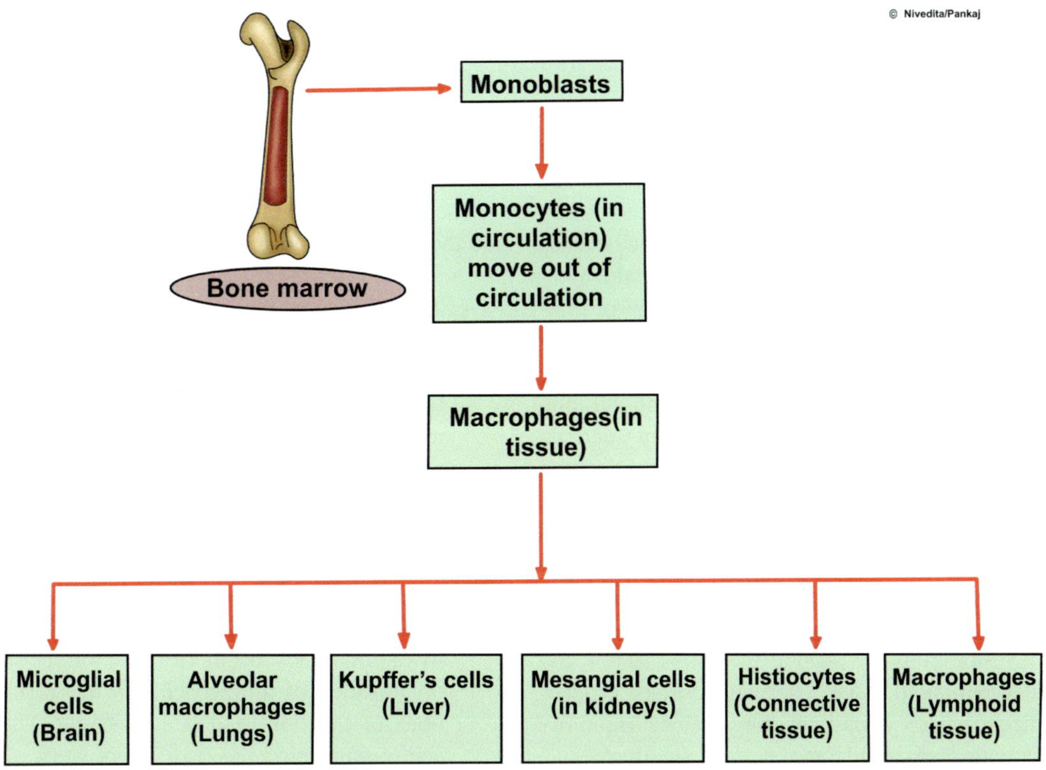

FIGURE 5.3 Formation of lymphocytes and generation adaptive immunity

diversity than T cells, forming millions of types of antibodies with different specific reactivities. The main types of B cells include:

1. **Plasma Cells:** These are the effector cells of B lymphocytes and are responsible for the production and secretion of antibodies. Once activated, B cells differentiate into plasma cells. Plasma cells produce large amounts of antibodies to fight specific antigens.
2. **Memory B Cells:** These long-lived B cells form following an initial immune response to an antigen. The immune system's long-term defence is aided by memory B cells, which quickly identify and respond to previously encountered antigens upon re-exposure.
3. **Naive B Cells:** Naive B cells are mature B cells that have not yet encountered their specific antigen. They circulate in the bloodstream and lymphoid organs, waiting to encounter antigens that will activate them.

Humoral immunity: B cells produce antibody molecules, which may be secreted or inserted into the plasma membrane. When an antigen activates a naïve or memory B cell, it proliferates and differentiates into an antibody-secreting effector cell, known as a plasmablast or plasma cell. In addition, B cells present antigens and secrete cytokines. B cells can recognise antigens and respond to the antigen in a specific way, expressing specific antibodies on the membrane.

The B-cell population of a specific activated clone significantly increases, and the new B lymphocytes are added to the original lymphocytes of the exact clone. These B cells also travel throughout the body to populate all the lymphoid tissue. However, they do not become active until they encounter another quantity of the same antigen that originally activated them. On subsequent exposure to the same antigen, the memory cells will cause a much more rapid and potent antibody response the second time because there are many more memory cells than the original B lymphocytes of the specific clone. The administration of multiple doses of antigen with intervals of several weeks or months between injections is usually required for effective immunisation because the increased potency and duration of the secondary response make immunisation successful. When naïve B lymphocytes encounter their associated antigens, become activated, and undergo clonal expansion, they differentiate into short-lived or long-lived plasma cells that produce large amounts of antibodies. The short-lived plasma cells provide rapid protection but are destroyed after a few days of antibody secretion. However, the long-lived plasma cells reside in tissues such as the bone marrow and gut-associated lymphoid tissue. They can continue producing antibodies for many years, providing lifelong immunity against infectious diseases like measles and smallpox.

Generating these lymphocytes is a complex process regulated by various factors, including chemokines, BCR signalling, and specific antigens. This process ensures a continuous replenishment of the B cell repertoire, essential for maintaining the body's defence against foreign substances.

5.2.4 Antibodies

Antibodies, also known as immunoglobulins (Igs), are a class of gamma globulins produced by B lymphocytes in response to the antigen. Antibodies recognise and bind to antigens to neutralise and remove them from the body.

There are five general classes of antibodies, each designated by a specific letter: IgA, IgD, IgE, IgG, and IgM. Among these five, 75 % of total antibodies are IgG. IgG is the only antibody class that significantly crosses the animal placenta, providing passive immunity to the foetus. Colostrum produced by cattle after parturition has a high concentration of antibodies, including immunoglobulins IgA, IgG, and IgM, which provide passive immunity to newborns during crucial phases of life. IgE has a role in allergic conditions. All these antibodies play various roles in the immune response, such as localised defence, recognition of antigens, and involvement in allergic reactions.

5.2.4.1 The Antibody's Structure

Comprises two pairs of chains, specifically one pair of heavy or long chains and one pair of light or short chains (Figure 5.4).

Each heavy chain contains about 400 amino acids, while each light chain contains about 200 amino acids. The antibody is divided into two halves, each by disulphide bonds. Each half of the antibody consists of one heavy chain and one light chain, and disulphide bonds join the two chains in each half. Each antibody molecule always has at least two and as many as ten such pairs. The combination of noncovalent and covalent bonds - disulphide holds the light and heavy chains together.

The antibody contains two regions: the constant region and the variable region.

A. **The constant region**, also known as the Fc (Fragment crystallisable) region, is similar in number and placement of amino acids in all antibodies of each type. The constant portion determines other properties of the antibody, such as antibody diffusivity in the tissues, adherence to specific structures in the tissues, attachment to the complement complex, and the ease with which the antibodies pass through membranes.
B. **The variable region** is also known as the Fab (Fragment antigen binding) region. The specificity of antibodies results from the unique structural organisation of amino acids in the variable portions of the light and heavy chains. The amino acid organisation has a different steric shape for each antigen specificity, allowing for rapid and tight bonding between the antibody and antigen.

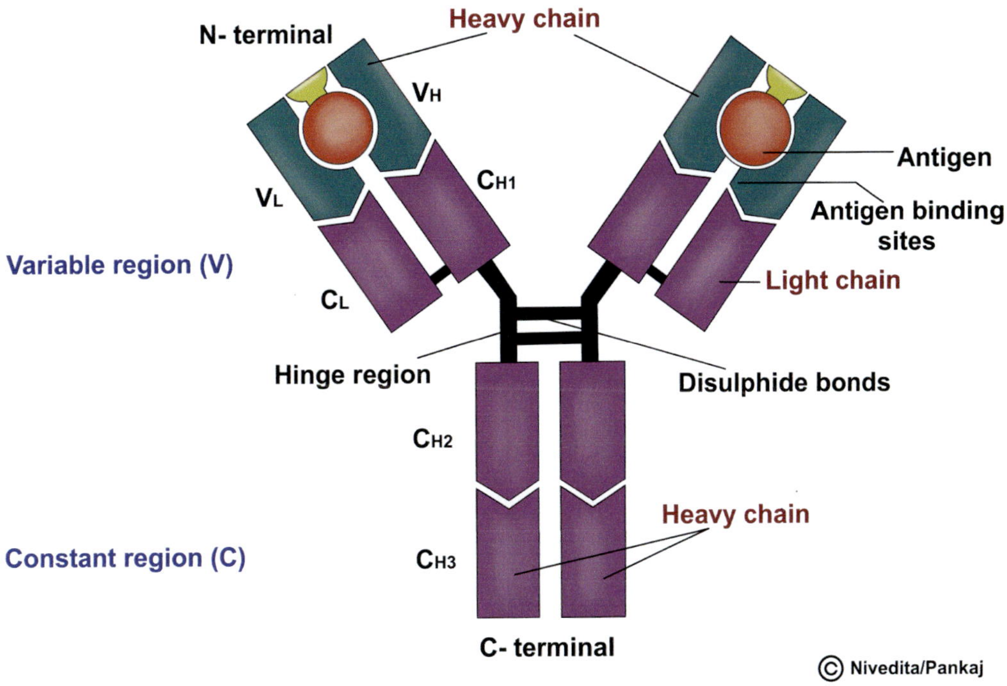

FIGURE 5.4 Structure of an antibody showing two heavy and two light chains

If an antibody is very specific, then there are many sites where it can bond with an antigen. This bonding is firm and is held together by different chemical forces such as hydrophobic bonding, hydrogen bonding, ionic attractions, and van der Waals forces.

5.2.5 Antigen-Presenting Cells

Antigen-presenting cells (APCs) play an important role in detecting, engulfing, and presenting antigens to T cells, triggering an immune response. There are three main types of APCs: macrophages, dendritic cells, and B lymphocytes.

1. **Macrophages** are large phagocytic cells that digest invading organisms to release antigens. Macrophages are highly versatile and heterogeneous, taking on different names according to their tissue location. Some of the different types of macrophages, based on their tissue location, include Lung-Alveolar macrophages or dust cells, Skin-Langerhans cells, Liver-Kupffer Cells, Central Nervous system-Microglial cells, Kidney Mesangial cells, and connective tissues-Histiocytes (Figure 5.5). They are present in various lymphoid tissues and are considered the major APCs.
2. **Dendritic cells**, conversely, are nonphagocytic and classified based on their location, such as in the spleen, lymph nodes, or skin. They are highly efficient in antigen presentation and are considered the most potent APCs.
3. **B lymphocytes**, traditionally known for their role in antibody production, have also been found to act as APCs, although they are the least efficient among the three types.

The role of APCs is to process and present antigens to T cells, specifically helper T cells. The major histocompatibility complex (MHC) proteins facilitate the interaction between APCs and T cells, which bind peptide fragments of antigen proteins degraded by APCs and transport them to the cell surface. T cells recognise major histocompatibility complexes using their T cell receptors (TCRs). The specific functions of cytotoxic and helper T cells depend on the presentation of antigens by APCs. MHC can be separated into two classes: class I MHC molecules, present on all nucleated cells of the body and responsible for presenting endogenous antigens to cytotoxic T cells, and class II MHC molecules, found on APCs and responsible for presenting exogenous antigens to helper T cells.

The process of antigen presentation by APCs is essential for the specificity and regulation of the adaptive immune response. It allows the immune system to differentiate between self and non-self antigens, thereby mounting an effective response against foreign invaders while preventing autoimmunity. The proper function of APCs and the presentation of antigens are critical for activating and regulating lymphocytes, which are instrumental in the adaptive immune response.

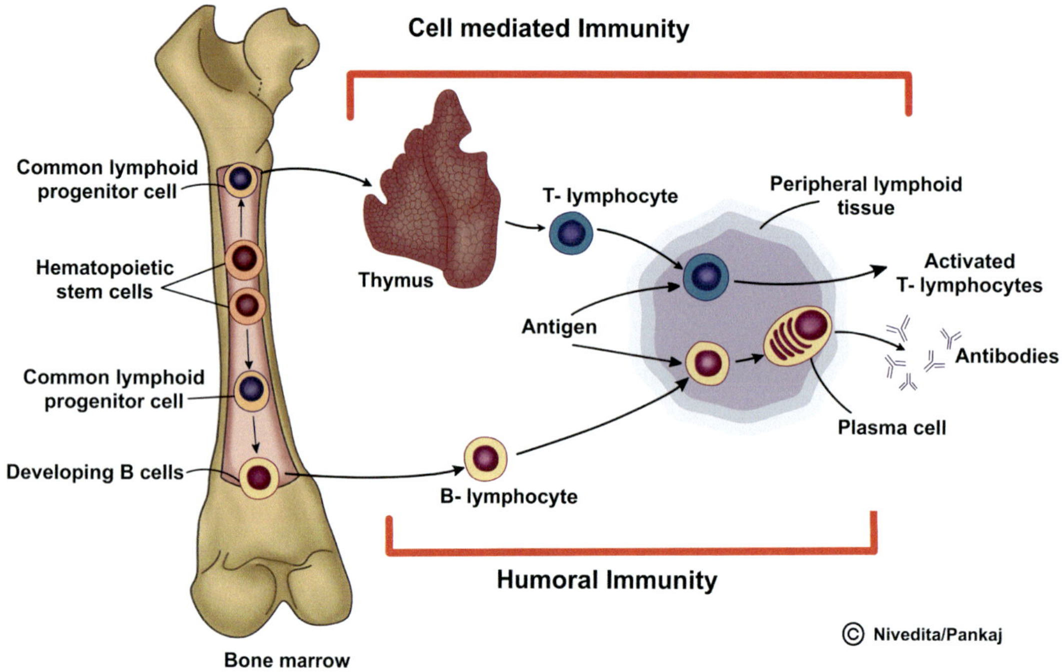

FIGURE 5.5 Diagram shows the formation of monocytes and, upon entry into different tissues, their synonyms

5.2.6 Toll-like Receptors

Toll-like receptors (TLRs) are a group of pattern recognition receptors that are crucial parts of the innate immune system. They recognise specific sequences or molecules in microbes, known as pathogen-associated molecular patterns (PAMPs). TLRs are mainly expressed by innate immune cells, particularly macrophages, and are the primary defence mechanism against invading pathogens. Upon recognition of PAMPs, TLRs trigger innate immunogenicity, leading to the generation of proinflammatory cytokines, chemokines, and antimicrobial peptides, as well as the upregulation of costimulatory molecules.

In mammals, 13 TLRs have been identified, each with a specific role in recognising distinct PAMPs. TLRs 1 to 6, except 3 are located on the cell surface membrane, where they identify a wide variety of microbial products, including bacterial lipoproteins, lipopolysaccharides (LPS), and flagellin. On the other hand, TLRs 3, 7, 8, and 9 are found intracellularly, where they recognise nucleic acids from viruses and bacteria. TLR4 responds to LPS present in tissues and the bloodstream, triggering proinflammatory signalling pathways. TLR4 has been shown to recognise endogenous molecules produced from tissue injury, indicating that TLR4 represents a key receptor on which both infectious and noninfectious stimuli converge to induce TLR4-mediated inflammation.

Toll-like receptors (TLRs) have a role in initiating the immune response to infections and tissue damage. They are also involved in several acute and chronic diseases, having a pivotal role as an amplifier of the inflammatory response. Their dysregulation can lead to chronic inflammation and contribute to the progression of various diseases.

5.2.7 Mode of Action of Antibodies

Antibodies can protect the animal body by two mechanisms: A. Direct actions and B. Complement pathway, as shown in Figure 5.6.

Direct actions: Because of the bivalent nature of the antibodies and the multiple antigen sites on most invading agents, when antibodies come into contact with these agents, they can deactivate them in various ways. These methods include:

1. **Agglutination:** Where multiple large particles having antigens on their body (RBCs and bacteria) are clubbed with each other to form a clump.
2. **Precipitation:** Where the molecular complex of a soluble antigen (e.g., tetanus toxin) and antibody complex is large enough that it becomes insoluble and precipitates.
3. **Neutralisation:** Where the antibodies cover the toxic sites of the antigenic agent.
4. **Lysis:** Where some strong antibodies can directly attack cellular agent membranes, causing them to rupture.

These direct actions of antibodies often are not strong enough to play a major role in protecting the body against the invader, but indirect actions provide more potent

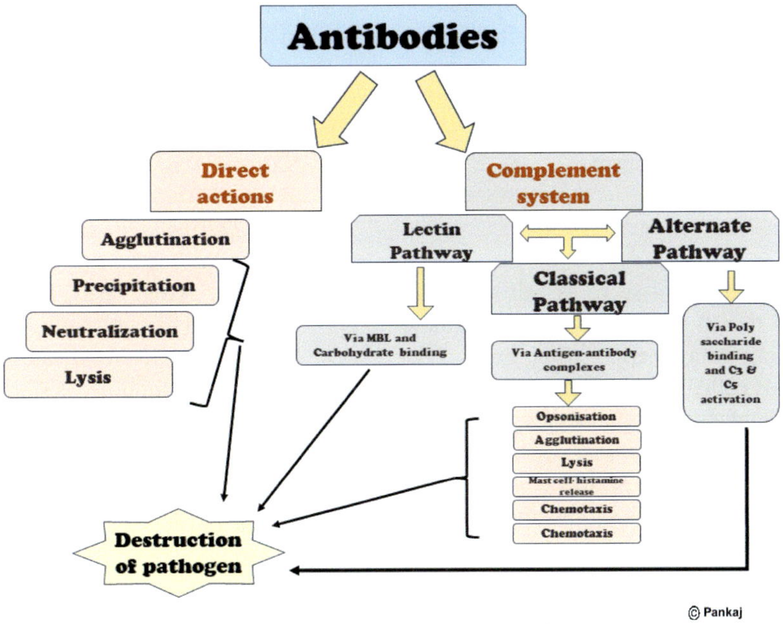

FIGURE 5.6 Mode of action of antibodies: Direct and complement system

protection by activating the complement pathways. A complement system enhances or accelerates various activities during the fight against invading organisms. It is a system of plasma enzymes identified by numbers from C1 to C9. Including the three subunits of C1 (C1q, C1r, C1s), there are 11 enzymes. Usually, these enzymes are inactive and are activated in three ways: a. Classical pathway, b. Lectin pathway, c. Alternate pathway.

a. **Classical pathway:** The classical pathway of the immune response begins when an antibody interacts with an antigen. This interaction exposes a specific reactive site on the antibody, which then binds with the C1 molecule of the complement system. This triggers a series of sequential reactions, starting with the activation of the proenzyme C1. The activated C1 enzymes then sequentially activate other enzymes in the system, leading to an amplified reaction from a small beginning. This results in multiple end products that help protect the body's tissues from damage caused by the invading organism or toxin. Key effects include:
   i. Opsonisation and phagocytosis, where the complement cascade product C3b significantly boosts phagocytosis by neutrophils and macrophages, engulfing bacteria attached to the antigen-antibody complexes. This process, known as opsonisation, can increase the number of bacteria destroyed several hundred times.
   ii. Lysis is another important effect, where the membrane attack complex (also known as the cytolytic complex), a combination of multiple complement factors C5b6789, inserts itself into the cell membrane, creating pores that lead to the osmotic rupture of the cell membranes of bacteria or other invaders.
   iii. The complement products also cause agglutination by changing the surfaces of the invading organisms, causing them to stick to each other. They can neutralise some viruses, rendering them nonvirulent.
   iv. Chemotaxis is initiated by fragment C5a, leading to a large influx of neutrophils and macrophages into the tissue area near the antigenic agent.
   v. When fragments C3a, C4a, and C5a are present, they trigger mast cells and basophils to release various substances, including histamine and heparin, into the nearby fluids. These substances increase regional blood flow and fluid leakage into the tissue, among other local tissue reactions that help inactivate or immobilise the antigenic agent.
   vi. Besides the inflammatory effects caused by the activation of mast cells and basophils, other complement products also contribute to local inflammation. These products increase blood flow, further increase capillary leakage of proteins, and cause interstitial fluid proteins to coagulate in the tissue spaces. This prevents the movement of the invading organism through the tissues.

b. **The lectin pathway** is activated when the serum protein mannose-binding lectin (MBL) binds to the mannose or fructose groups on the surface of bacteria, fungi, or viruses.

c. **The alternative pathway** is activated by a circulating protein called factor I. This protein binds to polysaccharides present in the cell membrane of invading organisms, leading to the activation of C3 and C5, which ultimately attack the antigenic products of the invading organism.

5.2.8 Cytokines

Cytokines are small proteins that serve as cell signalling molecules, aiding cell-to-cell communication in immune responses and regulating inflammation in the body. They are produced by various cells, including immune cells like macrophages, B lymphocytes, T lymphocytes, mast cells, endothelial cells, fibroblasts, and stromal cells. Cytokines significantly influence the balance between humoral and cell-based immune responses, regulating specific cell populations' maturation, growth, and responsiveness. They act through cell surface receptors and are involved in autocrine, paracrine, and endocrine signalling. Additionally, they are involved in various processes such as inflammation, pain, host immune responses to infection, trauma, sepsis, cancer, and reproduction.

5.2.8.1 Types of Cytokines

1. **Interleukins:** Produced by leukocytes, interleukins regulate cell growth, differentiation, and motility, affecting the immune system. Interleukins can be either inflammatory or anti-inflammatory, or both types.
2. **Interferons:** Interferons have antiviral, immunomodulatory, and antiproliferative effects, playing a key role in the immune response against viral infections.
3. **Tumour Necrosis Factors (TNF):** TNF plays a crucial role in inflammation and can induce fever, apoptotic cell death, and sepsis.
4. **Chemokines:** These cytokines play a crucial role in the immune response by stimulating the transfer of immune cells to the site of injury or infection.
5. **Lymphokines:** Lymphokines relay messages between immune cells, contributing to regulating the immune response.
6. **Monokines:** Monokines help mobilise an immune response and are produced by monocytes and macrophages.

5.2.9 Passive Immunity

Passive immunity refers to giving a person antibodies to a disease rather than producing them through their immune system. This kind of immunity offers instant protection, although it is not long-lasting and may only last for several weeks up to 3 or 4 months. Passive immunity can occur naturally, for example, when antibodies from the mother are passed to the foetus via the placenta or from breast milk to the baby's gut. It can also be produced artificially, such as when antibodies are given as a medication.

Passive immunity can be further classified into two types: natural and artificial.

1. **Natural Passive Immunity:** This immunity is obtained from the mother before and after childbirth. During pregnancy, the mother passes immunity to the foetus through IgG antibodies in the placenta. After birth, the antibodies (IgA) are transmitted through milk. The antibodies received from the mother are metabolised soon. Therefore, the passive immunity is short-lived.
2. **Artificial Passive Immunity:** This type of immunity is developed by injecting previously prepared antibodies using serum from humans or animals. Antibodies are obtained from the persons affected by the disease or from animals, particularly horses, which have been immunised artificially. The antibody (antiserum) serum is administered to animals who have developed the disease (therapeutic). It is also used as a prophylactic measure. This type of immunity helps provide immediate protection against acute infections like tetanus for poisoning by insects, snakes, and venom from other animals.

Unlike active immunity, which can take days or weeks to develop, passively administered antibodies can provide rapid and immediate protection. Passive antibodies are not dependent on the recipient's immune system, unlike vaccination, which relies on the host's immune response to create protective immunity. Passive immunisation may thus represent the therapy of choice in highly endemic areas where vaccine coverage is low or in individuals who cannot mount an effective immune response.

5.2.10 Allergy and Hypersensitivity

Immunity, while crucial for protecting the body from harmful pathogens, can sometimes lead to undesirable side effects such as allergies or hypersensitivity reactions. When the immune system overreacts to allergens, it causes allergic reactions. Allergens can be anything from food to physical agents like heat or light. There are several types of hypersensitivity reactions, some of which only occur in animals with a specific predisposition to allergies. One such reaction is anaphylaxis, a severe and widespread allergic reaction that can occur when an allergen enters the bloodstream. This reaction involves basophils in the blood and mast cells in the tissues, which, if sensitised by IgE reagins, can trigger a body-wide allergic response. This response includes the release of histamine, leading to widespread vasodilation and increased capillary permeability, which can result in significant plasma loss from the circulation. If not promptly treated with epinephrine, anaphylaxis can lead to circulatory shock and potentially death.

Another type of hypersensitivity reaction is urticaria, also known as hives, which results from antigens entering specific skin areas and causing localised anaphylactoid reactions. Histamine released in these areas causes vasodilation and increased capillary permeability, leading to localised skin swelling.

Hypersensitivity reactions can be classified into five types: Type I or anaphylactic reactions, Type II or cytotoxic reactions, Type III or antibody-mediated reactions, Type IV or cell-mediated reactions, and Type V or stimulatory/blocking reactions. Each type of reaction involves different immune system components and varies based on the type of antigen, the immune response generated, and the speed at which the response is produced.

Common allergens include food substances like wheat, egg, milk, and chocolate; inhalants like pollen grains, fungi, dust, and smoke; contactants like chemical substances, metals, animals, and plants; infectious agents like parasites, bacteria, viruses, and fungi; drugs like aspirin and antibiotics; and physical agents like cold, heat, light, pressure, and radiation.

It is important to note that the terms "allergy" and "hypersensitivity" are often used interchangeably. However, an allergic reaction typically refers to the signs and symptoms that an animal experiences. In contrast, a hypersensitivity reaction describes the immune response itself.

BIBLIOGRAPHY

Collivignarelli, F., Tamburro, R., Aste, G., Falerno, I., Del Signore, F., Simeoni, F., ... Vignoli, M. (2021). Lymphatic drainage mapping with indirect lymphography for canine mammary tumors. *Animals*, 11(4), 1115.

Gabriël, S., Geldhof, P., Phiri, I. K., Cornillie, P., Goddeeris, B. M., & Vercruysse, J. (2005). Placental transfer of immunoglobulins in cattle infected with Schistosoma mattheei. *Veterinary Immunology and Immunopathology*, 104(3–4), 265–272.

Gattinoni, L., Speiser, D. E., Lichterfeld, M., & Bonini, C. (2017). T memory stem cells in health and disease. *Nature Medicine*, 23(1), 18–27.

Hall, J. E., & Hall, M. E. (2020). *Guyton and hall textbook of medical physiology e-Book*. Elsevier Health Sciences.

Klein, T. B. G. (2012). *Cunningham's textbook of veterinary physiology-E-book*. Elsevier Health Sciences.

Lissoni, P., Messina, G., Pelizzoni, F., Rovelli, F., Brivio, F., Monzon, A., ... Di Fede, G. (2020). The fascination of cytokine immunological science. *Journal of Infectiology and Epidemiology*, 3(1), 18–28.

Louveau, A., Plog, B. A., Antila, S., Alitalo, K., Nedergaard, M., & Kipnis, J. (2017). Understanding the functions and relationships of the glymphatic system and meningeal lymphatics. *The Journal of Clinical Investigation*, 127(9), 3210–3219.

Tizard, I. R. (2017). *Veterinary immunology-e-book*. Elsevier Health Sciences.

von der Weid, P. Y., & Muthuchamy, M. (2010). Regulatory mechanisms in lymphatic vessel contraction under normal and inflammatory conditions. *Pathophysiology*, 17(4), 263–276.

Wiig, H., & Swartz, M. A. (2012). Interstitial fluid and lymph formation and transport: Physiological regulation and roles in inflammation and cancer. *Physiological Reviews*, 92(3), 1005–1060.

6 Physiology of the Haematological System

Rahul Singh Arya[1]*, Vasantha S. K. I.*[2]*, M. P. S. Tomar*[2]*, Naveen Swaroop M*[4] *and Tanmoy Rana*

[1]Central Agricultural University, Aizawl, India
[2]Sri Venkateswara Veterinary University, India
[3]Department of Veterinary Clinical Complex, West Bengal University of Animal & Fishery Sciences, Kolkata, India

6.1 INTRODUCTION

The blood has been considered the life force in a living animal since ancient times. It has been described as liquid connective tissue, transparent connective tissue, or the connective tissue without fibers. It flows through the body by the pumping action of the heart. However, the working of the musculature also contributes to circulation. The composition of blood can be divided into the fluid portion called the plasma and the cellular portion. The cellular portion comprises erythrocytes, various WBCs, and thrombocytes. They all have important functions in the body. The plasma is a complex liquid containing different proteins as colloids and various minerals and glucose in solution. The plasma proteins provide nutrition to the body, carry waste, maintain colloidal osmotic pressure, maintain blood pH, and provide immunity. Hormones and other biomolecules are also dissolved in the plasma. Most of the plasma proteins are produced by the liver. On the other hand, the cellular portion arises from the bone marrow. In some animals, hematopoiesis can also occur in other parts of the body called extramedullary hematopoiesis. The bone marrow can be divided into two portions of the red and yellow bone marrow. Red bone marrow is the active one, while the yellow bone marrow contains mostly fat cells.

Formation of blood cells is a continuous process and is tightly controlled by various hormones and other chemicals. It requires a proper supply of nutrients, ironically provided by the blood itself. The common nutrients required are proteins, minerals (iron, copper, etc.), vitamins (B complex), fats, and glucose. A deficiency of nutrients can precipitate the common condition of anemia. Cell division is sensitive and can be easily suppressed by persistent inflammatory and neoplastic conditions or acute or chronic toxicity. Toxicity of heavy metals like lead can affect hematopoiesis. Lesions like hematomas, abscesses, granulomas, cancers, and fibrosis can replace bone marrow, leading to loss of hematopoiesis. Poor hematopoiesis is clinically reflected by reduced total erythrocyte counts, total leucocyte count, hematocrit, and hemoglobin, however with considerable variation. Often, bone marrow malfunction is reflected by variation in shape and size of the erythrocytes. In veterinary science, this variation is of clinical significance in species like dogs and cats, while in cattle, the variation in size of RBCs, also called anisocytosis, is normal. The number of RBCs is more consistent than the leucocyte count, which keeps on varying with different diseases, inflammation, and even physiological conditions.

Hematopoiesis also depends on the normal functioning of other organs, particularly liver and kidneys. Poor liver function can suppress bone marrow function by reducing the supply of plasma proteins and accumulation of wastes. Poor renal function, particularly reduced GFR, eventually leads to anemia due to chronic exposure to retained urinary components. Certain drugs like chloramphenicol are known to suppress bone marrow function.

Like any other cells, erythrocytes arise from nucleated stem cells. It is to be noted that erythropoiesis is not only the division and differentiation of the stem cells but also the maturation of the RBC into a functional cell. During the process of maturation, they lose their nucleus, all cell organelles, all nucleic acid and most of the cytoplasm except for some portion responsible for energy production for the survival of the cell. During this process, the RBC gets supersaturated with hemoglobin, the oxygen carrying pigment. Consequent to this the RBC's are sensitive to osmatic imbalance as compared to other blood cells. Remnants of the nucleus or nucleic acids and even spindle fibers from improper cell division, are all clinical indicators of disturbance in hematopoiesis.

The Leucocytes are classified as granulocytes and agranulocytes. The granulocytes are those cells which have easily discernable and functional granules present in their cytoplasm. On the other hand, agranulocytes have none such granules in the cytoplasm. Granulocytes also have a peculiarity in their nuclear morphology, that it is multilobulated. Consequently, they are called as polymorphonuclear cells (PMNC). The agranulocytes have a large

DOI: 10.1201/9781003426851-6

nucleus with relatively consistent morphology and are called the Mononuclear cells (MNC). It is to be noted here that both the PMNC and MNC have only a single nucleus. The PMNC comprise of the neutrophils, eosinophils and basophils, classified according to the pH-based affinity of their cytoplasmic granules towards the dyes. They are synthesized in the bone marrow. The MNC comprise of the monocytes and lymphocytes. The monocytes are circulatory phagocytes and are synthesized in the bone marrow. The lymphocytes are synthesized in the bone marrow and also in the lymph nodes and spleen. In mammals they reach the normal blood circulation via the lymphatic flow.

The cells produced by the bone marrow are not all present in the circulation. Many leucocytes particularly polymorphonuclear cells are stored in the bone marrow reserve and are released when the need arises. While a significant portion of RBCs may be present in spleen. The spleen is the site for removal of old and senile RBC's. There is always a balance in the body in production of RBC's in the bone marrow and their removal from the circulation by the spleen. If destruction is more than the production then a clinical condition called splenic anemia is resulted. In case of acute loss of RBC's, the bone marrow cannot compensate the required number of cells needed immediately. So, the spleen releases the stored RBC's. Some immature nucleated RBC's are also released in the circulation by the bone marrow.

Thrombocytes are another important cell of blood arising from a special cell of the bone marrow called the megakaryocytes. They are required for preventing the hemorrhages. In birds they appear as nucleated cells while in mammals they are small fragments of original cell.

The process of hematopoiesis is a complex well-orchestrated function of the body. It is tightly controlled by various signaling molecules and hormones. It is robust and can show tremendous regeneration even after prolonged injury to the bone marrow. At the same time this process is also sensitive to injury especially toxicity, inflammation, systemic inflammatory cytokines and space occupying lesions like cancers.

6.2 COMPONENTS OF BLOOD

Transparent connective tissue called blood permeates every part of the body and helps maintain homeostasis by carrying hormones, nutrients, carbon dioxide, oxygen, and other substances to and from the body's cells aids in controlling the body's pH and temperature and, via phagocytosis and antibody synthesis, offers defence against illness. Whole blood is the term used to describe the blood found in the cardiovascular system. Peripheral blood is whole blood that travels through blood vessels carrying oxygen, nutrients, and waste products. When blood is drawn from an animal's vein or artery, it is referred to as peripheral blood. Blood is an opaque, horrifyingly red liquid. Under a microscope, whole blood appears as plasma, a transparent liquid in which different cell components are suspended. The primary constituent of plasma, which dissolves various solutes, is water. The components of a cell are platelets (thrombocytes), white blood cells (leukocytes), and red blood cells (erythrocytes). There are five types of white blood cells: neutrophils, basophils, eosinophils, monocytes, and lymphocytes. Using a variety of blood tests, clinicians regularly assess and evaluate the differences in order to try and identify the underlying cause of various diseases.

6.3 FUNCTION

Blood has three main functions: transportation, regulation, and defence.

6.4 TRANSPORTATION

All body cells receive oxygen through erythrocytes, or red blood cells, which contain haemoglobin. Nutrients and other essential elements are dissolved by blood plasma and subsequently transported to tissues via arteries and capillaries. Blood travels through veins to carry waste products from cellular metabolism to the kidneys and lungs, where they are eliminated from the body. Blood delivers hormones from endocrine glands to target organs and carries white blood cells to various sites of activity where they aid in the body's defence against infection. The process by which blood transports platelets to damaged blood vessel walls in order to form a plug that will stop the bleeding is known as haemostasis. Platelets also initiate the cascade that clots blood.

6.5 REGULATION

The blood has a part in regulating body temperature. The body temperature regulators in the brain are partially influenced by the blood's temperature when it passes over or through them. Blood is a fluid component of tissue. The composition of bodily tissue fluid is as uniform as feasible. If an animal becomes dehydrated or has low tissue fluid levels as a result of vomiting, diarrhoea, excessive perspiration, or another pathologic condition that causes fluid loss, some plasma will leave the circulation and enter the body tissues in an attempt to replenish lost fluid. Consequently, the blood has a higher concentration of cells (haemoconcentration) and less plasma. If an animal has too much bodily fluid – for example, after receiving subcutaneous fluids – it will enter the bloodstream. The cell count is diluted, or haemodilution, due to the excess fluid in the plasma. The blood's acid-base balance (pH) is maintained in part by blood. The normal range of blood pH is 7.35 to 7.45, with 7.4 being the ideal pH (slightly alkaline). The blood must remain within this precise range in order to maintain the animal's health. The pH needs to remain slightly alkaline in order to buffer the acidic waste products of cellular metabolism that it carries. Arterial blood has a slightly higher alkaline pH than venous blood because venous blood has a lower pH and carries most of the acidic waste products.

6.6 DEFENSE

Leukemic cells are transported by blood to locations where foreign invaders are present. These cells help sustain an animal's health by supporting its immune system. By delivering platelets to the sites of vascular damage, blood aids in haemostasis and prevents excessive bleeding in the animal.

6.6.1 Composition of Blood in Animals

Red blood cells, white blood cells, and platelets are formed elements combined with blood plasma, a liquid, to form blood, a connective tissue. The erythrocytes make up the largest population of all the cellular components, followed by platelets, and white blood cells, which make up the smallest. Basophils are the least common cell type, followed by eosinophils and monocytes, according to the differential leucocyte count. In carnivores and omnivores, the population of neutrophils exceeds that of lymphocytes; in ruminants, however, the opposite is true (Figure 6.1).

6.7 HEMATOPOIESIS (FORMATION OF BLOOD)

Few lymphocytes, out of all the cellular constituents, live for many years; all other cells have lifespans of hours to weeks and should be periodically replaced by newer cells. Hemopoiesis, also known as hematopoiesis, is the process that creates these cells and continues for the duration of an animal's life. The foetus's liver and spleen are where hematopoiesis takes place. The majority of an infant animal's body's bones contain red bone marrow, which is where this process mostly takes place. Many bones are impacted by the increased demand for blood cells that occurs during growth and development. The rate of production falls because the animal isn't growing as quickly as it is maturing. The highly vascularized connective tissue known as red bone marrow is found in the microscopic gaps between the trabeculae of spongy bone tissue. It is primarily found in the proximal epiphyses of the humerus and femur, pectoral and pelvic girdles, and bones of the axial skeleton. Hemocytoblasts, also known as pluripotent stem cells, make up about 0.05–0.1% of red bone marrow cells. They are derived from mesenchyme, the tissue from which practically all connective tissues develop. These cells have the ability to differentiate into a wide variety of cell types.

When certain red bone marrow becomes less active, yellow bone marrow – which is primarily made up of fat cells – is produced. However, increased physiological demands can cause yellow marrow to reactivate. The type of bone marrow is so named because of how ugly it looks. Red bone marrow is found in different places in the adult bodies of different species. In most species, the red marrow sites are the skull, ribs, sternum, vertebral column, pelvis, and the proximal ends of the femurs.

Red bone marrow stem cells self-replicate, multiply, and differentiate into cells that give rise to mast cells, reticular cells, adipocytes, blood cells, and macrophages. Certain stem cells have the capacity to differentiate into osteoblasts, chondroblasts, and muscle cells. These cells may find application in the replacement of bone, cartilage, and muscle tissue. The stroma, or framework, that holds up the red bone marrow cells is made of reticular fibers, which are produced by the reticular cells.

Red bone marrow's pluripotent stem cells divide into two additional types of stem cells, each of which has the potential to differentiate into a variety of cell types, in order to produce blood cells. These stem cells are referred to as lymphoid and myeloid stem cells. Red bone marrow is where myeloid stem cells start to develop into red blood cells, platelets, neutrophils, eosinophils, basophils, and monocytes. Red bone marrow is the starting point for lymphoid stem cell development, which is finished in lymphatic tissues where it gives rise to lymphocytes. Despite having unique cell identity markers in their plasma membranes, the different types of stem cells are not histologically distinct and instead resemble lymphocytes. These mycloid stem cells differentiate into precursor cells and progenitor cells

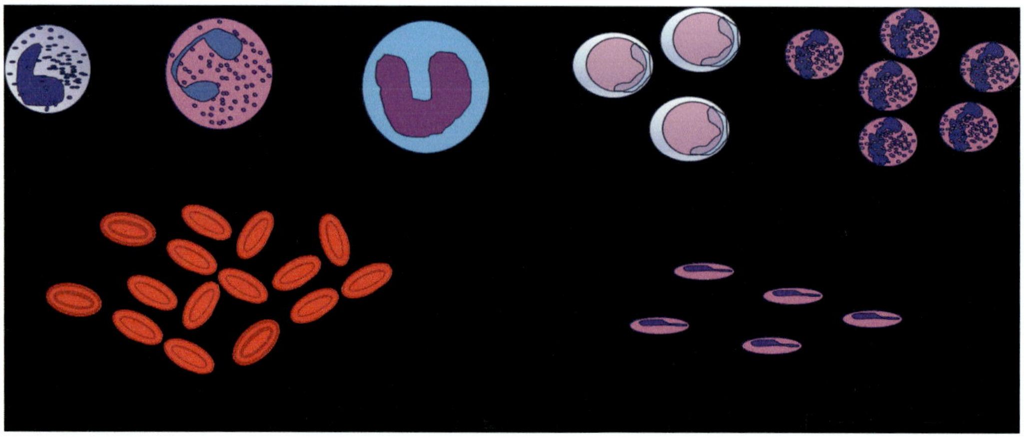

FIGURE 6.1 Diagrammatic presentation of formed components of blood. 1. White Blood Cells: Basophil (A), Eosinophil (B), Monocyte (C), Lymphocytes (D), Neutrophils (E). 2. Red Blood Cells. 3. Platelets

TABLE 6.1
Table Representing RBC Count in Different Species

Species	RBC (millions) or (volume mm^3)
Cattle, Pig, Dog, Cat	6–8
Sheep and Goat	10–14
Horse	7–12
Rabbit	5.5–6.5
Chicken	2.5–3.2

during hemopoiesis. The progenitor cells, which are dividing into CFU-E (erythrocyte), CFU-Meg (megakaryocyte, or platelets), and CFU-GM (granulocyte), are what form the colonies. Monocytes and eosinophils develop as a result of the precursor cells dividing and forming blasts.

Various hormones play important roles in all these mechanisms, e.g., hemopoietic growth factor, erythropoietin, thrombopoietin, cytokines, interleukins etc.

6.8 ERYTHROPOIESIS

Erythropoiesis is the process that creates red blood cells. In unipotent stem cells, proerythroblast differentiation is stimulated. The proerythroblasts undergo multiple stages of division, each of which produces a more advanced cell type. Eventually, the cells will stop dividing and will become nucleated. They also start producing haemoglobin. After this, red blood cells (RBCs) must go through three more phases of maturation before they are fully developed. For dogs, the entire process takes about one week; for cows, it takes about four to five days; and for birds, it takes about 36 hours.

Hormones, primarily erythropoietin (EPO), and the availability of iron, folic acid, vitamin B12, protein, and other components required to form red blood cells regulate the rate of erythropoiesis. The primary source of erythropoietin is the peritubular interstitial cells found in the kidney. Erythropoietin synthesis in the kidney is controlled by blood oxygen levels. Hypoxia is the cause of increased erythropoietin production. Erythropoietin stimulates erythropoiesis in the red bone marrow because a variety of diseases may result in tissues having less oxygen available to them. In cases of severe anaemia, nucleated red blood cells (NRBCs) may be released. Nucleated red blood cells (RBCs) are immature red blood cells that still contain their nuclei.They are not fully matured, so even though they are producing some haemoglobin, their efficiency is reduced (Table 6.1)

6.9 THROMBOPOIESIS

Thrombopoiesis, the process by which platelets are made, begins when a unipotent stem cell in the red bone marrow becomes differentiated into a megakaryocyte by means of a specific stimulant. Megakaryocytes are large multinucleated cells that remain within the bone marrow. Platelets made of fragments of megakaryocyte cytoplasm are released into peripheral blood. Platelets are also known as thrombocytes. Thrombopoiesis can take up to seven days to complete.

6.10 LEUKOPOIESIS

Leukopoiesis is the general term for the process of making white blood cells. Megakaryocytes and proerythroblasts are stem cells derived from the same pluripotent stem cell population. Different stimuli are used to promote the production of each distinct type of white blood cell. Granulopoiesis is the process by which a pluripotent stem cell differentiates into a neutrophil, eosinophil, or basophil. Early granulocytes are difficult to differentiate from one another because they all appear as large cells with lots of cytoplasm, large round nuclei, and a first set of nonspecific granules. Afterwards, the nonspecific granules are replaced by particular granules unique to each type of granulocyte.

After being activated by a specific stimulus, pluripotent stem cells give rise to both monocytes and lymphocytes. Since they lack specific cytoplasmic granules, they are known as agranulocytes. Some of the lymphocytes produced during the process known as lymphopoiesis proliferate outside of the bone marrow. Monopoiesis is the process by which monocytes grow and mature (Table 6.2).

TABLE 6.2
Table Representing WBC Count and Differential Leucocyte Count in Different Species

Species	Total WBC/UL(thousands)	Neutrophils	Lymphocytes	Monocytes	Eosinophils	Basophils
Cow	7–10	25–30	60–65	5	2–5	<1
Sheep	7–10	25–30	60–65	5	2–5	<1
Goat	8–12	35–40	50–55	5	2–5	<1
Horse	8–11	50–60	30–40	5–6	2–5	<1
Pig	15–22	30–35	55–60	5–6	2–5	<1
Dog	9–13	65–70	20–25	5–6	2–5	<1
Cat	10–15	55–60	30–35	5–6	2–5	<1
Chicken	20–30	25–30	55–60	10	3–8	1–4

BIBLIOGRAPHY

Dukes, H. H. (1955). *The physiology of domestic animals* (7th edition). Bailliere, Tindall & Cox.

Hall, J. E., & Hall, M. E. (2020). *Guyton and hall textbook of medical physiology e-book: Guyton and hall textbook of medical physiology e-book*. Elsevier Health Sciences.

Jain, N. C. (1986). *Schalm's veterinary hematology* (4th edition). Lea & Febiger.

Reece, W. O., Erickson, H. H., Goff, J. P., & Uemura, E. E. (Eds.). (2015). *Dukes' physiology of domestic animals*. John Wiley & Sons.

Schalm, O. W. (1962). Practical veterinary hematology. *The Canadian Veterinary Journal*, *3*(4), 116.

Thrall, M. A., Weiser, G., Allison, R. W., & Campbell, T. W. (Eds.). (2012). *Veterinary hematology and clinical chemistry*. John Wiley & Sons.

Tortora, G. J., & Derrickson, B. (2012). *Principles of anatomy and physiology*. John Wiley & Sons.

7 Physiology of the Muscular System

Pankaj Kumar Patel[1], *Virendra Kumar Singh*[2] *and Amrita Vasava*[3]

[1]College of Veterinary Science & A.H., Sardarkrushinagar, Kamdhenu University,Gujarat-India
[2]College of Veterinary Science & A.H., Navsari, Kamdhenu University, Gujarat-India
[3]College of Veterinary Science & A.H., Anand, Kamdhenu University, Gujarat-India

7.1 INTRODUCTION

Muscles are vital tissues that accomplish diverse functions by shortening and pulling on bones. The most visible aspect of muscle function is related to locomotion, which includes standing, lying down, grazing, running, and competing. Other functions, not so visual but necessary for overall body function, include muscles for respiration, digestion, parturition, blood and lymph circulation, swallowing, and generation of body heat. Muscles can receive and respond to stimuli, stretching and returning to their original shape after being stretched. The animal body has three types of muscle fibres: skeletal, smooth, and cardiac. Each is characterised by microscopic structural differences and their location, function, and innervation. Skeletal muscle tissue is essential for body movement and various physiological functions. It comprises a central, fleshy, contractile muscle "belly" flanked by two tendons at each extremity. This arrangement is strategically located within the body to span joints. When these muscles contract, they shorten the distance between the origin and insertion tendons, resulting in joint bending (Figure 7.1).

Skeletal muscles are under voluntary control and are crucial in locomotion, facial expressions, and posture. Smooth and cardiac muscles are associated with visceral structures, enabling activities such as organ emptying or content transportation. Skeletal muscles, except for sphincters, exert effects at a distance from their location, requiring one end to be anchored and the other attached to the movable part. An anatomical description of skeletal muscles typically involves specifying their origin (the less movable end) and insertion (the more movable end).

7.2 CLASSIFICATION OF MUSCLES

Muscles are classified based on various factors, including the presence or absence of cross-striations and the mode of control exerted over them.

7.2.1 Striated or Non-Striated

Figure 7.2 shows striated muscles, which encompass skeletal and cardiac muscles, are distinguished by cross-striations, while non-striated muscles, known as smooth muscles, are identified by the absence of cross-striations and are predominantly situated in the walls of visceral organs.

7.2.2 Control on Muscle

Control-wise, muscles fall into two types: voluntary and involuntary. Skeletal muscles, which are voluntary, can be consciously controlled and are innervated by somatic nerves. On the other hand, smooth and cardiac muscles are involuntary and cannot be controlled consciously. Autonomic nerves innervate them.

7.2.3 Muscle Movement

Skeletal muscles are also categorised based on the type of movement they induce and their strategic positions concerning the structures they affect. Flexors and extensors facilitate joint movement in different directions; adductors and abductors draw a limb toward or away from the median plane, respectively, while sphincters, arranged circularly, constrict body openings. Due to the high prevalence of movement abnormalities in veterinary patients with neuromuscular system disease, it is crucial to understand how skeletal muscle functions and is controlled by the nervous system. Abnormalities of cardiac muscle and smooth muscle also feature prominently in many other clinical disorders, and such muscle is often the target of pharmacological clinical intervention.

7.3 ORGANIZATION OF SKELETAL MUSCLE

The skeletal muscle is an intricately organised tissue consisting of bundles of muscle fibres called myofibres, illustrated in Figure 7.3.

Each myofibre is a cylindrical muscle cell ranging from 5 to 100 μm in diameter and around 10 to 30 centimetres in length. These muscle fibres are enveloped in endomysium, a connective tissue layer that hosts a rich capillary network. The endomysium is enclosed by perimysium, which encircles the muscle bundles, and epimysium, which encompasses the entire muscle (Figure 7.4).

The muscle fibre is the contractile unit that shortens and transmits a pull through the endomysium, perimysium, and epimysium to the tendon or aponeurosis attached to a

DOI: 10.1201/9781003426851-7

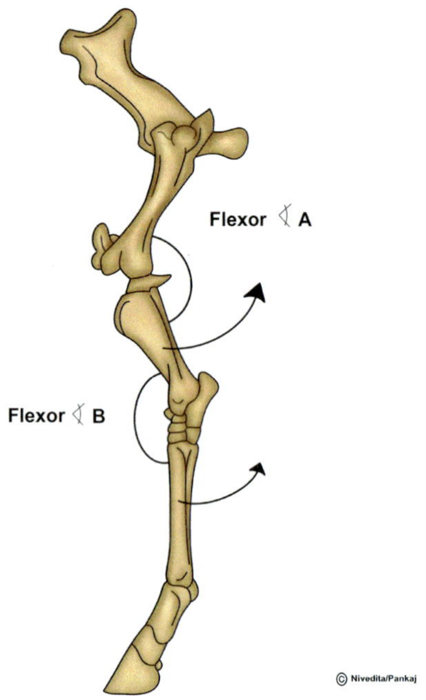

FIGURE 7.1 Control of muscle movement for posture adjustment. Arrows show the direction of joint movement.

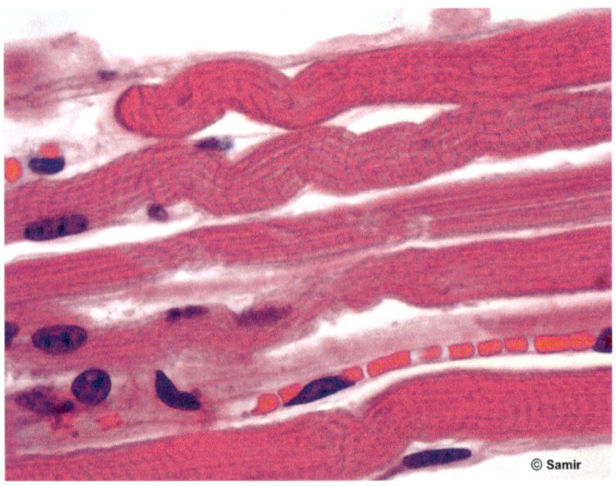

FIGURE 7.2 Diagram showing striations in the skeletal muscle.

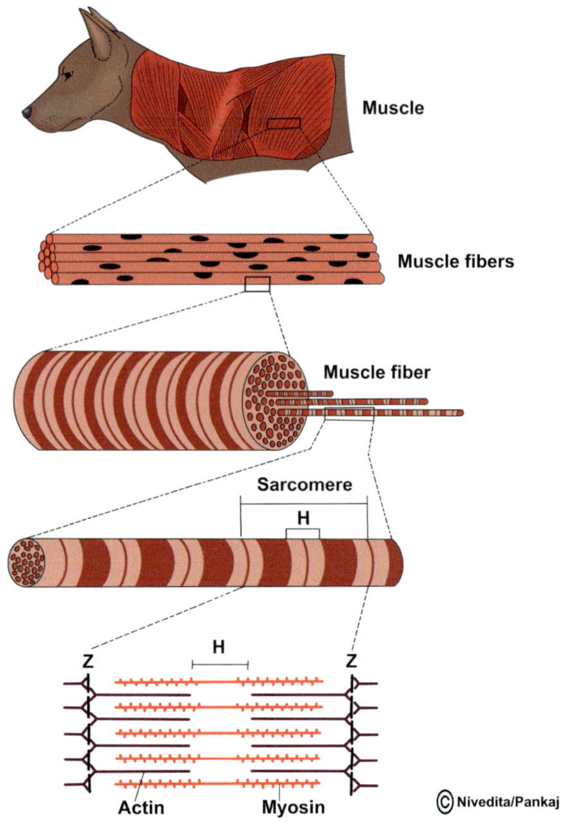

FIGURE 7.3 Diagram showing the organisation of skeletal muscle at multiple levels. The unit of muscle contraction – sarcomere also appears.

bone, thereby eliciting movement. Several hundred to several thousand myofibrils are arranged in parallel along the length of each muscle fibre; it looks like each layer is superimposed on another.

7.3.1 Arrangement of Myofibrils

Each myofibril comprises a linear sequence of repeating sarcomeres (Figure 7.5), which are the fundamental contractile units of the muscle fibre and can number in the tens of thousands.

The sarcomere contains a disk at each end known as the Z disk. The sarcomere comprises a range of large protein molecules responsible for muscular contraction, many of which are polymerised. Numerous thin protein filaments known as actin are attached to the Z disks and extend towards the centre of the sarcomere, resembling parallel fingers pointing at each other. These filaments are arranged in a regular spatial pattern, with a 2:1 ratio of actin to myosin. The actin filaments of two sarcomeres, common to the same Z line, form an I band or light bands. On the other hand, the myosin filaments are centrally located within a sarcomere and provide the dark bands (A band) of the characteristic striations along with the overlay of actin filaments. The myosin and actin filaments partially interdigitate, causing the myofibrils to have alternate light and dark bands. The dark bands or A bands are composed of myosin filaments and the overlapping ends of actin filaments, which exhibit anisotropic behaviour towards polarised light. This arrangement of actin and myosin filaments gives skeletal and cardiac muscle their striated appearance.

Each actin filament is made up of two helical strands of actin protein and two of tropomyosin protein, which are intertwined together as a larger helical complex. Troponin, complex globular protein molecules, are located intermittently along the tropomyosin-actin strand and can bind tropomyosin and actin. They have an affinity for calcium (Ca2+) ions. Myosin protein polymers in the form of thick filaments are present in suspension between the thin filaments of actin, running parallel to them. Approximately 500 myosin heads of a thick myosin filament form cross-bridges

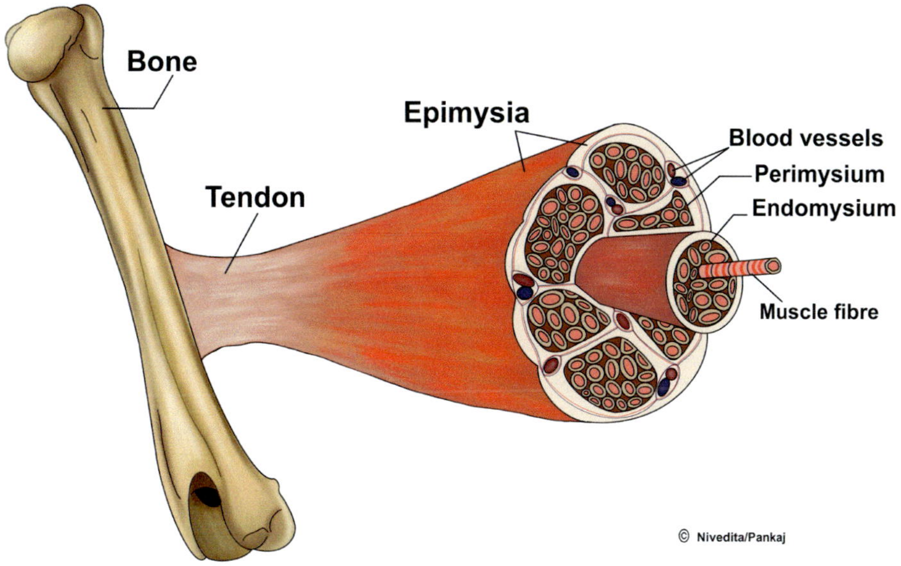

FIGURE 7.4 Skeletal muscle insertion in the bone and multi-level organization.

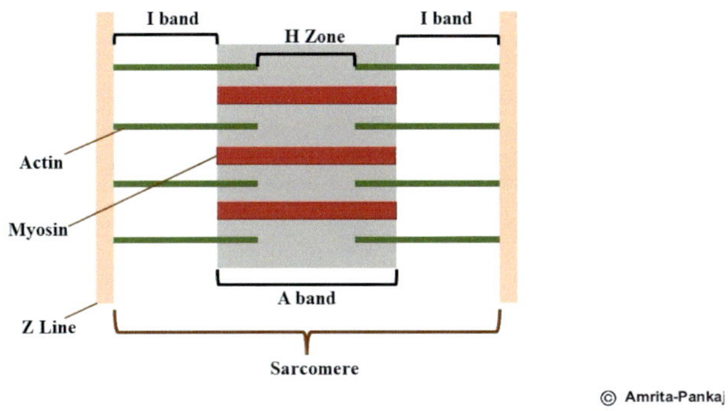

FIGURE 7.5 Diagram shows the sarcomere and the arrangement of actin and myosin filaments in it. Actin and myosin filaments are arranged parallel to one another.

that interact with actin to shorten the sarcomere as the myosin heads flex and relax.

7.3.2 Contractile Proteins

Myosin is a thick filament present in the sarcomere. Each myosin molecule has a molecular weight of approximately 480,000 and is composed of six polypeptide chains, including two heavy chains and four light chains, as seen in Figure 7.6.

The myosin molecule consists of two heavy chains, each weighing approximately 200,000, which twist around each other to form a double helix called the tail. At one end of the double helix, there are two free myosin heads, each consisting of a globular polypeptide structure and two light chains of molecular weight 20,000 each, which help control the function of the head during muscle contraction. The filament of myosin is constructed by bundling together over 200 individual myosin molecules, with the tails of these molecules forming the body of the filament. The heads of each myosin molecule, along with the protruding arms, form cross-bridges, which are flexible at two points called hinges. These hinged arms and heads play a crucial role in the contraction process, extending the heads far outward from the body of the myosin filament or bringing them close to the body. The myosin filaments are uniform in length, about 1.6 micrometres, and twisted to ensure cross-bridges extend in all directions. Additionally, there are no cross-bridge heads in the centre of the myosin filament for a distance of about 0.2 micrometres due to the extension of the hinged arms away from the centre or H zone. The myosin head plays a critical role in muscle contraction by functioning as an ATPase enzyme that hydrolyses ATP. This action allows the head to harness the energy released from the high-energy phosphate bond of ATP and use it to power the contraction process.

The structural core of the actin filament is a double helix of F-actin protein, depicted by the two lighter strands in the referenced Figure 7.7.

This helical structure is similar to that of the myosin molecule. The F-actin molecule comprises two helical strands with G-actin molecules with a molecular weight of approximately 42,000. Each G-actin molecule is bound to an ADP molecule, which is believed to be where myosin filament cross-bridges attach during muscle contraction. The binding sites are spaced every 2.7 nanometres along the actin filament and are offset between the two strands of the F-actin helix. The actin filament measures about 1 micrometre and is anchored firmly into the Z disks, extending into the gaps between myosin molecules in both directions. The actin filament is a complex structure that includes tropomyosin and two intertwined strands of actin protein, constituting a larger helical assembly. Troponin is another complex protein that is spaced periodically along

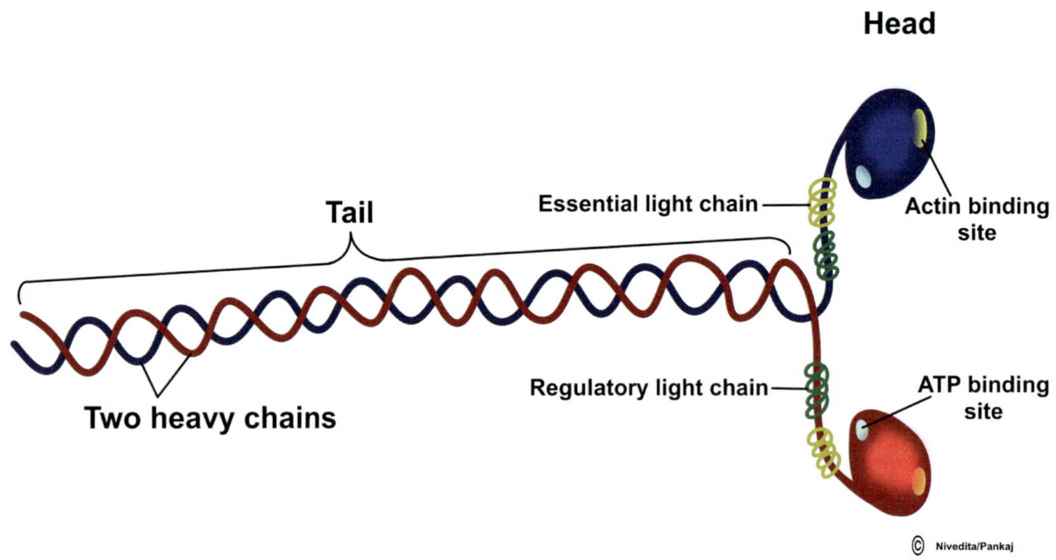

FIGURE 7.6 Molecular structure of myosin. Two heavy chains form the tail of myosin, while two light chains form the head along with each heavy chain.

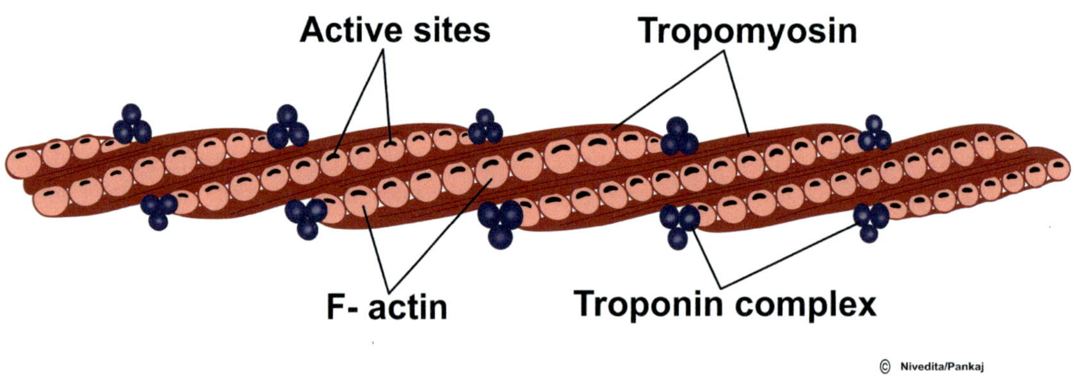

FIGURE 7.7 Molecular arrangement of actin is formed by two helical structures of F-actin. Tropomyosin strands fit in the groove formed by the helical structure of F-actin, which covers active binding sites.

the actin-tropomyosin strand and contains binding sites for tropomyosin, actin, and calcium ions.

7.3.3 Regulatory Proteins

Tropomyosin, a protein component of the actin filament, has a molecular weight of 70,000 and is 40 nanometres long. It spirals around the F-actin helix (See Figure 7.7). When the muscle is at rest, tropomyosin covers the active sites on the actin strands, preventing interaction with myosin and, thus, muscle contraction. Tropomyosin undergoes a structural change due to a signal, which consequently exposes the active sites on actin and initiates contraction.

Troponin intermittently attaches along the tropomyosin and is a complex protein of three subunits, each having a specific function in muscle contraction control. Troponin I binds to actin, troponin T binds to tropomyosin, and troponin C has a high affinity for calcium ions. This complex is crucial for securing tropomyosin to the actin and is believed to be the trigger for the contraction process when calcium ions bind to troponin.

Regulatory role of troponin and tropomyosin in muscle contraction: The troponin-tropomyosin complex inhibits the actin filament. Actin filaments, along with magnesium ions and ATP, swiftly and securely bind to the myosin molecule heads in the absence of the troponin-tropomyosin complex. However, when the troponin-tropomyosin complex is introduced to the actin filament, it prevents the binding between myosin and actin. In a relaxed muscle, the active sites on the normal actin filament are physically covered by the troponin-tropomyosin complex, which hinders them from attaching to the myosin filament heads and initiating muscle contraction. To initiate contraction, the inhibitory effect of the troponin-tropomyosin complex must be neutralised.

The inhibitory effect of the troponin-tropomyosin complex on the actin filaments is itself suppressed by a substantial quantity of calcium ions. The exact way in which the inhibition occurs is not fully comprehended. Still, one

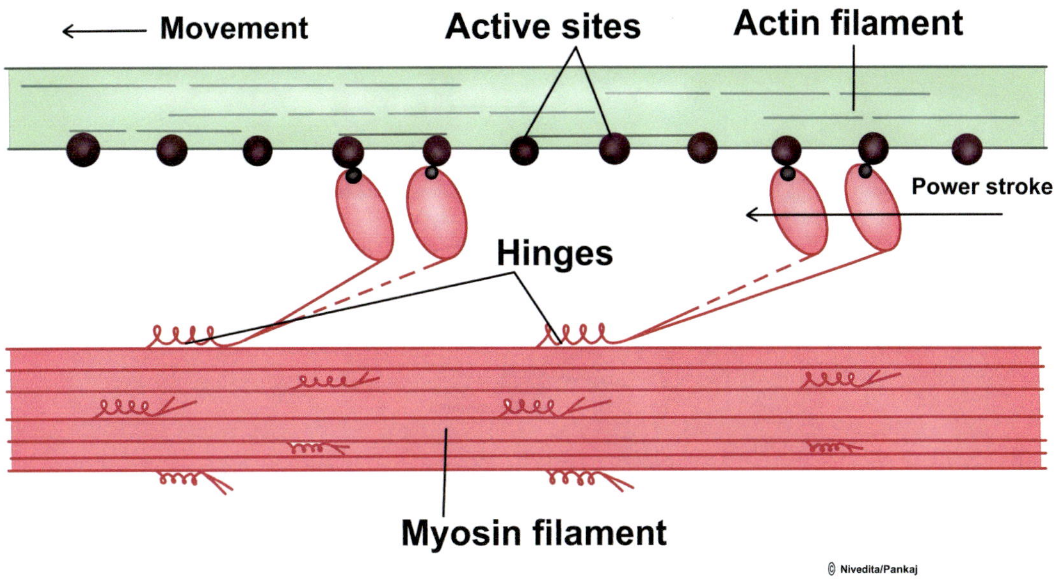

FIGURE 7.8 Diagram shows the sarcomere and the arrangement of the sarcotubular system. T Tubules are formed by invagination of the sarcolemma, and L tubules (sarcoplasmic reticulum) are parallel to sarcomere.

possible explanation is that when calcium ions attach to troponin C, the troponin complex undergoes a structural modification that leads to pulling on the tropomyosin molecule, causing it to move deeper into the gap between the two strands of actin. This movement exposes the active sites of the actin, allowing them to attract the myosin cross-bridge heads and facilitating the onset of contraction. While this mechanism is theoretical, it highlights the alteration of the normal interaction between the troponin-tropomyosin complex and actin by calcium ions, leading to a new state that triggers contraction.

7.3.4 Anchoring Proteins

Actinin is a structural protein that anchors actin filaments to the Z-discs, maintaining sarcomere integrity and actin alignment during muscle contraction. Desmin, a type III intermediate filament, supports muscle cells by linking Z-discs, myofibrils, and the cell membrane. Desmin is vital in preserving the structural stability of muscle cells and transmitting contractile forces across the muscle tissue. Nebulin, a large protein, is a molecular ruler, regulating actin filament length and organisation. The largest protein, Titin, provides passive elasticity, maintaining sarcomere integrity and regulating muscle tension.

7.3.5 Sarcotubular System

The skeletal muscle fibres harbour an intricate network of tubules known as the sarcotubular system, playing a crucial role in muscle function. Two distinct tubule sets compose this system, each with a unique arrangement among the myofibrils. The sarcoplasmic reticulum, encircling myofibrils, consists of tubules parallel to the myofibrils. In contrast, T tubules, arranged transversely, extend from one side of the fibre to the other, regularly spaced to serve each sarcomere, as shown in Figure 7.8.

The sarcotubular system facilitates impulse conduction from the muscle fibre surface to its core, with the sarcoplasmic reticulum serving as a vital calcium ion storage site, influencing muscle contraction initiation and termination. This intricate structure ensures anastomosing channel-like connectivity around each myofibril.

The system is further elucidated through the classification of two primary structures: T tubules and L tubules (sarcoplasmic reticulum). T tubules, formed by sarcolemma invaginations, penetrate the muscle fibre, allowing extracellular fluid flow through their lumen. In contrast, L tubules, or longitudinal tubules, constitute the sarcoplasmic reticulum, forming a closed tubular system around each myofibril. At regular intervals, L tubules dilate into lateral sacs called terminal cisternae, creating triads with adjacent T tubules. Triads are situated at the junction between 'A' and 'I' bands in skeletal muscle. L tubules store calcium ions, crucial for muscle contraction, and play a pivotal role in excitation-contraction coupling. The functions of the sarcotubular system are delineated through the roles of T and L tubules. T tubules ensure the rapid transmission of action potentials from the sarcolemma to myofibrils, facilitating efficient muscle stimulation. On the other hand, L tubules store a substantial quantity of calcium ions, releasing them into the sarcoplasm upon action potential arrival at the cisternae. This release of calcium ions initiates the processes essential for muscle contraction, highlighting the critical role of the sarcotubular system in the intricate mechanism of muscle function.

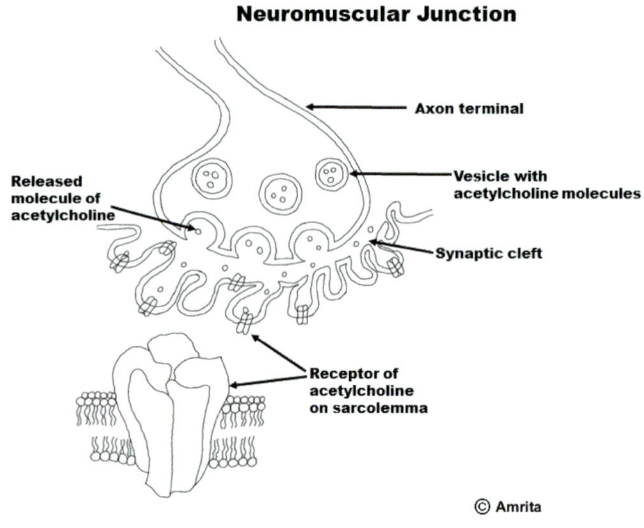

FIGURE 7.9 Diagram shows a schematic representation of the neuromuscular junction and receptors of acetylcholine on the sarcolemma.

7.4 MUSCLE CONTRACTION

Muscle contraction is a complex phenomenon involving the coordinated interaction between motor neurones and muscle fibres. Each muscle fibre is connected to multiple terminal branches of motor neurons, creating motor units that can vary in size. The size of each motor unit is unique and determined by the number of muscle fibres that are innervated by the terminal branches of the motor neurons. The larger motor units in the limbs and postural muscles are characterised by a single axon providing multiple muscle fibres. In contrast, the smaller motor units, linked with eye movements, have a single axon that may only offer a few muscle fibres.

7.4.1 Excitation Contraction Coupling

The point of contact between the end of a motor neurone and a muscle fibre is known as the neuromuscular junction. This junction serves as a connection between the two. Although the axon's terminal branch does not directly touch the muscle fibre, it communicates with it through a gap called the synaptic cleft. The neurotransmitter acetylcholine (ACh) is stored in vesicles within the axon's terminal branch, waiting to be released into the synaptic cleft when a signal arrives.

When an action potential arrives at the neuromuscular junction, it triggers the opening of voltage-gated Ca2+ and Na+ channels, causing an influx of calcium ions into the axon's terminal bulb. The DHP receptors, also known as voltage-gated Ca2+ channels, can be hindered by antihypertensive/antiarrhythmic drugs belonging to the class of Ca2+ channel blockers. The ryanodine receptors, also known as Ca2+ release channels, on the other hand, can be activated by caffeine. This rapid increase in calcium ion concentration of around 10 to 100 times triggers the binding of synaptic vesicles to the terminal bulb's plasma membrane, allowing the release of ACh into the synaptic cleft. The ACh diffuses into the muscle fibre sarcolemma's invaginations, binding to ACh receptors and initiating the process of depolarisation (Figure 7.9).

The enzyme acetylcholinesterase (AChE) hydrolyses ACh into acetic acid and choline almost immediately after its release. For the next action potential to be propagated to the muscle fibre, a new action potential at the neuromuscular junction must occur. AChE is found in high concentrations within the small size of the synaptic cleft. This, along with the limited diffusion distance of ACh in the synaptic cleft, accounts for the speedy hydrolysis of ACh.

Ach Receptor: The acetylcholine receptor is a key nervous system component, playing a crucial role in transmitting signals between nerve cells. It is a type of protein known as an ion channel, which forms a pathway across the cell membrane for ions to move in and out of the cell. The structure and function of the acetylcholine receptor are complex and finely tuned to allow for precise control over these ion movements. The receptor comprises five subunits, arranged in a ring to form a channel through the cell membrane. In foetuses, these subunits are two alpha, one beta, one delta, and one gamma. However, in adults, an epsilon subunit replaces the gamma subunit. Each of these subunits contributes to the overall structure and function of the receptor. The channel is normally narrow but can change shape and open when two acetylcholine molecules bind to the alpha subunits. This opening allows ions to pass through the channel. The channel is about 0.65 nanometres wide when open, which is large enough for positive ions like sodium (Na+), potassium (K+), and calcium (Ca2+) to pass through easily. Patch clamp studies, a type of experiment used to measure the movement of ions through individual channels, have shown that an open acetylcholine receptor channel can allow 15,000 to 30,000 sodium ions to pass in just one millisecond. This rapid movement of ions allows nerve cells to transmit signals quickly and efficiently. However, not all ions can pass through the acetylcholine receptor channel. Negative ions such as chloride are repelled and cannot pass through due to the negative charges at the channel's entrance. This selectivity is essential for maintaining the balance of ions inside and outside the cell, which is crucial for the proper functioning the nervous system.

7.4.2 Unloading of Calcium for Initiating Contraction

The depolarization propagates in all directions from the neuromuscular junction centrally located on the muscle fibre, starting with the T tubules. The T tubules act as communication links between the sarcolemma and the myofibrils within each muscle fibre. When a stimulus is received, the depolarisation continues in the T tubules, releasing calcium ions from the sarcoplasmic reticulum into the muscle fibre's cytosol, allowing the contraction to begin. With the entry of calcium ions, myosin filaments get attracted to unexposed active sites of actin filaments, allowing them to slide over each other, resulting in the reduction of sarcomere length. Just milliseconds after muscle contraction, the calcium membrane pump pumps calcium ions back into the sarcoplasmic reticulum, leading to the stoppage of muscle contraction.

7.4.3 Ratchet Theory of Muscle Contraction or Sliding Mechanism

The ratchet theory, also known as the walk-along theory or the sliding mechanism, provides a comprehensive framework for understanding the interaction between the activated actin filament and myosin cross-bridges during muscle contraction. When calcium ions activate the actin filament, the myosin cross-bridge heads are attracted to the active sites of the actin filament, initiating the contraction process. This theory describes the stepwise process of the myosin cross-bridges attaching to and disengaging from the active sites of the actin filament. When a myosin head attaches to an active site, it causes profound changes in the intramolecular forces between the head and arm of the cross-bridge, leading to a power stroke where the head tilts and drags the actin filament (Figure 7.10).

This process is repeated as the myosin cross-bridges stretch the actin filament, pulling actin filaments of two different ends toward the centre of the myosin filament or to the H zone. Each myosin cross-bridge is believed to operate independently; in a continuous cycle, there is a repeated process of attaching and pulling the actin filaments. The force of contraction is directly proportional to the number of myosin cross-bridges that are in touch with the actin filament at a particular moment. This theory provides valuable insights into the molecular events that underlie muscle contraction, shedding light on the coordinated interplay between the actin and myosin filaments. It is supported by considerable evidence and is a fundamental concept in muscle physiology. The walk-along theory has been a subject of extensive research, and recent studies have provided further evidence supporting its fundamental principles. Advanced imaging techniques, such as high-speed atomic force microscopy, have allowed researchers to directly observe the structural changes in myosin and actin filaments during the contraction process, providing visual confirmation of the mechanisms proposed by the walk-along theory. Additionally, studies utilising genetic and biochemical approaches have further elucidated the role of specific molecular components in the cross-bridge cycling process, supporting the independent and coordinated action of myosin cross-bridges during muscle contraction.

7.5 ROLE OF ATP IN THE MYOSIN CROSS-BRIDGE CYCLE AND MUSCLE CONTRACTION

The muscle contraction and relaxation process involves coordinated energy changes that enable the myosin cross-bridge heads to attach and detach with the mechanical alterations of the actin molecule. Before the initiation of muscle contraction, the heads of the cross-bridges bind with ATP. As soon as the myosin head attaches to ATP, the ATPase enzyme-like activity of the myosin head immediately breaks down the ATP molecule. However, the cleavage products, ADP + Pi, remain bound to the head. The energy released from ATP hydrolysis "cocks" the heads, increasing their angle of attachment to the cross-bridge arm and making them perpendicular to the active sites of the actin myofilaments. The tilting of the cross-bridge heads results in the release of ATP and Pi and the exposure of sites on the heads to bind new ATP. The binding of new ATP causes the detachment of the myosin cross-bridge heads from the actin myofilaments. The ATPase of the myosin cross-bridge heads then hydrolyses ATP as before, "cocking" the heads, and the process is repeated when the subsequent neuromuscular transmission causes the depolarisation of the sarcotubular system. The cyclic process continues until the Z disk is pressed against the ends of the myosin filaments. If the muscle is overloaded, no further pulling can occur. When a muscle contracts, a significant amount of ATP is cleaved to form ADP, and the more work the muscle performs, the more ATP is cleaved; this is known as the Fenn effect.

7.6 ENERGY SOURCES TO CARRY OUT MUSCLE CONTRACTION

Muscle contraction is powered by Adenosine Triphosphate (ATP), which is converted into Adenosine Diphosphate

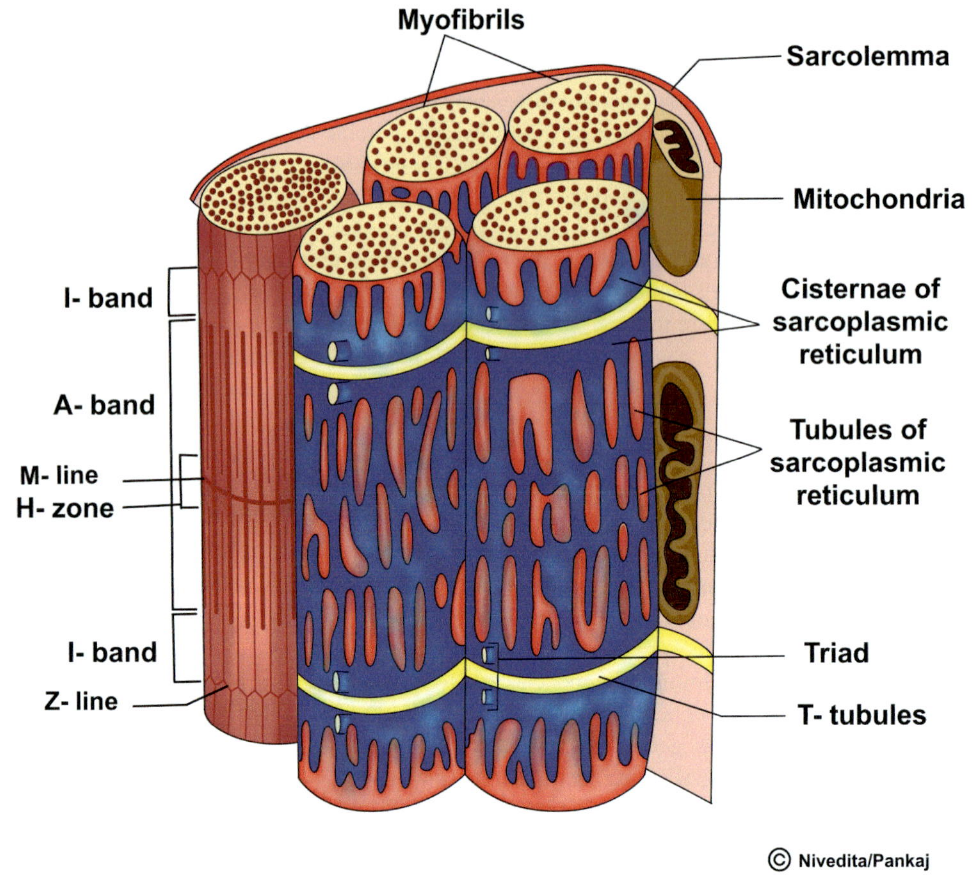

FIGURE 7.10 Sliding mechanism of muscle contraction.

(ADP) and inorganic phosphate (Pi). The concentration of ATP in muscle fibres is around four millimolars and can sustain full contraction for only a fraction of a second. Phosphocreatine is the first energy source for rephosphorylating ADP to form new ATP, but the amount is limited, only about five times greater than ATP.

7.6.1 Creatine Phosphate

Creatine phosphotransferase triggers Lohmann's reaction, releasing high-energy phosphate from creatine phosphate. The energy produced by the stored ATP and phosphocreatine is only enough to maintain the maximum level of muscle contraction for a brief period of 5 to 8 seconds. This process occurs through intermediary metabolism within the muscle fibre and oxidative phosphorylation in the mitochondria's electron transfer chain.

7.6.2 Glycolytic Pathway

Glycolysis is the second energy source, breaking down glucose in muscle cells to pyruvic acid and lactic acid, releasing energy to convert ADP to ATP. This process can continue even without oxygen, allowing muscle contraction to continue for many seconds or even more than a minute. The capacity of glycolysis to maintain maximum muscle contraction is restricted by the accumulation of end products in muscle cells, typically after approximately one minute.

7.6.3 Oxidative Phosphorylation

The third and final energy source is oxidative metabolism, which combines oxygen with cellular foodstuffs to produce ATP. Over 95% of all energy used by muscles for sustained long-term contraction comes from oxidative metabolism.

During prolonged endurance exercises, the predominant source of fuel for aerobic metabolism is fatty acids. These fatty acids undergo beta-oxidation to produce acetyl-CoA, which subsequently enters the citric acid cycle to generate ATP. Muscle contraction efficiency is approximately 50-70%, with excess energy dissipating as heat. This heat is crucial for regulating body temperature. When at rest and cooling, shivering involves muscle contraction and generates heat, which is helpful in maintaining body temperature.

7.7 FAST-CONTRACTING OR SLOW-CONTRACTING FIBRES

Each muscle in our body is composed of a combination of fast and slow muscle fibres, with other fibres falling between these two extremes. These muscle fibres exist on a continuum ranging from slow to fast. They can be categorised into two main types: slow-twitch (Type I) and fast-twitch (Type II) and vary in properties from each other as follows: 1. Contraction Speed: Slow-twitch (Type I) fibres contract slowly, suited for endurance activities, while fast-twitch (Type II) fibres contract rapidly, ideal for quick, powerful movements 2. Metabolism: Slow-twitch fibres primarily use aerobic respiration, making them fatigue-resistant, while fast-twitch fibres rely on anaerobic respiration, leading to quicker fatigue 3. Myosin ATPase Activity: Fast-twitch fibres, specifically Type IIx, have high myosin ATPase activity, which allows them to generate more force. However, the heavy reliance on anaerobic metabolism also makes them highly fatigable. 4. Force Production: Fast-twitch fibres provide bigger and more powerful forces for shorter durations, while slow-twitch fibres produce lower power contractions over longer periods 5.Fatigue Resistance: Slow-twitch muscle fibres have a higher resistance to fatigue as they primarily rely on aerobic respiration, while fast-twitch fibres fatigue more quickly, especially those primarily using anaerobic glycolysis 6. Capillary Supply and Myoglobin Content: Slow-twitch fibres have a rich capillary supply, numerous mitochondria, and a high concentration of myoglobin, making them red in colour, while fast-twitch fibres generally contain fewer blood vessels, mitochondria, and myoglobin, resulting in a paler colour. 7. Twitch Speed: Slow-twitch fibres possess slower twitch speeds, while fast-twitch fibres have higher twitch speeds. 8. Endurance vs Power: Slow-twitch fibres are designed for endurance activities that require long-term, repeated contractions, like maintaining posture or running a long distance. Fast-twitch fibres are ideal for quick contractions of short duration 9. Fibre Type Transitions: Muscle fibres can transition from slow to fast and vice versa with exercise training, impacting their characteristics and utility for specific activities 10. Colour: Slow-twitch fibres are often referred to as red fibres due to their high myoglobin content, while fast-twitch fibres are paler in colour due to lower myoglobin content.

7.8 MECHANICS BEHIND MUSCLE CONTRACTION

A motor unit refers to a cluster of muscle fibres that are triggered by a single motor neuron along with its axon terminals. The size of a motor unit can differ depending on the complexity of the movement that needs to be controlled. For instance, muscles involved in delicate and precise movements have fewer muscle fibres in a motor unit, whereas muscles that do not require fine control may have several hundred muscle fibres in a motor unit. Each motor neuron innervates multiple muscle fibres, and the number of fibres innervated depends on the type of muscle. The muscle fibres in each motor unit overlap with other motor units in micro-bundles of 3 to 15 fibres; they are not clustered together, which allows separate motor units to contract in support of one another instead of entirely as individual segments.

7.9 MUSCLE ACTION POTENTIAL AND ACTION POTENTIAL CURVE

The process of generating a muscle action potential shares similarities with the initiation and conduction of action potentials in nerve fibres, with some quantitative differences. In skeletal muscle fibres, the resting membrane potential is about -80 to -90 millivolts, which is more negative than in neurons. The action potential's duration ranges from 1 to 5 milliseconds, which is about five times longer than in large myelinated nerves. Additionally, the velocity of conduction is 3 to 5 m/sec, which is approximately 1/13th of the velocity of conduction in the large myelinated nerve fibres that stimulate skeletal muscle. As shown in Figure 7.11 in muscle tissue, the action potential curve has three major parts: the latent period, depolarisation, and repolarisation.

The latent period, which occurs immediately after applying a stimulus, is a brief phase lasting 0.5 to 1 millisecond, during which no change in the electrical potential is observed. Following the stimulus, there is a slight irregular deflection of the baseline for a very short period, known as the stimulus artefact. The current leakage from the stimulating electrode to the recording electrode creates a disturbance in the muscle, which results in the artefact. Subsequently, the stimulus artefact is followed by the latent period, which exists as a brief pause of 1–2 milliseconds after the action potential reaches the muscle before tension becomes noticeable. During the latent period, the action potential is propagated along the sarcolemma, and upon the release of calcium ions from the sarcoplasmic reticulum, the subsequent phases of muscle contraction occur.

7.10 EXCITABILITY CURVE OR STRENGTH-DURATION CURVE

The strength-duration curve, also called the excitability curve, is a visual representation of the correlation between the intensity of an electrical stimulus at a muscle's motor point and the pulse's duration (Figure 7.12).

This curve is used to assess nerve damage in lower motor neurone lesions. There are two main types of curves: the "nerve curve" and the "muscle curve." The muscle curve indicates a scenario where all nerve fibres supplying the muscle have degenerated, resulting in a steep and right-shifted curve. Two crucial parameters on the strength-duration curve are rheobase and chronaxie. The term "rheobase" refers to the smallest amount of electrical current necessary to trigger a neural response, regardless of how long the current is applied, while chronaxie is the time needed for a stimulus of twice the rheobase strength to elicit a response. Chronaxie is inversely proportional to

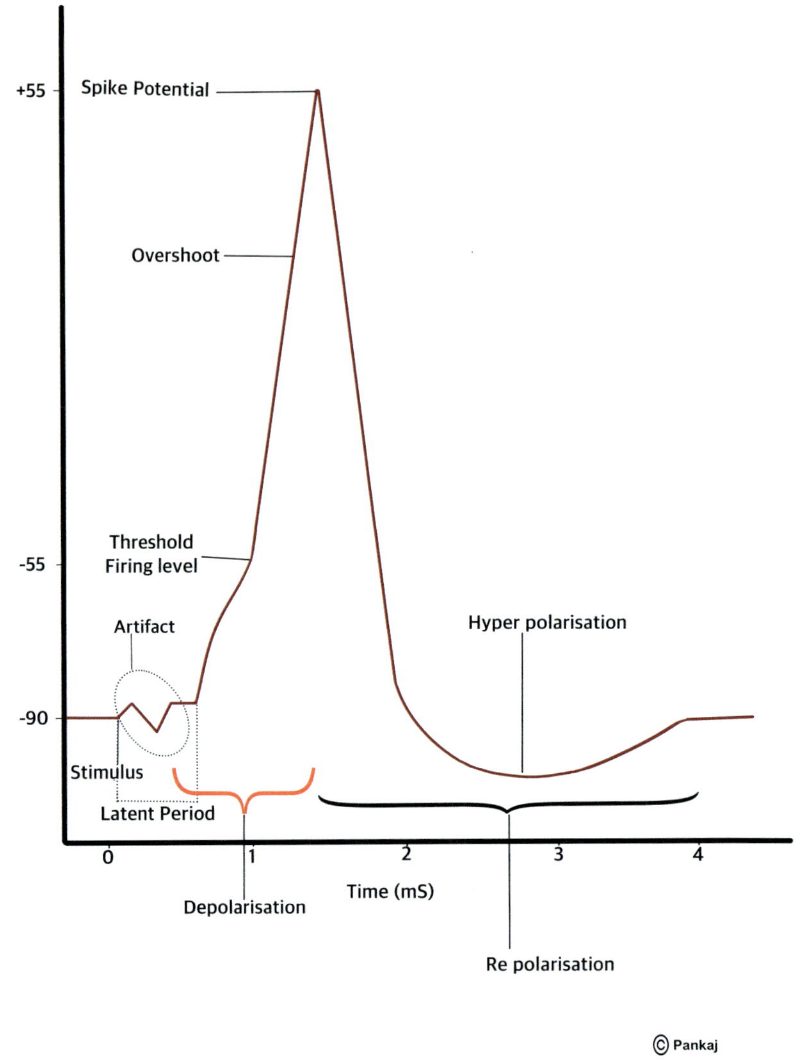

FIGURE 7.11 Graph shows the change in membrane potential after the application of the stimulus.

excitability, indicating that its value is higher for denervated muscles. The curve is influenced by various factors such as skin resistance, subcutaneous tissue, temperature, electrode characteristics, age, and fatigue. It is also used to evaluate sensory deficits in peripheral neuropathy and assess nerve injury.

7.11 TYPES OF MUSCLE CONTRACTION BASED ON CHANGES IN THE LENGTH OF MUSCLE FIBRES OR TENSION OF THE MUSCLE

Muscular contraction is divided into two types based on the change in the length of muscle fibres or tension of the muscle: 1. Isotonic Contraction: Isotonic contraction is a type of muscular contraction where the tension remains constant, and the length of the muscle fibre is changed. For instance, during the simple flexion of the arm, the muscle fibres shorten without a change in tension.2. Isometric Contraction: Isometric contraction is a type of muscular contraction where the length of muscle fibres remains the same, and the tension is increased. An example of muscle stiffness and strain occurs when a heavy object is pulled, causing increased tension in the muscles without changing their length. These contractions play a crucial role in understanding muscle function during various activities and exercises. They are essential in maintaining posture, generating force, and enabling movement.

7.12 SUMMATION, TETANIZATION, AND FORCE MODULATION IN MUSCLE CONTRACTION

Muscle contraction strength can be adjusted through a process known as summation, which involves the additive effect of individual muscle twitches. There are two main types of summation: multiple fibre summation and

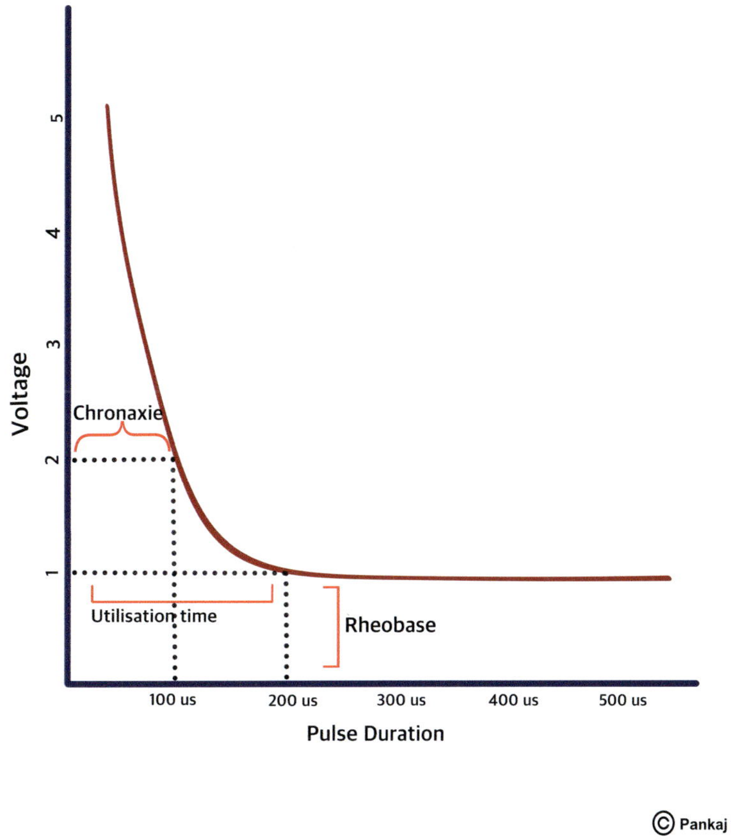

FIGURE 7.12 Graph shows the relationship between the application of strength and duration of stimuli.

frequency summation, both of which can lead to sustained contraction called tetanisation.

Multiple Fibre Summation: The central nervous system can selectively recruit motor units within a muscle, starting with smaller units and progressing to larger ones as needed. This recruitment strategy, known as the size principle, allows for precise control of muscle force, with small increments for light contractions and larger increments for stronger contractions. Smaller motor units are typically activated first due to their more excitable motoneurons, and motor units are activated asynchronously to ensure smooth muscle contractions.

Frequency Summation and Tetanisation: At lower frequencies of stimulation, individual muscle twitches are distinct. As the frequency increases, subsequent contractions begin before the previous ones have fully relaxed, leading to a progressive increase in overall muscle contraction strength. When the frequency reaches a certain threshold, the contractions merge into a smooth, sustained muscle contraction known as tetanisation. Beyond this frequency, further increases do not enhance contraction strength due to the high level of calcium ions in the muscle, which prevents relaxation between action potentials.

Staircase Effect (Treppe): When a muscle begins contracting after a period of rest, its initial strength may gradually increase to a plateau after several twitches, a phenomenon known as the staircase effect or treppe. This effect is believed to be due to the accumulation of calcium ions in the cytosol.

Strength of Stimulus: The force of muscle contraction is dependent on the intensity of the stimulus received. Various levels of stimuli can result in varying degrees of contraction force.

Effect of Number of Stimuli: The contractility of a muscle is affected by the number of stimuli it receives. A single stimulus results in a simple muscle twitch, while multiple stimuli can lead to beneficial effects, superposition, or summation effects, enhancing the contraction force and altering the contraction profile.

Effect of Load on Muscle Contraction: The muscle's work is assessed in two scenarios: free-loaded and after-loaded. The work done in the free-loaded condition is higher than in the after-loaded condition. This is consistent with the Frank-Starling law, as the initial length of the muscle fibre (or the degree of stretch) is greater in the free-loaded condition, leading to a stronger contraction and, thus, more work done. The free-loaded condition refers to a situation where the muscle contracts without any resistance, while the after-loaded condition refers to a situation where the muscle contracts against a load or resistance.

Muscle Fatigue: When a muscle contracts strongly for an extended period, it can become fatigued. This is primarily due to muscle glycogen depletion, which impairs the contractile and metabolic processes. In addition, the transmission of nerve signals through the neuromuscular junction can decrease after prolonged, intense muscle activity, which also contributes to muscle fatigue. Furthermore, suppose blood flow to a contracting muscle is interrupted. In that case, the muscle can become almost completely fatigued within a minute or two due to the loss of nutrient supply, particularly oxygen.

The coordination of agonist and antagonist muscles is essential to move body parts across a joint. This process, known as coactivation, is the brain and spinal cord's motor control centres regulate it. To position an animal's limb in the middle, the degree of contraction of the muscles that move the limb in opposite directions is balanced accordingly. By varying the activation of these muscle sets, the nervous system directs the precise positioning of the animal's body parts.

7.13 REFRACTORY PERIOD IN MUSCLE

The refractory period in muscles refers to a period during which a muscle cell is incapable of repeating an action potential, or in simpler terms; it cannot respond to a new stimulus. This period is crucial in preventing hyperactivity in excitable tissues and ensuring unidirectional propagation of action potentials. Refractory periods are classified into two types – absolute and relative.

The absolute refractory period corresponds to the depolarisation and repolarisation phases of the action potential, during which a second action potential cannot be initiated, regardless of the strength of the stimulus. The absolute refractory period for skeletal muscles lasts roughly 1–3 milliseconds. The prolonged absolute refractory period observed in cardiac muscle is primarily due to the elongated plateau phase of the cardiac action potential. This mechanism ensures that a complete contraction and relaxation cycle is achieved before the initiation of another contraction. The duration of the refractory period is crucial for the maintenance of rhythmic cardiac activity and prevents the occurrence of premature and/or irregular contractions as well as tetanus, which can compromise cardiac output.

The relative refractory period corresponds to the hyperpolarisation phase of the action potential. A second action potential can be initiated during this period, but it requires a greater stimulus than before. The relative refractory period for muscle fibres is approximately 5.99 ± 2.7 milliseconds.

In the context of smooth muscle, such as the ureter, the refractory period can be manipulated by altering the calcium content of the sarcoplasmic reticulum or by inhibiting certain channels.

7.14 RIGOR MORTIS

Muscle rigidity can manifest without neural stimulation, a condition known as rigor or physiological contracture, differing from typical muscle contraction. In this state, actin and myosin filaments are locked together due to an insufficient supply of ATP necessary for muscle relaxation. Postmortem, this lack of ATP leads to rigor mortis, where muscles remain rigid until lysosomal enzymes initiate decomposition after 12–24 hours. Muscles that were most active prior to death exhibit rigor mortis first due to their greater ATP depletion. Once death occurs, ATP synthesis ceases, preventing any further muscle relaxation.

BIBLIOGRAPHY

Andersson-Cedergren, E. (1959). Sarcotubular system. *Journal of Ultrastructure Research*, 2(Suppl. 1), 126.

Ariki, M., & Boyer, P. D. (1980). Characterization of medium inorganic phosphate-water exchange catalyzed by sarcoplasmic reticulum vesicles. *Biochemistry*, 19(9), 2001–2004.

Hall, J. E. (2021). *Guyton and hall textbook of medical physiology, 14th edition e-book*. Elsevier Health Sciences.

Klein, T. B. G. (2020). *Cunningham's textbook of veterinary physiology-E-book*. Elsevier Health Sciences.

Lin, B. L., Song, T., & Sadayappan, S. (2011). Myofilaments: movers and rulers of the sarcomere. *Comprehensive Physiology*, 2, 675–692.

Myhre, J. L., & Pilgrim, D. (2014). A Titan but not necessarily a ruler: Assessing the role of titin during thick filament patterning and assembly. *The Anatomical Record*, 297(9), 1604–1614.

Reece, W. O., Erickson, H. H., Goff, J. P., & Uemura, E. E. (Eds.). (2015). *Dukes' physiology of domestic animals*. John Wiley & Sons.

Schaeffer, P. J., & Lindstedt, S. L. (2013). How animals move: Comparative lessons on animal locomotion. *Comprehensive Physiology*, 3(1), 289–314.

Tsianos, G. A., & Loeb, G. E. (2017). Muscle and limb mechanics. *Comprehensive Physiology*, 7(2), 429–462.

Tsokos, M. (2005). Postmortem changesl overview. *Encyclopedia of Forensic and Legal Medicine*, 456–476.

Vandenboom, R. (2011). Modulation of skeletal muscle contraction by myosin phosphorylation. *Comprehensive Physiology*, 7(1), 171–212.

8 Physiology of the Nervous System

M. K. Vidya[1], C. N. Darshan[2] and N. R. Srikanth[3]

[1]Department of Physiology and Biochemistry, Veterinary College, Hassan, KVAFSU - Karnataka Veterinary, Animal and Fisheries Sciences University, India
[2]Department of Animal Husbandry and Veterinary Services, Karnataka, India
[3]Veterinary Biochemistry, Livestock Research Station, Lam, Guntur, Sri Venkateswara Veterinary University, India

8.1 INTRODUCTION

The nervous system is the body's communication network, consisting of the central nervous system (brain and spinal cord) and the peripheral nervous system (nerves outside the brain and spinal cord). It controls everything from basic functions like breathing to complex processes like decision-making and memory. It is a highly intricate network of neurons, the specialized cells that coordinate a complex interplay of functions within the organism. The nervous system is not just mere reflex responses; it dynamically processes sensory information to generate adaptive behaviour and form lasting memories. This remarkably sophisticated information processing allows organisms to react to their environment and also to learn from experiences, in a way, to adjust their future responses accordingly.

8.2 DIVISIONS OF NERVOUS SYSTEM

The nervous system is divided into two main parts: the central nervous system (CNS) and the peripheral nervous system (PNS) (Figure 8.1).

8.2.1 Central Nervous System (CNS)

The CNS, comprising the Brain and Spinal Cord, acts as the integrative control centre responsible for receiving, processing, and disseminating sensory information to coordinate complex cognitive processes, effector responses, and maintenance of physiological homeostasis. While the brain integrates sensory stimuli, generates thought and emotion, and coordinates motor output, the spinal cord serves as a communication conduit that facilitates the bidirectional flow of sensory information from the periphery to the brain and motor commands from the brain to the periphery.

8.2.2 Peripheral Nervous System (PNS)

The peripheral nervous system (PNS) consists of nerves and ganglia (clusters of nerve cell bodies) that extend from the central nervous system (CNS) to various parts of the body. It serves as a communication network between the CNS and the rest of the body, enabling sensory information to be transmitted to the brain and motor commands to be sent from the brain to muscles and glands. It governs the skeletal muscles, enabling the animal to consciously execute a wide range of movements with precision. It follows a descending motor pathway originating from the primary motor cortex in the brain. This cortical area in the brain involves specialized neurons responsible for generating voluntary movements. The signals generated descend from the brainstem and spinal cord, traversing into the synapse with the alpha motor neurons. These alpha motor neurons innervate skeletal muscle fibres, triggering their actions by contraction.

The PNS is further divided into two main functional components:

a. Somatic Nervous System (SNS): Controls voluntary movements and receives sensory information from the external environment. It includes sensory neurons that transmit information from the body's sensory receptors (e.g., skin, muscles) to the CNS, and motor neurons that convey signals from the CNS to skeletal muscles, enabling voluntary movements. Based on the site of recognition and exertion of its action, the neurons that are involved are classified as sensory and motor neurons.
   i. Sensory Neurons: These neurons detect stimuli from the external environment or within the body (such as touch, temperature, pain, and proprioception) and transmit this information to the CNS, particularly the brain and spinal cord.
   ii. Motor Neurons: These neurons transmit signals from the CNS to skeletal muscles, resulting in voluntary movements. Motor neurons carry commands from the brain or spinal cord to the muscles, allowing for coordinated movement and muscle contraction.

   The somatic nervous system plays a crucial role in various activities, including walking and performing other voluntary actions. It allows animals to interact with their environment and respond to sensory stimuli consciously.

DOI: 10.1201/9781003426851-8

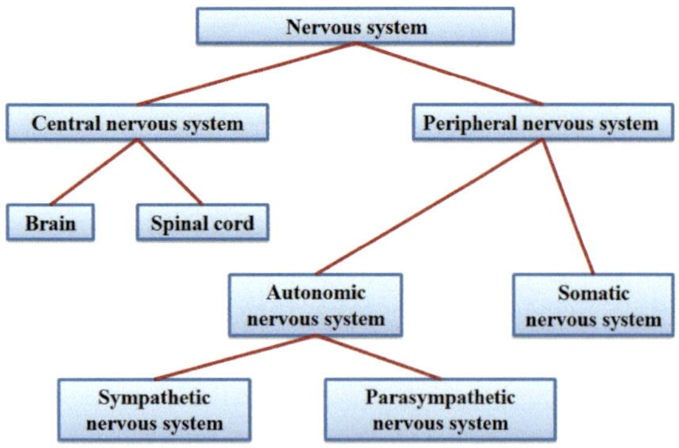

FIGURE 8.1 Divisions of the Nervous system

b. Autonomic Nervous System (ANS): Regulates involuntary bodily functions, such as heart rate, digestion, respiration, and glandular activity, to maintain internal homeostasis. The ANS operates without conscious control and is further divided into the sympathetic and parasympathetic divisions, which often have opposing effects on target organs, helping to balance physiological responses in different situations. The ANS consists of two main divisions with complementary functions:
   i. Sympathetic Division: Often referred to as the "fight or flight" response, the sympathetic division releases norepinephrine (noradrenaline) and epinephrine (adrenaline) from preganglionic neurons, which bind to adrenergic receptors on target organs to activate them to widespread physiological changes that prepare the body for exertion, such as increased heart rate, respiration rate, blood pressure, pupillary dilation, and blood sugar levels.
   ii. Parasympathetic Division: Known as the "rest and digest" response, the parasympathetic division shows opposing effects of the sympathetic nervous system, promoting relaxation, conserving energy, and facilitating digestion and other restorative processes. Acetylcholine is the primary neurotransmitter in this system. It slows heart rate, constricts airways, stimulates digestion, and promotes relaxation.

8.3 NERVOUS TISSUE

Nervous tissue is a vital part of the nervous system, consisting of a highly intricate network of neurons and glial cells. Neurons transmit signals, while glial cells support and protect them. It's essential for sensation, movement, cognition, and communication within the body. The nervous system is not just a mere reflex response; it dynamically processes sensory information to generate adaptive behaviour and form lasting memories. This remarkably sophisticated information processing allows organisms to react to their environment and also to learn from experiences, in a way, to adjust their future responses accordingly.

8.4 CELLS OF THE NERVOUS SYSTEM

The nervous system comprises various types of cells, each with specialized functions. The two important types are neurons and microglia.

1. Neurons: Neurons are the primary functional units of the nervous system, responsible for transmitting electrical impulses and processing information. They consist of a cell body (soma), dendrites (which receive signals from other neurons), and an axon (which transmits signals to other neurons, muscles, or glands).
2. Glial Cells: Glial cells, also known as neuroglia or glia, provide support and protection to neurons. There are several types of glial cells, like astrocytes, oligodendrocytes, microglia and ependymal which work together to ensure the proper functioning of the nervous system, facilitating communication, and maintaining homeostasis.

8.5 NEURONS

The nervous system is built upon a scaffold of neurons. Neurons are extremely specialized cells responsible for signal transduction and communication of the nervous system. They are unique in having long extensions called axons that transmit electrical signals and branching dendrites that receive signals from synapses that were established from adjacent neurons. They vary in shape, size, and function. The complex structure of the neurons allows complex communications within the nervous system. Information transfer occurs at specialized junctions referred to as synapses, where the neurotransmitters play a role in bridging the gap between the neurons enabling vast possibilities of signaling (Figure 8.2).

The neuronal cell body, also known as the soma or perikaryon, is the metabolic and structural centre of the neuron involved in integrating incoming signals from dendrites, processing the information, and generating outgoing

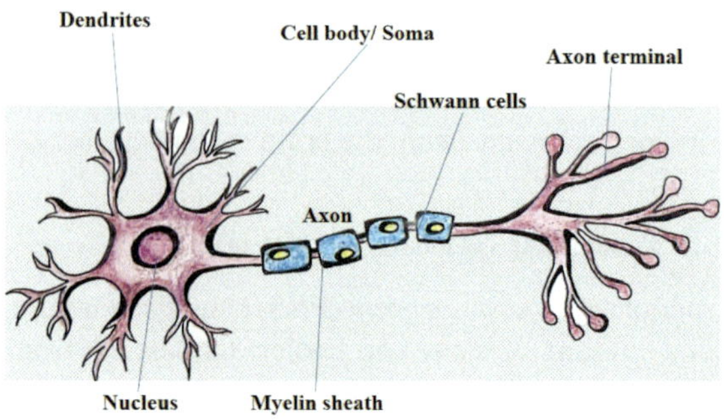

FIGURE 8.2 Structure of Neuron

signals along the axon. It consists of a nucleus that houses the cell's genetic material (DNA) and controls cellular activities by regulating gene expression. Organelles such as the endoplasmic reticulum, Golgi apparatus, mitochondria, and ribosomes, which are required for protein synthesis, metabolism, and cellular maintenance, are found in the cytoplasm. Along with them, Nissl bodies, also known as Nissl substances or Nissl granules, which are clusters of rough endoplasmic reticulum and ribosomes, are also found in the neuronal cell body. They are involved in protein synthesis, particularly the production of neurotransmitters and other proteins needed for neuronal function. A continuation of the body is a specialized region called the axon hillock, where the axon originates. It serves as the initial segment of the axon (a long, slender projection that conducts electrical impulses away from the cell body and toward other neurons, muscles, or glands) and features a high density of voltage-gated ion channels, which are essential for generating action potentials. These channels allow sodium ions to enter the neuron in response to depolarization, triggering the initiation of an action potential traveling at a speed of 0.5-120 m/s along the axon. Unlike the neuronal cell body, the axon hillock lacks Nissl bodies, contributing to the rapid conduction of action potentials along the axon. The axon hillock is also responsible for the integration of excitatory and inhibitory signals from dendrites and other synaptic inputs to initiate an action potential if the threshold for excitation is reached. Furthermore, the cell body has branching extensions that receive incoming signals from other neurons or sensory receptors, called dendrites. Dendrites have receptive surfaces that are covered with numerous small protrusions called dendritic spines, which provide surface area for synaptic connections with other neurons. These spines contain receptors for neurotransmitters released by axon terminals of neighbouring neurons. These chemical signals can be either excitatory, depolarizing the dendrite, or inhibitory, hyperpolarizing the dendrite. However, the dendrites integrate these signals from multiple synaptic inputs, summing up excitatory and inhibitory signals to determine whether the neuron will generate an action potential at the axon hillock. Though dendrites do not typically generate action potentials themselves, they propagate electrical signals, called graded potentials, toward the cell body.

Neurons can be classified based on various criteria, including their structure, function, and neurotransmitter characteristics.

Based on functionality the neurons are classified into Sensory (Afferent) neurons that transmit information from sensory receptors (such as those for touch, temperature, and pain) to the central nervous system, allowing for the perception of the external environment and internal bodily conditions, Motor (Efferent) neurons that convey signals from the central nervous system to muscles and glands, controlling voluntary and involuntary movements as well as glandular secretion, and Interneurons (Association neurons) that form connections between other neurons within the central nervous system, facilitating communication and integration of sensory, motor, and cognitive signals. They play a crucial role in information processing and neural circuitry.

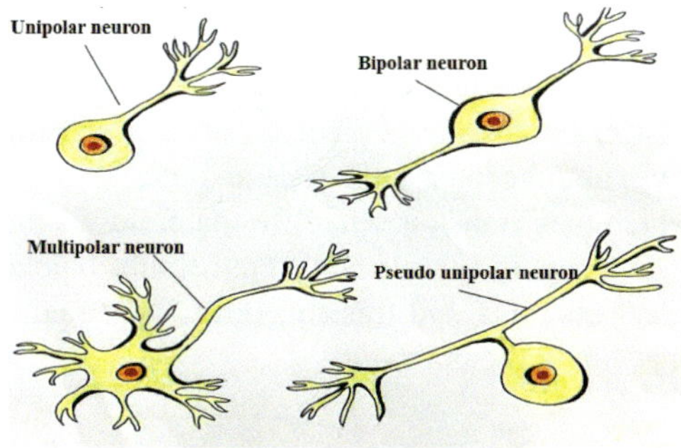

FIGURE 8.3 Different types of Neuron

Based on the neurotransmitter released by them, neurons are classified into Glutamatergic neurons (glutamate; excitatory), GABAergic neurons (GABA; inhibitory), Dopaminergic neurons (dopamine), Serotonergic neurons (serotonin), Noradrenergic neurons (norepinephrine), etc.

Structurally the neurons are classified into Multipolar, Bipolar, and Unipolar (Pseudounipolar) neurons (Figure 8.3).

8.5.1 Unipolar Neurons

Unipolar neurons, also known as pseudounipolar neurons, have a single process that extends from the cell body. This single process functions as both an axon that transmits signals to the central nervous system and a dendrite that receives sensory input. The single, elongated process of unipolar neurons allows for rapid conduction of sensory signals over long distances. This is facilitated by the myelin sheath, which insulates the axonal segment of the process and increases the speed of signal transmission. Their cell bodies are commonly found in the peripheral nervous system, particularly in sensory ganglia such as the dorsal root ganglia. They are primarily involved in sensory transmission, conveying sensory information from peripheral receptors (touch, temperature, and pain) to the central nervous system.

8.5.2 Bipolar Neurons

Unlike unipolar neurons, bipolar neurons have a distinct dendrite extending from one end of the cell body and also a single axon extending from the other end. This structural arrangement allows for separate reception of sensory input via the dendrite and transmission of signals to other neurons or sensory organs via the axon. Being relatively rare, located in sensory ganglia, these are primarily found in specialized sensory organs, such as the retina of the eye, the olfactory epithelium (nasal cavity), and the vestibular and cochlear apparatus of the inner ear. This proximity allows for efficient transmission of sensory information from the periphery to the central nervous system. Their specialized structure and function make them well-suited for relaying specific types of sensory information to higher brain centres for further processing and interpretation. In the retina, for example, bipolar neurons relay visual information from photoreceptor cells (rods and cones) to ganglion cells, which then transmit signals to the brain via the optic nerve. Similarly, in the olfactory epithelium, bipolar neurons detect odorants and transmit olfactory signals to the olfactory bulb in the brain.

8.5.3 Multipolar Neurons

These are the most common type of neuron in the central nervous system, characterized by having multiple dendrites and a single axon. The numerous dendrites increase the surface area available for synaptic connections, allowing the neuron to receive input from multiple sources. The single axon that carries electrical signals away from the cell body to other neurons, muscles, or glands may branch extensively as it extends from the cell body, allowing for communication with multiple target cells. These neurons integrate excitatory and inhibitory inputs from dendrites to determine whether to generate an action potential. They can be found in various regions of the brain and spinal cord, where they participate in sensory processing, motor control, cognitive functions, and other complex neural processes.

8.6 NEUROGLIA

Glia, also known as glial cells or neuroglia, are non-neuronal cells that provide support and protection for neurons in the nervous system. They outnumber neurons and include various types such as astrocytes, oligodendrocytes, microglia, Schwann cells, ependymal cells, and satellite cells (Figure 8.4), each with specific functions.

Glia are crucial for maintaining the structural integrity of the nervous system, regulating the extracellular environment, modulating synaptic transmission, and participating in immune responses and repair processes. They play essential roles in both normal brain function and various neurological disorders.

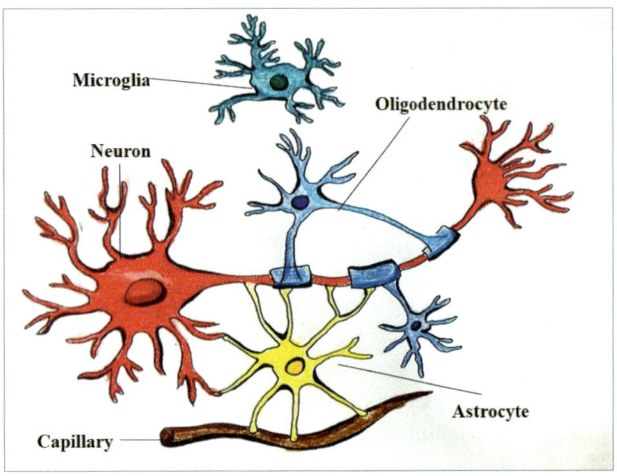

FIGURE 8.4 Various types of glial cells

8.6.1 Astrocytes

Astrocytes, the most abundant glial cells, are star-shaped and found in the brain and spinal cord. They provide structural support to neurons and help maintain the integrity of the blood–brain barrier, which regulates the passage of substances between the bloodstream and the brain. They also regulate the extracellular environment by taking up and releasing ions and neurotransmitters. Moreover, they provide energy substrates such as glucose to neurons and help regulate synaptic activity. Further, they help clear neurotransmitters from the synaptic cleft, thereby terminating synaptic transmission and preventing excessive neuronal excitation. Neurotrophically, astrocytes are involved in the secretion of growth factors and other molecules that promote neuronal survival and growth during development and in response to injury. They also play a major role in the modulation of synaptic function as they can influence synaptic transmission and plasticity through the release of signaling molecules such as glutamate, ATP, and D-serine. Additionally, they are involved in regulating the pH, ion concentration, and water balance of the extracellular space, contributing to the overall homeostasis of the CNS.

8.6.2 Ependymal Cells

Ependymal cells are a type of glial cell that lines the ventricles of the brain and the central canal of the spinal cord. These cells are cuboidal to columnar in shape and have cilia for the movement of CSF and microvilli on their surfaces. They are involved in the production and circulation of cerebrospinal fluid (CSF), which cushions the brain and spinal cord, provides nutrients, removes waste, and helps regulate the extracellular environment. They also form a barrier between the CSF-filled ventricles and the surrounding brain tissue. This barrier helps regulate the exchange of substances between the CSF and the brain, contributing to the overall homeostasis and protection of the CNS. Additionally, in some regions of the CNS, such as the lining of the lateral ventricles, ependymal cells possess stem cell-like properties and can give rise to new neurons and glial cells in response to injury or disease.

8.6.3 Microglia

Microglia are the highly dynamic and primary immune cells of the brain and spinal cord, constantly surveying their surroundings for signs of damage, infection, or inflammation. They can rapidly change their morphology and function in response to various signals. When activated by injury, infection, or inflammation, microglia can proliferate, migrate to the site of injury, and release cytokines and other signaling molecules to initiate an immune response. They thereby engulfing and digesting the cellular debris, pathogens, and dying neurons, helping to maintain tissue integrity and support neuronal survival. In addition to their role in immune defense, they also play roles in synaptic pruning, neurogenesis, and modulating neuroinflammation, thereby contribute to the maintenance of neuronal networks. Since the microglia play a crucial role in immune surveillance and neuronal homeostasis, dysfunction of microglia has been implicated in various neurological disorders, highlighting their importance in brain health and disease.

8.6.4 Oligodendrocytes

Oligodendrocytes have a relatively small cell body with a round or oval shape. The processes from the cell body wrap around the axons of neurons, forming a segment of the myelin sheath around multiple axons, creating a layered, insulating structure. The primary function of oligodendrocytes is to produce myelin, a fatty substance that insulates axons, allowing for efficient and rapid transmission of electrical signals along the neurons. They also provide structural support to neurons and contribute to the overall stability of the CNS. In the event of injury, oligodendrocytes can contribute to the repair and regeneration of damaged myelin.

8.6.5 Schwann Cells

Schwann cells are elongated glial cells that play a crucial role in the maintenance and function of peripheral nerves. Schwann cells wrap around the axons of neurons in the PNS to form the myelin sheath. Unlike oligodendrocytes, each Schwann cell typically myelinates only one segment of a single axon. Gaps between adjacent Schwann cells along the axon where the myelin sheath is absent are called Nodes of Ranvier. These gaps are essential for rapid signal transmission through saltatory conduction. The primary function of Schwann cells is to myelinate axons in the PNS, which speeds up the transmission of electrical impulses. Following a nerve injury, these cells can proliferate and prompt regrowth of axons, aiding in the repair of damaged nerves.

8.6.6 Satellite Cells

Satellite cells have a small and flattened cell body. They are found in the peripheral nervous system, specifically within ganglia. Within sensory, sympathetic, and parasympathetic ganglia, they surround the cell bodies and form a continuous layer around the neurons, providing a physical barrier between the neurons and the surrounding connective tissue. Satellite cells provide structural support to the neurons in the ganglia. They help regulate the exchange of nutrients and waste products between neurons and their surrounding environment. They play a role in modulating the microenvironment of the ganglia, including ion balance and neurotransmitter levels, which can affect neuronal excitability and function. In response to nerve injury, the activated satellite cells proliferate, produce growth factors and other molecules that support neuronal survival and regeneration. They can respond to changes in the neuronal environment, potentially playing a role in chronic pain conditions and neuropathies.

8.7 REACTION OF NEURAL TISSUE TO INJURY

The physiological response of neural tissue to injury is a complex, dynamic and multifaceted process that involves a balance between protective and detrimental mechanisms process (Figure 8.5).

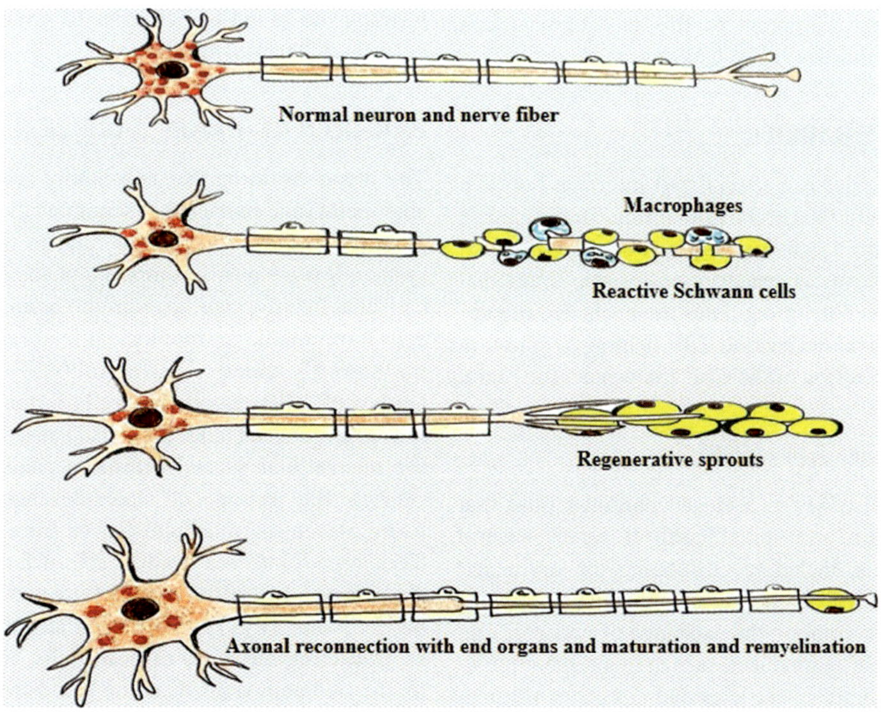

FIGURE 8.5 Reaction of neural tissue to injury

It involves a series of events that includes both immediate responses and long-term processes aimed at containing the damage and promoting repair. Primarily, following the injury, there is an excessive release of neurotransmitters, glutamate in particular, leading to over-activation of receptors (excitotoxicity), which disrupts the ionic balance and causes calcium influx and neuronal damage. Acutely, the activated microglia cells proliferate and migrate to the site of injury to release cytokines and chemokines, which attract other immune cells like astrocytes to form a glial scar around the injury site. This scar helps contain the injury but can also impede axonal regeneration. Secondarily, the injury can compromise the blood–brain barrier, allowing immune cells and potentially harmful substances like reactive oxygen species to enter the CNS, thereby exacerbating the inflammation process.

In the later stages, there may be attempts at neural regeneration, with axonal sprouting and formation of new connections. However, the extent of regeneration is often limited in the central nervous system due to factors such as inhibitory signaling molecules and the formation of scar tissue by reactive astrocytes. Ultimately, the response to neural injury can vary depending on factors such as the type and severity of the injury, the location within the nervous system, and the animal's age and overall health.

8.8 EXTRACELLULAR ENVIRONMENT TO THE CNS

The extracellular environment surrounding the central nervous system plays a crucial role in maintaining its function. It includes cerebrospinal fluid, which provides cushion and support, and the blood–brain barrier (BBB), which regulates the passage of substances between the bloodstream and the brain. Any disruption to this environment can impact the overall functioning of CNS.

8.9 BLOOD CSF BARRIER

The blood-CSF barrier is a specialized interface between the blood vessels and the cerebrospinal fluid in the brain's ventricles. It features a tight junction between the cells that line the choroid plexus, a structure within the brain that produces CSF. This barrier regulates the exchange of substances between the blood and the CSF, helping to maintain the chemical composition of the CSF and protect the brain.

8.9.1 Cerebrospinal Fluid (CSF)

Cerebrospinal fluid (CSF) is a clear, colourless fluid that surrounds the brain and spinal cord. It serves several important functions, including cushioning the brain and spinal cord against impact, providing buoyancy to support their weight, and facilitating the exchange of nutrients and waste products. CSF is produced primarily by the choroid plexus within the brain's ventricles and circulates through the ventricular system and around the brain and spinal cord before being absorbed back into the bloodstream through the arachnoid granulations. The production and absorption of CSF help regulate intracranial pressure. Its composition is tightly regulated to maintain a stable chemical environment for proper nervous system function. The CSF is about 99% water, providing the fluid medium necessary for its functions. The concentration of electrolytes in CSF is as follows: Sodium (138-150 mEq/L), Potassium (2.5-3.0 mEq/L), Calcium (2.1-2.4 mEq/L), Magnesium (1.0-1.5 mEq/L), Chloride (120-130 mEq/L), and Bicarbonate (23-30 mEq/L). The total protein concentration in CSF is low, around 15-45 mg/dL, which is significantly lower than in blood plasma. CSF glucose levels are typically about 50-80 mg/dL, which is about two-thirds of the blood glucose level. Normally, CSF contains very few cells, usually 0-5 WBCs per microliter, and is typically free of RBCs. Other components include Lactate (1.1-2.4 mmol/L), amino acids, neurotransmitters, and metabolites. The osmolarity of CSF is similar to that of blood plasma, which is around 290 mOsm/L. The pH of CSF is slightly lower than that of blood, typically around 7.3-7.4.

8.9.2 Choroid Epithelial Cells

Located within the lateral, third, and fourth ventricles, the choroid plexus has specialized cells called choroid epithelial cells. These are modified ependymal cells that have microvilli instead of cilia. Their primary function includes the production of cerebrospinal fluid (CSF), a clear fluid that surrounds and protects the brain and spinal cord. Additionally, they are responsible for actively transporting ions, nutrients, and other substances from the blood into the CSF, while also regulating the composition of the CSF and forming the blood-CSF barrier to protect the brain from potentially harmful substances.

8.10 BLOOD–BRAIN BARRIER (BBB)

The blood–brain barrier is a highly selective semipermeable membrane barrier that separates the circulating blood from the brain's extracellular fluid in the central nervous system. It is primarily composed of specialized endothelial cells that line the capillaries in the brain. These endothelial cells have unique morphological features that contribute to the barrier's selective permeability. The most distinctive feature of BBB endothelial cells is the presence of tight junctions between adjacent cells. These tight junctions seal the intercellular space, forming a continuous barrier that restricts the passage of molecules between cells, effectively preventing most substances from crossing from the bloodstream into the brain. Lack of fenestrations reduces the permeability of the endothelium and helps maintain the barrier's integrity. The BBB regulates molecular transport by restricting the entry of potentially harmful substances, toxins, and pathogens from the bloodstream into the brain.

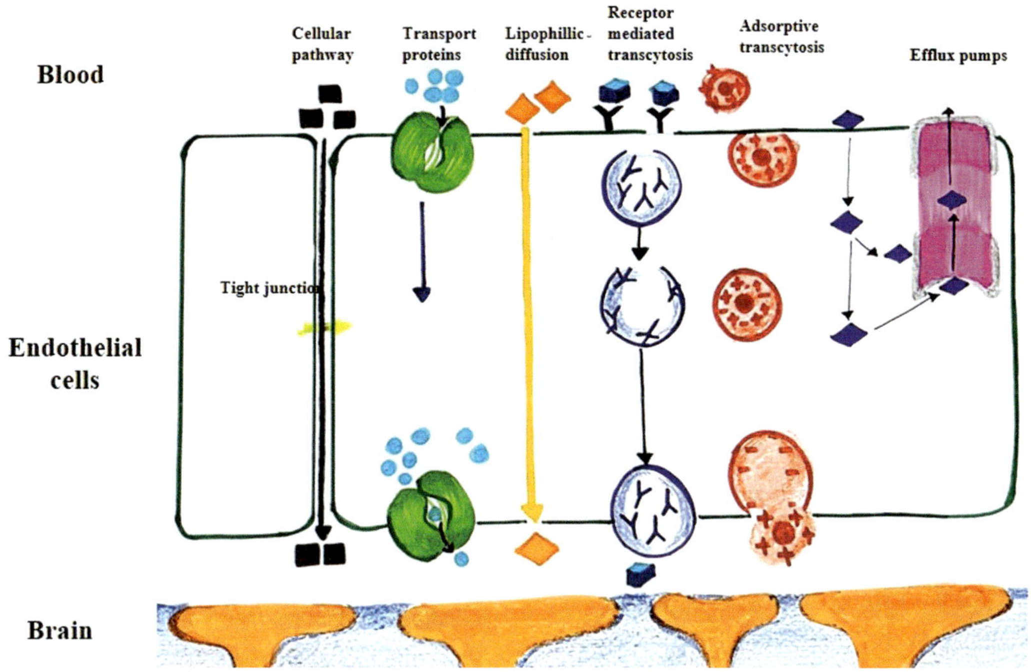

FIGURE 8.6 Transport across the blood–brain barrier

Transport across the blood–brain barrier occurs through several mechanisms, each serving specific functions (Figure 8.6):

1. Passive Diffusion: Small, lipophilic (fat-soluble) molecules, such as oxygen and carbon dioxide, can diffuse passively through the lipid bilayer of endothelial cells due to their small size and lipophilic nature. However, this route is limited for larger or polar molecules.
2. Transport Proteins: Specialized transport proteins embedded within the endothelial cell membrane facilitate the selective transport of essential molecules, such as glucose, amino acids, and neurotransmitters, across the BBB. These proteins include glucose transporters (e.g., GLUT1), amino acid transporters, and nucleoside transporters.
3. Receptor-Mediated Transcytosis: Certain molecules, such as insulin and transferrin, can cross the BBB via receptor-mediated transcytosis. In this process, molecules bind to specific receptors on the luminal (blood-facing) side of endothelial cells, triggering their internalization into vesicles. These vesicles then transport the molecules across the endothelial cell and release them on the abluminal (brain-facing) side.
4. Efflux Transporters: Efflux transporters, such as P-glycoprotein (P-gp), are located on the luminal side of endothelial cells and actively pump out various substances from the brain back into the bloodstream. This mechanism helps protect the brain by removing potentially harmful compounds and regulating the brain's exposure to drugs and toxins.
5. Paracellular Transport: Although minimal, some small molecules can pass through the tight junctions between endothelial cells via paracellular transport. However, this route is tightly regulated and allows only limited diffusion of substances.

8.11 ELECTROCHEMICAL BASIS OF NEURONAL FUNCTION

The electrochemical basis of neuronal function is rooted in the transmission of signals between neurons. Neurons communicate through electrical impulses called action potentials, which travel along the neuron's axon. These impulses are generated by changes in the neuron's membrane potential, primarily driven by the movement of ions such as sodium, potassium, chloride, and calcium across the cell membrane. Neurotransmitters released at synapses further modulate these electrical signals, allowing for complex neuronal communication and information processing in the brain and nervous system.

The neuronal membrane, also known as the plasma membrane, is a critical structure that surrounds and protects neurons, enabling them to function properly. It is primarily composed of a lipid bilayer interspersed with proteins and plays several crucial roles in neuronal activity. The neuronal membrane controls the movement of substances in and out of the cell, maintaining the necessary environment for neuronal function. It is selectively permeable, allowing some substances to pass while blocking others. This semi-permeable membrane maintains a resting membrane potential, typically around -70 mV, due to the differential distribution of ions (primarily sodium and potassium) across the membrane. This potential is essential for the generation and propagation of action potentials. Voltage-gated ion channels open in response to changes in membrane potential, allowing specific ions (e.g., Na+, K+, Ca2+) to flow in and out of the neuron causing rapid changes in membrane potential (depolarization and repolarization) that travel along the axon, allowing the neurons to transmit electrical signals over long distances. These signals are received by the membrane-bound receptors, thereby they trigger the intracellular signaling pathways that modulate neuronal activity and function.

8.12 ELECTROCHEMICAL GRADIENT IN NEURONS

The electrochemical gradient in neurons refers to the combined effects of both the concentration gradient and the electrostatic forces acting on ions across the cell membrane. It is a crucial concept in understanding how neurons generate and propagate electrical signals.

1. Concentration Gradient: This component of the electrochemical gradient is driven by differences in the concentration of ions inside and outside the neuron. For example, potassium ions (K+) have a higher concentration inside the neuron, while sodium ions (Na+) have a higher concentration outside. Chloride ions (Cl-) and calcium ions (Ca2+) also contribute to the gradient. These concentration differences establish a potential energy that drives the movement of ions down their concentration gradients. The approximate concentrations of ions inside and outside the neuron are tabulated in Table 8.1.
2. Electrostatic Forces: In addition to the concentration gradient, electrostatic forces influence the movement of ions across the cell membrane. Ions carry electric charges. A typical neuron is negatively charged on the inside compared to the outside. This is due to the presence of large negatively charged molecules leading to an unequal distribution of ions. This creates an electric field that attracts positively charged ions into the neuron and repels negatively charged ions.

TABLE 8.1
Concentrations of Ions Inside and Outside the Neurons

Ion	Intracellular Concentration (mM)	Extracellular Concentration (mM)
Potassium (K+)	140	5
Sodium (Na+)	15	150
Chloride (Cl-)	7-10	110
Calcium (Ca2+)	0.0001	2

The combined effect of these two forces determines the direction and magnitude of ion movement across the cell membrane. For example, during an action potential, sodium ions (Na+) move into the neuron driven by both the concentration gradient (higher outside) and the attraction to the negatively charged interior of the neuron. Understanding the electrochemical gradient is essential for grasping various neuronal processes, including resting membrane potential, action potential generation, synaptic transmission, and cellular signaling.

8.12.1 Concentration Gradient

The concentration gradients refer to the differences in the concentration of ions and molecules inside and outside the cell. Inside the neuron, the primary intracellular ions include negatively charged proteins and potassium (K+) that repolarizes the membrane after an action potential. Outside the neuron, in the extracellular space, the primary ions are sodium (Na+) ions which are involved in initiating action potentials by depolarizing the membrane, chloride (Cl-) ions that helps modulate the neuron's excitability and can contribute to inhibitory synaptic potentials, and calcium (Ca2+) which plays diverse roles in neurotransmitter release, synaptic plasticity, and intracellular signaling pathways. This asymmetric distribution of ions creates an electrochemical gradient across the cell membrane, which is crucial for various neuronal functions, including action potential generation, synaptic transmission, and cellular signaling. These concentration gradients are established and maintained by various active and passive transport mechanisms, such as the sodium-potassium pump (Na+/K+ ATPase), ion channels, and transporters.

8.12.2 Active Transport in Neurons

Active transport in neurons involves the movement of ions or molecules against their concentration gradients, requiring energy in the form of ATP. Two types of active transport mechanisms important in neurons are:

1. Sodium-Potassium Pump (Na+/K+ ATPase): This pump actively transports sodium ions (Na+) out of the neuron and potassium ions (K+) into the neuron against their concentration gradients. For each cycle, it moves 3 Na+ ions out of the neuron and 2 K+ ions into the neuron, utilizing ATP hydrolysis to power the process. This pump helps maintain the resting membrane potential and the concentration gradients of Na+ and K+ across the cell membrane.

2. Calcium Pump: Neurons also use active transport mechanisms to regulate intracellular calcium ion (Ca2+) levels. Calcium pumps located on the cell membrane and intracellular organelles, such as the endoplasmic reticulum and mitochondria, actively transport Ca2+ ions against their concentration gradient, helping to regulate intracellular Ca2+ concentrations and control various cellular processes such as synaptic transmission, neuronal excitability, and gene expression.

8.12.3 Passive Transport in Neurons

Passive transport in neurons refers to the movement of ions or molecules across the cell membrane without the input of metabolic energy (ATP). It occurs in response to concentration or electrochemical gradients and includes two main processes:

1. Diffusion: It is the movement of ions or molecules from a region of higher concentration to a region of lower concentration. In neurons, this process allows K+, Na+ and Cl- ions to move along their concentration gradients across the cell membrane. For example, during the resting state, potassium ions tend to diffuse out of the neuron, following their concentration gradient.
2. Facilitated Diffusion: Some molecules, such as glucose and certain ions, require assistance from membrane proteins to facilitate their movement across the cell membrane. These membrane proteins, such as ion channels and carrier proteins provide a pathway for specific molecules to move down their concentration gradients. Ion channels allow Na+, K+, Ca2+, and Cl- ions to pass through the membrane, facilitated by changes in the channel's conformation.

8.12.4 Ion Channels

The four main types of ion channels through which the ions exert their function in neurons are given in Table 8.2.

1. Voltage-Gated Ion Channels: These channels open or close in response to changes in the membrane potential.

TABLE 8.2
Differences among Voltage-Gated, Ligand-Gated, Mechanically-Gated and Leakage Ion Channels

Feature	Voltage-Gated Channels	Ligand-Gated Channels	Mechanically-Gated Channels	Leakage Channels
Opening trigger	Changes in the electrical membrane potential	Binding of a specific molecule (ligand), such as a neurotransmitter	Mechanical force or deformation of the cell membrane	Always open, allowing ions to leak through passive diffusion
Function	Initiate and propagate action potentials; critical in electrical signaling in neurons and muscles	Mediate synaptic transmission; convert chemical signals to electrical signals	Respond to physical changes like stretch, pressure, or vibration	Maintain the resting membrane potential; contribute to the steady-state ion flow
Examples	Sodium (Na+) channels, Potassium (K+) channels, Calcium (Ca2+) channels	Nicotinic acetylcholine receptors, GABA_A receptors, NMDA receptors	Stretch-activated ion channels, Piezo1 and Piezo2 channels	Potassium (K+) leak channels, Sodium (Na+) leak channels
Location	Found in neurons, muscle cells (e.g., skeletal, cardiac, smooth)	Located at synapses, neuromuscular junctions, various cell membranes	Present in sensory receptors (e.g., touch receptors, inner ear cells)	Ubiquitous in many cell types across different tissues
Role in physiology	Conduction of electrical signals, regulation of muscle contractions, and release of neurotransmitters	Facilitate synaptic transmission and neural communication	Sensory perception (e.g., hearing, touch, balance)	Stabilize the resting membrane potential; help in setting the membrane's baseline permeability
Selective permeability	Typically highly selective for specific ions (e.g., Na+, K+, Ca2+)	Selective to the ions that the ligand-gated receptor channel allows (e.g., Na+, Cl−)	Varies; can be selective or non-selective depending on the channel type	Generally selective for specific ions (e.g., K+ channels allow K+ to pass)
Activation dynamics	Fast activation and inactivation kinetics	Activation depends on ligand concentration and binding dynamics	Activation depends on the intensity and duration of mechanical force	Constantly active; does not require activation
Regulaion	Modulated by changes in membrane voltage, can be influenced by phosphorylation	Regulated by the presence and concentration of ligands, can be influenced by phosphorylation and allosteric modulators	Modulated by mechanical forces, such as pressure and stretch	Regulated by the electrochemical gradient and membrane potential

   a. Voltage-Gated Sodium (Na+) Channels: Crucial for the initiation and propagation of action potentials. When the membrane depolarizes, these channels open, allowing Na+ ions to enter the cell, further depolarizing the membrane.
   b. Voltage-Gated Potassium (K+) Channels: Important for repolarizing the membrane following an action potential. They open in response to depolarization, allowing K+ ions to exit the cell, restoring the resting membrane potential.
   c. Voltage-Gated Calcium (Ca2+) Channels: Play a key role in neurotransmitter release at synapses. When these channels open in response to depolarization, Ca2+ ions enter the cell and trigger the release of neurotransmitters.
2. Ligand-Gated Ion Channels: These channels open or close in response to the binding of a specific chemical messenger (ligand), a neurotransmitter.
   a. Ionotropic Receptors: Examples include the NMDA and AMPA receptors for glutamate, which are important for excitatory synaptic transmission, and the GABA(A) receptor for GABA, which mediates inhibitory transmission.
   b. Nicotinic Acetylcholine Receptors: These receptors respond to the neurotransmitter acetylcholine and are important for synaptic transmission at neuromuscular junctions and in the central nervous system.
3. Mechanically-Gated Ion Channels: These channels open or close in response to mechanical forces such as stretch, pressure, or deformation of the cell membrane. The channels have mechanoreceptors which are found in sensory neurons, involved in the sense of touch, hearing, and proprioception. For example, the channels in hair cells of the inner ear open in response to sound vibrations, leading to the perception of sound.
4. Leakage (Non-Gated) Ion Channels: These channels are always open and allow ions to flow continuously across the membrane, contributing to the resting membrane potential.
   a. Potassium Leak Channels: Allow K(+) ions to move out of the neuron, helping to maintain the resting membrane potential.
   b. Sodium Leak Channels: Allow Na(+) ions to move into the neuron, which also influences the resting membrane potential.

8.12.5 Electrostatic Forces in Neurons

Electrostatic forces play a vital role in neuronal function by influencing the movement and behaviour of ions within and around the neurons. These forces arise due to the electric charges carried by ions and the electric field created by those charges.

1. **Resting Membrane Potential:** The resting membrane potential of neurons typically ranges from -40 to -90 millivolts (around -70mV i.e., there is a difference of -70mV between the inside and outside environment of a neuron). This difference is primarily maintained by the unequal distribution of ions across the cell membrane wherein the inside of the neuron is negatively charged relative to the outside due to the presence of large negatively charged molecules and an unequal distribution of ions such as K+ and Na+. It is further influenced by the resulting electrostatic forces established by ion channels, pumps, and other membrane proteins. For example, the sodium-potassium pump (Na+/K+ ATPase) actively transports sodium ions out of the neuron and potassium ions into the neuron, helping to establish and maintain the resting potential.
2. **Graded Potential**: Graded potentials are local changes in membrane potential that occur in response to stimuli, but they are not large enough to trigger an action potential. These potentials are graded because their magnitude varies depending on the intensity of the stimulus. They can occur in various regions of the neuron, including the cell body, dendrites, and synaptic terminals. These potentials travel across the neurons and reach the axon hillock (trigger zone) where they will be summated to generate the action potential. In neurons, the graded potentials can be either depolarizing or hyperpolarizing, depending on whether they shift the membrane potential towards more positive or more negative side, respectively. There are two kinds of graded potentials: Excitatory Postsynaptic Potentials (EPSPs) and Inhibitory Postsynaptic Potentials (IPSPs)
   a. **EPSP**s are depolarizing graded potentials that occur in neurons in response to the release of neurotransmitters from presynaptic terminals. When neurotransmitters (ACh and glutamate) bind to their receptors on the postsynaptic membrane, it causes the opening of ion channels, leading to an influx of positive ions (Na+) into the neuron. This influx of positive charge depolarizes the postsynaptic membrane of a neuron. If the depolarization caused by EPSPs reaches the threshold at the axon hillock, it can trigger an initiation of action potential leading to the propagation of the signal along the axon. EPSPs are important in neuronal communication because they facilitate the transmission of excitatory signals from one neuron to another.
   b. **IPSP**s are hyperpolarizing graded potentials that occur in neurons in response to the release

of inhibitory neurotransmitters from presynaptic terminals. When inhibitory neurotransmitters (GABA and Glycine) bind to their receptors on the postsynaptic membrane, it causes the opening of ion channels that lead to an efflux of positive ions or an influx of negative (Cl-) ions. This results hyperpolarization of the postsynaptic membrane, making it more negative compared to the resting membrane potential. By hyperpolarizing the postsynaptic membrane, IPSPs reduce the likelihood of the neuron firing an action potential in response to excitatory inputs. IPSPs play a crucial role in neuronal communication by counteracting the effects of excitatory signals and regulating neuronal excitability.

The graded potentials, both excitatory and inhibitory, are integrated at the axon hillock of a neuron to determine whether an action potential is generated. There are two main types of summation:

1. Temporal summation which occurs when multiple graded potentials from the same presynaptic neuron arrive at the postsynaptic neuron's axon hillock in rapid succession, before the previous graded potential has fully decayed. These individual potentials summate temporally, increasing in amplitude and potentially reaching the threshold for action potential initiation.
2. Spatial summation which occurs when graded potentials from multiple presynaptic neurons synapse onto the same postsynaptic neuron. If these inputs arrive simultaneously, their effects summate spatially at the axon hillock.
   All the graded potentials (either excitatory or inhibitory inputs) summate at the axon hillock, and their net effect determines whether the membrane potential surpasses the threshold potential to generate an action potential. So it can be derived that the magnitude of action potential generated is directly correlated with the frequency of graded potentials.
3. **Action Potential**: An action potential is a rapid and transient change in membrane potential that is generated by the opening and closing of voltage-gated ion channels in the neuronal membrane. Voltage-gated sodium channels undergo several conformational states during the generation and propagation of an action potential. At resting membrane potential, the voltage-gated sodium channels remain closed until the membrane potential depolarizes to the threshold level. When the membrane potential reaches the threshold, voltage-gated sodium channels rapidly transit to an open state, allowing the influx of sodium ions, causing depolarization. After a brief period of time, typically around 1-2 millisecond, the voltage-gated sodium channels enter an inactive state. In this state, the channel is refractory to opening in response to further depolarization. This prevents the channel from reopening immediately after depolarization, ensuring that action potentials propagate in one direction. Following the inactive state, the channels transition to a closed state, but they are capable of reopening in response to subsequent depolarizations. This state allows the channels to reset and prepare for the generation of another action potential. These conformational changes are critical for the rapid and precise propagation of action potentials along the axon of a neuron.

The minimum level of depolarization that must be reached to open voltage-gated sodium channels and initiate an action potential is called Threshold potential. It is typically around -55 to -50 mV in most neurons, but this can vary depending on the neuron type and location. The threshold potential is influenced by several factors, including the balance of ion conductance (primarily sodium and potassium), which are affected by ion channel densities, ion gradients across the membrane, and the presence of modulators such as neurotransmitters. If the threshold potential is not reached, the action potential is not generated. Hence the concept of action potential is referred to as 'all or none phenomenon'.

The action potential consists of two distinct phases:

a. **Rising phase of action potential** occurs during depolarization when the membrane potential rapidly becomes less negative. This phase is initiated when the membrane potential reaches the threshold level, typically around -55 to -50 millivolts. During this phase the Voltage-gated sodium channels open in response to depolarization, causing a rapid influx of positive charge. The rising phase continues until the membrane potential reaches its peak, typically around +30 to +40 millivolts. The rising phase is essential for initiating and propagating the action potential along the neuron's membrane thereby allowing the transmission of electrical signals.

b. **Falling phase of action potential** occurs during repolarization, following the peak of depolarization. During this phase Voltage-gated sodium channels begin to close, reducing the influx of sodium ions into the cell. Voltage-gated potassium channels open, allowing potassium ions to flow out of the cell. As potassium ions leave the cell, the membrane potential begins to decrease, returning towards its resting state. The falling phase continues until the membrane potential reaches its resting level, typically around -70 millivolts. This repolarization process restores the negative charge inside the cell, preparing it for the next action potential. The falling phase is crucial for terminating the action potential and resetting the neuron's membrane potential.

Refractory period is a critical phase in the action potential cycle during which a neuron is temporarily unable to generate another action potential. The refractory period ensures the orderly propagation of action potentials along the axon and prevents backward propagation of action potentials. It also limits the frequency of action potentials, preventing excessive depolarisations that could lead to neural dysfunction or damage. There are two main types of refractory periods.

1. **Absolute Refractory Period**: During this phase, the neuron is completely unresponsive to any stimulus. It begins at the onset of the action potential and continues until the membrane potential begins to return to its resting state. This period is primarily due to the inactivation of voltage-gated sodium channels, which prevents the generation of another action potential.
2. **Relative Refractory Period**: Following the absolute refractory period, there is a period during which the neuron can respond to a strong stimulus, but it requires a greater-than-normal depolarization to reach the threshold for generating another action potential. This period occurs during the later phase of repolarization and early hyperpolarization when some of the voltage-gated sodium channels have recovered from inactivation. Additionally, voltage-gated potassium channels may still be open, contributing to membrane hyperpolarization.

Several factors affect the generation of action potentials across the neuron. Primarily, the stimulus applied to the neuron should be strong enough to reach the threshold potential and depolarize the membrane. The levels of excitatory and inhibitory neurotransmitters or their receptors also govern the neuronal excitability in an adverse way. Further, neuronal injury, disruptions in the extracellular or intracellular ionic concentrations can also affect the membrane potential and make it more difficult to generate action potentials.

8.13 PROPAGATION OF ACTION POTENTIAL

The trigger zone, also known as the axon hillock, is a specialized region of the neuron where the action potential is generated. Once this membrane potential reaches the threshold level (-55 to -50 millivolts), the voltage-gated sodium channels open, initiating the depolarization phase. As sodium ions enter the cell, the depolarization spreads along the axon due to local circuit currents. This influx of positive charge depolarizes the adjacent regions of the axon membrane, triggering the opening of next voltage-gated sodium channel. This process continues down the length of the axon with the uniform strength and speed all along. After an action potential propagates, the region of the axon becomes temporarily refractory to further stimulation. This prevents the action potential from propagating backward and ensures that the signal moves in one direction along the axon.

8.13.1 Conduction Speed of Action Potential

The conduction speed of an action potential is influenced by axon diameter, myelination, and temperature. The faster conduction of action potential in larger-diameter axons is attributed to the less internal resistance to the flow of ions as the action potential travels along a wider cross-sectional area. In myelinated axons it is attributed to saltatory conduction wherein the action potentials "leap" between the gaps called nodes of Ranvier, where the axon membrane is exposed. This saltatory conduction increases the speed of propagation by allowing the action potential to "skip" along the axon, rather than propagating continuously along its entire length. Higher temperatures in optimum range generally increase the conduction speed of action potentials as the elevated temperatures increase the rate of diffusion and enzymatic reactions involved in the propagation of the action potential.

8.14 SYNAPTIC TRANSMISSION

A synapse is a specialized junction between two neurons or between a neuron and a target cell (a muscle or gland cell) that allows communication between them (Figure 8.7).

Synaptic transmission is the process by which neurons communicate with each other in the nervous system. It occurs through the release of various neurotransmitters into the synaptic cleft. It involves the release of neurotransmitters from the presynaptic neuron, which then bind with receptors on the postsynaptic neuron, leading to the generation of electrical signals or impulses.

At chemical synapses, neurotransmitters are released from the presynaptic neuron and bind to receptors on the postsynaptic neuron. This binding can lead to the opening of ion channels, allowing the movement of Na+, K+, and Ca2+ ions in or out of the neuron. The movement of these ions is influenced by both concentration gradients and electrostatic forces.

An action potential travels down the axon of the presynaptic neuron to the axon terminal. The depolarization of the membrane at the axon terminal causes voltage-gated calcium channels to open, allowing calcium ions to flow into the presynaptic neuron. The influx of calcium ions triggers synaptic vesicles, which contain neurotransmitters, to move toward and fuse with the presynaptic membrane. This process is facilitated by proteins such as synaptotagmin and

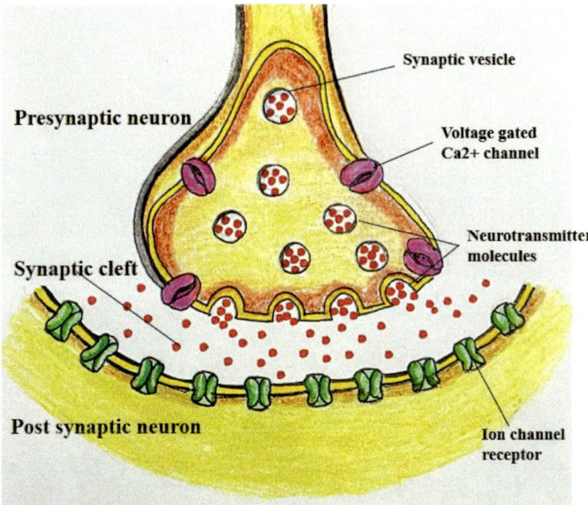

FIGURE 8.7 Synapse

SNARE proteins (SNAP-25, synaptobrevin, and syntaxin). Upon fusion, the vesicles release their neurotransmitter contents into the synaptic cleft through a process known as exocytosis. The neurotransmitters diffuse across the synaptic cleft and bind to specific receptor sites on the postsynaptic membrane. These receptors can be ionotropic (directly linked to ion channels) or metabotropic (indirectly linked to signaling pathways). Binding of neurotransmitters to receptors causes ion channels to open or close, leading to changes in the postsynaptic cell's membrane potential (Figure 8.8).

This can result in an excitatory postsynaptic potential (EPSP) or inhibitory postsynaptic potential (IPSP), depending on the type of neurotransmitter and receptor involved. The action of neurotransmitters is then terminated through various mechanisms, such as reuptake into the presynaptic neuron, enzymatic degradation, or diffusion away from the synaptic cleft. This process is fundamental for all aspects of

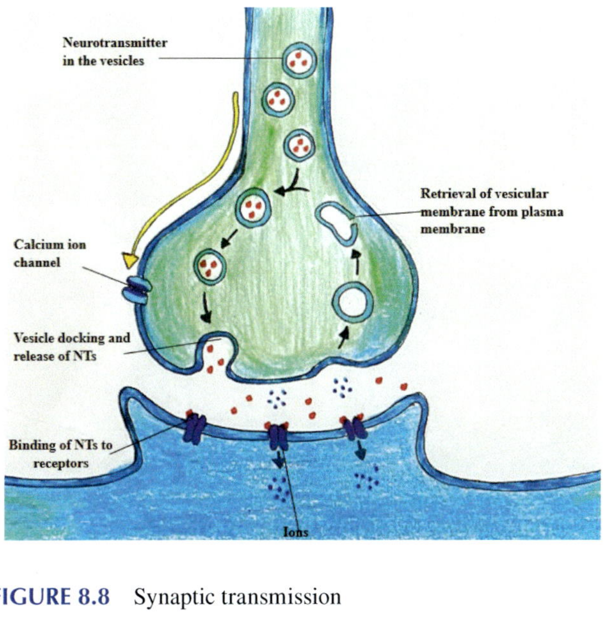

FIGURE 8.8 Synaptic transmission

brain function, including sensory perception, motor control, memory, and emotion.

There are two types of synapses: electrical and chemical synapses

1. **Electrical synapse:** Electrical synapses are less common than chemical synapses. In this type, the cytoplasm of two neurons is connected by gap junctions, allowing ions and small molecules to pass directly from one cell to another. This direct transmission results in rapid and synchronized signaling between neurons. Electrical synapses are found in various brain regions and are involved in processes like coordination of movement and synchronization of neural activity.
2. **Chemical synapse:** Chemical synapses are the most common type of synapse in the nervous system. They involve the release of neurotransmitters from the presynaptic neuron into the synaptic cleft, where they bind to receptors on the postsynaptic neuron, leading to changes in the postsynaptic membrane potential. This process allows for signal transmission between neurons and is essential for functions such as learning, memory, and sensory processing. Chemical synapses can exhibit various properties, including excitatory or inhibitory effects on the postsynaptic neuron, and can undergo processes like synaptic plasticity, which underlies learning and memory.

8.15 NEUROTRANSMITTERS OF NS

The nervous system (NS) uses various neurotransmitters (Acetylcholine, Dopamine, Serotonin, Glutamate, Gamma-aminobutyric acid, Norepinephrine, Endorphins) to transmit signals between neurons and other cells. Neurotransmitters can be grouped into two classes: Neurotransmitters of CNS & PNS and Excitatory & Inhibitory Neurotransmitters.

A brief information on the grouping of neurotransmitters based on their precursors and functions is provided in Table 8.3.

1. Neurotransmitters of PNS include Acetylcholine, Norepinephrine (noradrenaline), Epinephrine (adrenaline), Dopamine, and Serotonin.
2. Neurotransmitters of CNS include Glutamate, GABA, Acetylcholine, Dopamine, Serotonin, Norepinephrine, Histamine, Endorphins, and Enkephalins.

1. **Excitatory neurotransmitters:** Excitatory neurotransmitters are neurotransmitters that promote the generation of action potentials in postsynaptic neurons, thereby increasing neuronal activity. The most common and abundant excitatory neurotransmitter in the central nervous system is glutamate. When glutamate binds to its receptors on

TABLE 8.3
Grouping of Neurotransmitters Based on their Precursors

Precursor	Neurotransmitter(s)	Enzyme(s) Involved in Synthesis	Primary Locations	Receptors
Choline, Acetyl-CoA	Acetylcholine (ACh)	Choline acetyltransferase	Neuromuscular junctions, CNS, autonomic ganglia	Nicotinic, Muscarinic
Tyrosine	Dopamine (DA)	Tyrosine hydroxylase, Aromatic L-amino acid decarboxylase	CNS, particularly the substantia nigra, ventral tegmental area	D1-D5 (Dopaminergic receptors)
	Norepinephrine (NE)	Tyrosine hydroxylase, Dopamine β-hydroxylase	CNS, sympathetic nervous system	α1, α2, β1, β2, β3 (Adrenergic receptors)
	Epinephrine (E)	Tyrosine hydroxylase, Dopamine β-hydroxylase, Phenylethanolamine N-methyltransferase	Adrenal medulla, some CNS neurons	α1, α2, β1, β2, β3 (Adrenergic receptors)
Tryptophan	Serotonin (5-HT)	Tryptophan hydroxylase, Aromatic L-amino acid decarboxylase	CNS, particularly the raphe nuclei	5-HT1 to 5-HT7 (Serotonergic receptors)
Glutamate	GABA (γ-Aminobutyric Acid)	Glutamate decarboxylase	CNS	GABA_A, GABA_B
	Glutamate	Glutaminase	CNS, throughout the brain	NMDA, AMPA, Kainate, Metabotropic glutamate receptors
Serine	Glycine	Serine hydroxymethyltransferase	Spinal cord, brainstem	Glycine receptors
Histidine	Histamine	Histidine decarboxylase	CNS, particularly the hypothalamus	H1, H2, H3, H4 (Histaminergic receptors)
Arginine	Nitric Oxide (NO)	Nitric oxide synthase	CNS, endothelial cells	Diffuses to act on nearby cells, not through conventional receptors
Pro-opiomelanocortin (POMC)	Endorphins	Proteolytic enzymes	CNS, particularly the pituitary gland	μ, δ, κ (Opioid receptors)
Membrane phospholipids	Endocannabinoids	Enzymes like N-acyl-phosphatidylethanolamine-specific phospholipase D	CNS	CB1, CB2 (Cannabinoid receptors)

the postsynaptic neuron, it triggers depolarization of the neuron, leading to an action potential.

2. **Inhibitory neurotransmitters:** Inhibitory neurotransmitters are neurotransmitters that decrease the likelihood of a postsynaptic neuron firing an action potential, thereby reducing neuronal activity. The primary inhibitory neurotransmitter in the central nervous system is gamma-aminobutyric acid (GABA). When GABA binds to its receptors on the postsynaptic neuron, it typically increases the flow of negatively charged ions into the neuron, hyperpolarizing it and inhibiting the generation of an action potential.

Another important inhibitory neurotransmitter in the central nervous system is glycine. Like GABA, glycine also hyperpolarizes the postsynaptic neuron by increasing the flow of negatively charged ions into the neuron.

8.16 RECEPTORS OF NEUROTRANSMITTERS

Neurotransmitters of PNS and CNS, either excitatory or inhibitory, exert their effects by binding to specific receptors on the postsynaptic neuron or other target cells.

- Based on the structure, there are two main types of neurotransmitter receptors:
  1. **Ionotropic receptors (ligand-gated ion channels):** These receptors directly regulate the flow of ions across the cell membrane when bound by neurotransmitters. When a neurotransmitter binds to an ionotropic receptor, it causes a conformational change in the receptor protein, leading to the opening or closing of an ion channel. This rapid change in ion flow directly alters the membrane potential of the postsynaptic neuron, either depolarizing it (excitatory effect) or hyperpolarizing it (inhibitory effect). Examples include nicotinic acetylcholine receptors and certain glutamate receptors.
  2. **Metabotropic receptors (G protein-coupled receptors)**: These receptors indirectly influence ion flow and cellular processes through intracellular signaling pathways when activated by neurotransmitters. When a neurotransmitter binds to a metabotropic receptor, it activates a G protein, which then initiates a cascade of intracellular signaling events. These signaling pathways can modulate the ion channels, and typically induce slower and longer-lasting effects compared to ionotropic receptors. Examples include muscarinic acetylcholine receptors, dopamine receptors, and serotonin receptors.
- Based on the neurotransmitter that the receptors bind to, they are broadly classified into: Adrenergic, Cholinergic and Glutamate receptors. The salient features of adrenergic, Cholinergic and Glutamate receptors are tabulated in Table 8.4.

TABLE 8.4
Salient Features of Adrenergic, Cholinergic and Glutamate Receptors

Feature	Adrenergic Receptors	Cholinergic Receptors	Glutamate Receptors
Types/Subtypes	Alpha (α1, α2) and Beta (β1, β2, β3)	Nicotinic (nAChR) and Muscarinic (mAChR)	Ionotropic (AMPA, NMDA, Kainate) and Metabotropic (mGluR)
Primary Ligand	Norepinephrine (noradrenaline) and Epinephrine (adrenaline)	Acetylcholine (ACh)	Glutamate
Receptor Mechanism	G-protein coupled receptors (GPCRs)	nAChR: Ligand-gated ion channels, mAChR: GPCRs	Ionotropic: Ligand-gated ion channels, Metabotropic: GPCRs
Location	Various tissues, including heart, lungs, blood vessels, CNS	Neuromuscular junction, autonomic ganglia, CNS, various tissues	CNS, peripheral nervous system, various tissues
Primary Functions	Regulation of cardiovascular functions, smooth muscle tone, metabolic processes	Neuromuscular transmission, autonomic control, various CNS functions	Synaptic transmission, neural plasticity, excitatory neurotransmission
Response Time	Relatively slow (milliseconds to seconds)	nAChR: Fast (milliseconds), mAChR: Slow (milliseconds to seconds)	Ionotropic: Fast (milliseconds), Metabotropic: Slow (seconds)
Common Agonists	Epinephrine, Norepinephrine	Nicotine (nAChR), Muscarine (mAChR)	Glutamate, NMDA (for NMDA receptors), AMPA (for AMPA receptors)
Common Antagonists	Alpha-blockers (e.g., phentolamine), Beta-blockers (e.g., propranolol)	nAChR antagonists (e.g., curare), mAChR antagonists (e.g., atropine)	NMDA receptor antagonists (e.g., ketamine), AMPA receptor antagonists

TABLE 8.5
Differences Between Nicotinic and Muscarinic Receptors

Feature	Nicotinic Receptors	Muscarinic Receptors
Type	Ionotropic (ligand-gated ion channels)	Metabotropic (G-protein coupled receptors)
Primar location	Neuromuscular junctions, autonomic ganglia, CNS	CNS, parasympathetic nervous system (heart, smooth muscles, glands)
Subtypes	N1 (NM) - muscle type, N2 (NN) - neuronal type	M1, M2, M3, M4, M5
Agonists	Nicotine	Muscarine
Antagonists	Curare, hexamethonium	Atropine, scopolamine
Mechanism of action	Directly opens ion channels, allowing Na+ and K+ to pass	Activates G-proteins, leading to various intracellular effects
Response time	Fast (milliseconds)	Slow (seconds to minutes)
Physiological effects	Skeletal muscle contraction, autonomic ganglionic transmission	Modulation of heart rate, smooth muscle contraction, glandular secretion
Synaptic transmission	Excitatory	Can be excitatory or inhibitory
Clinical relevance	Used in anesthesia (neuromuscular blockers)	Used in treating various conditions like bradycardia, asthma, and gastrointestinal disorders

8.16.1 Cholinergic Receptors

Cholinergic receptors are receptors that bind the neurotransmitter acetylcholine and mediate its effects on target cells. There are two main types of cholinergic receptors. The differences between nicotinic and muscarinic cholinergic receptors are given in Table 8.5.

1. **Nicotinic acetylcholine receptors (nAChRs):** Nicotinic receptors are pentameric ion channels composed of five subunits arranged around a central pore. They are named after nicotine, a compound found in tobacco which mimics the effects of ACh on these receptors, leading to their activation. Binding of ACh or nicotine to the receptor causes conformational changes that open the channel, allowing the passage of cations Na+ and K+. This results in depolarization of the postsynaptic membrane, leading to excitatory effects. Nicotinic receptors are widely distributed in both CNS and PNS. In the CNS, they are primarily found at synapses between neurons and are involved in fast synaptic transmission. In the PNS, nicotinic receptors are found at neuromuscular junctions, where they mediate the effects of ACh on skeletal muscle contraction, as well as in autonomic ganglia, where they mediate synaptic transmission between preganglionic and postganglionic neurons of the autonomic nervous system. Activation of nicotinic receptors can lead to various physiological responses, including muscle contraction, release of neurotransmitters in the brain, and modulation of synaptic transmission. Their effects on different tissues are given in Table 8.6.
2. **Muscarinic acetylcholine receptors (mAChRs):** Muscarinic receptors are metabotropic receptors that bind with acetylcholine and are named after muscarine, a compound found in certain mushrooms that activates these receptors. Unlike nicotinic receptors, which are ionotropic and directly regulate ion flow across the cell membrane, muscarinic receptors indirectly modulate cellular processes through intracellular signaling pathways when activated by ACh. Muscarinic receptors are G protein-coupled receptors (GPCRs) having five subtypes, designated M1 through M5, which are encoded by different genes and exhibit distinct tissue distributions and signaling properties. In the CNS, muscarinic receptors play roles in regulating synaptic transmission, modulating neuronal excitability, and mediating various cognitive functions. In the PNS, muscarinic receptors are found on the effector organs of the parasympathetic nervous system, including smooth muscle, cardiac muscle, and glands. Activation of muscarinic receptors leads to various effects on different tissues which are given in Table 8.7.

TABLE 8.6
Action of Nicotinic Cholinergic Receptors in Different Tissues

Tissue/Location	Action of Nicotinic ACh Receptors
Neuromuscular Junction	Stimulation of skeletal muscle contraction
Autonomic Ganglia	Stimulation of sympathetic and parasympathetic ganglia
Adrenal Medulla	Stimulation of catecholamine release (epinephrine and norepinephrine)
Brain	Modulation of neurotransmitter release and neuronal excitability
Peripheral Nerves	Modulation of neurotransmitter release and neuronal excitability

TABLE 8.7
Action of Muscarinic Cholinergic Receptors in Different Tissues

Tissue/Location	Action of Muscarinic Cholinergic Receptors
Heart (Atria)	Decreased heart rate (negative chronotropic effect)
	Decreased conduction velocity (negative dromotropic effect)
	Decreased contractility (negative inotropic effect)
Smooth Muscle (Bronchi)	Constriction, leading to bronchoconstriction
Smooth Muscle (GI Tract)	Increased motility and peristalsis
Smooth Muscle (Bladder)	Contraction of detrusor muscle, leading to urination
Eye (Iris)	Constriction of the pupil (miosis)
Eye (Ciliary Muscle)	Contraction, leading to accommodation for near vision
Glands (Salivary, Sweat, Lacrimal)	Stimulation, leading to secretion
Brain (Central Nervous System)	Modulation of neurotransmitter release and neuronal excitability
Blood Vessels (Endothelial Cells)	Vasodilation

8.16.2 Adrenergic Receptors

Adrenergic receptors are a class of G protein-coupled receptors that bind the neurotransmitters norepinephrine (noradrenaline) and epinephrine (adrenaline). These receptors are named after adrenaline, also known as epinephrine, because it was initially identified as the ligand that activates them. There are two main types of adrenergic receptors:

1. Alpha-adrenergic receptors (α-adrenergic receptors): Alpha-adrenergic receptors are activated by norepinephrine, and to a lesser extent, adrenaline, released from sympathetic nerve terminals during the "fight or flight" response. Their activation leads to physiological responses such as increased blood pressure, pupil dilation, and vasoconstriction. They are classified into two subtypes: alpha-1 adrenergic receptors and alpha-2 adrenergic receptors.
   - α1 receptors: These receptors are primarily found on smooth muscle cells in blood vessels, particularly in peripheral arteries and veins. Activation of alpha-1 receptors leads to vasoconstriction, resulting in increased blood pressure. Alpha-1 receptors are also found on other tissues, such as the iris muscles of the eye, where their activation leads to pupil dilation (mydriasis), and on the smooth muscle of the urinary bladder, where their activation contributes to urinary retention (Table 8.8).

TABLE 8.8
Summary of the Actions of Alpha-1 Adrenergic Receptors in Different Tissues

Tissue/Location	Action of Alpha-1 Adrenergic Receptors
Vascular Smooth Muscle	Contraction, leading to vasoconstriction
Pupillary Dilator Muscle	Contraction, leading to pupil dilation
Intestinal Smooth Muscle	Contraction, leading to decreased motility
Genitourinary Smooth Muscle	Contraction, leading to urinary sphincter contraction and ejaculation control
Liver	Glycogenolysis, leading to increased blood glucose levels
Salivary Glands	Stimulation, leading to saliva production
Heart	Increased contractility (positive inotropic effect)
Central Nervous System	Modulation of neurotransmitter release and neuronal activity

   - α2 receptors: These receptors are found both presynaptically and postsynaptically in various tissues, including the central nervous system, peripheral nervous system, and certain peripheral tissues. Presynaptic alpha-2 receptors act as autoreceptors, regulating the release of norepinephrine from sympathetic nerve terminals. Postsynaptic alpha-2 receptors can mediate various effects, depending on the tissue. For example, activation of alpha-2 receptors in the brain can lead to sedation, while activation in peripheral tissues can lead to vasoconstriction or inhibition of insulin release (Table 8.9).

TABLE 8.9
Summary of the Actions of Alpha-2 Adrenergic Receptors in Different Tissues

Tissue/Location	Action of Alpha-2 Adrenergic Receptors
Presynaptic Nerve Terminals	Inhibition of neurotransmitter release (negative feedback mechanism)
Central Nervous System	Modulation of sympathetic outflow, leading to decreased sympathetic activity
Vascular Smooth Muscle	Vasodilation (inhibition of norepinephrine release)
Platelets	Inhibition of platelet aggregation
Pancreas (Islets of Langerhans)	Inhibition of insulin release
GI Tract	Inhibition of motility and secretion
Kidney	Renin release inhibition, leading to decreased renin levels and blood pressure regulation

TABLE 8.10
Summary of the Actions of Beta-1 Adrenergic Receptors in Different Tissues

Tissue/Location	Action of Beta-1 Adrenergic Receptors
Heart (Cardiac Muscle)	Increased heart rate (positive chronotropic effect)
	Increased contractility (positive inotropic effect)
	Increased conduction velocity (positive dromotropic effect)
Kidney (Juxtaglomerular Cells)	Increased renin release, leading to increased blood pressure
Fat Cells	Lipolysis, leading to increased free fatty acids in blood
Tissue/Location	Action of Alpha-2 Adrenergic Receptors
Presynaptic Nerve Terminals	Inhibition of neurotransmitter release (negative feedback mechanism)
Central Nervous System	Modulation of sympathetic outflow, leading to decreased sympathetic activity
Vascular Smooth Muscle	Vasodilation (inhibition of norepinephrine release)
Platelets	Inhibition of platelet aggregation
Pancreas (Islets of Langerhans)	Inhibition of insulin release
GI Tract	Inhibition of motility and secretion
Kidney	Renin release inhibition, leading to decreased renin levels and blood pressure regulation

2. Beta-adrenergic receptors (β-adrenergic receptors): Beta-adrenergic receptors bind with both noradrenaline and adrenaline. These receptors are classified into three main subtypes: beta-1 (β1) adrenergic receptors, beta-2 (β2) adrenergic receptors, and beta-3 (β3) adrenergic receptors.
    - β1 receptors: These receptors are primarily found in the heart, where they mediate the effects of sympathetic stimulation on cardiac function. Activation of beta-1 receptors leads to positive chronotropic effect and inotropic effect, and increased conduction velocity through the atrioventricular node (Table 8.10).
    - β2 receptors: Found in smooth muscle cells, bronchioles of the lungs, walls of blood vessels, and the uterus, activation of these receptors leads to relaxation of smooth muscle, resulting in bronchodilation, vasodilation, and relaxation of the uterine muscle during pregnancy. Beta-2 receptors are also found in hepatocytes, where their activation leads to glycogenolysis and gluconeogenesis (Table 8.11).
    - Beta-3 adrenergic receptors (β3 receptors): These receptors are found in adipose tissue and the urinary bladder. Their activation in adipocytes leads to lipolysis, while activation in the urinary bladder leads to relaxation of the detrusor muscle, promoting bladder filling (Table 8.12).

TABLE 8.11
Summary of the Actions of Beta-2 Adrenergic Receptors in Different Tissues

Tissue/Location	Action of Beta-2 Adrenergic Receptors
Smooth Muscle (Bronchioles)	Relaxation, leading to bronchodilation
Smooth Muscle (Blood Vessels)	Relaxation, leading to vasodilation
Liver	Glycogenolysis, leading to increased blood glucose levels
Skeletal Muscle	Glycogenolysis, leading to increased muscle strength
Pancreas (Islets of Langerhans)	Stimulation of insulin release
Uterus	Relaxation of uterine smooth muscle
Eye (Ciliary Body)	Stimulation of aqueous humor production
Platelets	Inhibition of platelet aggregation

TABLE 8.12
Summary of the Actions of Beta-3 Adrenergic Receptors in Different Tissues

Tissue/Location	Action of Beta-3 Adrenergic Receptors
Adipose Tissue	Lipolysis, leading to decreased fat storage and increased thermogenesis
Bladder	Relaxation of detrusor muscle, leading to increased bladder capacity
Brown Adipose Tissue	Stimulation of thermogenesis and energy expenditure
Colon	Inhibition of smooth muscle contraction, leading to decreased motility
Cardiovascular System	Vasodilation in some vascular beds, leading to decreased blood pressure

8.17 GLUTAMATE RECEPTORS

Glutamate receptors are a class of ionotropic and metabotropic receptors that bind the neurotransmitter glutamate. Glutamate is the primary excitatory neurotransmitter in the central nervous system and plays essential roles in synaptic transmission, neuronal development, and synaptic plasticity. There are two main types of glutamate receptors (ionotropic and metabotropic) and their differences are given in Table 8.13.

1. **Ionotropic glutamate receptors:** These receptors directly regulate flow of ions across the cell membrane when activated by glutamate. There are three subtypes of ionotropic glutamate receptors:

TABLE 8.13
Differences Between Ionotropic and Metabotropic Glutamate Receptors

Feature	Ionotropic Glutamate Receptors	Metabotropic Glutamate Receptors
Types/Subtypes	AMPA, NMDA, and Kainate receptors	mGluR1 to mGluR8
Receptor Mechanism	Ligand-gated ion channels	G-protein coupled receptors (GPCRs)
Primary Function	Mediate fast synaptic transmission	Modulate synaptic transmission and neuronal excitability
Ion Permeability	Permeable to Na+, K+, and Ca2+ (depending on subtype)	Not directly ion-permeable
Activation Requirements	Activated by glutamate binding	Activated by glutamate binding
Response Time	Fast (milliseconds)	Slow (seconds to minutes)
Role in Synaptic Transmission	Directly mediate excitatory postsynaptic currents (EPSCs)	Modulate neurotransmitter release and synaptic plasticity
Subunit Composition	AMPA: GluA1-4, NMDA: GluN1, GluN2 (A-D), GluN3, Kainate: GluK1-5	Various subunits form different mGluR types
Pharmacological Agents	AMPA agonists: AMPA, NMDA agonists: NMDA, Kainate agonists: Kainate	Various agonists and antagonists specific to mGluR subtypes
Functional Importance	Rapid excitatory neurotransmission	Modulation of synaptic plasticity, learning, memory, and neuroprotection

- **AMPA receptors** (α-amino-3-hydroxy-5 -methyl-4-isoxazolepropionic acid receptors): AMPA receptors are tetrameric protein complexes composed of different combinations of four subunits: GluA1, GluA2, GluA3, and GluA4. These subunits assemble to form functional receptor complexes, with the most common combination being heterotetramers containing GluA1 and GluA2 subunits. When glutamate binds to AMPA receptors located on the postsynaptic membrane, the receptors undergo a conformational change that leads to the opening of the ion channel pore, allowing the influx of Na+ and K+ ions into the postsynaptic neuron. This depolarization of the postsynaptic membrane results in the generation of EPSPs, which contribute to the integration of synaptic inputs and the generation of action potentials. AMPA receptors are responsible for the fast component of excitatory synaptic transmission and are critical for synaptic plasticity.
- **NMDA receptors** (N-methyl-D-aspartate receptors): NMDA receptors are involved in mediating long-lasting synaptic plasticity. These receptors are composed of GluN1, GluN2, and GluN3 subunits, which assemble to form functional receptor complexes. The GluN1 subunit is essential for receptor function, while the GluN2 and GluN3 subunits determine the functional properties of the receptor like its pharmacology and conductance. Activation of NMDA receptors requires both binding of glutamate to the receptor, and simultaneous depolarization of the postsynaptic membrane to remove the magnesium (Mg2+) ion that blocks the ion channel pore. This unique property of NMDA receptors, known as the voltage-dependent magnesium blockade, allows them to act as coincidence detectors, integrating both synaptic activity and postsynaptic membrane depolarization. Once activated, NMDA receptors allow influx of Ca2+ ions into the postsynaptic neuron, which triggers various intracellular signaling cascades that are responsible for synaptic plasticity and the neuronal function (Table 8.14 for differences between AMPA and NMDA receptors).
- **Kainate receptors**: Less understood compared to AMPA and NMDA receptors, these receptors are involved in modulating synaptic transmission and are present in various regions of brain.

2. **Metabotropic glutamate receptors (mGluRs):** These receptors indirectly modulate cellular processes through intracellular signaling pathways when activated by glutamate. There are eight subtypes of metabotropic glutamate receptors, which are classified into three groups based on their sequence homology, signal transduction mechanisms, and pharmacological properties. Metabotropic glutamate receptors play roles in modulating synaptic transmission, neuronal excitability, and synaptic plasticity.

TABLE 8.14
Differences Between AMPA and NMDA Receptors

Feature	AMPA Receptors	NMDA Receptors
Subunit Composition	Typically composed of GluA1-4 subunits	Composed of GluN1, GluN2 (A-D), and GluN3 subunits
Activation Requirements	Glutamate binding alone is sufficient	Requires glutamate binding, glycine (or D-serine) as a co-agonist, and depolarization to relieve Mg2+ block
Ion Permeability	Permeable to Na+ and K+ (some are also Ca2+ permeable)	Permeable to Na+, K+, and Ca2+
Magnesium Block	Not blocked by Mg2+ at resting membrane potential	Blocked by Mg2+ at resting membrane potential, relieved by depolarization
Kinetics	Fast activation and desensitization	Slower activation and desensitization
Role in Synaptic Transmission	Mediates fast excitatory postsynaptic currents (EPSCs)	Mediates slower EPSCs and plays a key role in synaptic plasticity (e.g., LTP and LTD)
Calcium Signaling	Limited role in calcium signaling (except Ca2+-permeable AMPA receptors)	Major role in calcium signaling due to Ca2+ permeability
Pharmacological Agents	Agonists: AMPA, Antagonists: CNQX, NBQX	Agonists: NMDA, Co-agonists: Glycine, D-serine, Antagonists: APV, MK-801
Functional Importance	Primary mediator of rapid synaptic transmission	Crucial for synaptic plasticity, learning, and memory

8.18 INACTIVATION OF NEUROTRANSMITTERS

The inactivation of neurotransmitters is a crucial step in regulating the duration and intensity of synaptic transmission, terminating their signaling effects and ensuring precise control of neuronal communication. Many neurotransmitters are taken back into the presynaptic neuron through specific transporter proteins located on the presynaptic membrane. Once inside the neuron, the neurotransmitter can be either recycled for future release or broken down by enzymes located in the synaptic cleft. For example, acetylcholine is rapidly broken down by the enzyme acetyl cholinesterase into choline and acetate. In another way, few neurotransmitters can diffuse away from the synaptic cleft, leading to their gradual dilution in the extracellular fluid. This process contributes to the termination of neurotransmitter signaling. Furthermore, in certain cases, neighbouring glial cells may take up neurotransmitters from the synaptic cleft, contributing to their removal from the extracellular space.

8.19 NEUROMUSCULAR SYNAPSE

The neuromuscular synapse, or neuromuscular junction (NMJ), is a highly specialized and efficient synapse that facilitates communication between motor neurons and muscle fibres, leading to muscle contraction (Figure 8.7 and Figure 8.8). The components of this junction include - motor neuron terminal, synaptic cleft, motor end plate, acetylcholine receptors, endplate potential (EPP), muscle action potential and neuromuscular transmission.

8.19.1 Synaptic Transmission at Neuromuscular Synapse

Synaptic transmission at the neuromuscular junction involves a series of events that lead to the transmission of signals from a motor neuron to a skeletal muscle fibre.

Structurally, NMJ has small membrane-bound vesicles that contain the neurotransmitter acetylcholine. These vesicles are called Synaptic Vesicles. When an action potential reaches the presynaptic terminal, voltage-gated calcium channels open, allowing Ca^{2+} ions to enter the terminal. In active zones, which are the specialized areas of the presynaptic membrane, the synaptic vesicles dock and release their contents into the synaptic cleft through exocytosis. The entry of Ca^{2+} ions triggers the fusion of synaptic vesicles with the presynaptic membrane, releasing ACh into the synaptic cleft, a narrow gap (about 20-50 nm wide) between the presynaptic terminal and the muscle fibre membrane (sarcolemma). Acetylcholine released from the presynaptic terminal diffuses across this cleft to reach the postsynaptic membrane. The postsynaptic membrane is highly folded (junctional folds), increasing the surface area to accommodate more acetylcholine receptors (nAChRs) on the motor end plate. These receptors bind to acetylcholine and open the channels, allowing Na^{+} ions to enter the muscle fibre and K^{+} ions to exit, resulting in depolarization of the muscle membrane (endplate potential). The endplate potential is a graded potential, meaning its magnitude varies depending on the amount of ACh released by the motor neuron and the number of nAChRs activated on the motor end plate. If the end plate potential is of sufficient magnitude to reach the threshold for excitation, it triggers an action potential that propagates along the sarcolemma (muscle fibre membrane) and into the T-tubules. This results in opening of voltage-gated calcium channels in the sarcoplasmic reticulum, a specialized organelle within the muscle fibre. This releases stored calcium ions into the cytoplasm of the muscle fibre, which bind with the protein troponin in the muscle fibre, causing a conformational change that exposes binding sites on the actin filament. Myosin heads bind to actin, forming cross-bridges, and ATP hydrolysis powers the sliding of actin filaments past myosin filaments, resulting in muscle contraction.

Within the synaptic cleft, the basal lamina is a layer of extracellular matrix involved in providing alignment and stability to NMJ. It contains acetylcholinesterase (AChE), the enzyme responsible for breaking down acetylcholine in the synaptic cleft, terminating the signal and ensuring that muscle contraction does not persist indefinitely.

The neuromuscular synapse plays a critical role in controlling skeletal muscle contraction and movement. Dysfunctions at the neuromuscular junction can lead to muscle weakness, paralysis, and various neuromuscular disorders, such as myasthenia gravis.

8.19.2 Boutons en Passage

The autonomic nerve fibres consists of bouton (or synaptic bouton), which is a swelling at the end of an axon terminal that forms a synapse with another neuron or a target cell. These boutons contain synaptic vesicles filled with neurotransmitters (adrenergic NTs in sympathetic nerve fibre and cholinergic NTs in parasympathetic nerve fibres) and are responsible for releasing neurotransmitters into the synaptic cleft to transmit signals to the postsynaptic cell. These are transient or temporary boutons that are observed during certain developmental stages or in response to changes in neuronal activity. These boutons may appear and disappear over time and may represent sites of active synaptogenesis or synaptic remodeling. They can form synapses with target cells or other neurons but may not be stable or long-lasting.

8.19.3 Sympathetic Signal Transduction

Sympathetic signal transduction refers to the process by which signals generated in the sympathetic division of the autonomic nervous system are transmitted from presynaptic neurons to postsynaptic target cells, leading to physiological responses.

The sequence of sympathetic signal transduction includes origination of signals, synaptic transmission in the sympathetic ganglia, postganglionic transmission to target tissues and activation of adrenergic receptors. The first and foremost step is origination of signals wherein the sympathetic signals originate in the cell bodies of preganglionic neurons followed by synaptic transmission in the sympathetic ganglia. The axons of preganglionic neurons project to sympathetic ganglia located outside the spinal cord. Upon reaching the ganglia, the preganglionic neurons synapse with postganglionic neurons. This results in binding of neurotransmitter released by preganglionic neurons (ACh) to nicotinic acetylcholine receptors on the postganglionic neurons. Postganglionic neurons of the sympathetic division project to various target tissues, including smooth muscle, cardiac muscle, glands, and adipose tissue. The neurotransmitter released by postganglionic sympathetic neurons is predominantly norepinephrine (noradrenaline), although some postganglionic neurons release acetylcholine as well. Norepinephrine binds to adrenergic receptors on the target cells, initiating cellular responses. Adrenergic receptors are G protein-coupled receptors located on the plasma membrane of target cells. Activation of these receptors either alpha-adrenergic receptors ($\alpha 1$, α 2) or beta-adrenergic receptors ($\beta 1$, $\beta 2$, $\beta 3$) leads to the activation or inhibition of intracellular signaling pathways, depending on the receptor subtype and the target tissue. The various physiological responses include increased heart rate and contractility, vasoconstriction, bronchodilation, pupil dilation, and mobilization of energy stores (e.g., glycogenolysis and lipolysis). These responses collectively prepare the body for "fight or flight" responses during stress or arousal situations.

The key molecules involved in Sympathetic Signal Transduction are:

1. cAMP (Cyclic AMP): cAMP is a second messenger that mediates the effects of neurotransmitters and hormones. Adrenergic receptors (β_1, β_2, and α_1 receptors) activate adenylate cyclase (AC), which converts ATP to cAMP. cAMP activates protein kinase A (PKA), which phosphorylates various target proteins, leading to cellular responses such as increased heart rate, glycogenolysis, and smooth muscle relaxation.
2. PKA (Protein Kinase A): PKA is a serine/threonine kinase that is activated by cAMP. Upon activation by cAMP, PKA phosphorylates specific proteins, altering their activity and cellular function. It is known to increase heart rate and force of contraction through phosphorylation of calcium channels and contractile proteins. Metabolically, it promotes glycogenolysis in liver and muscle cells. It induces relaxation of smooth muscle cells in bronchi and blood vessels through β_2 receptors.
3. Adenylate Cyclase: Adenylate cyclase (AC) is an enzyme that catalyzes the conversion of ATP to cAMP. It is stimulated by G-protein coupled receptors like β-adrenergic receptors. Found in the plasma membrane of cells, in sympathetic signal transduction, AC is activated by binding of norepinephrine to β-adrenergic receptors, leading to the production of cAMP and subsequent activation of PKA.

8.19.4 Parasympathetic Signal Transduction

Parasympathetic signal transduction refers to the process by which signals generated in the parasympathetic division of the autonomic nervous system (ANS) are transmitted from presynaptic neurons to postsynaptic target cells, leading to physiological responses. The parasympathetic division is often associated with the "rest and digest" response, promoting relaxation and conservation of energy.

The sequence of parasympathetic signal transduction includes origination of signals, synaptic transmission in the parasympathetic ganglia, postganglionic transmission to target tissues and activation of muscarinic receptors. The parasympathetic signals originate in the cell bodies of preganglionic neurons located in the brainstem and the sacral region of the spinal cord (craniosacral outflow). These preganglionic neurons project to parasympathetic ganglia located near or within the target organs. Upon reaching the parasympathetic ganglia, the axons of preganglionic neurons synapse with postganglionic neurons. The preganglionic neurons release acetylcholine which binds to nAChRs on the postganglionic neurons. Postganglionic neurons of the parasympathetic division project to various target tissues, including smooth muscle, cardiac muscle, glands, and visceral organs. The neurotransmitter released by postganglionic parasympathetic neurons is also acetylcholine. Unlike sympathetic neurons, the ACh binds to muscarinic acetylcholine receptors on the target cells in parasympathetic postganglionic neurons to initiate cellular responses. Activation of these muscarinic receptors are G protein-coupled receptors (M1, M2, M3, M4, and M5) leads to the activation or inhibition of intracellular signaling pathways, depending on the receptor subtype and the target tissue. The various physiological responses include decreased heart rate, increased gastrointestinal motility and secretion, contraction of smooth muscle in the bladder, constriction of the pupils (miosis), and stimulation of glandular secretion (e.g., saliva, tears, and digestive enzymes). These responses collectively promote relaxation, digestion, and conservation of energy.

8.20 REFLEXES OF NERVOUS SYSTEM

Living beings, from the simplest to the most complex, exhibit a remarkable ability to adapt to their environment. This adaptation occurs at two different levels. One is a conscious choice and the other is autonomous, subconscious response triggered by the body's regulatory systems. The conscious cognitive response to a physical threat involves skeletal muscle movement. However, there are numerous pre-programmed control mechanisms that enable muscular systems to respond involuntarily without the conscious decision of the subject.

Reflexes are involuntary and rapid actions of the system in response to a stimulus. They are essential for survival which help to protect from external harm and maintains internal homeostasis. Reflexes in addition to being involuntary, are stereotypic motor responses to sensory stimuli. These reflexes are mediated by neural circuits called reflex arcs which bypass the conscious processing centres of the brain, allowing rapid responses that are essential for survival.

There are four main categories of reflexes:

1. **Somatic reflexes:** These are quick, involuntary movements of the body's skeletal muscles controlled by the somatic nervous system integrated at the level of spinal cord. Though living beings constantly use these muscles for activities like walking, they don't consciously trigger them. These autonomous actions happen seamlessly in the background of the conscious thought of the beings.
2. **Autonomic reflexes:** These reflexes involve autonomic nervous system and control involuntary functions like digestion, heart rate and metabolic responses to stress. Living beings are unaware of these reflexes happening and are mediated at the level of midbrain and medulla.
3. **Somatic reflexes:** As introduced earlier these are involuntary movements controlled by spinal cord and are unusually fast. The key to this instantaneous response lies within a specialized neural pathway called the reflex arc. This pathway operates in such a way that the body reacts swiftly without any conscious cognitive thought to ensure that the critical reflex happen in a split-second.
4. **The Reflex Arc – A speedy-bypass circuit:** The cornerstone of the somatic reflex lies in the reflex arc. Reflex arc is a dedicated neural pathway that bypasses the brain for immediate response.

The reflex arc processes information at five different interconnected levels.

1. **Stimulus:** A stimulus is anything in the environment that can trigger a detectable change in an organism. Technically speaking, an event in the environment triggers a sensory receptor to send a signal. This change can be physical or chemical and it may happen internally or externally. External stimuli come from outside the organisms' environment for example light, temperature, sound, etc. Elevated blood pressure and altered sugar levels are examples of internal stimuli.
   Modus operandi of stimulus includes sensory reception, signal transduction, transmission and processing and response.
   Sensory reception: The stimulus comes into contact with sensory organs. Specialized cells called sensory receptors detect the stimuli form the environment and these cells are present not only in the sensory organs but are spread throughout the body.
   Signal transduction: Sensory receptors that receive the stimulus convert it into electrical signal.
   Transmission and processing: The electrical signals traverse through nerves to CNS for interpretation.
   Response: The CNS generates response based on the interpretation of the processed information. The response can be either physiological or behavioural. Physiological responses include the changes in the bodily functions while the behavioural responses display the actions of the organisms that withdrew them from the environmental dangers.

Stimuli are indispensable for organisms' survival. They allow interactions of the organisms with the environment, assess dangers, acquire resources and maintain homeostasis internally.

2. **Sensory neuron:** Sensory neurons are the porters of perception to the outer world. They are the afferent limbs of the somatosensory neural pathways. They are specialized nerve cells that detect and transmit information about changes in both the external and internal environments to the CNS. Based on the specialization to detect specific kind of stimulus, there are different types of sensory neurons.
   Mechanoreceptors: these are sensory neurons to detect touch, pressure and vibration.
   Thermoreceptors: These are sensory neurons to detect changes in temperature.
   Photoreceptors: These are sensory neurons in the retina of the eye to detect light.
   Chemoreceptors: These are sensory neurons to detect taste and smell.
   Propioceptors: These are sensory neurons to detect the position and movement.
   The primary function of the sensory neurons is to translate sensory stimuli into electrical signals that the CNS can comprehend. The process involves stimulus detection, sensory transduction, action potential generation, and signal transmission.
3. **Integration centre:** In a reflex arc, the integration centre is the decision maker. It is located within the CNS, either in spinal cord or brainstem. Function of the integration centre can be summarized as receiving of sensory input, processing of information, generating response and sending motor output.
4. **Motor neuron:** also referred to as motoneuron or efferent neuron is a nerve cell that transmits signal from CNS to the peripheral effectors.
5. **Effector muscle:** Skeletal muscle is the heart of the reflex arc. The effector muscle receives the message and contracts or relaxes depending on the reflex type.
   Uniqueness of the reflex arc: The reflex arc's elegance lies in its lightning speed. Bypassing the brain, it elicits responses in milliseconds. This is critical for situations where a quick reaction is vital.
   The role of Spinal Cord: The spinal cord plays main role in the reflex arc. It serves as an integration centre that receives sensory input, processing through interneurons and through the motor neurons finally triggering the motor response. This allows for a rapid, localized response without the involvement of extensive analysis of the brain.
   The interplay of the somatic reflex and the brain: Though the reflex arc operates independent for quick responses, but the brain is still informed about the reflexes. The brain does get the sensory information along a separate pathway. This makes the subject aware of the stimulus after the initial reflex. In addition, the brain can modulate and refine the future reflexes based on the situations. This interplay can be categorised into 3 main events:
   1. **Information relay:** Even the reflex arc stands alone, it still sends the information to the brain. This makes brain informed about the incidences which lead to informed decisions for later actions.
   2. **Learning and memory:** The brain learns the reflex actions and makes conscious decisions to avoid similar situations in future.
   3. **Modulating reflexes:** The brain can modulate the reflexes to some extent. For example, the Gag reflex triggers involuntary cough when the back of the throat gets stimulated. It is a preventive measure to avoid choking. However, during eating or drinking, this Gag reflex is modulated momentarily for smooth movement of the ingesta down the oesophagus.

The following are the somatic reflexes each with a specific function.

Stretch reflex: Stretch reflex is a fundamental reflex that plays vital role in maintaining posture and balance. It is an involuntary response maintained by reflex arc which helps muscles react quickly to the changes in length. It supports weight bearing of the muscles and helps prevent falling under the influence of gravity.

In skeletal muscles, the muscle spindle acts as the sensory organ, detecting when the muscle is stretched. The voluntary striated muscle, innervated by alpha motor neurons from the ventral root of the spinal cord, serves as the effector organ. Together, these components form a negative feedback loop that regulates muscle length and maintains posture. Muscle spindles are encapsulated structures found within the fleshy part of the skeletal muscles. They are spindle shaped and are much thinner than the surrounding muscles. Muscle spindles house specialized muscle fibres called intrafusal muscle fibres; while the most common contractile fibres of the muscle are called extrafusal muscle fibres which make up the bulk of the muscle and are contractile in nature.

Intrafusal muscle fibres are further divided into two types: nuclear bag fibres and nuclear chain fibres. Nuclear bag fibres are larger and fewer in number compared to nuclear chain fibres. About two nuclear fibres per spindle exist. Numerous nuclei are clustered in a bag-like region in the central part of the fibre, hence the name. The ends of the nuclear bag fibres are striated, containing contractile machinery just like regular muscle fibres. These fibres are densely innervated by large-diameter primary sensory neurons, which wrap around the central region. When a muscle fibre is stretched, the nuclear bag fibres contract along with the entire muscle. This stretch is detected by the sensory neurons wrapped around them. The primary sensory neurons fire action potentials, sending signals to the spinal cord. The efferent signals trigger stretch reflex, causing the stretched muscles to contract and there might be indirect inhibition of the opposing muscles through interconnected neurons within the spinal cord. This in turn helps in coordinated movement to maintain posture and balance. Nuclear chain fibres, alongside the nuclear bag fibres, are another critical players within the muscle spindle, acting as sensory organs for detecting changes in muscle length. These are smaller and more in number compared to nuclear bag fibres (3–9 per spindle). As in the name, the nuclei are arranged in a single, chain like row within the central region of the fibre. Similar to the nuclear bag fibres, the ends of nuclear chain fibres are striated and contain contractile machinery. In contrast to nuclear bag fibres, nuclear chain fibres respond by contracting slowly and continuously (tonic contraction), to the stretch. These are innervated by smaller diameter secondary sensory neurons compared to nuclear bag fibres. These secondary neurons fire action potentials that convey information about the static length of the muscle to the CNS. This continuous signal is critical for maintaining posture and body mindfulness.

In short, nuclear chain fibres update continuously, slowly and steadily on muscle length to the CNS. This is vital for maintaining posture and sense of body position. Nuclear bag fibres, on the other hand, function more like prompt responders, triggering the stretch reflex for immediate adjustments when a muscle is stretched too far. Together, these two types of intrafusal muscle fibres within the muscle spindle provide a comprehensive picture of muscle length, ensuring coordinated movement and proper posture.

Muscle spindles initiate two main types of large-diameter afferent (Sensory) fibres, Primary sensory fibres (Group Ia) and Secondary sensory fibres (Group II). Primary sensory fibres are associated with nuclear bag fibres. They are large in diameter for rapid transmission of the nerve impulses. Secondary sensory fibres are smaller in diameter than primary fibres and transmit signals slower.

The combined effect of large diameter afferent and efferent fibres coupled with monosynaptic pathway creates a direct line of communication between muscle spindle and the motor neuron. This makes the effective quick reflex response possible. Example is knee-jerk response in humans. On the other hand, if the muscle shortens, the spindles calm down and send fewer signals allowing the muscle to relax and lengthen further. Though, the negative feedback mechanisms are in place, they are slow. Major of the problem with the feedback mechanisms is they take into consideration the static sensitivity alone. When an animal loses its balance, the initial error in length of the muscle is too small to be effective. This error in feedback system is overcome by dynamic sensitivity of the spindles. The sensitivity of the spindles relies on change in the length of the muscle or the posture rather than the state. This dynamism comes from the afferent group of fibres arising from nuclear bag fibres, which are sensitive to the rate of lengthening of the muscle. Those originating from nuclear chain fibres are static in responses.

The stretch reflex, triggered by sensory fibres and utilizing efferent fibres, contracts a muscle *via* reciprocal innervation while the inverse stretch reflex relaxes it to prevent muscle damage.

1. **Reciprocal Innervation:** it ensures coordinated muscle movement. When a muscle (agonist) contracts due to a stretch reflex, the opposing muscle (antagonist) relaxes. This is achieved through inhibitory signals sent from sensory neuron to the motor neurons innervating the antagonist muscle. The stretch reflex on the agonist muscle and the reciprocal inhibition on the antagonist muscle happen simultaneously.
2. **Efferent Fibres:** Efferent fibres are motor neurons that carry signals from the CNS to muscles, causing them to contract. In stretch reflex, the motor neurons carrying the signal for muscle contraction are efferent fibres.
3. **Inverse Stretch Reflex (Golgi Tendon Reflex):** The inverse stretch reflex, also known as Golgi tendon reflex, acts as a safety mechanism to prevent excessive muscle force generation. Golgi tendon organs, located near muscle-tendon junctions, detect excessive tension and send inhibitory signals to the motor neurons of the contracting muscle, causing it to relax.
4. **Knee-jerk Reflex – A Classic Example of Stretch Reflex:** The knee-jerk reflex also called as patellar reflex, is a stretch reflex present in many animals including humans. It's a type of stretch reflex that tests the integrity of the reflex arc. When the tendon below the kneecap (patellar tendon) is given a blow, it stretches the quadriceps muscle, a muscle in the front of thigh. Stretch receptors in the quadriceps muscle, the muscle spindles, detect this stretch. Sensory neurons send a signal from the muscle spindles to the spinal cord. In the spinal cord, the sensory neurons synapse with motor neurons that control the quadriceps muscle. The motor neurons send a signal back to the quadriceps muscle, causing it to contract. The contraction of the quadriceps muscle straightens the leg at the knee joint. Doctors often use the knee jerk reflex as a way to assess the health of the subject's nervous system. A normal response to a tap on the patellar tendon is a quick extension of the lower leg. An abnormal response, such as a weak or absent reflex, could indicate a problem with the spinal cord or the nerves in the leg. Knee-jerk reflexes are

shown by dogs, cats, rabbits, horses, sheep, cows, pigs along with humans.

5. **The Withdrawal and Crossed Extensor Reflex:** The withdrawal reflex is considered pre-potent to the stretch reflex and inverse stretch reflex. This means that when there is a conflict between the reflexes, the withdrawal reflex takes priority. The main function of the withdrawal reflex is to protect the body from harmful stimuli like pain or extreme heat. It's an instantaneous response that allows the subject to pull away quickly without needing to analyse cognitively. This reflex uses polysynaptic connections (uses multiple connections in the spinal cord) to activate certain flexor muscles and deactivate extensor muscles. The withdrawal reflex is elicited by the detection of nociceptive stimuli at the cutaneous innervation level. There will be a flexion of the associated limb when the subject stands on a sharp or hot object. The magnitude of the response increases with the increase in stimulus so much so that if the stimulus is stronger enough, the whole weight bearing of the subject will be shifted to another/opposite (contralateral) limb for body support. This coordinated response involves activation of the ipsilateral extensor muscles (extensor reflex) in the supporting limb. Additionally, the upper limbs may exhibit a crossed extensor reflex, with extension of the ipsilateral arm (on the same side as the supporting leg) potentially serving a pushing motion away from the stimulus, while the contralateral arm flexes. Stretch reflex maintains posture with a momentary response. In contrast, the flexor reflex is a more prolonged action, protecting the body by keeping it withdrawn from danger for longer duration. Flexor reflex exhibits prolonged activation due to polysynaptic pathways. This multi-synaptic architecture allows for repeated stimulation of alpha-motor neurons through pathways of varying lengths. Additionally, positive feedback loops within the circuit (reverberating circuits) further extend alpha-motor neuron activation, resulting in after-discharge.
6. **Autonomic Reflexes:** As already discussed, the somatic nervous system controls voluntary striated skeletal muscles for cognitive interaction with the environment and situations where as, the autonomic nervous system (ANS) acts as an internal conductor, managing involuntary functions involving smooth muscles, like blood flow, digestion, etc., to maintain homeostasis. The ANS comprises three anatomically distinct divisions: Sympathetic nervous system (SNS) which orchestrates the fight-or-flight response *via* widespread activation; Parasympathetic nervous system (PNS) that promotes a rest-and-digest state through targeted activation of organs; Enteric nervous system (ENS) that functions as a semi-autonomous network governing gastrointestinal (GI) tract processes. ENS is strongly influenced by both the other divisions of the ANS.

The ANS, often referred to as the involuntary nervous system, plays a critical role in maintaining physiological equilibrium within the body. Autonomic reflexes, a fundamental component of the ANS, represent a complex interplay between sensory organs, the CNS, and effector tissues, ensuring the body's constant adaptation to internal and external stimuli.

The autonomic reflex arc shares similarities with its somatic counterpart but exhibits key distinctions. Sensory information from visceral organs and glands, relayed by visceral afferent neurons, reaches the brainstem or spinal cord. Within dedicated autonomic centres, complex integration of the sensory information occurs. These centres, located in the brainstem and thalamus, coordinate the appropriate autonomic response. For instance, the nucleus tractus solitarius (NTS) in the medulla oblongata integrates cardiovascular and respiratory information, while the pontine micturition centre in the pons controls bladder function. The response is then relayed through efferent neurons to target tissues, which include smooth muscle, cardiac muscle and glands.

Unlike, the single-synapse connection in the somatic system, the autonomic reflex arc involves two – a preganglionic neuron that synapses within an autonomic ganglion, and postganglionic neuron that innervates the effector tissue. This two-neuron arrangement allows for more intricate modulation and integration within the CNS.

Numerous autonomic reflexes ensure optimal physiological function. Here are a few key autonomic reflexes:

1. **Baroreceptor Reflex:** This reflex maintains blood pressure stability. When blood pressure rises, baroreceptors in the carotid sinus and aortic arch detect the change. Signals travel to the brainstem, leading to parasympathetic activation and sympathetic inhibition, resulting in vasodilation and decreased heart rate, ultimately lowering blood pressure. Conversely, a fall in blood pressure triggers the opposite response.
2. **Pupillary Light Reflex:** Light intensity is detected by retinal photoreceptors. In bright light, a reflex arc mediated by the parasympathetic nervous system constricts the pupil, reducing light intake. Conversely, in dim light, the sympathetic system dilates the pupil to allow for more light entry.
3. **Gastrointestinal Reflexes:** Autonomic reflexes govern digestion from food intake to waste elimination. Salivation, gastric acid secretion, and intestinal motility are all under autonomic control, ensuring proper food breakdown and nutrient absorption.

4. **The Cough Reflex:** This protective reflex safeguards the respiratory system. When irritants stimulate cough receptors in the airways, the brainstem initiates a forceful expulsion of air through the glottis, aiming to dislodge the irritant. Both the sympathetic and parasympathetic systems contribute to this reflex, with the sympathetic system causing bronchoconstriction to increase air pressure and the parasympathetic system coordinating the diaphragm and abdominal muscle contractions for forceful expulsion.
5. **The Micturition Reflex:** This complex reflex governs bladder emptying. As the bladder fills, stretch receptors in the bladder wall send signals to the pons in the brainstem. When filling reaches a certain threshold, the parasympathetic system triggers bladder muscle contraction, while simultaneously relaxing the urethral sphincter, allowing for urine flow. Conversely, the sympathetic system can inhibit bladder contraction and maintain urethral closure during socially inappropriate moments.
6. **The Thermoregulatory Reflex:** The body strives to maintain a constant core temperature (around 37°C). When skin thermoreceptors detect a decrease in ambient temperature, the sympathetic system activates. This response includes vasoconstriction to decrease blood flow to the skin, conserving body heat. Additionally, shivering, mediated by the sympathetic system, generates heat through involuntary muscle contractions. Conversely, in hot environments, the parasympathetic system promotes vasodilation, sweating, and increased blood flow to the skin to facilitate heat dissipation.
7. **The Oculocardiac Reflex:** This reflex helps maintain cardiovascular stability during certain manoeuvres. When pressure is applied to the eyeballs, such as during valsalva manoeuvre (straining with closed glottis), the vagus nerve (parasympathetic) is stimulated, leading to a transient decrease in heart rate. This reflex helps prevent excessive blood pressure spikes during straining activities.
8. **The Hering-Breuer Reflex:** This pulmonary reflex plays a role in regulating breathing depth and preventing overinflation of the lungs. Stretch receptors in the alveoli (air sacs) signal the brainstem when lung inflammation reaches a certain point. This triggers the parasympathetic system to inhibit the diaphragm and intercostal muscles, leading to exhalation and preventing lung overinflation.

BIBLIOGRAPHY

Bai, Liqun, Hua Xu, James F. Collins, and Fayez K. Ghishan. "Molecular and functional analysis of a novel neuronal vesicular glutamate transporter." *Journal of Biological Chemistry* 276, no. 39 (2001): 36764–36769.

Barnett, Mark W., and Philip M. Larkman. "The action potential." *Practical Neurology* 7, no. 3 (2007): 192–197.

Carnevale, Nicholas T., and Michael L. Hines. *The NEURON book*. Cambridge University Press, 2006.

Daneman, Richard, and Alexandre Prat. "The blood–brain barrier." *Cold Spring Harbor Perspectives in Biology* 7, no. 1 (2015): a020412.

Evans, Richard J., Victor Derkach, and Annmarie Surprenant. "ATP mediates fast synaptic transmission in mammalian neurons." *Nature* 357, no. 6378 (1992): 503–505.

Gomperts, Bastien D., Peter E. R. Tatham, and Ijsbrand M. Kramer. *Signal Transduction*. Gulf Professional Publishing, 2002.

Koop, L. K., and P. Tadi. "Neuroanatomy, sensory nerves." [Updated 2023 Jul 24]. In *StatPearls* [Internet]. Treasure Island, FL: StatPearls Publishing, 2024, January. Available from: https://www.ncbi.nlm.nih.gov/books/NBK539846

McCorry, Laurie Kelly. "Physiology of the autonomic nervous system." *American Journal of Pharmaceutical Education* 71, no. 4 (2007): 78.

Mueller, Jerel K., and William J. Tyler. "A quantitative overview of biophysical forces impinging on neural function." *Physical Biology* 11, no. 5 (2014): 051001.

Quraishe, Shmma, Lindsey H. Forbes, and Melissa R. Andrews. "The extracellular environment of the CNS: influence on plasticity, sprouting, and axonal regeneration after spinal cord injury." *Neural Plasticity* 2018 (2018): 2952386.

Reschechtko, Sergey, and J. Andrew Pruszynski. "Stretch reflexes." Current Biology 30, no. 18 (September 2020):R1025–R1030. doi: 10.1016/j.cub.2020.07.092. PMID: 32961152.

Sabatini, B. L., and W. G. Regehr. "Timing of synaptic transmission." *Annual Review of Physiology* 61, no. 1 (1999): 521–542.

Stuart, Greg, Jackie Schiller, and Bert Sakmann. "Action potential initiation and propagation in rat neocortical pyramidal neurons." *The Journal of Physiology* 505, no. 3 (1997): 617–632.

Takagi, Hiroshi. "Roles of ion channels in EPSP integration at neuronal dendrites." *Neuroscience Research* 37, no. 3 (2000): 167–171.

Uemura, Etsuro E. "Electrochemical basis of neuronal function." *Dukes' Physiology of Domestic Animals*, 13th edition, 2015.

Wang, Bin, and Olga K. Dudko. "A theory of synaptic transmission." *Elife* 10 (2021): e73585.

Waterhouse, James, and Iain Campbell. "Reflexes: Principles and properties." *Anaesthesia & Intensive Care Medicine* 15, no. 4 (2014): 201–206.

Westfall, Thomas C., and David P. Westfall. "Neurotransmission: The autonomic and somatic motor nervous systems." In Goodman and Gilman's The Pharmacological Basis of Therapeutics, 12th ed., edited by L. Bruton, 171–218. New York: McGraw Hill, 2011.

Zhang, Jiawei. "Basic neural units of the brain: Neurons, synapses and action potential." *arXiv preprint arXiv:1906.01703* (2019).

9 Physiology of the Visual System

Abbas Rabiu Ishaq
Paws and Claws Specialist Veterinary Clinic, Al-Sahafa District, Riyadh, Kingdom of Saudi Arabia

9.1 INTRODUCTION

Vertebrates and invertebrates have eyes that fulfill fundamental jobs of image creation using photoreceptors that extend to the neutrophil (Strausfeld, 2017). Advanced molecular studies elucidated the conservation of certain transcription factors that stand as the basis of the gene network that controls the development of the visual system of vertebrates and invertebrates (Gehring, 2011; Steinmetz et al., 2010). Invertebrate and vertebrate eyes share homologous cytological and neuroanatomical features already present in their common bilaterian ancestor (De Robertis & Sasai, 1996). Primary steps of visual processing are similar within vertebrates and differ from those of invertebrates. It is performed by the retina and optic tectum in vertebrates, while the lamina, medulla and lobula perform it in invertebrates. In both cases, neuronal networks show similar architectural designs (Wernet, Huberman, & Desplan, 2014; Strausfeld & Hirth, 2013).

9.2 TYPES OF ANIMAL EYES

Animal eyes fall under one of two types, based on morphology. These are simple and complex eyes (Erclik et al., 2009) (Figure 9.1).

9.2.1 Simple Eyes

Simple eyes consist of individual or small clusters of photoreceptors, typically accompanied by pigment cells and/or lens-forming cells that allow the organism to differentiate between light and dark environments, including the source direction. Organisms cannot use the eyes to create images (Arendt, 2003). Simple eyes comprise a large array of photoreceptors, pigment cells, and lens cells and can further be classified into two mutually exclusive classes called single-chambered and compound chambered (Treisman, 2004). Still on mechanism employed in capturing light rays, single-chambered eyes can be seen in pinhole eyes of the cephalopod *Nautilus*, which consist of a pigmented pit that relies on shadows to form an image, one that contains a lens to focus light (vertebrates and squids), or a mirror to reflect it (scallops) (Land, 2005).

Compound eyes found in insects and crustaceans employ a fundamentally different design. They are made up of ommatidial facets, each of which can be termed a 'mini-eye' containing a group of photoreceptors and, in most cases, a lens (Erclik et al., 2009).

9.2.1.1 Parts of the Eye

9.2.1.1.1 Cornea

The cornea is a highly transparent layer that forms the anterior aspect of the outer coat of the eye. It is continuous with the opaque sclera, which is the other component of the outer coat. The transitional zone, where the cornea becomes sclera, is known as the limbus or corneoscleral junction (Downie et al., 2021).

The opaque sclera covers most of the surface of the globe and shows a varying thickness, ranging from 0.3 mm immediately posterior to the insertion of the rectus muscles, up to 1.0 mm near the optic nerve (Dua & Augusto, 2000). It has been reported that the sclera is composed of collagen bundles that are approximately 64 nm in length, and branched fibrils that are randomly arranged and vary in diameter from 20 to 30 nm (Meek & Carlo, 2015).

9.2.1.1.2 Conjunctiva

The conjunctiva is a thin, translucent mucous membrane, consisting of a superficial epithelium overlying loose connective tissue called the *lamina propria* or stroma (Takahashi et al., 2013). Embryologically, the conjunctiva develops from the ocular surface ectoderm (Shumway et al., 2018). It covers the anterior region of the eyeball, extending from the limbus to the eyelid fornices and then back to the mucocutaneous junction. Major roles of the conjunctiva include immune surveillance (Gipson & Mary, 2022) and production of tear constituents (Doughty, 2012)

9.2.1.1.3 Lacrimal System

The lacrimal system comprises three major divisions. The first is the secretory component, consisting of the exocrine tubuloalveolar serous lacrimal gland and two types of accessory lacrimal glands (Krause and Wolfring glands), which contribute the aqueous components of the tear film (Takahashi et al., 2013). The second component consists of the tear meniscus and blinking. The third component is the system that mediates the drainage of these secretions (Murube, 2012).

Tear film is a protective fluid layer that contains lipids, proteins, and electrolytes, which overlies the surface of the eye (Giudice et al., 2021). The ocular surface and tear

DOI: 10.1201/9781003426851-9

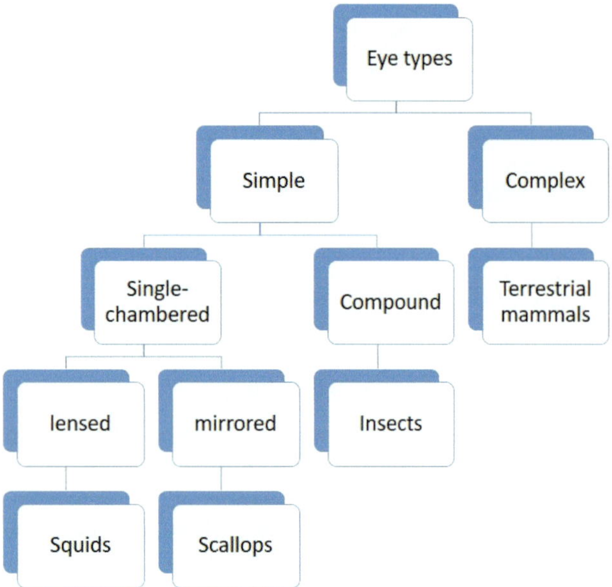

FIGURE 9.1 Schematic hierarchy showing the various eye types with an example of an organism in every case.

film are intimately related; a loss of homeostasis in either component can trigger a vicious circle of dry eye disease, a self-perpetuating cycle of tear instability, tear hyperosmolarity, ocular surface inflammation, and epithelial damage (McKee et al., 2014).

9.3 COMPLEX EYES

Complex eye types are found in more complex animals falling into the phylum Chordata. The image-forming mechanism employed includes structures for shadows, for refraction (lenses and corneas), and for reflection (mirrors) (Treisman, 2004) (Figure 9.2 & Figure 9.3).

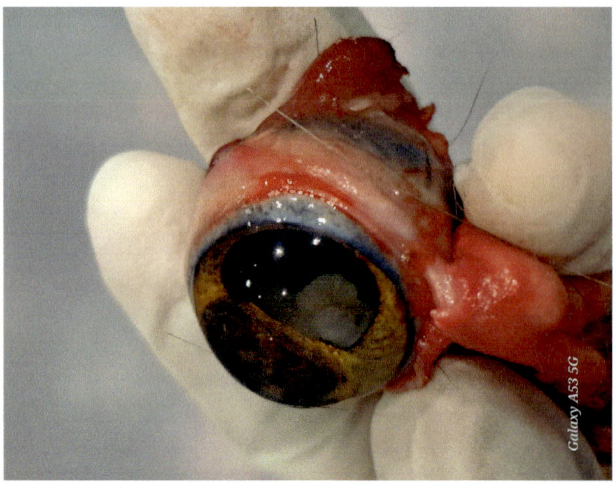

FIGURE 9.2 Structures for shadows, for refraction (lenses)

FIGURE 9.3 Structures for shadows, for refraction (corneas)

9.3.1 Photoreceptor Cell Types

The photoreceptors seen in both simple and complex eyes predominantly belong to either rhabdomeric or ciliary types, based on the mechanism used to maximize light sensitivity (Arendt, 2003). Rhabdomeric photoreceptors, found in the insect compound eye, fold the apical membrane into microvilli for the storage of photopigment. Ciliary photoreceptors, found in the vertebrate retina, fold the membrane of a modified cilium. These two photoreceptor types differ in at least four additional ways: (1) they employ distinct phototransduction cascades. Ciliary photoreceptors use cyclic GMP as a second messenger system, whereas rhabdomeric receptors employ inositol triphosphate (Morris et al., 2007). (2) The opsin photopigments and proteins involved in receptor function and response attenuation, while related, belong to distinct classes specific to each photoreceptor type. (3) The two receptors have opposite physiological responses to stimulation. Ciliary photoreceptors hyperpolarize in response to light, while rhabdomeric receptors depolarize. (4) Only ciliary photoreceptors express the *Retinal homeobox* (*Rx*) transcription factor (Arendt et al., 2004).

9.4 MUSCLES OF THE EYE

This can be either intrinsic or extrinsic (Sambasivan et al., 2011).

9.4.1 Extrinsic Muscles

These are also termed extra-ocular muscles and include the inferior rectus (IR); medial rectus (MR); superior oblique (SO); levator palpebrae (LP); superior rectus (SR); lateral rectus (LR); and retractor bulbi (RB). They are derived from the cranial mesoderm and are responsible for the bulk of ocular motions (Sambasivan et al., 2009).

9.4.2 Intrinsic Muscles

Sphincter pupillae, dilator pupillae, and ciliaris (Dimitrova et al., 2009).

9.4.3 Innervation of the Eye

The innervation for the eye can be for vision, supplied by the optic nerve (Cranial nerve II) or for ocular movements, carried out by oculomotor (Cranial nerve III), trochlear (Cranial nerve IV), and abducens (Cranial nerve VI) (Kur et al., 2012).

9.4.4 Blood Supply

Ophthalmic artery and vorticose vein, a branch of the ophthalmic vein (Flammer et al., 2013)

9.5 VISUAL PATHWAY

9.5.1 Accommodation

We are to make a comparative discussion on visualization with respect to certain media such as water. It is observed that it is difficult for terrestrial vertebrates to view under the water (Cronin, 1986). This is because the major refractive material in terrestrial vertebrates' eyes is the clear cornea. When the eyes are submerged in water (fresh, brackish, or marine), the aqueous and vitreous humor fluids lose their refractoriness to the water with similar specific gravity (Cronin, 1986). When the cornea does the necessary adjustments to produce direct rays presumed to be cast on the retina, the lens, serving a function similar to the smooth adjustment knob of a microscope, finishes the arrangement; this is termed accommodation. In situations where the lens fails to make adjustments, rays to be passed behind to produce blurred images behind the retina, a condition known as hyperopia (Sanes & Zipursky, 2010).

9.5.2 Accommodation in Amphibious Eyes

Aquatic animals or amphibious animals that are amphibious in nature have lenses that are so powerful that they do not require corneal aids for adjustment to produce sharp images underwater. This is achieved as a result of two distinct lens modifications in these groups of animals (Gislén et al., 2003). One is that the lens is spherical, contrary to the lozenge-shaped terrestrial animal lens forms. The second is the intrinsic gradient of the lens refractive index that decreases with distance (Land & Nilsson, 2012). When in free space (air), aquatic animals' eyes fail to make the necessary adjustment due to the combination of the spherical lens and the curved cornea combination. Rays are refracted by both the cornea and lens, making images focus in front of the retina. As a result, aquatic animals tend to have myopia in terrestrial environments (Martin, 1998). The question is, are all aquatic and amphibious animals myopic in terrestrial environments?

Some animals do show amphibious views, they maintain ultra-visual accommodation in aquatic environments and still have the ability to produce images with reasonable quality in terrestrial environments. This form of visual modification is seen in some marine mammals and aquatic birds (Martin, 1998). This is achieved by specialized modifications comprising the reduction of corneal refractive power in air through an increment in its radius, thus rendering it a bit myopic. Another way is by closing the iris to make the pupil shrink. The ultimate reduction in pupillary aperture brings about an increased depth of the field of view. These adjustments are believed to improve both hyperopic and myopic images, permitting the eye to retain some optimal visual functions in the strange environment (Gislén et al., 2003).

9.5.3 Avian Mechanisms of Accommodation

It is believed that avian species have more advanced accommodation techniques compared to other terrestrial vertebrates (Glasser & Howland, 1996). The process in avian is a holistic one and brings about good results. This comprises changes in the overall curvature and position of the lens, and involvement of the cornea or the iris (Glasser & Howland, 1996; Levy & Sivak, 1980). Though evolutionary terrestrial, birds still possess additional accommodation abilities seen in aquatic and amphibious animals. They have domed cornea (As in mammals) with a flattened lens, manipulated by the ciliary process and somewhat hemispherical retina. The striking difference is the presence of a ring of scleral ossicles (Os-sclera) surrounding the region of the iris and sometimes a highly muscular iris, incorporating independent central and peripheral sets of muscle. There exist variations from one avian species to another. For instance, albatrosses, penguins, and other seabirds, the lens is nearly spherical and the cornea flatter to reduce its refractive contribution in air (Martin, 1998). This makes refractive adjustment to be solely carried out by the lens. An amphibious ocular architecture is also seen in some avian species, this relies on the strengthening ring of scleral ossicles (Os-sclera), the extrinsic muscles of the iris, and the comparatively powerful ciliary body that squeeze the lens and shove it forward through the pupillary aperture when the eye enters water. The resulting bulge in the lens's anterior surface grants it increased refractive power, compensating for the loss of corneal refraction. This visual mechanism is seen in diving ducks (Levy & Sivak, 1980), cormorants (Katzir & Howland, 2003), other diving birds and to some extent penguins (Not avians though) (Howland & Sivak, 1984). A study conducted about the amphibious vision in australasian gannets (*Morus serrator*) (Machovsky-Capuska et al., 2012) using infrared video retinoscopy revealed that corneas of these birds are not unusually flattened, and eyes are properly focused in both air and water. It is revealed that it takes these species less than 100 milliseconds to achieve accommodation at air-water interphase (Schaeffel et al., 1987). In essence, vision underwater requires rapid adaptation to a dimmer, bluer light environment than in air (Litwiler & Cronin, 2001).

9.5.4 Accommodation in Rhesus Monkeys (*Macaca mulatto*)

Accommodation in most reptiles and birds (explained earlier) is aided by the iris which pinches the lens to produce a more highly refracting anterior surface in a condition known as iridogenic lenticonus (Duke-Elder, 1958; Walls, 1944). When the ciliary muscle is in a relaxed position, the iris is pulled posteriorly over the anterior surface of the lens. This possibly exerts enough posterior pressure on the malleable lens to account for the deeper anterior chamber and the thinner lens seen in the normal eye. During accommodation, there is possibility that contraction of the ciliary body could have an effect on corneal curvature (Bito et al., 1987). As the ciliary muscle contracts and its apex moves forward and inward, the ciliary ring narrows and the zonular fibers slacken, allowing the lens to become more spherical (Goodchild, Ghosh, & Martin, 1996). As the lens thickens and the anterior chamber shallows, the iris is stretched forward over the anterior lens surface. There is a practical loss of accommodation with ageing in the rhesus monkeys either in vivo or in vitro (Neider et al., 1990)

9.5.5 Accommodation in Racoon

The raccoon (*Procyon lotor*) is a small carnivore which eats in the upright position, using hand- and finger-like front paws and digits to wash, hold and examine its food at close range (Rohen et al., 1989). It has a prominent ciliary smooth muscle and zonular apparatus. The ciliary muscle and zonular apparatus exhibit a shift from longitudinal to reticular or circular orientation of some ciliary muscle bundles, anterior movement of the muscle as a whole, and more oblique crossing of the zonular fiber bundles in the zonular plexus (Kruse et al., 1990). During accommodation, the lens thickens slightly as it advances anteriorly, while the apparent cornea to retina distance increases slightly (Walls, 1944). These minimal accommodative modifications account for the greatest accommodative capability of racoons over any non-primate terrestrial mammal so far studied and the relative persistence of accommodative prowess even with advance age (Rohen et al., 1989).

9.5.6 Accommodation in Cats

Cats possess a relatively large eyeball with a broadly curved cornea, a disproportionately large pupil and a globular shaped lens situated rather deep posteriorly within the eye (Goodchild et al., 1996; Vakkur, Bishop, & Kozak, 1963). This offers cats great light-gathering capacity, reaching an estimated object intensity, retinal illumination 5.2 times greater in the cat than in the human (Goodchild et al., 1996). Tapetum lucidum located in the choroid just behind the retina acts as a diffuse reflecting surface (Coles, 1971; Weale, 1953), further enhances the efficiency of the cat's eye for dim-light vision by reflecting light back through the visual receptors. There is an attributed decrease in image sharpness, owing to the job of the tapetum lucidum to 'magnify' images (Walls, 1944). The pupil size is an important asset for image processing in cats. It assumes an elliptical shape with increasing light intensities, and at modest photopic levels the pupil is reduced in shape to just a narrow slit. The rostral superior colliculus (SC) plays an important part in the control of accommodation (Centrally). While the lateral suprasylvian (LS) area, the cortical area surrounding the middle suprasylvian sulcus (MSS) of the cat, is related to the control of lens accommodation (Bando et al., 1988; Wakabayashi et al., 2017). The LS area receives visual inputs. Some neurons in the mentioned area respond to changes in ocular disparity and target size and to motion in depth, which are important visual cues for accommodation (Ohtsuka & Sato, 2000). Some LS neurons also exhibited burst discharges preceding the onset of spontaneous accommodation (Bando et al., 1984). Accommodation-related signals from the cortex project mainly to the rostral SC (Wakabayashi et al., 2017). Changes in lens power could have only minimal influence on the refractive state of the eat's eye, because of the relatively large separation between the cornea and lens, as such, accommodation is more effectively produced by the forward and backward movement of the principal planes of the lens (Goodchild et al., 1996). The dynamic range of pupillary response in the cat, with over a 6 log-unit luminance range, and pupil diameter varying from 16 mm to less than 0.5 mm (Kappauf, 1943). The pupil size influences very significantly the quality of the image formed by the cat, with worst images produced using pupil fully dilated. The more sharp images are produced with an increasingly smaller pupil diameters (Abramov et al., 1972). optimal performance is achieved when the pupil reduces 2-3 mm in diameter, which gives an optical resolution limit exceeding 20 cycles/deg (Bonds, 1974). This level of accommodative quality is seen when kitten's eye develops to adult optical standards over the $1^1/_2$ months or so after birth (Bonds & Freeman, 1978).

9.5.6.1 Cat Retina Architecture

We are emphasizing on distribution of rods and cones within the central and peripheral retina of the cat. It is believed that rods are more in number compared to cones in cats' eyes even within the area centralis, the region attributed for highest cone density in most animals (Sterling, 1983). Steinberg, Reid, and Lacy (1973) used constructed detailed maps of receptor density over the entire cat retina from photomicrographs of sections through the inner segment of an excised retina. The outcome is that cone density is highest within the area centralis but that even here rods are more than 10 times as numerous, confirming the observations of Sterling (1983). Maximum rod density occurs within an anular region 10-15 degrees from the center of the area centralis and gradually decreases with increasing retinal eccentricity. By comparing these receptor densities with the ganglion cell counts of Stone (1972), Steinberg et al. (1973) estimated the degree of receptor/ganglion cell convergence for different portions of the retina. Within the area centralis, the ratio of cones to ganglion cells is at a minimum, suggesting that this would be the region of maximum photopic acuity. On the other hand, the minimum rod/ganglion cell ratio occurs 5 degrees into the periphery, where optimum scotopic acuity would likely occur. These ratio estimates may vary, depending upon the portion sampled, because of the horizontally elongated region of high ganglion cell density in the cat retina (Hughes, 1975; Stone, 1972).

The accounted convergence of receptors onto ganglion cells is believed to be mediated by an organised network of neural connections within the outer and inner plexiform layers, and the details of these connections in the cat, studied using the electron microscope. Cones in the cat retina synapse onto two types of bipolar cells, an invaginating type which connects with from 4 to 9 individual cones and a flat type which contacts 8-14 cones (Boycott & Kolb, 1973), there is no evidence for a bipolar, which contacts just a single cone, such as the midget bipolar cells seen in primate retina. These cone bipolars in the cat, in turn, synapse directly onto ganglion cells

(Kolb & Famiglietti, 1974). Rods, on the other hand, contact another type of bipolar, termed rod bipolars (Boycott & Kolb, 1973), and these do not synapse directly onto ganglion cells, but, instead, onto amacrine cells which then contact the ganglion cells (Kolb & Famiglietti, 1974). Also, the receptors are interconnected by horizontal cells, the dendritic terminals of which synapse only with cones and the axon terminal only with rods (Kolb, 1974). It is seen that rods can have direct input to cones, perhaps via electrically passive gap junctions (Nelson, 1977). Though the significance of this synaptic arrangement between receptors is not clear, but it could be that the horizontal cells are somehow involved in the control of retinal sensitivity at different background illuminations (Enroth-Cugell & Lennie, 1975; Enroth-Cugell & Shapley, 1973). Ganglion cells in the cat retina are believed not to have homogeneous morphology (Brown & Major, 1966; Honrubia & Elliot, 1970; Leicester & Stone, 1967). There are morphologically three forms of ganglion cells distinguished based on axon thickness and dendritic field size. These are α, β *and* γ, they are seen at all retinal eccentricities. Distribution wise, within the central region, β cells are predominant, and for each type, the spread of dendritic fields increases systematically with retinal eccentricity, but at any particular retinal locus the spread of *a* cells is usually greater than that of β cells. With respect to size, the y cells display large dendritic fields but possess small axons (Boycott & Wassle, 1974).

9.5.6.2 Cat Retina Physiology

We are to begin with photochemical and neural events which transpire prior to the ganglion cells. It is reported that even in the dark, the ganglion cells in the cat retina generate frequent, irregularly timed spikes which some believe are of receptor origin (Barlow & Levick, 1969). Spontaneous activity of the ganglion cells could be modulated by presentation of visual targets within a limited region of the retina, this infers that in all light-adapted ganglion cells this region (the receptive field), could be subdivided into two zones. That is, an approximately circular central region surrounded entirely by an annular area. Stimulation of one zone increases the cell's firing rate (ON response), while stimulation of the other causes a decrease in firing, followed by a burst of activity upon removal of the stimulus (OFF response). Some cells consisted of ON-centers, while others of OFF-centers. In each region, cells are always arranged in an antagonistic fashion (Kuffler, 1953). The receptive field of cat retina is composed of two distinct response mechanisms (center and surround) arranged concentrically, each with an approximately Gaussian sensitivity distribution and the response of the cell is determined by the subtractive interaction of these two mechanisms (Cleland & Enroth-Cugell, 1968; Enroth-Cugell & Robson, 1966; Rodieck, 1965; Rodieck & Stone, 1965). The majority of receptive fields for the cat ganglion cells follow this antagonistic center-surround arrangement and tend to display no selectivity for contour orientation or direction of movement. The receptive field centers of ganglion cells, (whether ON or OFF type), vary in diameter from 0.5 to 8-degree visual angle (Wiesel, 1960) and are smallest within the area centralis. The antagonistic surround portion of the field is several times larger than the field center, and there is evidence that the surround extends through the center (Cleland & Enroth-Cugell, 1968; Hammond, 1973; Rodieck & Stone, 1965). The inhibitory effect of this surround region on the center disappeared with dark adaptation, leaving only an ON or an OFF response throughout the field (Barlow, Fitzhugh, & Kuffler, 1957). In exceptional cases, some ganglion cells exhibit weak excitatory effects to visual stimulation many degrees removed from the conventional receptive field center (McIlwain, 1964).

Ganglion cells could also be classified along another dimension, such as based on different response properties, independent of the center-surround organization discussed earlier. One of such classifications is the finding that one class of ganglion cells, termed X cells, display approximately linear summation over the entire receptive field, while a second class, termed Y cells, which is clearly nonlinear. Evidence for linear summation consisted of finding a "null position" within the receptive field, such that repetitive phase reversal of a bipartite field (e.g., one complete light/dark cycle of grating) which does not produce a net change in background activity (Enroth-Cugell & Robson, 1966). Cells that fall under type X typically exhibit the following characteristics: field centers more of the ON type

(Not all); field centers tend to be smaller than Y type; maintained activity usually higher than Y type; respond only to slow or moderate target speeds; show no periphery effect; predominate within the area centralis; surround portion of field smaller diameter; small stationary targets evoke sustained response; resolve higher spatial frequencies than Y type and possess slow axonal conduction (Cleland et al., 1973). On the other hand, type Y cells: respond to very rapidly moved targets; show clear effects on the periphery; are insensitive to image defocusing; respond transiently to onset and offset of stimulus; possess fast conducting axons (Ikeda & Wright, 1972). There are proposals that these two classes of retinal ganglion cells constitute separate neuronal channels which subserve contrasting roles in visual sensation (Cleland et al., 1971). There are assumptions on possible involvement of X cells in analyzing the spatial features of stimuli, and possible abilities of Y cells to process information concerning movement and time varying stimulation (Ikeda & Wright, 1972). Other than the early discussed center-surround cell categories, Stone & Fukada, 1974; Stone & Hoffmann, 1972 described a third category tagged 'W'. These cells have field centers which are same in size to Y cell centers (at a given retinal eccentricity), and that the axons of the W cells conduct more slowly than either type X or type Y axons (Stone & Fukada, 1974). These W cells appear to comprise the bulk of the ganglion cells in the 'tagged' visual streak of the cat retina (Wilson, Rowe, & Stone, 1976). The fact that this third category of cells project heavily to the superior colliculus suggests that they may play some role in visuomotor behavior (Stone & Fukada, 1974). Some similarities exist between the physiological properties of the X, Y, and W ganglion cells as well as with those for α, β *and* γ.

9.5.6.3 Projection

Here, we are to discuss what happens to the axons after exiting the retina. Axons of the retinal ganglion cells, upon exiting the retina through the optic disc, partially cross at the optic chiasm. In all species, fibers from the nasal retina of each eye cross to the contralateral side of the brain (Crossing over) while fibers originating from the temporal retinae project ipsilaterally. An exception to this is seen in cats, where limited region around the area centralis which gives rise to bilateral projections showing a combination of crossed *and* uncrossed fibers (Stone & Fukada, 1974). With respect to size of the receptive field centers, this composite strip of crossed-uncrossed axons is wider horizontally for Y cells than for X cells (Kirk, Levick, Cleland, & Wassle, 1976). The complete/partial crossing-over is followed by the optic fibers projection to several subcortical sites in the brain (Garey & Powell, 1968; Laties & Sprague, 1966). It is proven that most of the of ganglion cell axons innervate cells of the lateral geniculate nucleus (LGN), which primarily consists of two anatomically distinct portions, the more prominent dorsal lateral geniculate and the smaller ventral lateral geniculate. The majority of cells in the dorsal LGN receive input from either X or Y cells (Wilson, Rowe, & Stone, 1976), while the ventral LGN innervation appears to be largely from W cells (Spear et al., 1975). Owing to their diffuse receptive-field properties and their projections to pretectal regions, it is strongly suggested that cells of the ventral LGN are involved in the pupillary response to light, vestibular and oculomotor reflexes. Cells in the dorsal LGN, on the other hand, display more well-defined receptive field properties and project to the visual cortex. The remaining ganglion cell axons project to structures in the midbrain, including the most prominent one, superior colliculus. There are strong suggestions that these midbrain structures are involved in visuomotor coordination while the geniculocortical pathways play the major role in spatial resolution and pattern recognition (Autrum et al., 1973).

9.6 LATERAL GENICULATE NUCLEUS (LGN)

9.6.1 Anatomy

To begin with, I am mentioning the presence of well-defined laminar structure. Cells of the geniculate are grouped into three major layers, with the optic tract fibers from the contralateral eye projecting exclusively to the dorsal and ventral layers. while the ipsilateral fibers project to the middle layer (Garey & Powell, 1968; Guillery, 1966; Hayhow, 1958; Laties & Sprague, 1966; Stone & Hansen, 1966). An additional sublamina of the ventral layer which also receives an ipsilateral projection was described by Guillery (1970). Neurons of the LGN are classified into 4 types based on peculiar morphological characteristics. These are classes 1, 2, 3, and 4. Class 1 neurons have large cell bodies and thick, straight dendrites punctuated by occasional spiney appendages. Class 2 neurons have relatively smaller somata and thin, arching dendrites that terminate in a cluster spheroid appendage. Class 3 neurons have smallest cell bodies and very thin, wavy dendrites along which appear complex arrays of ovoid specializations (Garey & Powell, 1967). Class 4 cells are small in size and possess dendrites resembling class 1 cells (Stone & Dreher, 1973). Based in distribution, class 1 and 2 cells are found in layers A, AI and, while class 4 cells are confined mainly to the C1 and C2 laminae of the LGN. Class 1 and 2 cells are exclusively relay cells (not interneurons), this ascertained through the technique of intracellular staining (Ogawa, Takimori, & Takahashi, 1978). There exist some morphological correlations between earlier mentioned X and Y cells with class 1, 2, 3 and 4 cells. Class 1 cells are morphologically associated with Y cells, class 3 with X type, while class 2 structural traits exhibited by some X cells, others for Y (Garey & Blakemore, 1977a, 1977b; Gilbert & Kelly, 1975; LeVay & Ferster, 1977). LeVay and Ferster (1977) suggested that class 1 neurons are Y cells, class 2 are X, and class 3 are simply interneurons. Afferent input to the cells of LGN arrive from numerous sources, with axons of retinal ganglion cells accounting for major excitatory input. These axons divide into several branches prior to terminating on geniculate cells (Guillery, 1966), suggesting that each individual optic fiber innervates several cells of the LGN. The projection of the fibers onto geniculate cells is retinotopically organized in a very precise manner (Stone & Hansen, 1966). The cells of the cat LGN also receive modified topographical projections from the visual cortex (Beresford, 1961; Guillery, 1966; Hollaänder, 1970; Kalil & Chase, 1970; Szentágothai, Harmori, & Tombol, 1966; Updyke, 1975), as well as a less organized projection from the superior colliculus (Altman, 1962). There is proposal that the LGN receives afferents from other nonvisual structures of the brain, this is because LGN activity can be modulated by stimulation of the midbrain reticular formation (Burke & Cole, 1978; Suzuki & Taira, 1961).

9.6.2 Physiology

Here, we are to elaborate how the LGN components bring about visualization possibilities. To begin with, geniculate neurons in the cat possess receptive fields which are arranged concentrically, with both ON-center and OFF-center types (Hubel & Wiesel, 1961). Receptive field centers of LGN cells have same diameters as the ganglion cells and show field sizes increase with retinal eccentricity (Hoffman, Stone, & Sherman, 1972; Hubel & Wiesel, 1961).

Besides rare exceptions (Daniels, Norman, & Pettigrew, 1977), LGN fields show virtually no specificity for orientation or direction of movement, which is seen in the retina too (Kozak, Rodieck, & Bishop, 1965). Note that there exist some peculiar differences between ganglion cell and geniculate fields and are not to be used interchangeably. For instance, at LGN level, the surround mechanism exerts more of inhibitory influence on the center, in comparison to surrounds at the retinal level, thus enhancing the sensitivity of geniculate neurons to luminance gradients (Hammond, 1973; Hubel & Wiesel, 1961). Also, the surround portion of the geniculate field show some levels of responsiveness, even at scotopic levels very much below those at which the effects of ganglion cell surrounds disappear (Hammond, 1972; Maffei & Fiorentini, 1972). LGN contains some cells which can be influenced via either eye, such that the cell is excited by stimulation of one eye and inhibited through the other. This is practically impossible for the retina (Sanderson, Bishop, & Darian-Smith, 1971). Another feature to be noted is binocular inhibitory interactions, which bring about absolute segregation of left- and right-eye afferents to the different laminae. It is evident that the receptive fields of LGN cells include a third component not seen in retinal fields, such as annular region, surrounding the prominent surround portion, which is of the same polarity as the field center (That is ON or OFF) (Hammond, 1973). Under normal circumstances, stimulation of outer surround reduces the effectiveness of the inner surround, thus disinhibiting the center (Maffei & Fiorentini, 1972). This third field component, called the suppressive field, based on the inhibitory nature, is a feature peculiar to the LGN (Cleland et al., 1971).

Most LGN neurons are innervated by few optic fibers and in some cases by only a single fiber (Cleland et al., 1971; Hammond, 1973; Hoffmann, Stone, & Sherman, 1972; Singer & Creutzfeldt, 1970). Geniculate cells with ON-center fields receive excitatory input only from ON-center ganglion cells, and OFF-center geniculate cells are excited just by OFF-center ganglion cells (Cleland et al., 1971). This is in the order of LGN synaptic organization (Singer & Creutzfeldt, 1970). Secondly, the X/V grouping described for retinal cells is also preserved at the LGN is the sense that X-type retinal cells innervate only X-type geniculate neurons and Y-retinal cells only Y-geniculate neurons (Cleland et al., 1971; Hoffmann et al., 1972). X-type geniculate neurons have slower conducting axons than the V-type neurons, as seen in the retina. A group of W type cells, which are strictly seen in the ventral portion of layer C (Cleland et al., 1976), while X and Y cells, in contrast, consist almost exclusively of the population of neurons in laminae A and Al (Wilson, Rowe, & Stone, 1976). These X cells in the LGN are more strongly subject to intrageniculate inhibition than are Y cells (Fukada & Stone, 1976). Bullier and Norton (1977) notice some cells of the LGN that exhibit some properties of both X and Y cells. The response properties of geniculate cells and other neurons in the visual pathways is strictly time-dependent (Stevens & Gerstein, 1976a, 1976b).

9.6.3 Lateral Geniculate Nucleus Projections

Axons of geniculate neurons project to four visual areas of the cortex, named areas 17, 18, 19, and the lateral wall of the suprasylvian gyrus in cats (Garey & Powell, 1967; Niimi & Sprague, 1970). Larger geniculate axons with faster conducting abilities project more heavily to area 18, while smaller axons with slower conduction terminate predominantly in areas 17 and 19 (Garey & Powell, 1967; Hollaänder & Vanegas, 1977). In general, fast conducting LGN cells are Y type while the slower conducting cells are X type. This mode of projections to visual cortex suggests that areas 17 and 18 may perform different visual functions in the cat, with termination of radiation fibers from the LGN upon several cortical areas within the geniculocortical pathways occurring in parallel, not series.

9.7 THE VISUAL CORTEX

9.7.1 Anatomical Configuration of the Visual Cortex (VC)

Cat visual cortex comprises three areas, named 17, 18, and 19, defined by cytoarchitectonic criteria (Otsuka & Hassler, 1962). Microelectrode techniques are used to map the visuotopic organization of these areas (Albus, 1975a; Bilge et al., 1967; Hubel & Wiesel, 1962, 1965a; Tusa, Palmer, & Rosenquist, 1978). An addition to the three prominent areas is the lateral waIl of the suprasylvian gyrus (Clare & Bishop, 1954; Hubel & Wiesel, 1969). This is an area that comprises six retinotopically organized regions arranged in three mirrorsymmetrical pairs (Palmer, Rosenquist, & Tusa, 1978). There are projections from the visual areas of one hemisphere to those of the contralateral cortex via the corpus caIlosum (Ebner & Meyers, 1965; Hubel & Wiesel, 1967). The amount of cortex devoted to central vision is disproportionately large in relation to the more peripheral visual fields (Whitteridge & Jung, 1973), with this dense distribution being the greatest in area 17 (Tusa et al., 1978). All the visual areas are interconnected, with area 17 sending projections to 18, 19, and the suprasylvian gyrus, and areas 18, and 19 projecting back onto 17 (Wilson, 1968).

9.7.2 Physiological Orientation of the Visual Cortex (VC)

The earlier mentioned concentric receptive field organization of retinal and geniculate levels gives way at the cortex (VC) to a different layout of excitatory and inhibitory areas. An exception is that antagonistic areas are arranged in elongated strips so that the most effective stimulus is a straight contour (or series of numerous contours) in a particular orientation. With the exception of very few cortical cells with non-oriented receptive fields (Spinelli & Barrett, 1969), orientation specificity is a fundamental property of neurons in the cat's visual cortex. Receptive field mapping techniques showed how cells with the same preferred orientation are grouped into columns that are on the order of 73-100 microns in thickness (Hubel & Wiesel, 1962, 1963). Neighboring columns have orientations that are quite similar, within 10-15 deg or less, and a complete, orderly orientation sequence of 180 deg, is contained within a cortical column that is 300-700 *microns* in thickness (Albus, 1975b). These functional columns are seen in the form of sheets, or slabs, the walls of which are perpendicular to the cortical surface. Individual cells tend to show some selectivity for orientation sharpness of tuning, with half-widths at half amplitude ranging from as little as 5 deg up to at least 70 deg (Henry, Bishop, & Dreher, 1974). Symmetrical tuning curves are not expected to be seen in every cell, with cells exhibiting asymmetries tending to be seen more frequently in area 18 (Hammond & Andrews, 1978). Responsiveness of a cortical cell may vary with time, but orientation preference and tuning of cells remains remarkably stable over time (Hammond, Andrews, & James, 1975; Henry et al., 1973).

There is a generalized belief that excitatory inputs from LGN to cortex establish a coarse orientation bias which is greatly refined by inhibitory connections between columns to yield the sharp tuning characteristic of cat cortical cells. Some cells in cat visual cortex are responsive to other forms of visual stimulation. Cortical cells will produce bursts of impulses when spots of light traverse their receptive fields (Pettigrew, 1974) and complex, but not simple, cells can be activated by movement of textured stimuli composed of random noise, i.e., a stochastically generated array of circular dots (Hammond & MacKay, 1977). Cortical cells will produce high amounts of impulses when spots of light traverse their receptive fields (Pettigrew, 1974) and complex (not simple) cells can be activated by movement of textured stimuli composed of random noise, that is, a stochastically generated array of circular dots (Hammond & MacKay, 1977).

9.7.3 The Cell Types in Visual Cortex

The cells in cat visual cortex can be categorized into distinct groups, that is simple, complex, and hypercomplex, on the basis of other receptive field properties (Henry, 1977). Simple cells possess receptive fields that consist of adjacent ON and OFF regions within which spatial summation occurs and can be mapped using stationary flashing targets (Hubel & Wiesel, 1961).

Simple cells in general possess little or no spontaneous activity (Pettigrew, Nikara, & Bishop, 1968a), frequently respond more vigorously to one direction of movement than multidirectional (Bishop, Coombs, & Henry, 1971), prefer slowly moving stimuli (Movshon, 1975; Orban, Kennedy, & Maes, 1978; Pettigrew et al., 1968a), respond only within a narrow range or orientations (Hammond & Andrews, 1978; Henry et al., 1974; Hubel & Wiesel, 1962; Watkins & Berkley, 1974), commonly display linear spatial summation when tested with sinusoidal gratings (Movshon, Thompson, & Tolhurst, 1978a), possess small receptive Fields (In comparison to the other cell types) (Hubel & Wiesel, 1962), and are located predominantly in layer IV (Hubel & Wiesel, 1962).

On the other hand, complex cells respond very weakly to flashed stimuli anywhere within the receptive field and do not show spatial summation (Hubel & Wiesel, 1962). They have comparatively large receptive fields (Hubel & Wiesel, 1962), respond towards a broader range of orientations (Rose & Blakemore, 1974; Watkins & Berkley, 1974; Wilson & Sherman, 1976), are highly nonlinear in their response to sinusoidal gratings (Movshon, Thompson, & Tolhurst, 1978b), have a higher level of spontaneous activity (Lund, 1987; Pettigrew et al., 1968a), prefer relatively high speeds of movement (Movshon, 1975; Pettigrew et al., 1968a; Pollen & Ronner, 1975); and are concentrated in the cortical layers on either side of IV (Lund, 1987).

Shared features amongst the two types of cells include possession of larger receptive fields, less orientation selectivity, and higher preferred speeds with increasing retinal eccentricity. Except that these variations are more pronounced in the complex cells (Wilson & Sherman, 1976).

The last category, hypercomplex cell; are sensitive to the length of an oriented stimulus (Hubel & Wiesel, 1965a), have comparatively low spontaneous activity (Rose & Blakemore, 1974), vary widely in their degree of orientation tuning; and are encountered most often in areas 18 and

19, usually within layers II and III (Hubel & Wiesel, 1965a; Lund, 1987).

There are variable explanations on how the three cell types operate. One of these is that complex cells receive their input from simple cells, and hypercomplex from complex (Hubel & Wiesel, 1965a). A contrasting one states that the complex cells are working in parallel, not in series, with the simple cells. Simple and complex cells represent the cortical terminations of the X and Y projections, respectively, from the LGN (Hoffmann & Stone, 1971; Maffei & Fiorentini, 1973; Stone, 1972). Thus, fast-conducting axons (Y type) project to area 18, in which complex cells predominate, and slow-conducting axons to project to area 17, the cortical region in which simple cells are concentrated.

Neurons in areas 17 and 18 differ considerably in their spatial and temporal response properties, in a manner consistent with the belief that X cells project to area 17, while Y cells project predominantly to area 18 (Movshon, Thompson, & Tolhurst, 1978b). The area 17 neurons (whether simple or complex) preferred higher spatial frequencies and lower temporal frequencies relative to neurons in area 18. Thus, within area 17, simple and complex cells (Distinguished by their linearity of summation) did not differ in terms of the distribution of their preferred spatial frequencies, but simple cells did tend to be more narrowly tuned for spatial frequency (Movshon, Thompson, & Tolhurst, 1978b).

9.7.4 Binocularity in Visual Cortex Neurons

Here, we are discussing how the cortical neurons play a major role in making binocular visualization a possibility in cats. It is found that most cortical neurons possess *two* receptive fields, one believed to be associated with each eye. Activation of these cortical cells through either eye (Left and right simultaneously) (Hubel & Wiesel, 1962), thus, binocular cells simultaneous stimulation of both eyes yields a burst of activity which is greater than that produced by stimulation of either eye alone (Burns & Pritchard, 1968). A point to note is that, not all binocular cells show equal responsiveness through the two eyes. Rather, different cells exhibit varying degrees of ocular dominance, with the extremes being cortical cells which can be activated through one eye only. As in the case of orientation, these binocular cells are arranged in ocular dominance columns which extend from pial surface to white matter. Within a given column, cells display much the same eye preference, and the ocular dominance shifts systematically from column to column in a gradient (LeVay, Stryker, & Shatz, 1978; Shatz, Lindstrom, & Wiesel, 1977). Autoradiography technique shows that within layer IV (the terminal site of LGN afferents) of area 17 a complete cycle of ocular dominance occupies a band of tissue roughly 0.5 mm wide. It is evident that monocular cells predominate within the cortical projection area of central vision (Albus, 1975c). In terms of dimension, some binocular neurons have receptive fields which cover corresponding retinal areas.

For others, there exists various degrees of retinal disparity between the pair of monocular receptive fields, and these disparities are typically more pronounced in the horizontal direction (Barlow, Blakemore, & Pettigrew, 1967; Nikara, Bishop, & Pettigrew, 1968). Another form of binocular modification in receptive fields of cortical neurons exists in area 17, that allows cyclorotational eye movements, and interocular orientation disparities as large as 15 degrees (Blakemore, Fiorentini, & Maffei, 1972; Nelson, Kato, & Bishop, 1977).

Another stunning difference is seen in area 18 of the cat, which has a class of binocular cells that have different preferred directions of motion between the two eyes (Cynader & Regan, 1978; Pettigrew, 1973), and another class which has different preferred speeds of movement (Cynader & Regan, 1978). There are suggestions that such binocular neurons may encode information about objects moving in depth toward or away from the organism.

9.7.5 Role of Auxiliary Visual Areas

The lateral supersylvian area is one of the major auxillary visual areas that are of interest. Cells in this area tend to show similar properties that are exhibited by the complex cells in areas 17 and 18 (Hubel & Wiesel, 1969). Some astonishing features that are unique to the cells in lateral supersylvian area include lack of orientation selectivity, presence of extremely large receptive fields and paucity of fields representing the upper visual field (Spear & Baumann, 1975).

9.7.6 Visually Induced Potentials in The Cortex

This measures the combined activity from a large collection of cortical neurons. This is achieved by determining the spatial frequency at which the evoked potential has fallen to zero (Berkley & Watkins & Berkley, 1974), or by estimating the contrast level at which the evoked potential is zero (Campbell, Maffei, & Piccolino, 1973).

9.7.7 Visual Defects

This section will summarise visual deficiencies in relation to age. There is a level of visual deprivation in animals (cats) during the first two to three months of life. This is mostly affecting visual areas, excluding the retina (Wiesel & Hubel, 1963a, 1963b). This defect can be both unilateral and bilateral on X and Y mechanisms (Garey & Blakemore, 1977a; Sherman, Hoffmann, & Stone, 1972; Sherman & Stone; 1973; Hirsch & Spinelli 1970). All instances are described below, with respect to different parts of the visual system.

9.7.8 Monocular and Binocular Deprivation on the Lateral Geniculate Nucleus.

Monocular deprivation leads to a significant reduction in the proportion of Y cells sampled from the deprived laminae, while that of X cells remains relatively normal (Sherman et al., 1972; Sherman, Wilson, & Guillery, 1975). These physiological changes are accompanied by morphological effects which are relatively specific to the Y system (Garey & Blakemore, 1977a, 1977b; Lin & Sherman, 1978; Sherman, Guillery, Kaas, & Sanderson, 1974). Central to the LGN; that is at the optic radiation, there is a normal complement of Y neurons following monocular deprivation (Eysel, Grusser, & Hoffman, 1979). This can be attributed to greater reduction in the size of somas of Y cells, compared to X cells (Eysel et al., 1979). On the other hand, X cells tend to be normal throughout all laminae, in both monocular and binocular instances.

In a nutshell, binocular deprivation produces a less severe, though still significant, reduction in Y-cell frequency throughout the lateral geniculate nucleus, monocular crescent included, leaving X-cell frequency to be relatively normal (Sherman et al., 1972).

9.7.9 Deprivation at the Visual Cortex

Visual deprivation effects are so obvious at the level of visual cortex, even minor surgical interventions such as third eyelid flapping can show elicit some manifestations (Schechter & Murphy, 1976). Cells in cortical layer IV of eyes subjected to some levels of visual deprivation tend to shrink as compared to those innervating undisturbed eye (Shatz & Stryker, 1978). These effects of monocular deprivation tend not to be uniformly distributed. Incidence of cortical cells responsive through the deprived eye increases in more peripheral regions of the visual field, with both simple and complex, display abnormal receptive field properties (Wilson & Sherman, 1977).

Ultimately, physiological effects of monocular deprivation are found in both areas 17 and 18, except that in area 18, the contralateral pathway (That is, the hemisphere contralateral to the deprived eye) is much more resistant to the effects of monocular deprivation than is the ipsilateral pathway. This asymmetry between crossed and uncrossed projections is not observed in area 17 (Singer, 1978).

Other instances that can cause binocular visual defects include strabismus and alternating monocular occlusion (Blakemore, 1976; Hubel & Wiesel, 1965b).

On the other hand, binocular deprivation effects are relatively ess severe, except that there is a large increase in the number of visually unresponsive cells (Blakemore, Fiorentini, & Maffei, 1972; Imbert & Buisseret, 1975; Pettigrew, 1974). Among these are response of cells over a wider range of orientations (Hubel & Wiesel, 1965a), retinal disparities (Pettigrew, 1974), presence of fewer directionally selective cortical cells (Singer & Tretter, 1976) and an unusual number of cells with exceedingly large receptive fields which appear to lack inhibitory sidebands (Singer & Tretter, 1976).

9.7.10 Recovery from Monocular Deprivation Effects

Are there possible practical ways to reverse of decrease the gross effects mentioned above?

Binocularity in some cells may be temporarily restored following prolonged monocular deprivation by intravenous administration of bicuculline, which is an antagonist of the putative inhibitory neurotransmitter gamma-aminobutyric acid (GABA) (Duffy et al., 1976). Permanent, though not practical way of reversal is the enucleation of the non-deprived eye, thus, inducing a compensatory hypertrophy in the deprived eye (Kratz, Spear, & Smith, 1976). Another way is intraventricular injection (Kasamatsu & Pettigrew, 1976) or local perfusion (Pettigrew et al., 1968a) of 6-hydroxydopamine (6-HDA) prevented the usual loss of binocularity following monocular deprivation.

BIBLIOGRAPHY

Abramov, I., P. O'Bryan, C. G. Bernhard, M. G. F. Fuortes, G. Gemne, P. Gouras, H. K. Hartline *et al.* "Optical properties of vertebrate eyes." In *Physiology of Photoreceptor Organs* Volumes 1536, 7/2, pp. 449–482, Springer Berlin, Heidelberg, 1972.

Albus, K. "A quantitative study of the projection area of the central and the paracentral visual field in area 17 of the cat: I. The precision of the topography." *Experimental Brain Research* 24 (1975a): 159–179.

Albus, K. "A quantitative study of the projection area of the central and the paracentral visual field in area 17 of the cat: II. The spatial organization of the orientation domain." *Experimental Brain Research* 24 (1975b): 181–202.

Albus, K. "Predominance of monocularly driven cells in the projection area of the central visual field in cat's striate cortex." *Brain Research* 89 (1975c): 341–347.

Altman, Joseph. "Some fiber projections to the superior colliculus in the cat." *Journal of Comparative Neurology* 119, no. 1 (1962): 77–95.

Arendt, Detlev. "Evolution of eyes and photoreceptor cell types." *International Journal of Developmental Biology* 47, no. 7–8 (2003): 563–571.

Arendt, Detlev, Kristin Tessmar-Raible, Heidi Snyman, Adriaan W. Dorresteijn, and Joachim Wittbrodt. "Ciliary photoreceptors with a vertebrate-type opsin in an invertebrate brain." *Science* 306, no. 5697 (2004): 869–871.

Autrum, H., P. O. Bishop, V. Braitenberg, K. L. Chow, R. L. De Valois, R. B. Freeman, W. A. van de Grind *et al.* "Neurophysiological mechanisms in the visual discrimination of form." In *Central Processing of Visual Information A: Integrative Functions and Comparative Data* Volume 7 / 3 / 3A, pp. 153–207, Springer Berlin, Heidelberg, 1973.

Bando, T. A. K. E. H. I. K. O., K. E. N. J. I. Tsukuda, N. O. B. U. H. I. K. O. Yamamoto, J. Maeda, and N. A. K. A. A. K. I. R. A. Tsukahara. "Physiological identification of midbrain neurons related to lens accommodation in cats." *Journal of Neurophysiology* 52, no. 5 (1984): 870–878.

Bando, Takehiko, Haruo Toda, and Takeo Awaji. "Lens accommodation-related and pupil-related units in the lateral suprasylvian area in cats." *Progress in Brain Research* 75 (1988): 231–236.

Barlow, H. B., Colin Blakemore, and John D. Pettigrew. "The neural mechanism of binocular depth discrimination." *The Journal of Physiology* 193, no. 2 (1967): 327.

Barlow, H. B., Roo Fitzhugh, and S. W. Kuffler. "Change of organization in the receptive fields of the cat's retina during dark adaptation." *The Journal of Physiology* 137, no. 3 (1957): 338.

Barlow, H. B., and W. R. Levick. "Changes in the maintained discharge with adaptation level in the cat retina." *The Journal of Physiology* 202, no. 3 (1969): 699–718.

Beresford, W. A. "Fibre degeneration following lesions of the visual cortex of the cat." In *Neurophysiologie und Psychophysik des Visuellen Systems/The Visual System: Neurophysiology and Psychophysics: Symposion Freiburg/Br., 28.8.–3.9. 1960*, R. Jung, and H. Kornhuber (eds.) pp. 247–255. Berlin Heidelberg: Springer, 1961.

Berlucchi, G., G. S. Brindley, B. Brooks, O. D. Creutzfeldt, E. Dodt, R. W. Doty, H-J. Freund *et al.* "Projection of optic pathways to the visual cortex." In *Visual Centers in the Brain* Volumes 1536, 7 / 3 / 3B, pp. 247–268, Springer Berlin, Heidelberg, 1973.

Bilge, M., A. Bingle, K. N. Seneviratne, and D. Whitteridge. "A map of the visual cortex in the cat." *The Journal of Physiology* 191, no. 2 (1967): 116P–118P.

Bishop, P. O., J. S. Coombs, and G. H. Henry. "Responses to visual contours: spatio-temporal aspects of excitation in the receptive fields of simple striate neurones." *The Journal of Physiology* 219, no. 3 (1971): 625–657.

Bito, L. Z., P. L. Kaufman, M. Neider, O. C. Miranda, and P. Antal. "The dynamics of accommodation (ciliary muscle contraction, zonular relaxation and lenticular deformation) as a function of stimulus strength and age in iridectomized rhesus eyes." *Investigative Ophthalmology & Visual Science* 28, no. ARVO suppl (1987): 318.

Blake, Randolph. "Spatial vision in the cat." In *Frontiers in Visual Science: Proceedings of the University of Houston College of Optometry Dedication Symposium, Houston, Texas, USA, March, 1977*, vol. 8, pp. 209–219. Berlin, Heidelberg: Springer Berlin Heidelberg, 1978.

Blakemore, Colin. "The conditions required for the maintenance of binocularity in the kitten's visual cortex." *The Journal of Physiology* 261, no. 2 (1976): 423–444.

Blakemore, Colin, Adriana Fiorentini, and Lamberto Maffei. "A second neural mechanism of binocular depth discrimination." *The Journal of Physiology* 226, no. 3 (1972): 725–749.

Bonds, Alfred Bryan. "Optical quality of the living cat eye." *The Journal of Physiology* 243, no. 3 (1974): 777–795.

Bonds, A. B., and R. D. Freeman. "Development of optical quality in the kitten eye." *Vision Research* 18, no. 4 (1978): 391–398.

Boycott, B. B., and H. Kolb. "The connections between bipolar cells and photoreceptors in the retina of the domestic cat." *Journal of Comparative Neurology* 148, no. 1 (1973): 91–114.

Boycott, B. B., and H. Wässle. "The morphological types of ganglion cells of the domestic cat's retina." *The Journal of Physiology* 240, no. 2 (1974): 397–419.

Brown, J. E., and Diane Major. "Cat retinal ganglion cell dendritic fields." *Experimental Neurology* 15, no. 1 (1966): 70–78.

Bullier, Jean H., and Thomas T. Norton. "Receptive-field properties of X-, Y-and intermediate cells in the cat lateral geniculate nucleus." *Brain Research* 121, no. 1 (1977): 151–156.

Burke, William, and Andrew Malcolm Cole. "Extraretinal influences on the lateral geniculate nucleus." *Reviews of Physiology, Biochemistry and Pharmacology* 80 (1978): 105–166.

Burns, B. Delisle, and R. Pritchard. "Cortical conditions for fused binocular vision." *The Journal of Physiology* 197, no. 1 (1968): 149–171.

Campbell, F. W., L. Maffei, and M. Piccolino. "The contrast sensitivity of the cat." *The Journal of Physiology* 229, no. 3 (1973): 719.

Clare, Margaret H., and G. H. Bishop. "Responses from an association area secondarily activated from optic cortex." *Journal of Neurophysiology* 17, no. 3 (1954): 271–277.

Cleland, B. G., M. W. Dubin, and W. R. Levick. "Sustained and transient neurones in the cat's retina and lateral geniculate nucleus." *The Journal of Physiology* 217, no. 2 (1971): 473–496.

Cleland, B. G., and Christina Enroth-Cugell. "Quantitative aspects of sensitivity and summation in the cat retina." *The Journal of Physiology* 198, no. 1 (1968): 17–38.

Cleland, B. G., W. R. Levick, R. Morstyn, and H. G. Wagner. "Lateral geniculate relay of slowly conducting retinal afferents to cat visual cortex." *The Journal of Physiology* 255, no. 1 (1976): 299–320.

Cleland, B. G., W. R. Levick, and K. J. Sanderson. "Properties of sustained and transient ganglion cells in the cat retina." *The Journal of Physiology* 228, no. 3 (1973): 649–680.

Coles, J. A. "Some reflective properties of the tapetum lucidum of the cat's eye." *The Journal of Physiology* 212, no. 2 (1971): 393–409.

Cronin, Thomas W. "Optical design and evolutionary adaptation in crustacean compound eyes." *Journal of Crustacean Biology* 6, no. 1 (1986): 1–23.

Cynader, Max, and D. Regan. "Neurones in cat parastriate cortex sensitive to the direction of motion in three-dimensional space." *The Journal of Physiology* 274, no. 1 (1978): 549–569.

Daniels, J. D., Joyce L. Norman, and J. D. Pettigrew. "Biases for oriented moving bars in lateral geniculate nucleus neurons of normal and stripe-reared cats." *Experimental Brain Research* 29 (1977): 155–172.

De Robertis, E. M., and Yoshiki Sasai. "A common plan for dorsoventral patterning in Bilateria." *Nature* 380, no. 6569 (1996): 37–40.

Dimitrova, Diana M., Brian L. Allman, Mary S. Shall, and Stephen J. Goldberg. "Polyneuronal innervation of single muscle fibers in cat eye muscle: inferior oblique." *Journal of Neurophysiology* 101, no. 6 (2009): 2815–2821.

Doughty, Michael J. *Comparative Anatomy and Physiology of the Cornea and Conjunctiva.* Baton Rouge, LA: CRC Press, 2012.

Downie, Laura E., Stefan Bandlitz, Jan P. Bergmanson, Jennifer P. Craig, Debarun Dutta, Carole Maldonado-Codina, William Ngo, Jaya S. Siddireddy, and James S. Wolffsohn. "BCLA CLEAR - anatomy and physiology of the anterior eye." *Contact Lens and Anterior Eye 44*, no. 2 (2021): 132–156. Accessed December 3, 2023. https://doi.org/10.1016/j.clae.2021.02.009.

Dua, Harminder S., and Augusto Azuara-Blanco. "Limbal stem cells of the corneal epithelium." *Survey of Ophthalmology 44*, no. 5 (2000): 415–425. Accessed December 5, 2023. https://doi.org/10.1016/S0039-6257(00)00109-0.

Duffy, Frank H., S. Robert Snodgrass, James L. Burchfiel, and Janet L. Conway. "Bicuculline reversal of deprivation amblyopia in the cat." *Nature* 260, no. 5548 (1976): 256–257.

Duke-Elder, Stewart. *System of Ophthalmology Vol. 1 The Eye in Evolution*. Henry Kimpton, 1958.

Ebner, Ford F., and Ronald E. Myers. "Distribution of corpus callosum and anterior commissure in cat and raccoon." *Journal of Comparative Neurology* 124, no. 3 (1965): 353–365.

Enroth-Cugell, Christina, and P. Lennie. "The control of retinal ganglion cell discharge by receptive field surrounds." *The Journal of Physiology* 247, no. 3 (1975): 551–578.

Enroth-Cugell, Christina, and John G. Robson. "The contrast sensitivity of retinal ganglion cells of the cat." *The Journal of Physiology* 187, no. 3 (1966): 517–552.

Enroth-Cugell, Christina, and R. M. Shapley. "Flux, not retinal illumination, is what cat retinal ganglion cells really care about." *The Journal of Physiology* 233, no. 2 (1973): 311–326.

Erclik, Ted, Volker Hartenstein, Roderick R. McInnes, and Howard D. Lipshitz. "Eye evolution at high resolution: the neuron as a unit of homology." *Developmental Biology* 332, no. 1 (2009): 70–79.

Eysel, U. Th, O -J. Grüsser, and K -P. Hoffmann. "Monocular deprivation and the signal transmission by X-and Y-neurons of the cat lateral geniculate nucleus." *Experimental Brain Research* 34 (1979): 521–539.

Flammer, Josef, Katarzyna Konieczka, Rosa M. Bruno, Agostino Virdis, Andreas J. Flammer, and Stefano Taddei. "The Eye and the Heart." *European Heart Journal 34*, no. 17 (2013): 1270–1278. Accessed December 6, 2023. https://doi.org/10.1093/eurheartj/eht023.

Fukuda, Yutaka, and Jonathan Stone. "Retinal distribution and central projections of Y-, X-, and W-cells of the cat's retina." *Journal of Neurophysiology* 37, no. 4 (1974): 749–772.

Fukuda, Yutaka, and Jonathan Stone. "Evidence of differential inhibitory influences on X-and Y-type relay cells in the cat's lateral geniculate nucleus." *Brain Research* 113, no. 1 (1976): 188–196.

Gallar, Juana, M. Carmen Acosta, and Carlos Belmonte. "Activation of scleral cold thermoreceptors by temperature and blood flow changes." *Investigative Ophthalmology & Visual Science* 44, no. 2 (2003): 697–705.

Garey, L. J., and Colin Blakemore. "Monocular deprivation: morphological effects on different classes of neurons in the lateral geniculate nucleus." *Science* 195, no. 4276 (1977a): 414–416.

Garey, L. J., and C. Blakemore. "The effects of monocular deprivation on different neuronal classes in the lateral geniculate nucleus of the cat." *Experimental Brain Research* 28 (1977b): 259–278.

Garey, L. J., and T. P. Powell. "The projection of the retina in the cat." *Journal of Anatomy* 102, no. Pt 2 (1968): 189.

Garey, L. J., and Thomas Philip Stroud Powell. "The projection of the lateral geniculate nucleus upon the cortex in the cat." *Proceedings of the Royal Society of London. Series B. Biological Sciences* 169, no. 1014 (1967): 107–126.

Gehring, Walter J. "Chance and necessity in eye evolution." *Genome Biology and Evolution* 3 (2011): 1053–1066.

Gilbert, C. D., and J. P. Kelly. "The projections of cells in different layers of the cat's visual cortex." *Journal of Comparative Neurology* 163, no. 1 (1975): 81–105.

Gipson, Ilene K., and Mary Ann Stepp. "Anatomy and cell biology of the cornea, superficial limbus, and conjunctiva." *Albert and Jakobiec's Principles and Practice of Ophthalmology* (2022): 3–30.

Gislén, Anna, Marie Dacke, Ronald HH Kröger, Maths Abrahamsson, Dan-Eric Nilsson, and Eric J. Warrant. "Superior underwater vision in a human population of sea gypsies." *Current Biology* 13, no. 10 (2003): 833–836.

Giudice, Chiara, Laura Nordio, Micol Cadonici, Maria N. Perelli, and Mario Caniatti. "Epithelial lacrimal gland tumors in dogs and cats: is the human WHO classification appropriate for animals?" *Veterinary Pathology* (2021). Accessed December 6, 2023. https://doi.org/10.1177/03009858211025791.

Glasser, Adrian, and Howard C. Howland. "A history of studies of visual accommodation in birds." *The Quarterly Review of Biology* 71, no. 4 (1996): 475–509.

Goodchild, Ann K., Krishna K. Ghosh, and Paul R. Martin. "Comparison of photoreceptor spatial density and ganglion cell morphology in the retina of human, macaque monkey, cat, and the marmoset Callithrix jacchus." *Journal of Comparative Neurology* 366, no. 1 (1996): 55–75.

Guillery, R. W. "A study of Golgi preparations from the dorsal lateral geniculate nucleus of the adult cat." *Journal of Comparative Neurology* 128, no. 1 (1966): 21–49.

Guillery, R. W. "The laminar distribution of retinal fibers in the dorsal lateral geniculate nucleus of the cat: a new interpretation." *Journal of Comparative Neurology* 138, no. 3 (1970): 339–367.

Hammond, P. "Spatial organization of receptive fields of LGN neurones." *The Journal of Physiology* 222, no. 1 (1972): 53P–54P.

Hammond, P. "Contrasts in spatial organization of receptive fields at geniculate and retinal levels: centre surround and outer surround." *The Journal of Physiology* 228, no. 1 (1973): 115–137.

Hammond, P., and D. P. Andrews. "Orientation tuning of cells in areas 17 and 18 of the cat's visual cortex." *Experimental Brain Research* 31 (1978): 341–351.

Hammond, P., D. P. Andrews, and C. R. James. "Invariance of orientational and directional tuning in visual cortical cells of the adult cat." *Brain Research* (1975).

Hammond, P., and D. M. MacKay. "Differential responsiveness of simple and complex cells in cat striate cortex to visual texture." *Experimental Brain Research* 30 (1977): 275–296.

Hayhow, W. Ro. "The cytoarchitecture of the lateral geniculate body in the cat in relation to the distribution of crossed and uncrossed optic fibers." *Journal of Comparative Neurology* 110, no. 1 (1958): 1–63.

Henry, Geoffrey H. "Receptive field classes of cells in the striate cortex of the cat." *Brain Research* 133, no. 1 (1977): 1–28.

Henry, G. H., P. O. Bishop, and B. Dreher. "Orientation, axis and direction as stimulus parameters for striate cells." *Vision Research* 14, no. 9 (1974): 767–777.

Henry, G. H., P. O. Bishop, R. M. Tupper, and B. Dreher. "Orientation of specificity and response variability of cells in the striate cortex." *Vision Research* 13 (1973): 1771–1779.

Hirsch, Helmut V. B., and D. N. Spinelli. "Visual experience modifies distribution of horizontally and vertically oriented receptive fields in cats." *Science* 168, no. 3933 (1970): 869–871.

Hoffmann, K-P., and Jonathan Stone. "Conduction velocity of afferents to cat visual cortex: a correlation with cortical receptive field properties." *Brain Research* 32, no. 2 (1971): 460–466.

Hoffmann, K. P., J. Stone, and S. Murray Sherman. "Relay of receptive-field properties in dorsal lateral geniculate nucleus of the cat." *Journal of Neurophysiology* 35, no. 4 (1972): 518–531.

Holländer, Horstmar. "The projection from the visual cortex to the lateral geniculate body (LGB) an experimental study with silver impregnation methods in the Cat." *Experimental Brain Research* 10 (1970): 219–235.

Hollaänder, Horstmar, and Horacio Vanegas. "The projection from the lateral geniculate nucleus onto the visual cortex in the cat. A quantitative study with horseradish-peroxidase." *Journal of Comparative Neurology* 173, no. 3 (1977): 519–536.

Honrubia, Francisco M., and James H. Elliott. "Dendritic fields of the retinal ganglion cells in the cat." *Archives of Ophthalmology* 84, no. 2 (1970): 221–226.

Howland, Howard C., and Jacob G. Sivak. "Penguin vision in air and water." *Vision Research* 24, no. 12 (1984): 1905–1909.

Hubel, David H., and Torsten N. Wiesel. "Integrative action in the cat's lateral geniculate body." *The Journal of Physiology* 155, no. 2 (1961): 385.

Hubel, David H., and Torsten N. Wiesel. "Receptive fields, binocular interaction and functional architecture in the cat's visual cortex." *The Journal of Physiology* 160, no. 1 (1962): 106.

Hubel, David H., and Torsten N. Wiesel. "Shape and arrangement of columns in cat's striate cortex." *The Journal of Physiology* 165, no. 3 (1963): 559.

Hubel, David H., and Torsten N. Wiesel. "Receptive fields and functional architecture in two nonstriate visual areas (18 and 19) of the cat." *Journal of Neurophysiology* 28, no. 2 (1965a): 229–289.

Hubel, David H., and Torsten N. Wiesel. "Binocular interaction in striate cortex of kittens reared with artificial squint." *Journal of Neurophysiology* 28, no. 6 (1965b): 1041–1059.

Hubel, David H., and Torsten N. Wiesel. "Cortical and callosal connections concerned with the vertical meridian of visual fields in the cat." *Journal of Neurophysiology* 30, no. 6 (1967): 1561–1573.

Hubel, David H., and Torsten N. Wiesel. "Visual area of the lateral suprasylvian gyrus (Clare—Bishop area) of the cat." *The Journal of Physiology* 202, no. 1 (1969): 251–260.

Hughes, Auatin. "A quantitative analysis of the cat retinal ganglion cell topography." *Journal of Comparative Neurology* 163, no. 1 (1975): 107–128.

Ikeda, Hisako, and M. J. Wright. "Receptive field organization of 'sustained'and 'transient'retinal ganglion cells which subserve different functional roles." *The Journal of Physiology* 227, no. 3 (1972): 769–800.

Imbert, Michel, and Pierre Buisseret. "Receptive field characteristics and plastic properties of visual cortical cells in kittens reared with or without visual experience." *Experimental Brain Research* 22 (1975): 25–36.

Kalil, Ronald E., and Ronald Chase. "Corticofugal influence on activity of lateral geniculate neurons in the cat." *Journal of Neurophysiology* 33, no. 3 (1970): 459–474.

Kappauf, William E. "Variation in the size of the cat's pupil as a function of stimulus brightness." *Journal of Comparative Psychology* 36, no. 2 (1943): 125.

Kasamatsu, Takuji, and John D. Pettigrew. "Depletion of brain catecholamines: failure of ocular dominance shift after monocular occlusion in kittens." *Science* 194, no. 4261 (1976): 206–209.

Katzir, Gadi, and Howard C. Howland. "Corneal power and underwater accommodation in great cormorants (Phalacrocorax carbo sinensis)." *Journal of Experimental Biology* 206, no. 5 (2003): 833–841.

Kirk, D. L., W. R. Levick, B. G. Cleland, and H. Wässle. "Crossed and uncrossed representation of the visual field by brisk-sustained and brisk-transient cat retinal ganglion cells." *Vision Research* 16, no. 3 (1976): 225–231.

Kolb, Helga. "The connections between horizontal cells and photoreceptors in the retina of the cat: electron microscopy of Golgi preparations." *Journal of Comparative Neurology* 155, no. 1 (1974): 1–14.

Kolb, Helga, and E. V. Famiglietti. "Rod and cone pathways in the inner plexiform layer of cat retina." *Science* 186, no. 4158 (1974): 47–49.

Kozak, W., R. W. Rodieck, and P. O. Bishop. "Responses of single units in lateral geniculate nucleus of cat to moving visual patterns." *Journal of Neurophysiology* 28, no. 1 (1965): 19–47.

Kratz, Kenneth E., and Peter D. Spear. "Postcritical-period reversal of effects of monocular deprivation on striate cortex cells in the cat." *Journal of Neurophysiology* 39, no. 3 (1976): 501–511.

Kruse, Friedrich E., J. J. Chen, R. J. Tsai, and S. C. Tseng. "Conjunctival transdifferentiation is due to the incomplete removal of limbal basal epithelium." *Investigative Ophthalmology & Visual Science* 31, no. 9 (1990): 1903–1913.

Kuffler, Stephen W. "Discharge patterns and functional organization of mammalian retina." *Journal of Neurophysiology* 16, no. 1 (1953): 37–68.

Kur, Joanna, Eric A. Newman, and Tailoi Chan-Ling. "Cellular and physiological mechanisms underlying blood flow regulation in the retina and choroid in health and disease." *Progress in Retinal and Eye Research* 31, no. 5 (2012): 377–406.

Land, Michael F. "The optical structures of animal eyes." *Current Biology* 15, no. 9 (2005): R319–R323.

Land, Michael F., and Dan-Eric Nilsson. *Animal Eyes*. Oxford: Oxford University Press, 2012.

Laties, Alan M., and James M. Sprague. "The projection of optic fibers to the visual centers in the cat." *Journal of Comparative Neurology* 127, no. 1 (1966): 35–70.

Leicester, Jonathan, and Jonathan Stone. "Ganglion, amacrine and horizontal cells of the cat's retina." *Vision Research* 7, no. 9–10 (1967): 695–IN3.

LeVay, Simon, and David Ferster. "Relay cell classes in the lateral geniculate nucleus of the cat and the effects of visual deprivation." *Journal of Comparative Neurology* 172, no. 4 (1977): 563–584.

LeVay, Simon, Michael P. Stryker, and Carla J. Shatz. "Ocular dominance columns and their development in layer IV of the cat's visual cortex: a quantitative study." *Journal of Comparative Neurology* 179, no. 1 (1978): 223–244.

Levy, Brian, and J. G. Sivak. "Mechanisms of accommodation in the bird eye." *Journal of comparative Physiology* 137, no. 3 (1980): 267–272.

Lin, Chia-Sheng, and S. Murray Sherman. "Effects of early monocular eyelid suture upon development of relay cell classes in the cat's lateral geniculate nucleus." *Journal of Comparative Neurology* 181, no. 4 (1978): 809–831.

Litwiler, Tricia L., and Thomas W. Cronin. "No evidence of accommodation in the eyes of the bottlenose dolphin, Tursiops truncatus." *Marine Mammal Science* 17, no. 3 (2001): 508–525.

Lund, Jennifer S. "Local circuit neurons of macaque monkey striate cortex: I. Neurons of laminae 4C and 5A." *Journal of Comparative Neurology* 257, no. 1 (1987): 60–92.

Machovsky-Capuska, Gabriel E., Howard C. Howland, David Raubenheimer, Robin Vaughn-Hirshorn, Bernd Würsig, Mark E. Hauber, and Gadi Katzir. "Visual accommodation and active pursuit of prey underwater in a plunge-diving bird: the Australasian gannet." *Proceedings of the Royal Society B: Biological Sciences* 279, no. 1745 (2012): 4118–4125.

Maffei, L., and A. Fiorentini. "Retinogeniculate convergence and analysis of contrast." *Journal of Neurophysiology* 35, no. 1 (1972): 65–72.

Maffei, Lamberto, and Adriana Fiorentini. "The visual cortex as a spatial frequency analyser." *Vision Research* 13, no. 7 (1973): 1255–1267.

Martin, Graham R. "Eye structure and amphibious foraging in albatrosses." *Proceedings of the Royal Society of London. Series B: Biological Sciences* 265, no. 1397 (1998): 665–671.

McKee, Hamish D., Luciane C. Irion, Fiona M. Carley, Arun K. Brahma, Mohammad R. Jafarinasab, Mohsen Rahmati-Kamel, Mozhgan R. Kanavi, and Sepehr Feizi. "Re: Dua Et Al.: Human corneal anatomy redefined: a novel pre-descemet layer (dua's layer) (ophthalmology 2013;120:1778–85)." *Ophthalmology 121*, no. 5 (2014): e24–e25. Accessed December 5, 2023. https://doi.org/10.1016/j.ophtha.2013.12.021.

McIlwain, James T. "Receptive fields of optic tract axons and lateral geniculate cells: peripheral extent and barbiturate sensitivity." *Journal of Neurophysiology* 27, no. 6 (1964): 1154–1173.

Meek, Keith M., and Carlo Knupp. "Corneal structure and transparency." *Progress in Retinal and Eye Research* 49 (2015): 1–16. Accessed December 5, 2023. https://doi.org/10.1016/j.preteyeres.2015.07.001.

Montero, V. M., and R. W. Guillery. "Degeneration in the dorsal lateral geniculate nucleus of the rat following interruption of the retinal or cortical connections." *Journal of Comparative Neurology* 134, no. 2 (1968): 211–241.

Morris, Joshua, Albert Cardona, Maria Del Mar De Miguel-Bonet, and Volker Hartenstein. "Neurobiology of the basal platyhelminth Macrostomum lignano: map and digital 3D model of the juvenile brain neuropile." *Development Genes and Evolution* 217 (2007): 569–584.

Movshon, J. Anthony. "The velocity tuning of single units in cat striate cortex." *The Journal of Physiology* 249, no. 3 (1975): 445–468.

Movshon, J. Anthony, Ian D. Thompson, and David J. Tolhurst. "Receptive field organization of complex cells in the cat's striate cortex." *The Journal of Physiology* 283, no. 1 (1978a): 79–99.

Movshon, J. Anthony, I. D. Thompson, and D. J. Tolhurst. "Spatial and temporal contrast sensitivity of neurones in areas 17 and 18 of the cat's visual cortex." *The Journal of Physiology* 283, no. 1 (1978b): 101–120.

Murube, Juan. "The Origin of Tears. The Aqueo-serous Component in the XIX and XX Centuries." *The Ocular Surface* 10, no. 2 (2012): 56–66. Accessed December 6, 2023. https://doi.org/10.1016/j.jtos.2012.01.006.

Neider, Michael W., Kathryn Crawford, Paul L. Kaufman, and Laszlo Z. Bito. "In vivo videography of the rhesus monkey accommodative apparatus: age-related loss of ciliary muscle response to central stimulation." *Archives of Ophthalmology* 108, no. 1 (1990): 69–74.

Nelson, J. I., H. Kato, and P. O. Bishop. "Discrimination of orientation and position disparities by binocularly activated neurons in cat straite cortex." *Journal of Neurophysiology* 40, no. 2 (1977): 260–283.

Niimi, Kahee, and James M. Sprague. "Thalamo-cortical organization of the visual system in the cat." *Journal of Comparative Neurology* 138, no. 2 (1970): 219–249.

Nikara, T., P. O. Bishop, and J. D. Pettigrew. "Analysis of retinal correspondence by studying receptive fields of rinocular single units in cat striate cortex." *Experimental Brain Research* 6 (1968): 353–372.

Ogawa, Tetsuro, Toru Takimori, and Yasuo Takahashi. "Intracellular recording and staining of cat's lateral geniculate neurons." *Brain Research* 139, no. 1 (1978): 35–42.

Ohtsuka, K., and A. Sato. "Neuronal connectivity between accommodative and active fixation systems." In *Accommodation and Vergence Mechanisms in the Visual System* O. Franzén, H. Richter, and L. Stark (eds.), pp. 21–30. Basel: Birkhäuser Basel, 2000.

Orban, Guy A., Henry Kennedy, and Hugo Maes. "Influence of eccentricity on velocity characteristics of area 18 neurones in the cat." *Brain Research* 159, no. 2 (1978): 391–395.

Otsuka, Ryosaku, and Rolf Hassler. "Über Aufbau und Gliederung der corticalen Sehsphäre bei der Katze." *Archiv für Psychiatrie und Nervenkrankheiten* 203 (1962): 212–234.

Palmer, L. A., A. C. Rosenquist, and R. J. Tusa. "The retinotopic organization of lateral suprasylvian visual areas in the cat." *Journal of Comparative Neurology* 177, no. 2 (1978): 237–256.

Pettigrew, J. D. "Binocular neurones which signal change of disparity in area 18 of cat visual cortex." *Nature New Biology* 241, no. 108 (1973): 123–124.

Pettigrew, J. D. "The effect of visual experience on the development of stimulus specificity by kitten cortical neurones." *The Journal of Physiology* 237, no. 1 (1974): 49–74.

Pettigrew, J. D., T. Nikara, and P. O. Bishop. "Responses to moving slits by single units in cat striate cortex." *Experimental Brain Research* 6 (1968a): 373–390.

Pollen, D. A., and S. F. Ronner. "Periodic excitability changes across the receptive fields of complex cells in the striate and parastriate cortex of the cat." *The Journal of Physiology* 245, no. 3 (1975): 667–697.

Ramón y Cajal, Santiago, and Domingo Sanchez. "Contribución al conocimiento de los centros nerviosos de los insectos." *Madrid: Imprenta de Hijos de Nicolás Moya* 2 (1915): 161–164.

Rodieck, Robert W. "Quantitative analysis of cat retinal ganglion cell response to visual stimuli." *Vision Research* 5, no. 12 (1965): 583–601.

Rodieck, Robert W., and Jonathan Stone. "Analysis of receptive fields of cat retinal ganglion cells." *Journal of Neurophysiology* 28, no. 5 (1965): 833–849.

Rohen, Johannes W., Paul L. Kaufman, Michael Eichhorn, Patrick A. Goeckner, and Laszlo Z. Bito. "Functional morphology of accommodation in the raccoon." *Experimental eye Research* 48, no. 4 (1989): 523–537.

Rose, David, and Colin Blakemore. "An analysis of orientation selectivity in the cat's visual cortex." *Experimental Brain Research* 20 (1974): 1–17.

Rutishauser, Rolf, and Philip Moline. "Evo-devo and the search for homology ("sameness") in biological systems." *Theory in Biosciences* 124 (2005): 213–241.

Sambasivan, Ramkumar, Barbara Gayraud-Morel, Gérard Dumas, Clémire Cimper, Sylvain Paisant, Robert G. Kelly, and Shahragim Tajbakhsh. "Distinct regulatory cascades govern extraocular and pharyngeal arch muscle progenitor cell fates." *Developmental Cell* 16, no. 6 (2009): 810–821.

Sambasivan, Ramkumar, Shigeru Kuratani, and Shahragim Tajbakhsh. "An eye on the head: the development and evolution of craniofacial muscles." *Development* 138, no. 12 (2011): 2401–2415.

Sanderson, K. J., P. O. Bishop, and I. Darian-Smith. "The properties of the binocular receptive fields of lateral geniculate neurons." *Experimental Brain Research* 13 (1971): 178–207.

Sanes, Joshua R., and S. Lawrence Zipursky. "Design principles of insect and vertebrate visual systems." *Neuron* 66, no. 1 (2010): 15–36.

Schaeffel, Frank, Leslie Farkas, and Howard C. Howland. "Infrared photoretinoscope." *Applied Optics* 26, no. 8 (1987): 1505–1509.

Schechter, P. B., and E. H. Murphy. "Brief monocular visual experience and kitten cortical binocularity." *Brain Research* 109 (1976): 165–168.

Shatz, Carla J., and Michael P. Stryker. "Ocular dominance in layer IV of the cat's visual cortex and the effects of monocular deprivation." *The Journal of Physiology* 281, no. 1 (1978): 267–283.

Shatz, Carla J., S. Lindstro, and Thorsten N. Wiesel. "The distribution of afferents representing the right and left eyes in the cat's visual cortex." *Brain Research* 131, no. 1 (1977): 103–116.

Sherman, S. Murray, R. W. Guillery, J. H. Kaas, and K. J. Sanderson. "Behavioral, electrophysiological and morphological studies of binocular competition in the development of the geniculo–cortical pathways of cats." *Journal of Comparative Neurology* 158, no. 1 (1974): 1–18.

Sherman, S. Murray, K. P. Hoffmann, and J. Stone. "Loss of a specific cell type from dorsal lateral geniculate nucleus in visually deprived cats." *Journal of Neurophysiology* 35, no. 4 (1972): 532–541.

Sherman, S. Murray, and Jonathan Stone. "Physiological normality of the retina in visually deprived cats." *Brain Research* 60, no. 1 (1973): 224–230.

Sherman, S. Murray, J. R. Wilson, and R. W. Guillery. "Evidence that binocular competition affects the postnatal development of Y-cells in the cat's lateral geniculate nucleus." *Brain Research* 100 (1975): 441–444.

Shumway, Caleb L., Mahsaw Motlagh, and Matthew Wade. "Anatomy, head and neck, eye conjunctiva." 5, no. 1 (2018): 1–18.

Singer, W. "The effect of monocular deprivation on cat parastriate cortex: asymmetry between crossed and uncrossed pathways." *Brain Research* 157, no. 2 (1978): 351–355.

Singer, Wolf, and O. D. Creutzfeldt. "Reciprocal lateral inhibition of on-and off-center neurones in the lateral geniculate body of the cat." *Experimental Brain Research* 10 (1970): 311–330.

Singer, W., and F. Tretter. "Receptive-field properties and neuronal connectivity in striate and parastriate cortex of contour-deprived cats." *Journal of Neurophysiology* 39, no. 3 (1976): 613–630.

Spear, Peter D., and Thomas P. Baumann. "Receptive-field characteristics of single neurons in lateral suprasylvian visual area of the cat." *Journal of Neurophysiology* 38, no. 6 (1975): 1403–1420.

Spinelli, D. N., and T. W. Barrett. "Visual receptive field organization of single units in the cat's visual cortex." *Experimental Neurology* 24, no. 1 (1969): 76–98.

Steinberg, Roy H., Miriam Reid, and Paula L. Lacy. "The distribution of rods and cones in the retina of the cat (Felis domesticus)." *Journal of Comparative Neurology* 148, no. 2 (1973): 229–248.

Steinmetz, Patrick RH, Rolf Urbach, Nico Posnien, Joakim Eriksson, Roman P. Kostyuchenko, Carlo Brena, Keren Guy, Michael Akam, Gregor Bucher, and Detlev Arendt. "Six3 demarcates the anterior-most developing brain region in bilaterian animals." *EvoDevo* 1 (2010): 1–9.

Sterling, Peter. "Microcircuitry of the cat retina." *Annual Review of Neuroscience* 6, no. 1 (1983): 149–185.

Stevens, John King, and George L. Gerstein. "Spatiotemporal organization of cat lateral geniculate receptive fields." *Journal of Neurophysiology* 39, no. 2 (1976a): 213–238.

Stevens, John K., and George L. Gerstein. "Interactions between cat lateral geniculate neurons." *Journal of Neurophysiology* 39, no. 2 (1976b): 239–256.

Stone, Jonathan. "Morphology and physiology of the geniculocortical synapse in the cat: The question of parallel input to the striate cortex." *Investigative Ophthalmology* 11, no. 5 (1972): 338–346.

Stone, J., and Fukuda Y. "Properties of cat retinal ganglion cells: a comparison of W-cells with X-and Y-cells." *Journal of Neurophysiology* 37, no. 4 (1974): 722–748.

Stone, Jonathan, and Sandra M. Hansen. "The projection of the cat's retina on the lateral geniculate nucleus." *The Journal of Comparative Neurology* 126, no. 4 (1966): 601–624.

Stone, Jonathan, and Klaus-Peter Hoffman. "Very slow-conducting ganglion cells in the cat's retina: a major, new functional type?." *Brain Research* 43, no. 2 (1972): 610–616.

Stone, J. O. N. A. T. H. A. N., and B. O. G. D. A. N. Dreher. "Projection of X-and Y-cells of the cat's lateral geniculate nucleus to areas 17 and 18 of visual cortex." *Journal of Neurophysiology* 36, no. 3 (1973): 551–567.

Strausfeld, Nicholas J. "The insect visual system: correspondence with vertebrates and with olfactory." *Handbook of Brain Microcircuits* (2017): 293.

Strausfeld, Nicholas J., and Frank Hirth. "Homology versus convergence in resolving transphyletic correspondences of brain organization." *Brain, Behavior and Evolution* 82, no. 4 (2013): 215–219.

Suzuki, Hisao, and Norio Taira. "Effect of reticular stimulation upon synaptic transmission in cat's lateral geniculate body." *The Japanese Journal of Physiology* 11, no. 6 (1961): 641–655.

Szentágothai, J., J. Hámori, and Therese Tömböl. "Degeneration and electron microscope analysis of the synaptic glomeruli in the lateral geniculate body." *Experimental Brain Research* 2, no. 4 (1966): 283–301.

Takahashi, Yasuhiro, Akihide Watanabe, Hiromichi Matsuda, Yasuhisa Nakamura, Takashi Nakano, Ken Asamoto, Hiroshi Ikeda, and Hirohiko Kakizaki. "Anatomy of secretory glands in the eyelid and conjunctiva: a photographic review." *Ophthalmic Plastic & Reconstructive Surgery* 29, no. 3 (2013): 215–219.

Treisman, Jessica E. "How to make an eye." *The Company of Biologist* 131, no. 16 (2004): 3823–3827.

Tusa, R. J., L. A. Palmer, and A. C. Rosenquist. "The retinotopic organization of area 17 (striate cortex) in the cat." *Journal of Comparative Neurology* 177, no. 2 (1978): 213–235.

Updyke, B. V. "The pattern of projection of cortical areas 17, 18, and 19 onto the laminae of the dorsal lateral geniculate nucleus in the cat." *Journal of Comparative Neurology* 163, no. 4 (1975): 377–395.

Vakkur, GJ, PO Bishop, and W. Kozak. "Visual optics in the cat, including posterior nodal distance and retinal landmarks." *Vision Research* 3, no. 7–8 (1963): 289–314.

Wakabayashi, Taku, Tatsuhiko Sato, Chikako Hara-Ueno, Yoko Fukushima, Kaori Sayanagi, Nobuhiko Shiraki, Miki Sawa, Yasushi Ikuno, Hirokazu Sakaguchi, and Kohji Nishida. "Retinal microvasculature and visual acuity in eyes with branch retinal vein occlusion: imaging analysis by optical coherence tomography angiography." *Investigative Ophthalmology & Visual Science* 58, no. 4 (2017): 2087–2094.

Walls, Gordon Lynn. "The vertebrate eye and its adaptive radiation." *The Journal of Nervous and Mental Disease* 100, no. 3 (1944): 332.

Watkins, David W., and Mark A. Berkley. "The orientation selectivity of single neurons in cat striate cortex." *Experimental Brain Research* 19 (1974): 433–446.

Weale, Robert Alexander. "The spectral reflectivity of the cat's tapetum measured in situ." *The Journal of Physiology* 119, no. 1 (1953): 30.

Wernet, Mathias F., Andrew D. Huberman, and Claude Desplan. "So many pieces, one puzzle: cell type specification and visual circuitry in flies and mice." *Genes & Development* 28, no. 23 (2014): 2565–2584.

Whitteridge, David, and R. Jung. *Projection of Optic Pathways to the Visual Cortex*. Vol. 7. Berlin: Springer, Handbook of Sensory Physiology, 1973.

Wiesel, T. N. "Receptive fields of ganglion cells in the cat's retina." *The Journal of Physiology* 153, no. 3 (1960): 583.

Wiesel, Torsten N., and David H. Hubel. "Effects of visual deprivation on morphology and physiology of cells in the cat's lateral geniculate body." *Journal of Neurophysiology* 26, no. 6 (1963a): 978–993.

Wiesel, Torsten N., and David H. Hubel. "Single-cell responses in striate cortex of kittens deprived of vision in one eye." *Journal of Neurophysiology* 26, no. 6 (1963b): 1003–1017.

Wilson, J. R., and S. Murray Sherman. "Receptive-field characteristics of neurons in cat striate cortex: changes with visual field eccentricity." *Journal of Neurophysiology* 39, no. 3 (1976): 512–533.

Wilson, Jo R., and S. M. Sherman. "Differential effects of early monocular deprivation on binocular and monocular segments of cat striate cortex." *Journal of Neurophysiology* 40, no. 4 (1977): 891–903.

Wilson, M. E. "Cortico-cortical connexions of the cat visual areas." *Journal of Anatomy* 102, no. Pt 3 (1968): 375.

Wilson, P. D., M. H. Rowe, and J. Stone. "Properties of relay cells in cat's lateral geniculate nucleus: a comparison of W-cells with X-and Y-cells." *Journal of Neurophysiology* 39, no. 6 (1976): 1193–1209.

Zeki, Semir. "A century of cerebral achromatopsia." *Brain* 113, no. 6 (1990): 1721–1777.

10 Physiology of the Auditory System

Tanmay Mondal and Gurpreet Singh

Department of Veterinary Physiology and
Biochemistry, College of Veterinary Science, Rampura Phul
Guru Angad Dev Veterinary and Animal Sciences University, India

10.1 INTRODUCTION

The intricate physiology of hearing varies widely across the animal kingdom, prompting a nuanced exploration of auditory capabilities. Among humans, birds, and select mammals, the ability to hear is well-documented, enabling them to articulate sensations and respond to trained sounds. However, the perception of sound in other vertebrates and invertebrates remains a subject of debate. The presence of a histologic apparatus similar to that of humans suggests potential auditory experiences in some species, while others may interpret sound as tactile or vibratory stimuli. Recent studies challenging assumptions about fish hearing, particularly by von Frisch, underscore the complexity of auditory phenomena. Furthermore, the presence of a cochlea in amphibians adds to the ambiguity, as some reptiles with auditory structures display limited sound responsiveness. In contrast, birds and mammals, exemplified by dogs and cats, demonstrate not only auditory sensitivity but also the ability to distinguish subtle variations in tone. As this comparative examination unfolds, the intricate relationship between anatomy, behavior, and sound perception becomes increasingly apparent, urging a deeper understanding of the diverse auditory systems across the animal kingdom.

10.2 THE SOUND RECEIVING APPARATUS

10.2.1 The Pinna

The auricle, also known as the pinna, constitutes the visible part of the outer ear, serving as the initial collector of sound waves. Its primary function is to channel these sound waves into the ear canal, officially called the external auditory meatus, where the incoming sound is amplified. As the waves progress along the ear canal, they approach the eardrum, or tympanic membrane, a flexible oval membrane situated at the canal's termination.

The outer part of the ear, referred to as the pinna, has distinctive characteristics despite its limited mobility and relatively weak muscles. Unlike certain animals capable of orienting their ears for improved sound perception, our pinna doesn't contribute significantly to enhancing our hearing or determining the source of a sound. However, the pinna does have a purpose. Even though it may seem less functional, blocking the hollow parts of the ear or altering its size can affect hearing. For instance, making the pinna larger with your hand can improve the perception of certain sounds, as larger, cup-shaped ears that project at an angle are more effective for hearing than smaller, flatter ears close to the head. Additionally, the pinna has the capability to conduct sound through specific bone structures in a unique manner.

10.2.2 The External Auditory Meatus

The external acoustic meatus, alternatively known as the external auditory canal or ear canal (Latin: meatus acusticus externus), is an air-filled tubular space that extends from the auricle of the external ear into the temporal bone, ultimately leading to the tympanic membrane. Despite its curviness, the size and length of the external auditory meatus are generally inconsequential, except for very high-pitched sounds where it may contribute to amplification. This curvy structure prevents sound waves from directly reaching any part of the eardrum. The meatus is equipped with hair and wax, providing protective functions by preventing dust and bugs from penetrating deep into the ear. Moreover, its sensitivity to touch, governed by nerves, prompts defensive or evasive reactions even to slight touches, enhancing the overall defensive mechanisms associated with this crucial part of the outer ear (Figure 10.1).

10.3 THE SOUND-CONDUCTING APPARATUS

The main function of the middle ear is seen more as a pressure regulator than a conductor of sound waves. Recent theoretical perspectives suggest that the entire middle ear, including the tympanic membrane, ossicles, and muscles, is more focused on equalizing pressures rather than transmitting sound to the inner ear. Essentially, these structures work to protect the inner ear from excessive stress caused by sound waves, contributing to a more effective hearing process. While the middle ear may also play a role in sound conduction, its destruction or removal does not result in complete deafness, emphasizing the significance of bone and tissue conduction, along with the direct excitability of the auditory nerve to sound waves.

Comparatively, some animals lack or have underdeveloped middle ear structures, such as the tympanic membrane

DOI: 10.1201/9781003426851-10

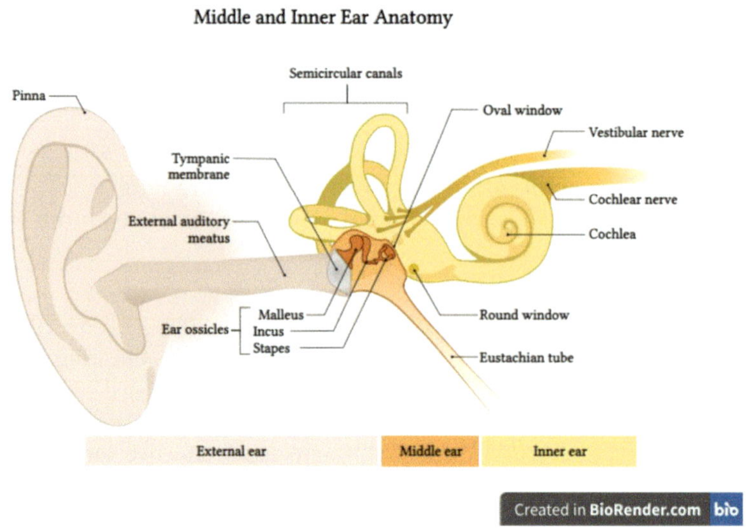

FIGURE 10.1 Structural components of the external, middle, and inner ear. The tympanic membrane separates the external ear from the middle ear.

or middle ear cavity, yet still exhibit good hearing. This discrepancy raises questions about the correlation between anatomical observations in lower animals and physiological studies. The absence or rudimentary state of these structures in amphibians, reptiles, and certain animals closer to humans might be associated with less advanced hearing abilities. Experimental evidence from studies on animals like frogs, turtles, and snakes suggests a potential lack of a true sense of hearing in species with underdeveloped middle ear structures.

10.3.1 The Tympanic Membrane

The tympanic membrane, or eardrum, possesses a compound nature attributed to its irregular funnel-like shape and the uneven length and tension of its fibers. Each of its individual radial fibers vibrates in response to specific sounds, their resonance determined by length and tension. The uncertainty surrounding whether the membrane oscillates as a whole has prompted extensive scrutiny over the past 70 years, particularly during physiological activities like swallowing and breathing, seemingly unrelated to direct hearing. Various techniques, including direct microscopy, manometry, and even photographic methods employing mirrors, have been utilized to study its movements. Despite these efforts, the results, as noted by Mangold, have not provided profound insights into explaining the unique qualities of different sounds. This has led to the intriguing suggestion that the middle ear may play a relatively minor role in the direct transmission of sound to the inner ear. Functionally, the tympanic membrane, a thin cone-shaped structure present in humans and several tetrapods, serves to transmit sound vibrations from the outer air to the ossicles in the middle ear and further to the cochlea.

10.3.2 The Ossicles

The auditory ossicles, including the malleus, incus, and stapes, function as a unit, forming a bent lever that serves to transfer and amplify air vibrations into the inner ear for processing as sound. Acting as a protective mechanism, the ossicular lever system, while debated, may prevent damage to the inner ear during forceful movements caused by loud sounds. Despite variations in opinions regarding certain protective aspects, the ossicular system's main purpose is to transmit vibrations efficiently. This lever system operates in a 3:2 ratio, significantly increasing pressure at the oval window, resulting in a thirtyfold amplification of sound. Furthermore, the ossicular system, in collaboration with the tympanic membrane, facilitates impedance matching, allowing optimal energy utilization in sound waves. The intricate interplay of the auditory ossicles and the tensor tympani and stapedius muscles ensures the delicate balance required for effective sound transmission while safeguarding the inner ear from potential damage caused by loud noises.

10.3.3 Round Window

The function of the Round Window is a subject with varying perspectives. Common ground among researchers is the acknowledgment that the membrane covering the Round Window permits the temporary displacement of perilymph, moving inward as transmitted sound waves tilt the stapes towards the organ of Corti. In the absence of the ossicular chain and equal exposure of both fenestrae to sound waves, perilymph movement would not occur. Functioning as one of the two openings from the middle ear into the inner ear, the Round Window is sealed by the secondary tympanic membrane, vibrating with an opposite phase to vibrations entering through the oval window. This movement of fluid in the cochlea stimulates hair cells on the basilar membrane, ensuring the occurrence of audition and serving as a vital communication point between the middle and inner ear.

10.4 THE SOUND-PERCEIVING APPARATUS

The inner ear has two main functions: the vestibular portion, which helps with position and balance, and the cochlear portion, which is responsible for hearing. The cochlear portion is connected to the cochlear nerve, a branch of the vestibulocochlear nerve (cranial nerve VIII), while the vestibular portion is connected to the vestibular nerve, another branch of the same cranial nerve. The inner ear is housed in a bony structure called the osseous labyrinth, which is like a complex network of passages. The cochlea, responsible for hearing, is coiled to fit within the limited space available. If it were uncoiled, it would extend into the brain.

10.4.1 Vestibular Structure and Function

The vestibular system helps us with balance and orientation. It's located in the osseous labyrinth, specifically in the vestibule and three semicircular canals (front, side, and back). Each canal is positioned at right angles to the others. There's also a cochlear part in the osseous labyrinth. Inside this labyrinth, there's a membranous structure with fluid called endolymph, which is high in potassium and low in sodium. Another fluid, perilymph, is outside the membranous labyrinth. It can circulate with cerebrospinal fluid through the vestibular aqueduct, connecting to the brain's subarachnoid space. This connection raises the risk of inner ear infections leading to meningitis (Figure 10.2).

In the vestibular portion, the membranous labyrinth has three semicircular ducts and two sacs called the utricle and saccule. The utricle connects to the saccule, and both connect to the cochlear portion. The cochlear sac, situated between layers of meninges, absorbs endolymph and allows immune cells to clear debris. Each semicircular canal has an enlarged section called an ampulla, housing equilibrium receptors called cristae. The utricle and saccule have sensory receptors called maculae, consisting of hair cells embedded in an otolithic membrane. This membrane has a gelatinous material with otoliths, calcium carbonate crystals. Due to gravity, otoliths can stress hair cells, helping sense head position. The weight of otoliths also allows sensing linear head acceleration or deceleration. Rotational acceleration or deceleration is detected by the ampullary cristae in semicircular ducts. Hair cells in these structures respond to head movements and play a role in maintaining balance and orientation.

10.4.2 Cochlear Structure and Function

The cochlea has a part called the cochlear duct or scala media, which runs through the center of the cochlea, dividing it into two parts: the scala vestibuli above and the scala tympani below. These parts are mostly separate, except for a small opening called the helicotrema at the tip of the cochlea. Inside the scala media, there are many structures known as organs of Corti. These structures convert sound waves into nerve impulses, which are then sent to the brain for us to hear. The position of each organ of Corti in the scala media, from the base (near the middle ear) to the top of the cochlea, and its individual nerve connection, determines the frequency of the sound we perceive (Figure 10.3).

Different frequencies of sound have different paths along the basilar membrane, which is the thicker base where nerve entries are located. A weak sound wave strengthens when it reaches the part of the basilar membrane with a resonant frequency similar to its own. High-frequency sounds travel a short

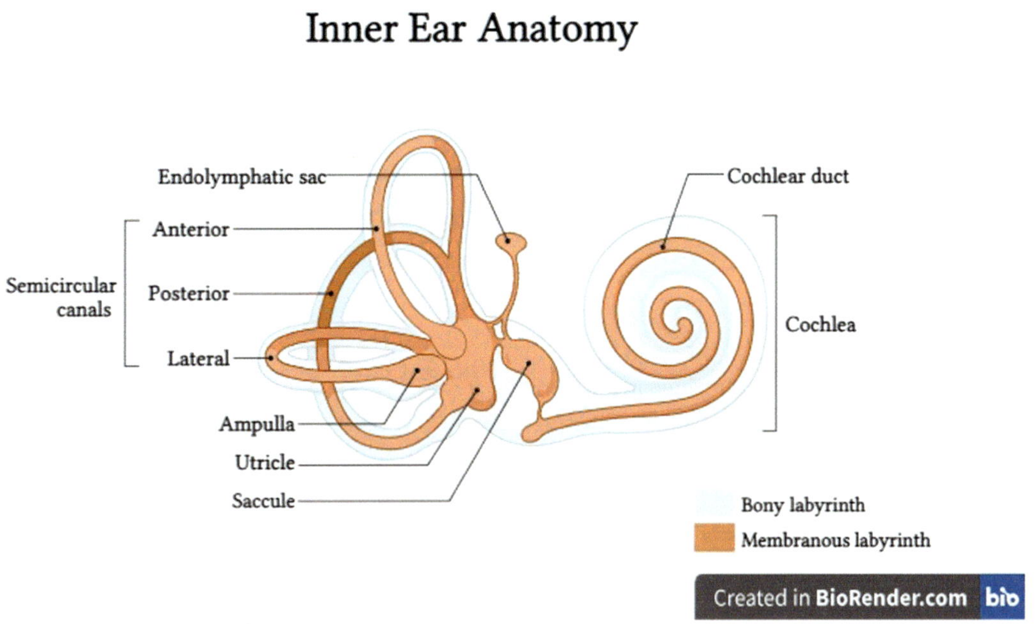

FIGURE 10.2 Structural component of the inner ear.

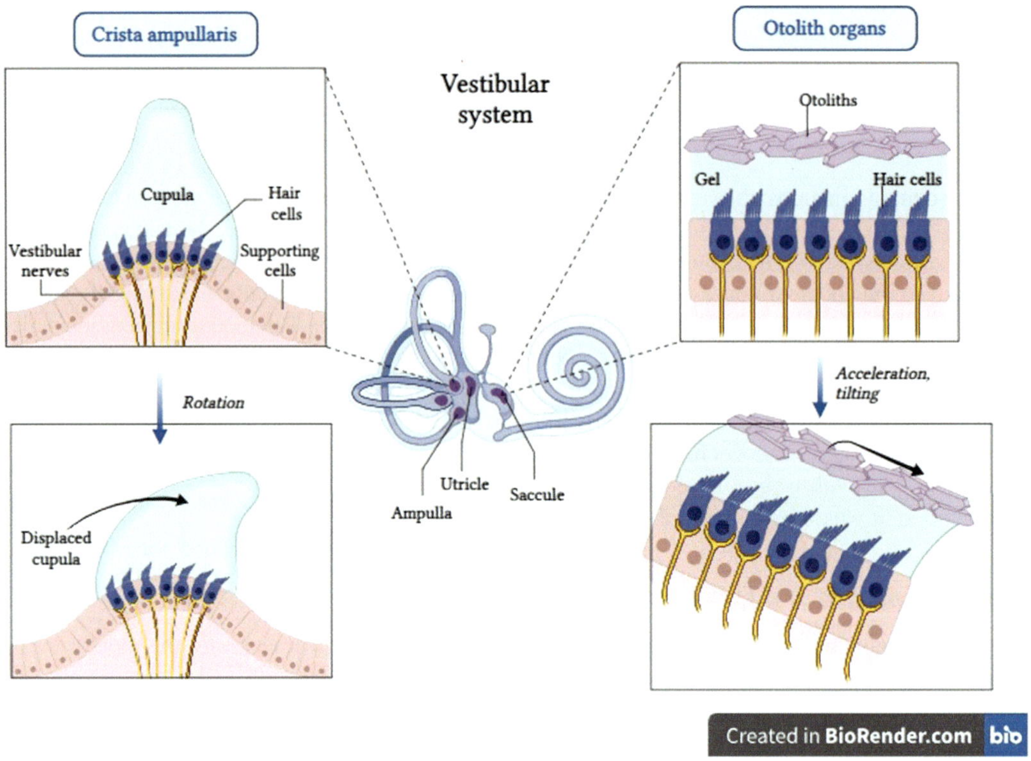

FIGURE 10.3 Structural component of the vestibular system of the inner ear.

distance before dying out, while low-frequency sounds travel a longer distance before doing the same. Every frequency between high and low is represented at different points along the basilar membrane. The organs of Corti are mainly made up of sensory receptors called hair cells with hairs reaching towards the tectorial membrane. When the basilar membrane vibrates due to sound waves, the hair cell cilia move against the tectorial membrane. This movement causes the hair cells to create a nerve impulse, which is then sent to the brain's auditory cortex through the vestibulocochlear nerve (Figure 10.4).

10.5 MECHANISM OF HEARING

The pinna, the visible part of the outer ear, helps collect and direct sound waves into the external auditory meatus (ear canal).

10.5.1 Eardrum Activation

The incoming sound waves strike the tympanic membrane or eardrum, causing it to vibrate in response to the changing air pressure.

10.5.2 Ossicle Transmission

The vibration of the eardrum is transmitted through the middle ear by three small bones known as the auditory ossicles (malleus, incus, and stapes).

The malleus (hammer) is connected to the eardrum and transfers vibrations to the incus (anvil), which, in turn, transmits them to the stapes (stirrup).

The stapes then pushes against the oval window, a membrane-covered opening that leads to the inner ear.

10.5.3 Inner Ear Activation

The movement of the stapes against the oval window sets the fluids inside the inner ear in motion, particularly the perilymph in the vestibule.

10.5.4 Perilymph Transmission

Perilymph, a fluid in the inner ear, is incompressible and thus efficiently transmits the sound wave through the scala vestibuli, a spiral-shaped chamber within the cochlea.

10.5.5 Organ of Corti Stimulation

As the sound wave travels through the cochlea, it reaches the scala media and scala tympani, stimulating the organ of Corti, a structure located along the basilar membrane in the cochlea.

The specific hair cells in the organ of Corti are activated based on the frequency and intensity of the sound wave, initiating a response.

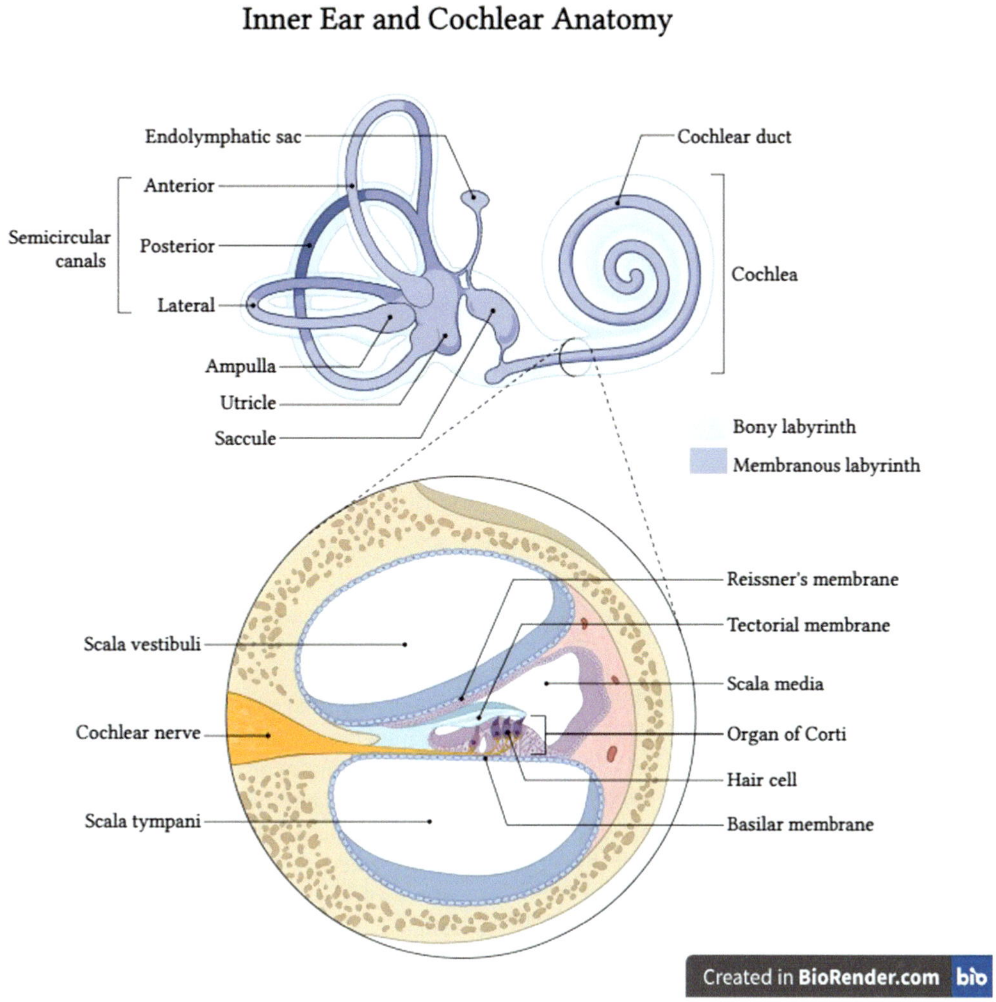

FIGURE 10.4 Structural component of the cochlear portion of the inner ear.

10.5.6 Liquid Movement Compensation

The movement of perilymph in the scala tympani is compensated by an outward movement of the cochlear (round) window, preventing excessive pressure build-up in the inner ear.

10.5.7 Hair Cell Stimulation

The stimulation of hair cells in the organ of Corti occurs as the basilar membrane vibrates in response to the fluid movement. Hair cells convert mechanical vibrations into electrical signals.

10.5.8 Nerve Transmission

Nerve impulses generated by the stimulated hair cells are transmitted via the cochlear branch of the vestibulocochlear nerve (auditory nerve) to the brain.

These electrical signals travel to the auditory cortex, where they are interpreted as sound, allowing us to perceive and understand the auditory stimulus.

BIBLIOGRAPHY

Borg, E. and Counter, S.A. 1989. The middle-ear muscles. *Scientific American*, 261(2), pp. 74–81.

Fettiplace, R. and Fuchs, P.A. 1999. Mechanisms of hair cell tuning. *Annual Review of Physiology*, 61(1), pp. 809–834.

Freeman, D.M. and Weiss, T.F. 1988. The role of fluid inertia in mechanical stimulation of hair cells. *Hearing Research*, 35(2–3), pp. 201–207.

Getty, R., Foust, H.L., Presley, E.T. and Miller, M.E. 1956. Macroscopic anatomy of the ear of the dog. *American Journal of Veterinary Research*, 17(64), pp. 364–375.

Gummer, A.W., Hemmert, W. and Zenner, H.P. 1996. Resonant tectorial membrane motion in the inner ear: its crucial role in frequency tuning. *Proceedings of the National Academy of Sciences USA*, 93, pp. 8727–8732.

Harrison, J.M. and Irving, R. 1964. Nucleus of the trapezoid body: Dual afferent innervation. *Science*, 143(3605), pp. 473–474.

Heine, P.A. 2004. Anatomy of the ear. *Veterinary Clinics: Small Animal Practice*, 34(2), pp. 379–395.

11 Physiology of the Somatosensory System

Sriti Pandey[1], Manoj Kumar Ahirwar[1], Pramod Kumar[2] and Rajesh Kumar[3]
[1]Department of Veterinary Physiology and Biochemistry, College of Veterinary Science and Animal Husbandry, Mhow, Indore, MP, India
[2]Department of Veterinary Physiology and Biochemistry, College of Veterinary Science and Animal Husbandry, Acharya Narendra Deva University of Agriculture and Technology, Ayodhya,UP, India
[3]Department of Veterinary Gynaecology and Obstetrics, College of Veterinary Science and Animal Husbandry, Acharya Narendra Deva University of Agriculture and Technology, Ayodhya,UP, India

11.1 INTRODUCTION

The somatosensory system informs the body about tactile, thermal, proprioceptive (body position, movement), and nociceptive (pain) information. This system also monitors the temperature of the body, external objects, and environment, and provides information about painful, itchy, and tickling stimuli. In general, the somatosensory systems inform us about objects in our external environment through touch (i.e., physical contact with the skin) and about the position and movement of our body parts (proprioception) through the muscle and joints.

11.2 COMPONENTS OF SOMATOSENSORY SYSTEM

The somatosensory systems process information about, and represent, several modalities of somatic sensation (i.e., pain, temperature, touch, proprioception).

11.2.1 Sensory Receptors

Receptors are specialized neuronal structures to detect diverse stimuli from the internal and external environment. Specialized sensory receptor cells (e.g., the photoreceptors of the eye) are located in specialized receptor organs, produce receptor potentials, contain synaptic specializations, and release neural transmitters. Specialized sensory receptors may be modified neurons (e.g., the photoreceptors and olfactory receptors) or modified epithelial cells (e.g., taste receptors and the auditory and vestibular hair cells). The vast majority of somatosensory receptors are formed by the endings of the somatosensory 1° afferent peripheral axon and its adjacent tissue. The adjacent tissue does not generate receptor potentials. There is only one type of sensory receptor cell in the somatosensory system, the Merkel cells, and they are found only in the skin. Instead of ending on specialized receptors, most peripheral axons of somatosensory 1° afferents travel to the skin, muscle or joint, branch near their terminal sites, and end in the skin, muscle, tendon or joint tissue. All the peripheral terminal branches of a 1° somatosensory axon end in a specific type of tissue (e.g., skin) and not in multiple types of tissue (i.e., not in skin and muscle). All the peripheral terminal branches of a 1° axon form only one type of somatosensory receptor.

Many of the 1° somatosensory afferent terminals are enveloped in a connective tissue capsule along with surrounding muscle, tendon, or cutaneous cells, or end on hair follicles. The hair follicles and the encapsulated tissue adjacent to the 1° afferent terminals (i.e., skin, muscle, tendon, and joint tissues) contain no synaptic specializations and do not generate receptor potentials or release neural transmitters. The complex of encapsulated tissue and afferent endings and the complex of hair follicle and afferent endings play a role in the receptor transduction process, and each complex is considered to form a "somatosensory receptor". Many other 1° somatosensory axons branch and terminate in skin, muscle, or joint as free nerve endings. These endings are bare of myelin, are not encapsulated, and are not associated with a specific type of tissue. The sensitivity of the receptors to specific stimuli (e.g., touch versus muscle stretch) is determined by the location of the receptor and by the non-neural tissue surrounding the 1° afferent terminal.

Receptors have the ability to convert other forms of energy into electrical energy i.e. action potential, causes transduction if adequate stimuli is applied. They have differential sensitivities to different stimuli. The permeability of the receptor is modified and ends up in graded receptor potential. They show different adaptability to a stimulus and speed of adaption varies with receptors. Each receptor has specific acuity i.e. discriminative ability.

Sensory receptors are classified based on function of the receptors, location of receptors, source of the stimulus, and type of stimulus or modality.

DOI: 10.1201/9781003426851-11

11.2.1.1 Classification of Sensory Receptors

Based on the function of the receptors

1. Phasic receptors: Produce action potential during part of stimulation hence they initiate response. It does not provide clue for duration of stimulus.
2. Tonic receptors: Generate action potential upon the application of stimulus and propagation of action potential goes on till the stimulus is removed. This gives information regarding the duration of stimulus. In this type, the possibility of weakening of the impulse irrespective of the maintenance of constant strength of stimulus due to prolonged exposure leading on to a phenomenon known as **receptor adaptation**.

Based on the location of receptors

1. Exteroceptors: Detect stimuli from outside the body in animals' micro or macro environment (change in temperature, air pressure). They can be divided into:
   1a. Superficial cutaneous sensations: Touch, light superficial pressure, temperature, cold, warmth, and pain.
   1b. Deep sensations: Receptors in the muscle tendon and joints together known as kinesthetic sensation mediated through the muscle spindle receptor and Golgi tendon organ to sense position and movement e.g. Special Sensory receptors and Skin receptors.
2. Interoceptors: Detect stimuli within the body (Blood pressure, O2 uptake). They are Proprioceptors: Convey information from the muscles, tendons, and joints, Visceroceptors: Carrying information from the viscera, Specialized receptors: Baroceptors, osmoceptors, chemoceptors, and volume receptors, Baroceptors: Present in carotid sinus and aortic arch to detect BP change, Osmoreceptors: Present in hypothalamus to detect osmotic change, and Chemoreceptors: Sensitive to chemical change. Present in tongue to detect taste and in nose to detect smell. Hypothalamus central chemoreceptors detect pH change, blood glucose, and hormonal levels.

Based on the source of the stimulus

Telereceptors: Detect the stimuli at a distance (Photo and auditory receptors)

Based on the stimulus or modality

1. Photoreceptors: Detect change in intensity of light mediating sense of vision.
2. Thermoreceptors: To detect change in temperature.
3. Electroceptors and Magnetoceptors: (Mostly in insects and birds). They detect electrical and magnetic field respectively.
4. Mechanoreceptors: Involved in sensing touch, pain, hearing, balance (equilibrium).
5. Chemoreceptors: Detect chemical signals, sense of gustation, olfaction, blood O2, tension, and change in pH (internal environment).

Based on the location of receptors sensations can also be classified as

1. Superficial cutaneous sensations: Touch, light superficial pressure, temperature, cold, warmth, and pain.
2. Deep sensations: Receptors in the muscle tendon and joints together known as kinesthetic sensation mediated through the muscle spindle receptor and Golgi tendon organ to sense position and movement.

11.2.1.2 Types of Somatosensory Receptor

11.2.1.2.1 Cutaneous Receptors

The somatosensory receptors in skin are known as the cutaneous receptors. These are classified as encapsulated receptors which include Meissner corpuscles, Pacinian corpuscles and Ruffini corpuscles. Other cutaneous receptors are unencapsulated i.e. do not end on or near specialized tissue. They may be mechanoreceptors, nociceptors or thermoreceptors. It includes hair follicle receptor (the 1° afferent ends on hair follicles) and the Merkel complex (the 1° afferent ends at the base of a specialized receptor cell called the Merkel cell) e.g. sensory receptors of the crude touch, pain and temperature.

a. Meissner Corpuscle- It is an elongated, encapsulated stack of flattened epithelial (laminar) cells found in glabrous (i.e., hairless) skin, within the dermal papillae.
b. Pacinian corpuscle – It is football-shaped, encapsulated, and contains concentrically layered epithelial (laminar) cells and in cross section it looks like a slice of onion. These receptors can be cutaneous, proprioceptive or visceral receptors, depending on their location. They are located in subcutaneous tissue beneath the dermis and in the connective tissues of bone, the body wall and body cavity.
c. Ruffini Corpuscle – It is cigar-shaped, encapsulated, and contains longitudinal strands of collagenous fibers that are continuous with the connective tissue of the skin or joint. These receptors can be cutaneous or proprioceptive receptors depending on their location. The receptors run parallel to the surface of the skin and are most sensitive to skin stretch.

d. Hair Follicle – It is an unencapsulated cutaneous receptor which run parallel to the hair shaft forming a lattice-like pattern. These receptors respond best to moving objects and signal the direction and velocity of the movement of a stimulus brushing against hairy skin.
e. Merkel Complex – It is unencapsulated cutaneous receptor consists of a specialized receptor cell, the Merkel cell, and a nerve ending, the Merkel disk. The Merkel complex is found in both hairy and non-hairy skin and is located in the basal layer of the epidermis.
f. Free Nerve Endings – Free nerve endings are found throughout the body, in skin, muscles, tendons, joints, mucous membranes, cornea, body mesentery, the dura, the viscera, etc. The free nerve endings in skin are stimulated by tissue-damaging (nociceptive) stimuli that produce the sensation of pain or by cooling of the skin or the warming of skin or by touch. Free nerve endings are considered to be the somatosensory receptors for pain, temperature and crude touch.

11.2.1.2.1.1 Mechanism of Action of Cutaneous Receptors

1. Thermoreceptors- The hypothalamus of the animals has central thermoceptors to monitor their internal temperature. Two different receptors to mediate warmth and cold.

The warmth receptors are situated near deep blood vessels. They are Ruffini end organ which can also serve to detect tactile sensation (Poly modal receptor). Ruffini's corpuscles are located in the connective tissue of the skin in association with collagen fibers. Hence detect skin stretch and are involved in proprioception. Warmth receptors detect change in temperature of 20° C to 47° C. The frequency of action potential increases with increase in the temperature.

The cold sensation is mediated by the end organ of Krause. The Cold receptors detect and are stimulated at the temperature between 10° C to 35° C. These are extremely sensitive and can detect change in even 0.5° C. There are also another thermoceptor which detect painful hot stimuli. The temperature ranges above 45° C. The frequency of action potential generated is in accordance with the amount of pain sensation.

Free nerve ending of the thermoreceptor neurons have specific thermoreceptor proteins. They are known as Thermo TRPs. The thermo TRPs are many, variable to different temperature range, and some are to detect distinct temperature. They get stimulated and act through ion channels and gating mechanism by modifying the protein configuration.

2. Nocioceptors- These detect pain sensation by naked nerve endings. This sensation does not have any special structure to mediate. These nociceptors are available in the superficial layers of dermis, which are represented by non-medullated fibers (Figure 11.1). They act as protective response to an injurious stimulus. These receptors have vibratory sense which is mediated by tactile and pressure receptors namely Merkel's discs and Pacinian corpuscles and Itching and tickling sense which is mediated by stimulation of both tactile and pain receptors (Table 11.1).

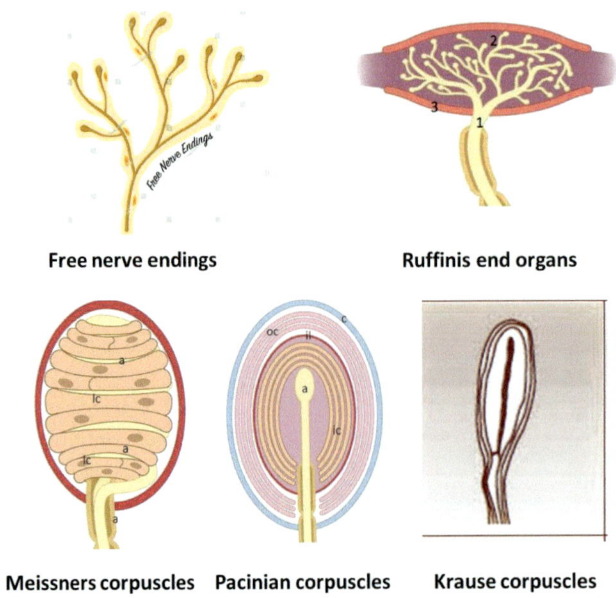

FIGURE 11.1 Cutaneous receptors of somatosensory system

TABLE 11.1
Cutaneous Receptors

Receptor	Type	Sensation	Signals	Adaptation
Meissner corpuscle	Encapsulated & layered	Touch: Flutter & Movement	Frequency/Velocity & Direction	Rapid
Pacinian corpuscle	Encapsulated & layered	Touch: Vibration	Frequency: 100-300 Hz	Rapid
Ruffini corpuscle	Encapsulated collagen	Touch: Skin Stretch	Direction & Force	Slow
Hair follicle	Unencapsulated	Touch: Movement	Direction &Velocity	Rapid
Merkel complex	Specialized epithelial cell	Touch, Pressure, Form	Location & Magnitude	Slow
Free nerve ending	Unencapsulated	Pain, Touch, or Temperature	Tissue damage, Contact, or Temperature change	Depends on information carried

11.2.1.2.2 Proprioceptive Receptors

The somatosensory receptors in muscles, tendons, joint ligaments, and joint capsules are known as proprioceptors. There are no specialized sensory receptor cells for body proprioception. In skeletal muscle, these are muscle spindles and Golgi tendon organs, as well as numerous free nerve endings. In joints, these are encapsulated endings similar to those in the skin, as well as numerous free nerve endings.

a. Muscle Spindles – They are encapsulated and consist of small muscle fibers, called intrafusal muscle fibers, and afferent and efferent nerve terminals. Muscle spindles are found in nearly all striated muscles. Muscle spindles are most numerous in muscles that carry out fine movements, such as the extraocular muscles and the intrinsic muscles of the hand. The muscle spindles are proprioceptors specialized to monitor muscle length (stretch) and signal the rate of change in muscle length by changing the discharge rate of afferent action potentials.

There are two types of intrafusal fibers. One of them is known as the nuclear bag, where their central portion is interrupted and dilated, which houses many nuclei. The nuclear bag region tapers towards the pole. Another intrafusal fiber is known as the nuclear chain. They have a uniform diameter throughout, but near the middle, a small but appreciable increase in diameter due to the increased number of cell nuclei. It is innervated by the gamma motor neurons or intrafusal fibers. The cell bodies of the gamma motor neurons are located in the ventral horn of the gray matter of the corresponding spinal segment. The skeletal muscles are innervated by α-motor neurons. These are referred to as extrafusal fibers. The gamma motor neurons are of two types: gamma plate neurons and gamma trail motor neurons.

The Gamma plate neurons innervate nuclear bag intrafusal fibers and are tonic neurons generating sustained action potential. Gamma trail motor neurons supply nuclear chain intrafusal fibers. These are phasic neurons and produce action potentials of short duration with high frequency.

b. Golgi Tendon Organ – They are encapsulated and contain intertwining collagen bundles, which are continuous with the muscle tendon, and fine branches of afferent fibers that weave between the collagen bundles. They resemble Ruffini corpuscles and are found in the tendons of striated extrafusal muscles near the muscle-tendon junction.

The Golgi tendon organ is a proprioceptor that monitors and signals muscle contraction against a force (muscle tension), whereas the muscle spindle is a proprioceptor that monitors and signals muscle stretch (muscle length).

c. Joint Receptors – These are encapsulated endings that resemble the Ruffini and Pacinian corpuscles and the Golgi tendon organs and are located within the connective tissue, capsule, and ligaments of joints.
d. Free Nerve Endings – These are abundant in muscles, tendons, joints, and ligaments. These free nerve endings are considered to be the somatosensory receptors for pain resulting from muscle, tendon, joint, or ligament damage and are not considered to be part of the proprioceptive system.

11.2.1.2.2.1 Mechanism of Action of Proprioceptors

These receptors are located in the skeletal muscle, tendons, and joints are known as the kinesthetic receptors. These inform the CNS about the movement and position of the limbs to maintain posture and equilibrium. The joint sense includes the sense of movement and the sense of position. The sense of movement is the ability to feel the movement with closed eyes. The sense of position is to feel the position of the body in relation to space. These two are mediated by receptors in and around the joint. Receptors involved are the spray nerve endings, Pacinian corpuscles, touch receptors, and free nerve endings located in the synovial membrane, ligaments, and tissues near the joint.

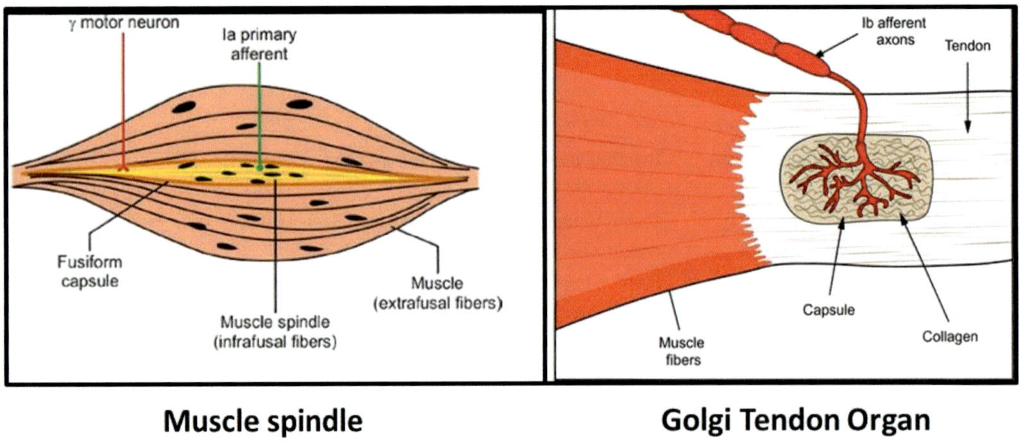

FIGURE 11.2 Propioceptive receptors of somatosensory system (Blumer et al., 2023)

The impulses generated from these reach the cerebral cortex through dorsal tracts, medial lemniscus, and thalamus to end in conscious perception of the position of the body in relation to space. The impulses from the muscle spindle organ, Golgi tendon organ, and vestibular apparatus of the internal ear do not reach the cerebral cortex. They are concerned with the reflex adjustment of muscle tone with the involvement of the spinal cord to maintain posture and equilibrium. These are known as non-sensory proprioceptive impulses as they never bring out conscious sensations.

Some of these proprioceptors send signals to the respiratory neurons about changes in the force exerted by the respiratory muscles and the movement of the ribcage. The function of the diaphragm with respiratory control is also mediated by one of the tendon organs in the external intercostals, which inhibit inspiration upon stimulation by a small change in the force of contraction of these muscles. The stretching of the diaphragm results in the inhibition of its contraction and inspiration acting through the tendon organs. The intercostal reflexes are significant due to the presence of numerous muscle spindle organs in the external intercostals. The muscle spindles help to coordinate breathing during changes in the posture and stabilize ribcage during reduction in lung compliance.

11.2.1.2.2.2 Mechanism of Action of Muscle Spindle Gamma Motor Neuron Excitation of these neurons ends up in contraction of intrafusal muscles. This in turn stretches the nuclear bag fibers and nuclear chain fibers. Sensory receptor of the muscle spindle in limited to these areas and can detect the tension upon these receptors. Stretch initiates depolarization in the sensory receptor membrane. Intrafusal fibers have no control on the tension developed in the extrafusal fibers. The nuclear bag fibers are tonic and sarcolemma of these are not excitable. No appreciable action potential is generated, hence no muscle twitch occur. On contrary nuclear chain fibers respond to neural excitation and are phasic in nature. It is excitable when stimulated and generates action potential in response to gamma trail motor neurons. This creates a twitch type of contraction.

Afferent fibers from nuclear bag spiral around the intrafusal muscle and form Annulospiral or Nuclear bag endings of the muscle spindle. These fibers are large and myelinated, responsible for initiation of myotatic reflex. Sensory fibers associated with nuclear chain intrafusal fibers terminate in plaque like ending namely the Floewer Spray ending. These fibers are small and are myelinated and are responsible for flexor reflex (Figure 11.2). These system functions to maintain posture of an animal and muscle tone of the skeletal muscles. The effect exerted by Gamma plate motor neurons is referred to as Gamma loop mechanism (Table 11.2).

11.2.2 Somatosensory Neurons

The cell bodies of the first-order somatosensory afferent neurons are located in posterior root or cranial root ganglia and are pseudounipolar cells. The cell body gives rise to a single process that divides to form a peripheral axon and a central axon. The peripheral axon travels to and ends in the skin, muscle, tendon or joint, and the central axon travels to and ends in the central nervous system (Table 11.3).

11.2.3 Somatosensory Receptor Organ

The sensory receptor organs contain the somatosensory receptors and deliver the somatosensory stimuli to the receptors. The receptors of most sensory systems are located in specialized sensory receptor organs (e.g., the photoreceptors in the eye and the auditory and vestibular hair cells in the inner ear) or within a restricted part of the body (e.g., the taste buds in the mouth and the olfactory receptors in the olfactory mucosa of the nose). For the tactile component of the somatosensory system, the skin covering the entire body, head, and face functions as the touch receptor organ, whereas joint tissues, muscles, and tendons act as the proprioception receptor organs.

TABLE 11.2
Proprioceptive Receptors

Receptor	Type	Sensation	Signals	Adaptation
Muscle Spindle	Encapsulated annulospiral and flower spray endings	Muscle stretch	Muscle length & velocity	Rapid initial transient and slow sustained
Muscle: Golgi Tendon Organ	Encapsulated collagen	Muscle tension	Muscle contraction	Slow
Joint: Pacinian	Encapsulated & layered	Joint Movement	Direction & velocity	Rapid
Joint: Ruffini	Encapsulated collagen	Joint pressure	Pressure & Angle	Slow
Joint: Golgi Organ	Encapsulated collagen	Joint torque	Twisting force	Slow

TABLE 11.3
Somatosensory Receptors and their Peripheral Axons

Receptor Type	Axon3 Group	CAP Peak	Conduction Velocity	Axon Diameter	Information Processed
Muscle Spindle: Annulospiral endings	Ia	Aα	70-120 m/sec	1-20 μM	Muscle length and velocity
Muscle Spindle: Flower Spray endings	II	Aβ	30-70 m/sec	6-12 μM	Muscle length
Golgi Tendon Organ	Ib	Aα	70-120 m/sec	12-20 μM	Muscle tension
Joint: Pacinian	II	Aβ	30-70 m/sec	6-12 μM	Joint movement
Joint: Ruffini	II	Aβ	30-70 m/sec	6-12 μM	Joint angle
Joint: Golgi Tendon Organ	II	Aβ	30-70 m/sec	6-12 μM	Joint torque
Meissner corpuscle	II	Aβ	30-70 m/sec	6-12 μM	Touch, flitter or movement
Pacinian corpuscle	II	Aβ	30-70 m/sec	6-12 μM	Vibration
Ruffini corpuscle	II	Aβ	30-70 m/sec	6-12 μM	Skin stretch
Hair follicle	II & III	Aβ & Aδ	10-70 m/sec	2-12 μM	Touch movement
Merkel complex	II	Aβ	30-70 m/sec	6-12 μM	Fine touch
Free Nerve endings	III	Aδ	5-30 m/sec	1-6 μM	Sharp pain or cool/cold
Free nerve endings	IV	C	0.5-2 m/sec	<1.5 μM	Dull or aching pain, or touch or warm

11.2.4 Somatosensory Cortex

The somatosensory information converges in the parietal lobe of the cerebral cortex where it is processed to provide a cohesive perception of the body and its physical environment. The postcentral gyrus includes the primary somatosensory cortex (Brodmann areas 3, 2, and 1) collectively referred to as primary somatosensory cortex S1.

1. Primary Cortical Receiving Area - The primary somatosensory cortex includes the postcentral gyrus and the posterior paracentral lobule of the parietal lobe. It processes the information about crude touch, pain and temperature.
2. Secondary Cortical Receiving Area - The secondary somatosensory cortex is located inferiorly in the pars opercularis of the parietal lobe, which forms part of the upper lip of the lateral sulcus. It is involved with specific touch perception and is thus integrally linked with the amygdala and hippocampus to encode and reinforce memories
3. Association Cortical Area - The somatosensory association cortex is located in the superior parietal lobe, which is posterior to the primary somatosensory cortex. The highest degree of convergence of somatosensory information occurs in the posterior parietal cortex.
4. Cortical Areas for Pain Sensation - Pain information is processed in multiple pathways.

11.3 RECEPTOR SPECIFICITY

Adequate stimulus is preferred if large magnitude of stimuli can provoke response from many receptors. Some receptors are designated by nature to respond to different stimuli, for example, nociceptors to detect pain, which may be due to temperature, pressure, or chemicals. These are termed polymodal receptors. The region in the sensory surface of the afferent neuron which detects the stimuli to generate a response upon stimulation is termed as Receptive field.

The somatosensory systems process information about, and represent, several modalities of somatic sensation (i.e., pain, temperature, touch, proprioception). Each of these modalities can be divided into sub-modalities, as shown in Table 11.1 (e.g., pain into sharp, pricking, cutting pain; dull, burning pain; and deep aching pain). Discriminative touch is also subdivided into touch, pressure, flutter, and vibration. Each of these sensations (i.e., sub-modalities) is represented by neurons that exhibit modality specificity. That is, when a somatosensory neuron is stimulated naturally (e.g., by skin warming) or artificially (e.g., by electrical stimulation of the neuron), the sensation perceived is specific to the information normally processed by the neuron (i.e., warm skin). Consequently, a "warm" somatosensory neuron will not respond to the cooling of the skin or to a touch stimulus that does not "warm" the skin. The somatosensory receptor and its central connections determine the modality specificity of the neurons forming a somatosensory pathway.

The sensory information necessary for discriminative touch, proprioception, pain, and thermal sensations are kept separate within the somatosensory pathways. To ensure the fidelity of stimulus representation in the discriminative touch-proprioceptive pathways, there is minimal convergence of modality specific information along the pathways. That is, within the medial lemniscal and main sensory trigeminal pathways, there is little mixing of information from different receptor types and from afferents with different adaptive properties up to the level of the cerebral cortex. For example, a cuneate nucleus neuron (2° medial lemniscal afferent) will synapse only with one type of posterior root neuron (e.g., 1° afferents with small "touch" receptive fields and rapidly adapting discharges). Within VPL and VPM, the medial lemniscal and ventral trigeminal lemniscal fibers terminate in different regions based on the sensory information they are carrying. Fibers carrying cutaneous information terminate within the core of the nucleus, whereas those carrying proprioceptive information terminate in the surrounding, peripheral shell of the nucleus. Within the cerebral cortex, there is a convergence of modality specific information and the response properties of cortical neurons become more complex. The primary somatosensory cortex is responsible for the first stage of cortical processing. Within the primary sensory cortex, discriminative touch and proprioceptive information from overlapping areas may be combined such that a cortical neuron may respond to both cutaneous and proprioceptive stimulation of a digit.

Each subcortical somatosensory neuron responds to modality-specific stimuli applied to a specific region of the body or face i.e. each somatosensory neurons have receptive fields. For example, an axon in the medial lemniscus (i.e., the fiber tract) that responds to tactile stimulation of the right index finger pad will not respond to tactile stimulation of any other area in the hand, body or face. The stimulated area producing the response is called the neuron's receptive field (Figure 5.3). The neuron's receptive field can also be defined anatomically as that area of the sense organ (i.e., skin, muscles or joints) innervated directly or indirectly by the neuron. Consequently, a somatosensory neuron can be described to channel information about stimulus location - as well as stimulus modality. Furthermore, the size of a neuron's receptive field is related to the body area innervated/represented. The receptive fields of neurons innervating/representing the finger pads, lips, and tongue are the smallest, whereas those of neurons innervating/representing the shoulders, back and legs are the largest. For greater accuracy in locating the point of stimulus contact or movement, smaller cutaneous receptive fields are required. For fine motor control, as in playing the piano or speaking, small proprioceptive receptive fields are required.

11.4 PATHWAYS OF SOMATOSENSORY SYSTEM

Sensory pathways consist of the chain of neurons, from receptor organ to cerebral cortex, that are responsible for the perception of sensations. The sensory information processed by the somatosensory systems travels along different anatomical pathways depending on the information carried. For example, the lemniscal pathway carries discriminative touch and proprioceptive information whereas, the spinothalamic pathways carry crude touch, pain and temperature information from the body. The various pathways of somatosensory system are-

11.4.1 Medial Lemniscal Pathway: Body Discriminative Touch and Proprioception

The pathway is important in identifying objects and features of surfaces by touch even though small shapes can be perceived with information solely from tactile receptors, most discriminations involve an active process of tactile exploration and multiple contacts on the skin, and an integration of cutaneous and proprioceptive information as well as efferent control. The medial lemniscus is a second-order neuron of the dorsal column-medial lemniscus pathway which commences at the nucleus gracilis and nucleus cuneatus at the caudal medulla; the arcuate fibers decussate at the caudal medulla and ascend via the medial lemniscus contralaterally in the brainstem until synapsing at the ventral posterolateral nucleus of the thalamus, the point at which the third-order neuron of the DCML pathway commences. The second-order neuron with the somatotopic arrangement, transports the sensory spinothalamic information of conscious proprioception, vibration, fine touch, and 2-point discrimination of skin and joints of the body and head; from the caudal medulla to the ventral posterolateral nucleus of the thalamus, and subsequently the primary somatosensory cortex. The third-order neuron ascends from the ventral posterolateral nucleus until synapsing with the primary somatosensory region of the brain cortex.

The primary function of the medial lemniscus is as a second-order neuron of the dorsal column-medial lemniscus pathway is to transport the sensory spinothalamic information of conscious proprioception, fine touch, 2-point discrimination of skin and joints of the body and head and vibration, from the caudal medulla to the ventral posterolateral nucleus of the thalamus, and subsequently the primary somatosensory cortex. Due to the unique somatosensory information and location of this pathway, the dorsal column-medial lemniscus tract may be used as an important clinical clue to pinpoint a possible site of injury and manage the patient accordingly.

The blood supply of the medial lemniscus varies according to the level of the brainstem. The nuclei gracilis and cuneatus are located in the medial posterior aspect of the medulla and receive vascular supply from the posterior spinal artery. The arcuate fibers decussate in the medial-anterior aspect of the medulla and are irrigated predominantly by the anterior spinal artery. The medial lemniscus begins ascending contralaterally in the brainstem in the medial aspect of the medulla and is supplied by the anterior spinal artery. As the medial lemniscus continues to ascend, it begins shifting posteriorly and laterally; at the level of the pons, the medial lemniscus receives supply from the basilar artery. By the time the medial lemniscus makes it to the midbrain, it is predominantly in the posterior lateral aspect of the midbrain, and the posterior cerebral artery irrigates it.

11.4.2 Main Sensory Trigeminal Pathway: Face Discriminative Touch and Proprioception

The main sensory trigeminal pathway, the cranial homologue of the medial lemniscal pathway, mediates discriminative touch and proprioception from the face. Somatosensory information from the face and dura travels via four cranial nerves: the trigeminal (V) nerve, facial (VII) nerve, glossopharyngeal (IX) nerve and the vagus (X) nerve. Recall that some afferent fibers in the facial, glossopharyngeal and vagus nerves carry somatosensory information from the ear (pinna, external auditory meatus, tympanic membrane, and middle ear cavity) and posterior dura. The central processes of the 1° afferent neurons carrying information processed for discriminative touch and proprioception from the face and dura terminate on 2° afferents in the main sensory trigeminal nucleus. Most of the 2° afferent axons from the main sensory trigeminal nucleus decussate and join the contralateral ventral trigeminothalamic tract (also called the ventral trigeminal lemniscus). The ventral trigeminothalamic tract runs in very close association with the medial lemniscus. The 2° afferent fibers ascend in the ventral trigeminothalamic tract to the thalamus where they terminate on 3° afferents in the ventral posteromedial nucleus (VPM) of the thalamus. The axons of these 3° afferents, ascend in the posterior limb of the internal capsule and terminate in the postcentral gyrus of the parietal lobe.

11.4.3 Neospinothalamic Pathway: Body – Sharp Prickling Pain and Cool/Cold

There are two well-defined spinothalamic pathways, chiefly concerned with pain and temperature sensations and with crude touch. The "fast" conducting neospinothalamic pathway is involved in conveying the "sharp/cutting" pain elicited at the time tissue is damaged. The "slower" conducting paleospinothalamic pathway is involved in conveying the "dull/burning" pain that accompanies the later inflammatory reaction in the damaged tissue as well as temperature and crude touch information. The archeospinothalamic pathway is a poorly defined pathway involved in a generalized sense of discomfort and diffuse pain.

The 1° afferents of the spinothalamic pathways send A-delta (neospinothalamic) and C (paleospinothalamic) fibers to the periphery where they form free nerve endings in skin, muscle, tendon, joint capsules and viscera. The central processes of these 1° afferent neurons are the pseudounipolar cells of the posterior root ganglia and enter the spinal cord in the lateral division of the posterior root. Recall that fibers entering via the medial division of the posterior root are concerned with tactile, pressure, proprioception, and vibration sensations. The lateral division axons branch and send fibers to the gray matter at the segment of entry and into the tract of Lissauer. The 1° afferents of the spinothalamic systems may end in the segment of root entry or one or two segments up. The neospinothalamics end on 2° afferents in the nucleus posteromarginalis. The paleospinothalamics end on 2° afferents in the substantia gelatinosa.

a. Neospinothalamic Pathway – The 2° neospinothalamic afferents (nucleus posteromarginalis axons) cross in the anterior white commissure to collect in the spinothalamic tract within the contralateral anterior and lateral (predominantly) funiculi. These 2° neospinothalamic afferent axons ascend the spinal cord and brain stem in the spinothalamic tract to terminate on 3° afferents in the ventral posterolateral (VPL) nucleus of the thalamus. Note that these 3° neospinothalamic VPL afferents differ from the VPL neurons synapsing with 2° afferent axons of the medial lemniscus. The 3° neospinothalamic VPL neurons send their axons to the primary somatosensory cortex (i.e., the postcentral gyrus of the parietal lobe).
b. Paleospinothalamic Pathway – The 2° paleospinothalamic afferents (axons of the substantia gelatinosa) travel a short distance to terminate in or near the ipsilateral nucleus proprius on 3° afferents. The axons of most paleospinothalamic 3° (nucleus proprius) afferents cross in the anterior white commissure to collect bilaterally in the spinothalamic tracts of the anterior (predominantly) and lateral funiculi. Along their ascending course to the thalamus, many of the paleospinothalamic 3° (nucleus proprius) afferents leave the spinothalamic tracts

to terminate in the brain stem reticular formation or midbrain periaqueductal gray. The remaining paleospinothalamic 3° afferents remain in the spinothalamic tracts and ascend to terminate on 4° paleospinothalamic afferents in the intralaminar nuclei of the thalamus (IL in the figure). The projections of the intralaminar nuclei axons (4° paleospinothalamic afferents) are to diffuse areas of the cerebral cortex which are believed to play a role in poorly localized sense of pain. For example, destruction of the primary somatosensory cortex (i.e., the postcentral gyrus of the parietal lobe) does not seem as detrimental to the appreciation of painful stimuli as it is to the appreciation of other somatic sensations.

11.4.4 Spinal Trigeminal Pathway: Face Pain, Temperature, and Crude Touch

The spinal trigeminal pathway, the cranial homologue of the spinothalamic pathway, mediates crude touch, pain and temperature from the face. Somatosensory information from the face and dura travels via four cranial nerves: the trigeminal (V) nerve, facial (VII) nerve, glossopharyngeal (IX) nerve and the vagus (X) nerve. Recall that some afferent fibers in the facial, glossopharyngeal and vagus nerves carry somatosensory information from the ear (pinna, external auditory meatus, tympanic membrane, and middle ear cavity) and posterior dura. The central processes of the 1° afferent neurons carrying information about simple touch, pain and temperature from the face and dura enter the spinal trigeminal tract and terminate on 2° afferents in the spinal trigeminal nucleus. Most of the 2° afferent axons from the spinal trigeminal nucleus decussate and form the contralateral ventral trigeminothalamic tract (also called the ventral trigeminal lemniscus). The 2° afferent fibers ascend in the ventral trigeminothalamic tract to the thalamus where they terminate on 3° afferents in the ventral posteromedial nucleus (VPM) or intralaminar nuclei of the thalamus. The axons of these 3° afferents ascend in the posterior limb of the internal capsule and terminate in the postcentral gyrus of the parietal lobe (VPM neurons only) or in more diffuse areas of the cerebral cortex (intralaminar nucleus neurons).

11.5 CONCLUSION

The somatosensory systems provide accurate information about the location and temporal features of stimuli and about sharp pain, tactile stimuli and the position and movement of body parts. The somatosensory systems process information about, and represent, several modalities of somatic sensation via its receptors. These receptors are located in specialized receptor organs, produce receptor potentials, contain synaptic specializations, and release neural transmitters. Somatosensory receptors send information via pathways consists of the chain of neurons, from receptor organ to cerebral cortex, that are responsible for the perception of sensations. Although it is convenient to subdivide somatosensory receptors and pathways for didactic, clinical and research purposes, it is important to keep in mind that most somatosensory stimuli act simultaneously and in varying degrees on all somatosensory receptors in the body part stimulated.

BIBLIOGRAPHY

Thomas E, Puget S, Valentin D, Songer P (2017). "Sensory evaluation—Profiling and preferences". In *The craft and science of coffee* B. Folmer (ed.), (pp. 419–456). Academic Press.

Patrick D (2020). *Somatosensory systems. Chapter 2. Neuroscience online*. McGovern Medical School.

Patrick D (2020). *Somatosensory pathways. Chapter 4. Neuroscience online*. McGovern Medical School.

Sherman C (2019). *The senses: The somatosensory system*. Dana Foundation.

Hashim IH, Kumamoto S, Takemura K, Maeno T, Okuda S, Mori Y (2017). "Tactile evaluation feedback system for multi-layered structure inspired by Human Tactile perception mechanism". *Sensors* (Basel, Switzerland) 17 (11): 2601.

Buccino G, Binkofski F, Fink GR, Fadiga L, Fogassi L, Gallese V, Seitz RJ, Zilles K, Rizzolatti G, Freund HJ (2001). "Action observation activates premotor and parietal areas in a somatotopic manner: An fMRI study". *The European Journal of Neuroscience* 13 (2): 400–4.

Seelke AM, Padberg JJ, Disbrow E, Purnell SM, Recanzone G, Krubitzer L (2012). "Topographic Maps within Brodmann's area 5 of macaque monkeys". *Cerebral Cortex* 22 (8): 1834–50.

Blumer R, Carrero-Rojas G, Calvo PM, Streicher J, de la Cruz RR, Pastor AM (2023). *Proprioceptors in extraocular muscles*. Experimental Physiology.

Romanowski CA, Hutton M, Rowe J, Yianni J, Warren D, Bigley J, Wilkinson ID (2011). "The anatomy of the medial lemniscus within the brainstem demonstrated at 3 tesla with high resolution fat suppressed T1-weighted images and diffusion tensor imaging". *Neuroradiology Journal* 24 (2): 171–6.

Somatosensory, Viscerosensory and Spinocerebellar Pathways (1997). *Neuroanatomy online*. McGovern Medical School.

12 Physiology of the Olfactory System

Tanmay Mondal[1] and Gurpreet Singh
Department of Veterinary Physiology and Biochemistry, College of Veterinary Science, Rampura Phul Guru Angad Dev Veterinary and Animal Sciences University

12.1 INTRODUCTION

Olfaction, commonly known as the sense of smell, is a remarkable chemosensory process crucial for the survival and well-being of organisms. In the animal kingdom, the ability to detect and differentiate thousands of diverse odorants at low concentrations has played a pivotal role in navigating the environment and ensuring sustenance. In humans and animals, the olfactory system, orchestrated by the olfactory nerve and various neuroanatomical structures, is intricately linked to complex functions such as taste and involuntary memory formation. From an evolutionary perspective, the intact sense of smell has been essential for evaluating the safety of ingestible substances, assessing potential dangers, and recognizing social relationships.

As life evolved, the need to sense chemicals in the environment became increasingly crucial. Single-celled organisms, like Escherichia coli, developed chemical receptors to survive by distinguishing between nourishing and harmful substances. The sense of smell, present ubiquitously in the animal kingdom, is particularly advanced in humans. Humans possess an extensive array of odorant receptor types, surpassing other sensory systems, allowing them to perceive a wide range of stimuli. Importantly, the olfactory system serves as a unique window into overall brain health. Smell loss is often an early indicator of neurodegenerative diseases such as Alzheimer's or Parkinson's, providing valuable insights into brain development. Remarkably, smell loss stands out as one of the most reliable predictors of future mortality in older populations, surpassing other health markers. This introduction sets the stage for a detailed exploration of the anatomy, physiology, of the olfactory pathway.

12.2 ANATOMY OF THE OLFACTORY SYSTEM

In Figure 12.1, a microscopic cross-section of the olfactory region offers insights into the intricate structure of olfactory receptor cells. Each cell comprises a cell body with two extending nerve fibers – one functioning as a dendrite and the other as an axon. Examining this microscopic view, we observe the dendritic process of the olfactory cell extending beyond the olfactory region membrane, nestled within crevices between sustentacular cells. These sustentacular cells play a vital role in providing substantial support to the dendritic processes, effectively shielding the olfactory cell bodies from direct exposure to the nasal cavity. Notably, within this location, hair-like structures known as olfactory cilia project into the nasal cavity from the olfactory vesicles, which represent the expanded parts of dendrites. Typically covered with a thin secretion from Bowman's glands (subepithelial glands), these cilia play a crucial role in the olfactory process. The ducts of Bowman's glands traverse through the epithelium to reach the surface, ensuring a constant supply of secretions that refresh the thin layer of fluid bathing the olfactory hairs on the surface of the olfactory region.

The mechanism of sniffing becomes integral in this process, facilitating the back-and-forth movement of air. This action increases the likelihood of the substance being smelled going into solution, acting as the stimulus for the subsequent transmission of impulses to the brain. The coordinated activity of the olfactory cells' axons, joining with others to form fibers and branches of the olfactory nerves, further contributes to the efficient transmission of olfactory signals. Additionally, the paragraph underscores the importance of basal cells, which undergo division and differentiation into sustentacular or olfactory cells, acting as a protective mechanism against the potential loss of smell resulting from nasal mucosal diseases (Figure 12.1).

12.3 OLFACTION

The olfactory system is a complex sensory apparatus encompassing the olfactory bulb, olfactory tract, lateral olfactory gyrus, and piriform lobe, as illustrated in Figure 12.2. Olfaction, a vital special sense, plays a pivotal role in the localization of food, reflex-stimulated secretion of digestive enzymes, and the detection of potential dangers. Key components of this system include olfactory cells, situated within the specialized olfactory epithelium on the ethmo-turbinate bones of the nasal cavity. Interestingly, the olfactory mucosa occupies a considerably larger area in dogs (100 cm^2) compared to humans (about 5 cm^2). These olfactory cells give rise to olfactory nerve fibers that terminate in the olfactory bulb. Olfaction, identified as the sensation of smell, is a result of detecting aerosolized odorous substances in the environment. As one of the special senses, humans rely on the olfactory system's various components, including the olfactory epithelium, olfactory nerves, olfactory bulb and tract, and multiple areas of the olfactory

DOI: 10.1201/9781003426851-12

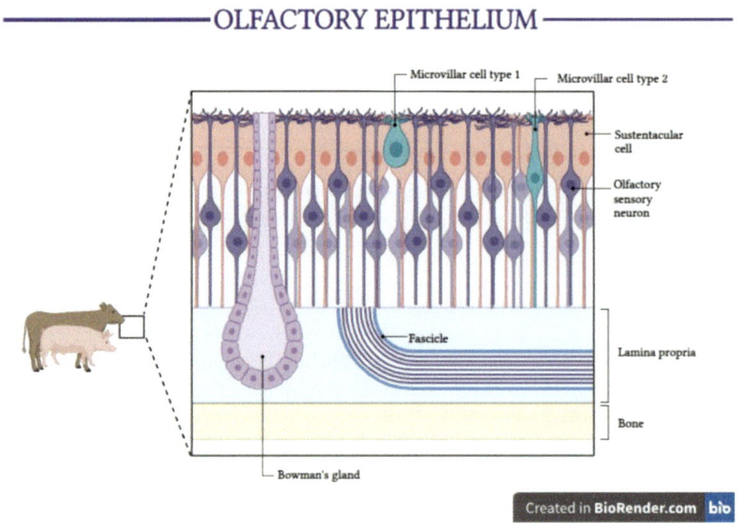

FIGURE 12.1 The olfactory region of the domestic animal and the cells associated with smell.

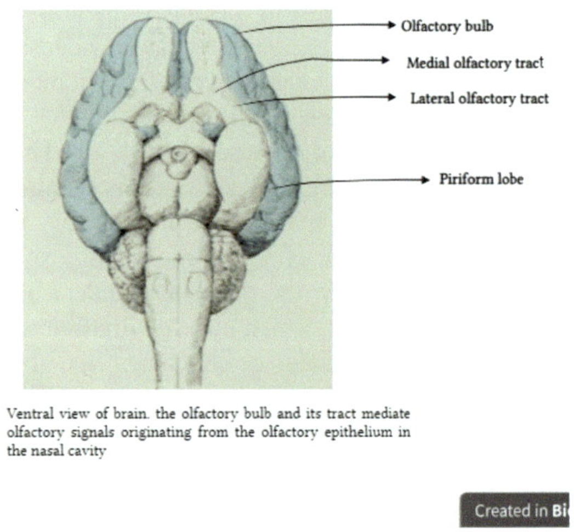

FIGURE 12.2 Ventral view of the brain. the olfactory bulb and its tract mediate olfactory signals originating from the olfactory epithelium in the nasal cavity.

cortex. Notably, the olfactory nerve boasts a unique feature – direct connectivity to the cortex, bypassing the thalamus, a distinction from other special senses whose axons relay in the thalamus. Another intriguing aspect is the continuous replacement of olfactory nerves throughout life, facilitated by the differentiation of basal stem cells. This intricate system underlines the importance of olfaction in shaping our sensory experiences and interactions with the environment (Figure 12.2).

12.4 RECEPTORS FOR ODORANTS

As fragrant molecules grace the nasal cavity, they gracefully dissolve within the fluid exquisitely secreted by olfactory glands nestled in the olfactory mucosa. An intricate dance unfolds as these molecules, enchanting olfactory receptors, find their partner in a protein ballet called olfactory-binding protein, gracefully present in the mucus. Crafted by the olfactory glands in the nasal mucosa, this protein acts as a virtuoso, thought to both carry and concentrate the enchanting molecules. The olfactory cells, true connoisseurs, possess an extraordinary ability to discern a myriad of scents even at the most minuscule concentrations. Within the olfactory mucosa reside bipolar sensory neurons, their opulent cell bodies resting just below the ethereal cribriform plate, a delicate sheet of the ethmoid bone. A sensory neuron, akin to a poetic muse, extends a single dendrite, culminating in an expansive olfactory knob gracing the surface of the olfactory mucosa. Each olfactory knob, a symphony in itself, gives birth to a symphony of 10–20 cilia, delicately spreading over the surface like a lustrous veil, embraced by the caress of mucus. For the poetry of odors to unfold, molecules must waltz through this mucosal layer, where the cilia, adorned with sensory receptors, stand ready to transduce the olfactory sonnet. In this olfactory ballet, groups of sensory neurons express their olfactory receptors, forming harmonies that resonate with specific odors, each note carrying a distinct affinity for the aromatic molecules. At the opposite end of this sensory ballet, a solitary non-myelinated axon emerges, not in solitude but in a gathering of elegance. Unlike their cranial counterparts, these axons converge, creating petite fascicles known as the fila olfactoria. Through ethereal foramina in the cribriform plate, they embark on a journey, an aromatic pilgrimage, entering the olfactory bulb. Here, in the grand ballroom, they engage in a refined dance, synapsing with tufted and mitral cells, contributing to the opulent symphony of the olfactory experience. In this fragrant tapestry, the olfactory system weaves an exquisite narrative, an eloquent ode to the intricacies of scent transduction and transmission.

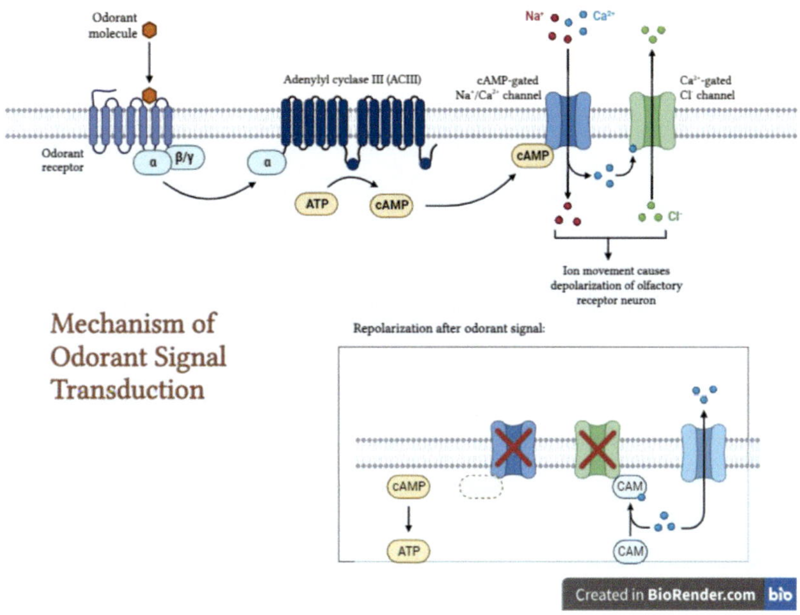

FIGURE 12.3 Mechanism of odorant signal transduction.

12.5 TRANSDUCTION OF OLFACTORY STIMULUS

The cilia membrane unfolds a tapestry adorned with G protein-coupled olfactory receptors, as depicted in the elegant tableau of Figure 12.3. The union of an odorant molecule with the receptor on the cilia sets in motion the activation of a G protein, known as Golf, forming a graceful partnership with guanosine triphosphate (GTP). The ballet of transduction commences, guided by two distinguished second-messenger systems: inositol trisphosphate (IP3) and cyclic adenosine monophosphate (cAMP), elegantly portrayed in Figure 12.3. The GTP-Golf complex takes the lead, orchestrating a sequence of events. First, phospholipase C is activated, giving rise to IP3, which, in turn, opens Ca2+ channels. Simultaneously, adenylyl cyclase is stirred to action, producing cAMP, which unfurls Na+ and Ca2+ channels in the membrane, allowing for the graceful entry of Na+ and Ca2+ (primarily Ca2+), culminating in the subsequent depolarization of olfactory cells. This cascade of events, akin to a symphony, engenders excitatory postsynaptic graded potentials (EPSPs) within the cilia. Embarking on a journey, these ciliary EPSPs traverse from cilia to the trigger zone, the illustrious axon hillock, where their strength, when surpassing the threshold potential, begets the generation of action potentials. These action potentials, like heralds, traverse the axons of the olfactory cells, wending their way to the olfactory bulb. Herein lies the nexus from which impulses radiate, cascading through extensive realms of the central nervous system, a route to the piriform lobe, where the perception of smell takes its throne. A poetic uniqueness graces olfactory transmission – a swift adaptation to stimuli. An initial surge of axonal discharge in response to stimulation gracefully yields to a steady-state discharge of diminished amplitude, adding a nuanced cadence to the olfactory symphony (Figure 12.3).

12.6 CENTRAL PATHWAY FOR OLFACTION

Olfactory nerve fibers, known as fila olfactoria, gracefully culminate in the ipsilateral olfactory bulb, a realm inhabited by tufted cells and mitral cells. As revealed in the captivating tableau, within this sensory symphony, dendrites of tufted and mitral cells intricately intertwine with the terminal ends of olfactory nerve fibers, giving birth to the mesmerizing glomeruli of the olfactory bulb. The alchemy of neurotransmitters, likely peptides, emanates from the terminal realms of olfactory axons, sparking a ballet of excitement in mitral and tufted cells, their movements finely tuned by the artistry of inhibitory periglomerular interneurons. Embarking on a grand journey, the axons of mitral and tufted cells venture beyond the olfactory bulb, crafting the majestic tapestry of the large lateral olfactory tract. The hallowed sanctuaries that receive these olfactory signals include the amygdala and entorhinal cortex, which, in turn, dispatch olfactory signals to the esteemed hippocampus and frontal cortex. In a departure from the norm, olfactory signals, rather than traversing the thalamus, navigate a direct course to evoke emotional responses. The saga continues as mitral and tufted cells extend their influence to the ipsilateral septal nucleus, while their emissaries traverse the enigmatic realms of the medial olfactory tract and the anterior commissure to reach the distant shores of the contralateral olfactory bulb. This intricate projection of olfactory signals unveils a narrative where emotional resonance finds its home in the entorhinal cortex, hippocampal formation,

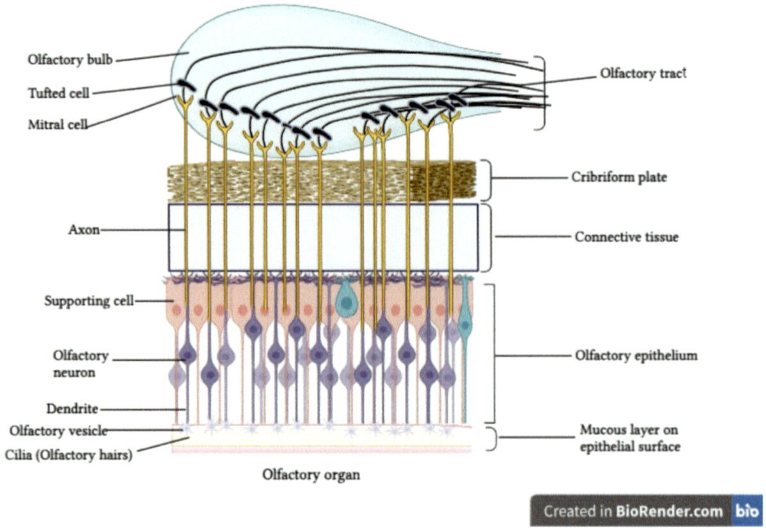

FIGURE 12.4 Lateral view of the olfactory neurons, olfactory nerve, olfactory bulb, and olfactory tract. Dendrites of the olfactory bipolar neurons bear 10-20 short cilia. Axons of the olfactory neurons form olfactory fibers, which terminate in the olfactory bulb to synapse with tufted and mitral cells

septal nuclei, and amygdala of the limbic system. Moreover, the autonomic responses to olfaction find their maestros in the hypothalamus and periaqueductal gray of the midbrain, integral components of the autonomic nervous system intertwined with the limbic symphony. In this orchestration, the olfactory signals transcend mere detection, becoming the very essence of olfactory memories and the evocation of profound emotional reactions (Figure 12.4).

12.7 PHEROMONES

Animals talk to each other using smells. Black-tailed deer and Rocky Mountain mule deer use special glands on their hind legs to send smells. These smells help them know if others are friends or strangers. The smell is put on their skin and hair, and they sniff each other about once an hour. A special substance in the deer's gland helps them talk through smells. It was the first time scientists studied this in mammals. Other animals also have scent glands. Rabbits have them between their foot pads, and cats have them on their chin and lips. Cats mark things by rubbing their heads on them. These special smells are called pheromones. They do a lot of jobs, like marking paths, saying who is who in a group, showing where to find food, and telling others if there's danger. Animals have a cool way of talking with smells.

12.8 CLINICAL CORRELATIONS

Anosmia, which means the loss of the sense of smell, and hyposmia, referring to a reduced ability to smell, are not commonly encountered clinical issues. The reasons behind anosmia and hyposmia are not always clear. Anosmia often stems from severe inflammation of the olfactory mucosa or bilateral lesions affecting the olfactory nerve or olfactory bulbs. Viral infections, like distemper and parainfluenza, can also lead to anosmia or hyposmia by causing damage to the olfactory mucosa. Head trauma, in some instances, may displace the olfactory bulb concerning the cribriform plate, causing harm to the axons of the fila olfactoria that pass through this plate en route to the olfactory bulb. Such displacement can result in a loss or reduction of the sense of smell. Evaluating olfactory deficiencies in animals can be done by observing their responses to smells, such as turning the head away, facial muscle contractions, or sniffing. However, it's important to avoid using irritating substances during these assessments, as the animal's reactions may be triggered not necessarily by an intact olfactory system but rather by the stimulation of sensory nerve endings of the trigeminal nerve in the nasal mucosa.

BIBLIOGRAPHY

Bieri, S., Monastyrskaria, K. and Schilling, B. 2004. Olfactory receptor neuron profiling using sandalwood odorants. *Chemical Senses* 29: pp. 483–487.

Breer, H., Krieger, J., Meinken, C., Kiefer, H. and Strotman, J. 1998. Expression and functional analysis of olfactory receptors. *Annals of the New York Academy of Sciences* 855: pp. 175–181.

Chaudhari, N. and Roper, S.D. 2010. The cell biology of taste. *Journal of Cell Biology* 190: pp. 285–296.

Pevsner, J., Sklar, P.B. and Snyser, S.H. 1986. Odorant-binding protein: localization to nasal glands and secretions. *Proceedings of the National Academy of Sciences USA* 83: pp. 4942–4946.

Vogt, R.G., Prestwich, G.D. and Lemer, M.R. 1991. Odorant-binding-protein subfamilies associate with distinct classes of olfactory receptor neurons in insects. *Developmental Neurobiology* 22: pp. 78–84.

13 Physiology of the Gustatory System

Mohamed Hasif G, Monalisha Debbarma and Ankit Jena
Siksha 'O' Anusandhan University, India

13.1 INTRODUCTION

Receptors are found in the taste buds of all the animals.

Canines as example of taste receptors: Table 13.1 describes three major types of receptors, as shown in the table, which are present in animals.

There is a protrusion on the dorsal and lateral surface of the tongue, the same receptors on human tongues. Dogs sense sweet, salt, sour and bitter tastes. They don't have a liking for salt due to the absence of sensitivity to salt receptors.

Taste buds of different species: Different species have different numbers of taste buds, with humans having the highest number of taste buds compared to dogs, as shown in the Table 13.2. Cats have the lowest number of taste buds compared to dogs.

Different tastes are located in different parts of the animal's tongue, as shown in Table 13.3.

13.2 SALIVARY INFLUENCE IN GUSTATION

Saliva in animals dissolves the chemicals that stimulate taste sensation. These chemicals enter the buds by forming pores. The columnar cells are present in about 40–120 in range for the taste buds (similar to a cluster of bananas).

TABLE 13.1
Reception of Taste Buds and its Location

Receptors of taste buds (types)	Location of tongue
Fungiform	Dorsal surface of rostral 2/3 rd and lateral margin & tip .
Vallate	Caudal portion of lateral side.
Foliate	Dorsolateral of caudal side.

TABLE 13.2
Species and Number of Taste Buds

Species	Number of taste buds
Dog	1700
Human	9000
Cat	470

TABLE 13.3
Taste and Location in the Tongue

Taste	Location in tongue
Meaty	Rostral 2/3rd of dorsal region
Sweet	Rostral & lateral region
Salty	Lateral region
*(very low in dogs)	

Small "taste pores," i.e., modified receptors in taste buds, have an extension of "microvilli" from these taste pores.

Life span: Replacement is initiated every 10 days.

Taste sensation is triggered by a large number of molecules and a very little number of cells responsible for taste. Each flavor receptor pertains to only one type of taste cell. Every cell recognizes sweet, sour, and bitter tastes, but very little recognition for salty tastes.

13.3 COMMUNICATIONS OF STIMULUS BY THE GUSTATORY SYSTEM

Saliva carries chemical substances that enter taste buds through pores at their top, where they bind to receptors located on the membrane of microvilli within taste cells. This binding triggers depolarization of the taste cell membrane. The specific mechanisms of membrane depolarization vary depending on the type of taste molecules binding to their respective receptors. For instance, salty taste transduction involves the influx of Na+ ions through amiloride-sensitive Na+ channels (ENaC), leading to membrane depolarization (Figure 13.1). However, the exact transduction pathways for salty tastes remain incompletely understood. One hypothesis suggests that the transduction process may involve the opening of voltage-gated Ca2+ channels, resulting in the influx of Ca2+ ions and subsequent release of neurotransmitters. Notably, taste cells responsible for detecting salty tastes lack well-defined synapses, and the neurotransmitter involved in their signaling remains unidentified. Additional research is needed to further elucidate the intricate mechanisms underlying salty taste perception.

Elevated intracellular Ca2+ levels, combined with membrane depolarization, trigger the release of ATP via gap junction channels in the plasma membrane. Released ATP acts on sensory nerve endings, initiating generator

DOI: 10.1201/9781003426851-13

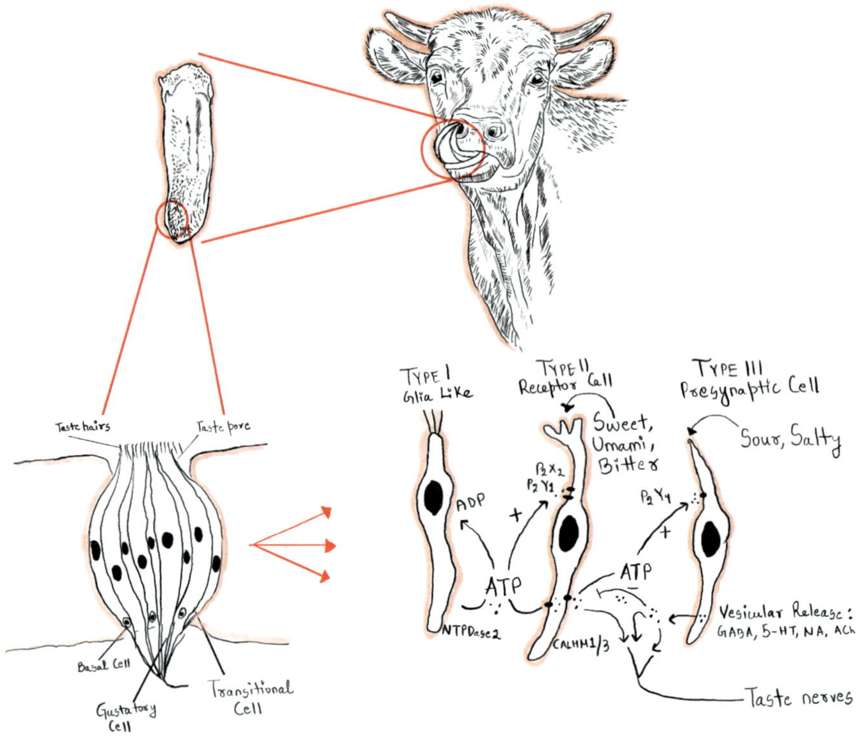

FIGURE 13.1 The gustatory organ of cattle for taste sensation is located in the tongue.

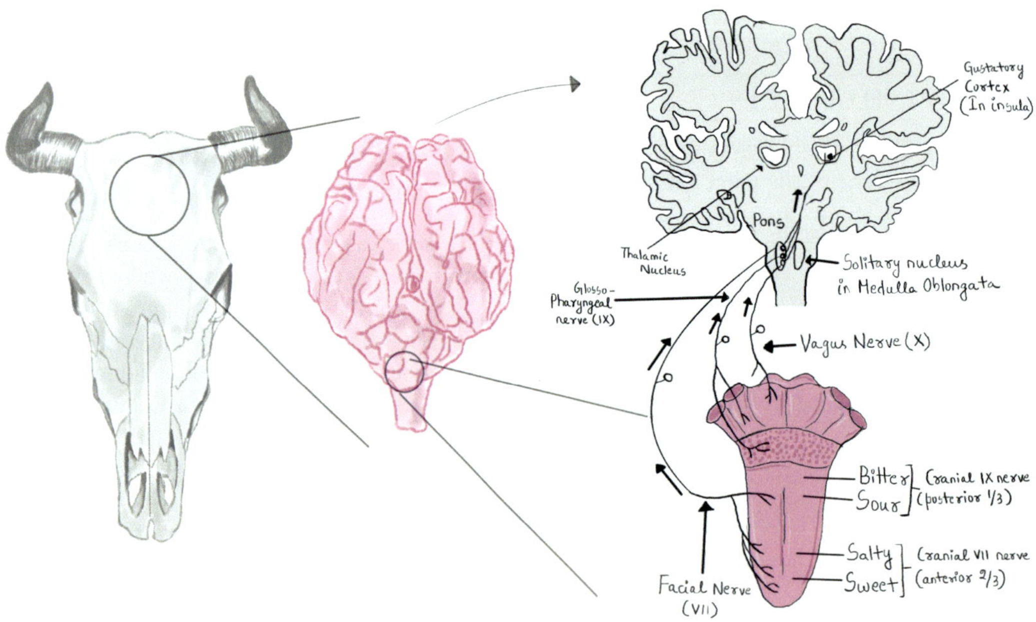

FIGURE 13.2 Central control of nerves for gustatory physiology of cattle.

potentials. If these generator potentials reach the threshold potential, action potentials are generated. Notably, taste cells responsible for sweet taste detection lack well-defined synapses, adding complexity to the transmission of signals from taste buds to sensory nerves. Additional research is needed to fully elucidate the intricate mechanisms involved in sweet taste transduction. Sweet taste transduction involves G protein-coupled receptors that activate phospholipase C (PLC). Upon activation, PLC generates two intracellular second messengers, inositol trisphosphate (IP3) and diacylglycerol (DAG), by hydrolyzing phosphatidylinositol 4,5-bisphosphate (PIP2). IP3 binds to calcium channels on the endoplasmic reticulum, causing them to open and allow Ca2+ to flow into the cytosol. There is speculation that taste-selective cation channels (TrpM5) are sensitive to changes in calcium levels, with IP3 playing a crucial role in their activation. This activation leads to the influx of Na+ ions through TrpM5, resulting in the generation of a depolarizing receptor potential in taste receptor cells (Figure 13.1).

13.4 CENTRAL NERVOUS SYSTEM CONTROL OF GUSTATION

Taste cells located on the tongue receive innervation from bipolar neurons, which contribute axons to two cranial nerves: the facial (VII) and glossopharyngeal (IX) nerves. The cell bodies of these bipolar neurons reside in the geniculate ganglion of the facial nerve and the distal ganglion of the glossopharyngeal nerve. Axons from the geniculate ganglion (VII) exit the facial nerve outside the cranium, forming the chorda tympani nerve, which traverses the middle ear cavity. The chorda tympani nerve then merges with the lingual nerve, ultimately innervating taste buds on the anterior two-thirds of the tongue. Meanwhile, central axons originating from the distal ganglion (IX) travel to the tongue via the lingual nerve, entering the base of the tongue and providing sensory fibers to taste buds in the posterior portion.

The central processes of bipolar neurons from the geniculate and distal ganglia extend into the nucleus of the solitary tract in the medulla oblongata. Figure 13.2 explains efferent fibers originating from the nucleus of the solitary tract ascend as the solitario-thalamic tract, terminating in the ventral posteromedial nucleus of the thalamus. Thalamic neurons then project to the ipsilateral cerebral cortex, facilitating the perception of taste. Additionally, the nucleus of the solitary tract projects to the amygdala within the limbic system, linking taste perception with emotional and behavioral responses. This neural pathway highlights the complex connectivity between taste receptors on the tongue, cranial nerves, brainstem nuclei, thalamus, and cerebral cortex, as well as their integration with regions associated with emotional processing (Figure 13.2).

BIBLIOGRAPHY

Chandrashekar, J., Hoon, M. A., Ryba, N. J. P., & Zuker, C. S. (2006). The receptors and cells for mammalian taste. *Nature*, 444(7117), 288–294.

Huang, A. L., Chen, X., Hoon, M. A., Chandrashekar, J., Guo, W., Tränkner, D., Ryba, N. J. P., & Zuker, C. S. (2006). The cells and logic for mammalian sour taste detection. *Nature*, 442(7105), 934–938.

Mukherjee, J., Das, P. K., & Banerjee, D. (2023). Lactation physiology. In: P. K. Das, V. Sejian, J. Mukherjee, & D. Banerjee (Eds.), *Textbook of Veterinary Physiology*. Springer. https://doi.org/10.1007/978-981-19-9410-4_25Chrome-extension://efaidnbmnnnibpcajpcglclefindmkaj/https://ndvsu.org/images/StudyMaterials/Physio/Lactation-Growth--Environment.pdf.

Reece, W. O., Erickson, H. H., Goff, J. P., & Uemura, E. E. (Eds.) (2015). *Dukes' Physiology of Domestic Animals*. John Wiley & Sons, 2015.

Roper, S. D. (2013). Taste buds as peripheral chemosensory processors. *Seminars in Cell & Developmental Biology*, 24(1), 71–79.

14 Physiology of the Bone

Abhishek Kalundia
Cornerstone Pet Clinic Hyderabad - 89, India

14.1 INTRODUCTION TO BONE PHYSIOLOGY

14.1.1 Definition of Bones

Bones are rigid organs that form the skeletal system of vertebrates, including mammals, birds, reptiles, amphibians, and fish. They serve as the structural framework of the body, providing support, protection, and anchorage for muscles. Bones are primarily composed of connective tissue, primarily collagen, and minerals such as calcium and phosphorus. They vary in shape, size, and density depending on their function and location in the body (Figure 14.1).

14.1.2 Importance of Bones in Animals

a. **Structural Support:**

Bones provide structural support for the body, giving it shape and stability. They form the framework that holds the body upright and allows animals to move.

b. **Protection of Vital Organs:**

Bones serve as protective coverings for vital organs such as the brain, spinal cord, heart, and lungs. The skull protects the brain, while the ribcage shields the heart and lungs from injury.

c. **Muscle Attachment:**

Bones serve as attachment sites for muscles, tendons, and ligaments. Muscles attach to bones via tendons, allowing movement of body parts when muscles contract.

d. **Facilitation of Movement:**

Bones, along with muscles and joints, enable animals to move. They form levers that muscles pull on to produce movement. Different shapes and arrangements of bones facilitate various types of movement, from running and jumping to flying and swimming.

e. **Blood Cell Production:**

Within the cavities of certain bones, such as the marrow cavity of long bones, hematopoiesis occurs. Hematopoiesis is the process of blood cell formation, including red blood cells, white blood cells, and platelets. These blood cells play vital roles in oxygen transport, immune response, and blood clotting.

f. **Mineral Storage and Homeostasis:**

Bones store essential minerals such as calcium and phosphorus, which are important for bone strength and integrity. During times of dietary deficiency, bones release these minerals into the bloodstream to maintain proper mineral balance in the body.

14.1.3 Overview of Bone Structure

1. **Macroscopic Structure:**
   a. Compact Bone: Also known as cortical bone, compact bone forms the dense outer layer of bones. It consists of multiple layers called lamellae, which are arranged in concentric circles around central canals called Haversian canals. These canals contain blood vessels, lymphatic vessels, and nerves that supply nutrients to the bone tissue.
   b. Spongy Bone: Also called cancellous or trabecular bone, spongy bone forms the inner layer of bones. It consists of a network of bony trabeculae or spicules separated by spaces filled with bone marrow. Spongy bone provides structural support while reducing the overall weight of the bone.
2. **Microscopic Structure:**
   a. Osteocytes: Osteocytes are mature bone cells embedded within the bone matrix. They maintain bone tissue by regulating mineral content and responding to mechanical stress.
   b. Osteoblasts: Osteoblasts are bone-forming cells responsible for synthesizing and secreting organic components of the bone matrix, such as collagen. They play a crucial role in bone growth and repair.
   c. Osteoclasts: Osteoclasts are multinucleated cells responsible for bone resorption, the process of breaking down bone tissue. They secrete enzymes and acids that dissolve

DOI: 10.1201/9781003426851-14

Physiology of Bone in brief		
1. Bone Formation (Osteogenesis):	Osteoblasts	These are bone-forming cells responsible for synthesizing and depositing new bone matrix, primarily composed of collagen and minerals like calcium and phosphate.
	Osteocytes	Osteoblasts that become embedded in the bone matrix. They play a role in bone remodeling and mineral homeostasis.
	Bone Matrix	A combination of organic components (mainly collagen) and inorganic components (primarily hydroxyapatite crystals), providing strength and rigidity to bone.
	Bone Remodeling	Continuous process involving resorption (by osteoclasts) and formation (by osteoblasts) of bone tissue, crucial for maintenance and repair.
2. Bone Resorption (Osteoclastogenesis)	Osteoclasts	Specialized cells responsible for breaking down bone tissue. They secrete acids and enzymes that dissolve the mineral components of bone, releasing calcium and phosphate into the bloodstream.
	RANK/RANKL/OPG System	Regulation of osteoclastogenesis involves a signaling pathway between osteoblasts (which express RANKL - Receptor Activator of Nuclear Factor Kappa B Ligand) and osteoclast precursors (which express RANK - Receptor Activator of Nuclear Factor Kappa B). Osteoprotegerin (OPG) acts as a decoy receptor, inhibiting RANKL and thus osteoclastogenesis.
3. Hormonal Regulation:	Parathyroid Hormone (PTH)	Released in response to low blood calcium levels, PTH stimulates osteoclast activity and bone resorption, leading to increased calcium release into the bloodstream.
	Calcitonin	Released by the thyroid gland in response to high blood calcium levels, calcitonin inhibits osteoclast activity, reducing bone resorption.
4. Bone Growth and Repair:	Endochondral Ossification	The process by which most bones develop, involving the replacement of cartilage by bone tissue.
	Intramembranous Ossification	Formation of bone directly from mesenchymal cells without the intermediate step of cartilage formation, seen in flat bones like the skull.

FIGURE 14.1 Physiology of bones in brief

mineralized bone matrix, releasing calcium and phosphorus into the bloodstream.

d. Bone Matrix: The bone matrix consists of organic and inorganic components. The organic matrix is predominantly composed of collagen fibers, which provide tensile strength and flexibility to the bone. The inorganic matrix consists primarily of hydroxyapatite crystals, composed of calcium phosphate, which provides compressive strength and rigidity to the bone.

e. Bone Marrow: Bone marrow is a soft, gelatinous tissue found within the medullary cavities of long bones and in the spaces between trabeculae of spongy bone. It consists of two types: red bone marrow, responsible for hematopoiesis (blood cell formation), and yellow bone marrow, primarily composed of adipocytes (fat cells).

14.1.4 Terminologies and Key Words

The following are a few basic terminologies which a veterinary orthopedic surgeon should be aware of.

1. **Bone Composition**: Bones are composed of organic components (e.g., collagen) and inorganic components (e.g., hydroxyapatite) that provide strength, flexibility, and rigidity to bone tissue.
2. **Bone Development**: Bones develop through processes such as intramembranous ossification and endochondral ossification, with growth occurring at epiphyseal growth plates during childhood and adolescence.
3. **Bone Remodeling:** Bone remodeling involves the continuous balance between osteoclast-mediated bone resorption and osteoblast-mediated bone formation, which is regulated by various factors including hormones, mechanical stress, and nutritional status.
4. **Mechanical Properties:** Bones exhibit mechanical properties such as strength, elasticity, and toughness, which are influenced by factors such as bone density, structure, and mineralization.
5. **Factors Affecting Bone Health**: Nutrition, exercise, hormonal balance, genetics, age, and developmental stage are important factors affecting bone health in animals.
6. **Bone Diseases and Disorders**: Common bone diseases and disorders in animals include osteoporosis, osteoarthritis, rickets, osteomalacia, and bone fractures, which can impact mobility, quality of life, and overall health.
7. **Comparative Bone Physiology**: Bone physiology varies among different animal species based on factors such as size, habitat, locomotor behavior, and evolutionary adaptations.

14.2 COMPOSITION OF BONES

14.2.1 Organic Components

a. **Collagen**: Collagen is the primary organic component of bone tissue, accounting for approximately 90% of the organic matrix. It provides tensile strength and flexibility to bones, allowing them to withstand bending and stretching forces without breaking. Collagen fibers are arranged in a crisscross pattern, forming a scaffold for mineral deposition.

b. **Non-Collagenous Proteins**: In addition to collagen, bone tissue contains various non-collagenous proteins that contribute to its structure and function. These proteins include osteocalcin, osteopontin, bone sialoprotein, and others. They regulate mieralization, cell signaling, and bone remodeling processes.

14.2.2 Inorganic Components

a. **Hydroxyapatite**: Hydroxyapatite is the primary inorganic component of bone tissue, accounting for approximately 70% of bone mass. It is a crystalline form of calcium phosphate, represented by the chemical formula Ca10(PO4)6(OH)2. Hydroxyapatite crystals provide compressive strength and rigidity to bones, making them resistant to deformation and fracture.

b. **Other Minerals**: In addition to calcium and phosphate, bones contain trace amounts of other minerals essential for bone health, including magnesium, potassium, sodium, and fluoride. These minerals play various roles in bone metabolism, such as regulating mineral deposition and bone remodeling processes.

14.2.3 Water

Water constitutes approximately 25–30% of bone tissue by weight. It fills the spaces within the bone matrix and facilitates nutrient transport, waste removal, and cellular communication within the bone tissue. Water content influences the mechanical properties of bones, affecting their strength, stiffness, and resilience.

14.3 BONE DEVELOPMENT

14.3.1 Embryonic Bone Formation

a. **Intramembranous Ossification**: In intramembranous ossification, bones develop directly from mesenchymal (undifferentiated) connective tissue. This process occurs primarily in flat bones, such as the skull bones and certain facial bones. Mesenchymal cells differentiate into osteoblasts, which secrete osteoid (organic bone matrix) that subsequently mineralizes to form bone tissue.

b. **Endochondral Ossification:** In endochondral ossification, bones develop from a hyaline cartilage precursor. This process occurs in most of the bones in the body, including long bones such as the femur and humerus. Chondrocytes (cartilage cells) within the cartilage model undergo hypertrophy and secrete factors that attract osteoblasts to the developing bone. Osteoblasts then deposit osteoid around the cartilage model, which eventually becomes mineralized, replacing the cartilage with bone tissue.

14.3.2 Bone Growth in Juveniles

a. **Epiphyseal Growth Plate**: During childhood and adolescence, long bones grow in length primarily through the activity of the epiphyseal growth plate, also known as the growth plate or physis. The growth plate is a layer of cartilage located near the ends of long bones. Chondrocytes within the growth plate undergo mitosis, leading to the proliferation of cartilage cells. As new cartilage cells are produced, older cells undergo hypertrophy and mineralization, eventually forming bone tissue. The growth plate closes when growth ceases, marking the end of longitudinal bone growth.

b. **Closure of Growth Plates**: Or Closure and Epiphyseal Line Formation – As individuals reach skeletal maturity, the epiphyseal growth plates close, signaling the end of longitudinal bone growth. Closure occurs when chondrocytes within the growth plate stop dividing and undergo apoptosis (programmed cell death). Once the growth plate is completely replaced by bone tissue, it leaves behind a thin line of compact bone known as the epiphyseal line or metaphysis.

14.4 BONE REMODELING

14.4.1 Bone Remodeling in Animals

Bone remodeling is a dynamic process that involves the continuous resorption (breakdown) and formation of bone tissue throughout an animal's life. It is essential for maintaining bone integrity, repairing micro-damage, and adapting bone structure to changing mechanical demands. Bone remodeling is regulated by a balance between osteoclast-mediated bone resorption and osteoblast-mediated bone formation.

14.4.2 Osteoclasts and Bone Resorption

a. **Osteoclast Activation**: Osteoclasts are large, multinucleated cells derived from hematopoietic stem cells. They are responsible for breaking down bone tissue through the secretion of enzymes and acids that dissolve mineralized bone matrices.

b. **Resorption Lacunae Formation**: Osteoclasts attach to the bone surface and form specialized structures called resorption lacunae. Within these lacunae, osteoclasts create an acidic microenvironment that dissolves hydroxyapatite crystals and digests organic bone matrix, releasing calcium and phosphorus into the bloodstream.

c. **Regulation of Osteoclast Activity**: Osteoclast activity is tightly regulated by various factors, including parathyroid hormone (PTH), calcitonin, cytokines, growth factors, and mechanical stress. PTH stimulates osteoclast activity and bone resorption, whereas calcitonin inhibits osteoclast activity and promotes bone deposition.

14.4.3 Osteoblasts and Bone Formation

a. **Osteoblast Differentiation**: Osteoblasts are bone-forming cells derived from mesenchymal stem cells. They synthesize and secrete organic components of the bone matrix, including collagen and other proteins.
b. **Bone Matrix Deposition**: Osteoblasts deposit osteoid, an unmineralized organic matrix, onto the surface of bone tissue. Osteoid serves as a scaffold for mineral deposition, facilitating the formation of hydroxyapatite crystals.
c. **Mineralization**: Once osteoid is deposited, mineralization occurs as calcium and phosphate ions precipitate onto the collagen fibers, forming hydroxyapatite crystals. This process gradually transforms osteoid into mineralized bone tissue.
d. Regulation of Osteoblast Activity: Osteoblast activity is regulated by various hormonal and local factors, including bone morphogenetic proteins (BMPs), insulin-like growth factors (IGFs), transforming growth factor-beta (TGF-β), mechanical stress, and the presence of mineral ions. These factors promote osteoblast differentiation and bone formation.

14.4.4 Hormonal Regulation of Bone Remodeling

Bone remodeling involves the coordinated action of osteoclasts and osteoblasts, known as coupling. Osteoclast-mediated bone resorption creates microenvironments conducive to osteoblast activity by releasing growth factors and mineral ions from the bone matrix. Osteoblasts subsequently migrate to resorption sites and deposit new bone tissue, restoring bone structure and function.

14.5 MECHANICAL PROPERTIES OF BONES

14.5.1 Strength

a. **Tensile Strength**: Tensile strength refers to the ability of bones to resist pulling or stretching forces applied along their length. This property is primarily due to the presence of collagen fibers in the bone matrix, which provide resistance to tensile forces. Collagen fibers are arranged in a crisscross pattern, allowing bones to withstand tensile stress without fracturing.
b. **Compressive Strength**: Compressive strength refers to the ability of bones to resist crushing or compression forces applied perpendicular to their surface. This property is primarily attributed to the mineralized component of bone tissue, particularly hydroxyapatite crystals, which provide rigidity and support to the bone structure. Compressive strength enables bones to withstand the weight of the body and external loads without collapsing.

14.5.2 Elasticity

a. **Elastic Deformation**: Bones exhibit elastic deformation, meaning they can temporarily deform under stress and return to their original shape once the stress is removed. This property allows bones to absorb energy and deform slightly without sustaining permanent damage. Elastic deformation is primarily attributed to the collagen fibers in the bone matrix, which can stretch and recoil.
b. **Plastic Deformation**: In addition to elastic deformation, bones can undergo plastic deformation, where they deform permanently under excessive stress. Plastic deformation occurs when the applied stress exceeds the bone's elastic limit, causing irreversible changes in bone shape or structure. Prolonged or repetitive loading can lead to micro-fractures and remodeling of bone tissue.

14.5.3 Toughness

a. **Toughness**: Toughness is a measure of a material's ability to absorb energy and deform plastically before fracturing. In the context of bones, toughness refers to their ability to withstand impact and loading without fracturing. Toughness is influenced by the combination of tensile strength, compressive strength, and elasticity. Collagen fibers provide toughness by absorbing energy and dissipating it throughout the bone matrix, reducing the risk of fracture.

14.5.4 Adaptations for Different Animal Lifestyles

a. **Bone Density and Shape**: Animals exhibit variations in bone density and shape based on their lifestyle and activity level. For example, terrestrial mammals often have denser bones with thick cortical bone layers to support body weight and resist bending forces. In contrast, flying birds have lightweight bones with thin cortical layers and extensive pneumatic cavities to reduce weight and facilitate flight.

b. **Wolff's Law**: Wolff's Law states that bone architecture adapts to the mechanical demands placed upon it. Bones remodel in response to changes in loading patterns, with increased mechanical stress leading to bone deposition and decreased stress leading to bone resorption. This adaptive process helps optimize bone structure and strength based on an animal's specific requirements.

14.6 BONE DISEASES AND DISORDERS

14.6.1 Osteoporosis

a. Definition: Osteoporosis is a metabolic bone disease characterized by a decrease in bone density and mass, resulting in increased bone fragility and susceptibility to fractures.
b. Causes: Osteoporosis in animals can result from various factors, including hormonal imbalances (e.g., decreased estrogen levels), nutritional deficiencies (e.g., calcium, vitamin D), aging, genetics, and certain medical conditions (e.g., hyperparathyroidism).
c. Symptoms: Animals with osteoporosis may exhibit signs of lameness, reluctance to move, decreased activity level, and fractures with minimal trauma.
d. Treatment: Treatment strategies for osteoporosis in animals may include dietary supplementation with calcium and vitamin D, hormone replacement therapy (e.g., estrogen), bisphosphonate medications to inhibit bone resorption, and exercise to promote bone strength and density.

14.6.2 Osteoarthritis

a. **Definition**: Osteoarthritis is a degenerative joint disease characterized by progressive deterioration of articular cartilage, inflammation of the synovial membrane, and formation of osteophytes (bone spurs) at joint margins.
b. **Causes**: Osteoarthritis in animals can result from various factors, including aging, obesity, joint trauma or injury, abnormal joint conformation, genetic predisposition, and chronic overuse.
c. **Symptoms**: Animals with osteoarthritis may exhibit signs of joint stiffness, swelling, pain, lameness, decreased range of motion, and reluctance to engage in physical activity.
d. **Treatment**: Treatment options for osteoarthritis in animals may include weight management, non-steroidal anti-inflammatory drugs (NSAIDs) for pain relief, joint supplements (e.g., glucosamine, chondroitin sulfate), physical therapy, joint injections (e.g., corticosteroids, hyaluronic acid), and surgical interventions (e.g., arthroscopy, joint replacement).

14.6.3 Rickets and Osteomalacia

a. **Definition**: Rickets and osteomalacia are metabolic bone diseases characterized by defective mineralization of bone tissue, leading to soft, weak bones.
b. **Causes**: Rickets and osteomalacia in animals are typically caused by nutritional deficiencies, particularly inadequate intake or absorption of calcium, phosphorus, and vitamin D. Other contributing factors may include renal disease, liver dysfunction, and certain medications (e.g., anticonvulsants).
c. **Symptoms**: Animals with rickets and osteomalacia may exhibit signs of bone deformities, growth retardation, skeletal pain, lameness, and fractures with minimal trauma.
d. **Treatment**: Treatment of rickets and osteomalacia in animals involves addressing the underlying nutritional deficiencies through dietary supplementation with calcium, phosphorus, and vitamin D. In severe cases, veterinary intervention may be necessary to correct skeletal deformities and manage associated complications.

14.6.4 Bone Fractures and Healing Process

a. **Definition**: Bone fractures occur when bones break or crack due to excessive force or trauma. The healing process involves a series of biological events aimed at repairing the fracture and restoring bone integrity.
b. **Causes**: Bone fractures in animals can result from various causes, including falls, collisions, blunt trauma, repetitive stress, and pathological conditions (e.g., bone tumors).
c. **Symptoms**: Animals with bone fractures may exhibit signs of pain, swelling, deformity, instability, and reluctance to bear weight on the affected limb.
d. **Treatment**: Treatment of bone fractures in animals may involve immobilization with splints, casts, or external fixation devices, surgical stabilization with internal fixation (e.g., plates, screws, pins), pain management, and rehabilitation therapy. The healing process typically involves inflammation, callus formation, remodeling, and eventual restoration of bone structure and function.

14.7 FACTORS AFFECTING BONE HEALTH

14.7.1 Nutrition

a. **Calcium and Phosphorus**: Adequate intake of calcium and phosphorus is essential for maintaining bone mineralization and strength in animals. These minerals serve as the building blocks of hydroxyapatite crystals, which provide rigidity to bone tissue.
b. **Vitamin D**: Vitamin D plays a crucial role in calcium absorption and utilization by promoting the expression of proteins involved in intestinal calcium transport. Animals deficient in vitamin D may experience impaired bone mineralization and increased risk of metabolic bone diseases.
c. **Protein**: Protein is necessary for collagen synthesis, which contributes to the organic matrix of bone tissue. Inadequate protein intake can impair bone formation and repair processes in animals.

14.7.2 Exercise and Physical Activity

a. **Weight-Bearing Exercise**: Weight-bearing exercise and physical activity stimulate bone formation and remodeling by subjecting bones to mechanical stress. This stress triggers osteoblast activity and deposition of new bone tissue, leading to increased bone density and strength.
b. **Impact and High-Intensity Exercise**: Impact and high-intensity exercise, such as running, jumping, and resistance training, are particularly effective in promoting bone health by inducing greater mechanical loading on bones.

14.7.3 Hormonal Balance

a. **Estrogen**: Estrogen plays a crucial role in maintaining bone density and inhibiting bone resorption in animals. Decreased estrogen levels, such as those seen during menopause in females or following ovariectomy (spaying), can lead to accelerated bone loss and increased risk of osteoporosis.
b. Parathyroid Hormone (PTH): PTH regulates calcium and phosphorus metabolism by stimulating osteoclast activity and bone resorption. Abnormalities in PTH secretion or function can disrupt bone homeostasis and lead to metabolic bone diseases, such as hyperparathyroidism or hypoparathyroidism.

14.7.4 Genetics

a. **Bone Structure and Density**: Genetic factors influence bone structure, density, and mineralization in animals. Certain genetic mutations or variations may predispose animals to osteoporosis, osteogenesis imperfecta (brittle bone disease), or other bone disorders.
b. **Bone Remodeling**: Genetic factors can also affect the regulation of bone remodeling processes, including osteoblast and osteoclast activity. Variations in genes encoding bone-related proteins may influence bone turnover rates and susceptibility to bone diseases.

14.7.5 Age and Developmental Stage

a. **Growth and Development**: Bone health is influenced by age and developmental stage in animals. During growth and development, adequate nutrition and hormonal signaling are critical for achieving optimal bone mineralization and skeletal growth.
b. **Senescence**: Aging is associated with changes in bone structure and composition, including decreased bone density, impaired bone remodeling, and increased susceptibility to fractures. Aging-related hormonal changes and chronic diseases can further exacerbate bone loss in older animals.

14.8 COMPARATIVE BONE PHYSIOLOGY

14.8.1 Variation in Bone Structure among Different Animal Groups

a. **Variations in Bone Structure**: Bones exhibit structural variations among different animal species based on factors such as size, shape, and biomechanical requirements. For example, terrestrial mammals often have dense cortical bone with thick trabeculae to support body weight and resist bending forces, while birds have lightweight bones with thin cortical layers and extensive pneumatic cavities to facilitate flight.
b. **Composition of Bone Tissue**: While the basic components of bone tissue (e.g., collagen, hydroxyapatite) are similar across vertebrate species, there may be variations in the relative proportions and organization of these components. For instance, marine mammals may have specialized adaptations to cope with the challenges of diving, such as increased bone density and mineralization to withstand hydrostatic pressure.

14.8.2 Growth Pattern and Adaptations

a. **Differences in Growth Patterns**: Animals exhibit diverse growth patterns and rates depending on factors such as lifespan, metabolic rate, and reproductive strategy. For example, fast-growing species may have accelerated skeletal growth during early development to reach sexual maturity quickly, while slow-growing species may have more prolonged growth periods.

b. **Specialized Adaptations**: Some animal groups have evolved specialized skeletal adaptations to suit their ecological niche or locomotor strategy. For instance, arboreal animals may have elongated limb bones and flexible joints for climbing, while cursorial animals may have shortened limb bones and robust skeletons for running.

14.8.3 Adaptations for Locomotion

The mechanical properties of bones are closely linked to an animal's locomotor behavior and habitat. Animals with specialized locomotor modes (e.g., running, swimming, flying) have bones adapted to withstand the specific mechanical stresses associated with their movement patterns. For example, the wings of birds and bats are lightweight yet strong to support flight, while the limbs of terrestrial mammals are optimized for weight-bearing and propulsion.

14.8.4 Bone Remodeling and Adaptation

a. **Environmental Influences**: Bone remodeling and adaptation are influenced by environmental factors such as diet, activity level, and habitat. Animals living in challenging environments (e.g., deserts, high altitudes) may undergo skeletal adaptations to cope with limited resources or extreme conditions. For example, desert-dwelling animals may have adaptations for water conservation, leading to changes in bone structure and mineral metabolism.

b. **Functional Demands**: Bone remodeling is driven by functional demands placed on the skeleton, such as mechanical loading and stress distribution. Animals engaged in repetitive or strenuous activities may undergo bone remodeling to strengthen and optimize their skeletal structure for enhanced performance and durability.

14.9 CLINICAL APPLICATIONS OF PHYSIOLOGY OF BONE

14.9.1 Diagnosis and Management of Bone Diseases

a. **Osteoporosis**: Understanding bone physiology is crucial for diagnosing and managing osteoporosis in animals. Veterinarians utilize techniques such as dual-energy X-ray absorptiometry (DEXA) to assess bone density and identify animals at risk of osteoporotic fractures. Management strategies may include nutritional supplementation, exercise prescription, and pharmacological interventions to prevent bone loss and reduce fracture risk.

b. **Osteoarthritis**: Knowledge of bone physiology informs the diagnosis and treatment of osteoarthritis in animals. Veterinarians may recommend joint supplements, physical therapy, and pain management strategies to alleviate symptoms and improve joint function. In severe cases, surgical interventions such as arthroscopy or joint replacement may be necessary to restore mobility and quality of life.

c. **Rickets and Osteomalacia**: Understanding the role of calcium, phosphorus, and vitamin D in bone metabolism is essential for diagnosing and treating rickets and osteomalacia in animals. Veterinary professionals may prescribe dietary supplements and monitor serum mineral levels to correct deficiencies and promote proper bone mineralization.

14.9.2 Orthopedic Surgery and Rehabilitation:

a. **Fracture Repair**: Knowledge of bone physiology guides orthopedic surgeons in the treatment of fractures in animals. Surgical techniques such as open reduction and internal fixation (ORIF) or external fixation may be employed to stabilize fractured bones and facilitate healing. Post-operative rehabilitation programs, including controlled exercise, physical therapy, and pain management, play a critical role in optimizing bone healing and restoring function.

b. **Joint Surgery:** Understanding bone physiology is essential for performing joint surgeries such as arthroscopy, osteotomy, and joint replacement in animals. Surgeons must consider factors such as bone density, mechanical loading, and joint biomechanics when planning and executing surgical procedures to achieve optimal outcomes and minimize complications.

14.9.3 Nutritional Management

a. **Dietary Recommendations:** Knowledge of bone physiology informs nutritional recommendations for animals at risk of bone diseases or disorders. Veterinary professionals may advise on appropriate calcium, phosphorus, and vitamin D supplementation, as well as the use of joint supplements containing glucosamine and chondroitin sulfate to support bone and joint health.

b. **Weight Management**: Maintaining a healthy body weight is essential for bone health in animals. Veterinary professionals may provide dietary counseling and weight management strategies to prevent obesity-related bone disorders and reduce the risk of orthopedic conditions such as osteoarthritis and joint strain.

14.9.4 Preventive Medicine and Wellness Care

a. **Health Screening:** Regular health screenings and wellness exams allow veterinary professionals to assess bone health and detect early signs of bone diseases or disorders in animals. Diagnostic tests such as radiography, DEXA scanning, and blood chemistry panels may be utilized to evaluate bone density, mineral metabolism, and overall musculoskeletal function.
b. **Exercise Prescription**: Veterinary professionals may prescribe tailored exercise programs to promote bone health and prevent musculoskeletal injuries in animals. Exercise interventions may include activities such as walking, swimming, and agility training, which help maintain bone density, strengthen muscles, and improve joint mobility.

14.10 FUTURE DIRECTIONS IN BONE PHYSIOLOGY RESEARCH

a. **Molecular Mechanisms**: Further elucidation of the molecular mechanisms regulating bone remodeling, mineralization, and mechanotransduction pathways in animals.
b. **Genetic Studies**: Investigation of genetic factors influencing bone development, metabolism, and susceptibility to bone diseases and disorders in various animal species.
c. **Biomechanical Studies:** Integration of biomechanical modeling and simulation techniques to study bone structure-function relationships and predict bone response to mechanical loading and stress.
d. **Nutritional Interventions**: Evaluation of novel nutritional strategies and dietary supplements for optimizing bone health and preventing bone diseases in animals.
e. **Therapeutic Approaches**: Development of novel therapeutic interventions, including pharmacological agents, gene therapies, and tissue engineering strategies, for the treatment of bone diseases and disorders in animals.
f. **Translational Research**: Translation of findings from basic research in bone physiology to clinical applications and veterinary practice, with a focus on improving diagnostic techniques, treatment options, and rehabilitation protocols for bone-related conditions in animals.

To summarize, we can say that continued research in bone physiology is essential for advancing our understanding of skeletal biology, improving animal welfare, and developing innovative strategies for the prevention and treatment of bone diseases and disorders in animals. By addressing key knowledge gaps and exploring new avenues of investigation, we can ultimately enhance the health and longevity of animals across diverse species and environments.

14.11 CONCLUSION

In conclusion, bone physiology plays a critical role in the structural integrity, locomotion, and overall health of animals. Throughout this chapter, we have explored various aspects of bone physiology, including bone composition, development, remodeling, mechanical properties, and factors influencing bone health. Bones serve as dynamic organs that undergo constant remodeling in response to mechanical, hormonal, and environmental cues. Understanding the mechanisms underlying bone physiology is essential for the prevention, diagnosis, and treatment of bone diseases and disorders in animals.

BIBLIOGRAPHY

Alexander, R. McN. (2003). *Principles of animal locomotion*. Princeton University Press.

Carter, D. R., & Beaupré, G. S. (2001). *Skeletal function and form: Mechanobiology of skeletal development, aging, and regeneration*. Cambridge University Press.

Currey, J. D. (2002). *Bones: Structure and mechanics*. Princeton University Press.

Fossum, T. W. (2018). *Small animal surgery*. Elsevier Health Sciences.

Hall, J. E., & Guyton, A. C. (2015). *Guyton and hall textbook of medical physiology*. Saunders/Elsevier.

Kanis, J. A., & Johnell, O. (2005). Requirements for DXA for the management of osteoporosis in Europe. *Osteoporosis International*, 16(3), 229–238.

Kardong, K. V. (2018). *Vertebrates: Comparative anatomy, function, evolution*. McGraw-Hill Education.

Marieb, E. N., & Hoehn, K. (2019). *Human anatomy & physiology*. Pearson Education.

Martin, R. B., Burr, D. B., & Sharkey, N. A. (1998). *Skeletal tissue mechanics*. Springer Science & Business Media.

Seibel, M. J. (2006). Molecular markers of bone turnover: Biochemical, technical and analytical aspects. *Osteoporosis International*, 17(4), 429–435.

Thrall, D. E. (2015). *Textbook of veterinary diagnostic radiology*. Elsevier Health Sciences.

Tobias, K. M., & Johnston, S. A. (2018). *Veterinary surgery: Small animal*. Elsevier Health Sciences.

15 Physiology of the Digestive System of Monogastric Animals

S. K. I. Vasantha[1], M. P. S. Tomar[1], M. Naveen Swaroop[1] and Devendra Pathak[2]
[1]Sri Venkateswara Veterinary University, Tirupati, India
[2]Guru Angad Dev Veterinary & Animal Sciences University, Ludhiana, India

15.1 INTRODUCTION

Cattle and horses predominantly consume plant matter. They possess voluminous gastrointestinal tracts, often featuring complex structures like multiple-chambered stomachs, as seen in ruminants. Digestion in herbivores primarily relies on symbiotic microbial fermentation processes to break down fibrous plant material. Carnivores exhibit comparatively simpler and smaller digestive systems optimized for rapid enzymatic breakdown of proteins and fats derived from animal sources. Species such as pigs and certain avian varieties display dietary versatility, consuming both plant and animal matter. Their digestive physiology encompasses a combination of enzymatic processes akin to carnivores and microbial fermentation similar to herbivores, facilitating the digestion of a diverse range of foodstuffs. Animals exhibit varied digestive adaptations aligned with their dietary preferences, ranging from specialized herbivores to carnivores and adaptable omnivores. Although animals consume food in its natural state, the conversion of nutrients from a non-absorbable to an assimilable form is imperative for proper utilization by the organism. This transformative process, known as digestion, is orchestrated by the gastrointestinal tract (GIT). Key facets of digestion include

- **Movement of Food:** Food traverses the alimentary tract through peristaltic contractions, ensuring its sequential passage from the oral cavity to the stomach and onward through the intestines.
- **Secretion of digestive juices**: Specialized glands within the GIT secrete digestive juices replete with enzymes and other requisite compounds essential for the enzymatic breakdown of food constituents.
- **Enzymatic/ microbial digestion**: Upon arrival in the stomach and intestines, food undergoes enzymatic degradation facilitated by the enzymes present in digestive secretions. In certain instances, microbial communities also contribute to the breakdown of complex substrates.
- **Absorption of nutrients:** The intestinal mucosa, furnished with villi and microvilli, orchestrates the absorption of digested nutrients into the circulatory and lymphatic systems for subsequent distribution to cells and tissues throughout the organism.
- **Control of digestive function:** Regulatory mechanisms, encompassing neural reflexes and hormonal signaling pathways, orchestrate the precise coordination and modulation of digestive processes to optimize nutrient assimilation and waste elimination.

15.2 PHYSICAL FEATURES OF THE GIT IN DIFFERENT ANIMALS

The digestive tracts of various animals exhibit remarkable diversity in structure and function. Dogs possess a relatively short and uncomplicated digestive tract, highlighted by an ileocecal valve and a small cecum. In contrast, pigs feature a simple stomach similar to dogs, yet their small and large intestines are comparatively longer. Notably, the pig's cecum and colon showcase sacculations known as haustrae, enhancing the retention of digesta. Equines stand out for their exceptionally large colon, facilitating efficient digestion. Among domestic and laboratory animals, sheep boast the longest small intestines in proportion to their body size. However, sheep lack sacculations or significant voluminosity in their cecum and colon, distinguishing them from other species (Table 15.1). These variations underscore the adaptation of digestive systems to different dietary and physiological needs across animal taxa (Lentle and Janssen, 2008)

15.3 GASTROINTESTINAL BLOOD FLOW

The hepatic portal system serves as a vital conduit for blood originating from the gastrointestinal tract, spleen, and pancreas, directing it exclusively to the liver before it enters systemic circulation. This selective routing enables the liver to fulfill its pivotal role as a metabolic regulator and detoxifier. By subjecting blood borne substances to hepatic processing prior to their dissemination throughout the body, this mechanism ensures that potentially harmful agents, such as bacteria and toxins absorbed from the gastrointestinal tract, are intercepted and neutralized before they can exert systemic effects, thereby safeguarding overall

DOI: 10.1201/9781003426851-15

TABLE 15.1
Endocrine Activity of the GIT

Name of the hormone	Source	Stimulus for secretion	Biological action	Reference
Gastrin	G-cells in pyloric glands	Peptides and amino acids in stomach	Stimulation of gastric secretion and mucosal growth	Duan et al., 2022
Cholecystokinin (CCK)	I-cells in the mucosa of duodenum and jejunum	Fats, Fatty acids and amino acids	Stimulation of Pancreatic enzyme secretion, Gall bladder contractions, and small intestinal secretion and moderate inhibition of gastric motility	Wang et al., 2019
Secretin	S-cells in the mucosa of the duodenum	Acid chyme	Stimulation of Pancreatic electrolyte secretion and small intestinal secretions and mild inhibition of gastric motility and secretion	Ma & Lee, 2020
Gastric inhibitory peptide (GIP)	Upper and lower intestine	Glucose, amino acids and fatty acids	Inhibition of gastric motility and secretion and stimulation of intestinal secretion and insulin	Reimann et al., 2020
Enteroglucagon	Upper and lower intestine	Glucose and fatty acids	Stimulation of blood flow and mitotic activity	Nightingale & Spiller, 2023
Motilin	Upper and lower intestine	variable	Stimulation of MMC in antroduodenum	Deloose et al., 2019
Neurotensin	Upper and lower intestine and colon	Fat	Stimulation of vasodilatation and inhibition of smooth muscles	Janig, 2022.
Pancreatic poly peptide	Pancreas	Protein	Stimulation of pancreatic secretion	Chandra & Liddle, 2020
Vasoactive intestinal polypeptide (VIP)			Moderate inhibition of gastric secretion	Iwasaki et al., 2019
Neuropepetide Y			Stimulates feeding	Assan et al., 2021
Somatostatin	D cells in the gastric glands		Inhibits the release of gastrin and hydrochloric acid	Chen et al., 2023
Incretin	Duodenum	Glucose	Stimulates insulin secretion	Holst et al., 2021
Ghrelin	Stomach		Stimulates feeding	Davis, 2018

physiological integrity. Majority of non-fat, water-soluble nutrients absorbed from the gastrointestinal tract are transported via the portal venous system directly to the liver sinusoids. Consequently, both the reticuloendothelial cells and the parenchymal cells of the liver play pivotal roles in the absorption and temporary storage of approximately 50-75% of these absorbed nutrients.

Moreover, blood flow within various segments and layers of the gastrointestinal tract fluctuates in response to physiological activity. During heightened activity, such as digestion, blood flow to the gastrointestinal tract increases. This augmentation is mediated by a complex interplay of gastrointestinal hormones including cholecystokinin (CCK), vasoactive intestinal peptide (VIP), gastrin, and secretin, alongside vasodilatory agents like kinins (kallidin and bradykinin), diminished oxygen levels, and the release of adenosine. These factors collectively induce vasodilation, thereby facilitating enhanced blood flow to the gastrointestinal tract, which is essential for optimal digestive and absorptive processes (Wyatt & Haugk, 2018).

15.4 NEURAL CONTROL OF GIT

The gastrointestinal tract (GIT) boasts its own intrinsic nervous system known as the enteric nervous system (ENS), which operates independently of the central nervous system. However, the GIT also undergoes regulation via the autonomic nervous system. The ENS comprises two primary plexuses: the myenteric plexus (or Auerbach's plexus), positioned between the longitudinal and circular muscle layers, and the submucosal plexus (or Meissner's plexus), situated within the submucosa. The myenteric plexus predominantly governs gastrointestinal movements, while Meissner's plexus orchestrates secretory activities and regulates local blood flow. In terms of neural innervation, parasympathetic input to the GIT is facilitated by the vagus and pelvic nerves, whereas sympathetic innervation occurs via the post-ganglionic abdominal plexus. Broadly, parasympathetic nerves promote contractile activity and secretion within the GIT, whereas sympathetic nerves tend to inhibit these functions. This intricate interplay between the enteric, parasympathetic, and sympathetic nervous systems ensures coordinated regulation of gastrointestinal motility, secretion, and blood flow, contributing to digestive efficiency and homeostasis (Sharkey & Mawe, 2023).

15.5 MOTOR CONTROL OF THE GIT

The gastrointestinal tract (GIT) orchestrates a continuous process of mixing ingested food with digestive secretions and propelling it through coordinated movements. Within the GIT, two fundamental types of functional movements are discernible.

Mixing movements: These movements, characterized by contractions of the gastrointestinal musculature, serve to thoroughly intermingle ingested food with digestive juices and enzymes. By fostering the fragmentation of large food particles and promoting homogenous dispersion of digestive secretions, mixing movements optimize the efficiency of subsequent digestion and absorption processes.

Propulsive movements: These movements, facilitated by rhythmic contractions of the gastrointestinal smooth muscle, propel food forward along the length of the GIT at an appropriate velocity conducive to effective digestion and absorption. By sequentially advancing ingested material through successive compartments of the alimentary canal, propulsive movements ensure a continuous flow of nutrients, facilitating their optimal utilization by the body. Collectively, these intricate mixing and propulsive mechanisms within the GIT contribute to the comprehensive breakdown and assimilation of ingested nutrients, thereby supporting essential physiological functions and metabolic processes (Tellez-Isaias, 2023).

15.5.1 Prehension

Prehension, the process of seizing and conveying food to the mouth, is facilitated by specialized organs in domestic animals, primarily the lips, teeth, and tongue. In animals such as cats and dogs, the forelimbs play a crucial role in holding food, while head and jaw movements aid in guiding the food into the mouth. Domestic animals create a negative pressure within the oral cavity to drink liquids effectively. Cats and dogs, for instance, utilize the mobile tip of their tongue to form a ladle-like shape, enabling them to scoop and carry liquids into their mouths. This intricate combination of anatomical adaptations and coordinated movements ensures efficient prehension and ingestion of food and liquids in domestic animals (Davis & Davis, 2019).

15.5.2 Mastication

Mastication, the mechanical breakdown of food in the oral cavity, serves to reduce particle size and increase surface area, facilitating subsequent enzymatic digestion in the GIT. Additionally, mastication promotes thorough mixing of food with saliva, aiding in the process of deglutition. Herbivores exhibit adept masticatory capabilities, characterized by the efficient grinding of plant material, while carnivores typically lack such specialization. Incisors are utilized by organisms to initiate food fragmentation, while molars are employed for further comminution and grinding (Priyanka et al., 2023).

15.5.3 Deglutition

Deglutition, the physiological process of swallowing, entails the coordinated contraction and relaxation of muscles in the pharynx and esophagus, facilitating the movement of ingested material from the oral cavity to the stomach. Both mastication and deglutition are intricately regulated reflex actions governed by the central nervous system (CNS). The swallowing center, situated within the medulla oblongata of the brainstem, plays a pivotal role in orchestrating the sequential muscle contractions and relaxations essential for efficient swallowing. Sensory information from receptors in the oral cavity, pharynx, and esophagus is transmitted to the swallowing center, which in turn coordinates the efferent signals to the relevant muscle groups involved in the swallowing process. This sophisticated neural control mechanism ensures the timely and coordinated progression of food boluses through the upper gastrointestinal tract (McQuilken, 2021).

15.6 COMPOSITION OF SALIVA AND ITS MAJOR FUNCTIONS

Water content: Salivary fluid is predominantly water, constituting approximately 99% of its volume. This aqueous base facilitates various physiological functions within the oral cavity.

Electrolytes: Saliva contains a variety of electrolytes essential for maintaining electrolyte balance and pH regulation. These include sodium, potassium, calcium, chloride, magnesium, bicarbonate, and phosphate ions.

Proteins: Saliva contains a diverse range of proteins, including enzymes, immunoglobulins (antibodies), and other antimicrobial factors. These proteins play crucial roles in digestion, immune defense, and maintaining oral health. Examples of enzymes present in saliva include salivary amylase for carbohydrate digestion and lysozyme for antibacterial action.

Mucosal Glycoproteins: Saliva also contains mucosal glycoproteins, which contribute to the lubricating and protective properties of saliva, helping to maintain oral mucosal integrity.

Traces of Albumin and polypeptides/oligopeptides: Small amounts of albumin, along with various polypeptides and oligopeptides, are present in saliva. These substances may have roles in oral health and immune function.

Glucose and Nitrogenous products: Saliva may contain glucose, along with nitrogenous products such as urea and ammonia. These substances may originate from systemic circulation or local metabolic processes within the salivary glands.

The interactions among these components contribute to the multifunctional nature of saliva, which includes lubrication, digestion, antimicrobial defense, pH regulation, and maintenance of oral health. Saliva's diverse composition reflects its importance in oral physiology and overall well-being (Ekstrom et al., 2019).

The stomach is lined with mucus-secreting cells, but it also contains tubular gland types: the pyloric and oxyntic glands. HCL, pepsinogen, intrinsic factor, and mucus are secreted by the oxyntic glands. Pepsinogen and the hormone gastrin are also secreted by the pyloric glands, but mucus predominates.

15.6.1 Pyloric Glands

These glands release gastrin and mucus. These cells secrete a mucus that facilitates the passage of digesta into the small intestine. Furthermore, it neutralizes the acid and shields the stomach wall from the hydrochloric acid's corrosive effects. Mucus-secreting cells that release thick, viscous mucus that both protects the stomach wall and facilitates food movement line the whole luminal surface of the stomach. Because this mucus has an alkaline composition, it shields the stomach wall from the powerful hydrochloric acid.

15.6.2 Oxyntic Glands

The three cell types that make up this structure are the neck cells that secrete mucus, the peptic or chief cells that secrete pepsinogen, and the parietal cells that secrete hydrochloric acid and intrinsic factor (Figure 15.1). The acid solution secreted by activated parietal cells is isotonic with bodily fluids and has a pH of 0.8. When gastric acid and previously synthesized pepsin come into contact with the inactive form of pepsinogen, it gets activated. An active proteolytic enzyme with a pH range of 1.8 to 3.5 is called pepsin. Gastric lipase is a tributyrase; which breaks down butterfat but is not lipolytically active toward other types of fat. There is no statistical significance even with the tributyrase activity. While gelatinase aids in the liquefaction of some of the meat's proteoglycans, gastric amylase has little effect on the digestion of carbohydrates. Along with hydrochloric acid, the parietal cells release the chemical called intrinsic factor, which is necessary for the absorption of vitamin B12. It is commonly recognized that the lack of intrinsic factor causes pernicious anemia because vitamin B12 is needed to stimulate erythropoiesis in the bone marrow (Paxton, 2020).

15.6.3 Pancreatic Secretions

Chyme, a mixture of partially digested food, gastric, and salivary secretions, exits the stomach and is assisted in its further digestion in the small intestine by biliary and pancreatic secretions. Pancreatic juice is a transparent, alkaline secretion that is divided between an organic phase that contains enzymes and an aqueous phase with a high bicarbonate level and low chloride content. The preservation of a pH that is appropriate for the pancreatic enzymes' activity, as well as the partial neutralization of the acid chyme, depends on the bicarbonate ion. Enzymes for the digestion of the three main nutritional categories – carbohydrates, proteins, and fats – are present in the organic phase. Important proteolytic enzymes in the pancreatic juice are trypsin, chymotrypsin, and carboxypeptidase, although several elastases and nucleases are also present. Trypsin and chymotrypsin split the whole and partially digested protein into peptides of various sizes but do not release individual amino acids. Carboxypeptidase, on the other hand, splits peptides into individual amino acids.

As procarboxypeptidase, chymotrypsinogen, and trypsinogen, pancreatic proteolytic enzymes are secreted in an inactive state. They need to be secreted into the intestinal system before they become active. Trypsinogen is activated when the intestinal mucosa comes into contact with an enzyme known as enterokinase, which stimulates the chyme. It is also possible for trypsin to autocatalytically activate trypsinogen. Trypsin stimulates the enzymes chymotrypsinogen and procarboxypeptidase. The primary enzyme in the pancreatic juice that breaks down carbohydrates is pancreatic amylase. With the exception of cellulose, it hydrolyzes starches, glycogen, and most other carbs to produce disaccharides and a small number of trisaccharide (Agrawal & Aoun, 2014).

Water and bicarbonate ions are secreted into the pancreatic juice by the epithelial cells of the ductules and ducts that lead from the acini. The concentration of bicarbonate ions in pancreatic juice can rise five times higher than that in blood plasma when the pancreas is stimulated to release large amounts of its juice.

15.6.4 Bile

Bile is an alkaline secretion that has a greenish or greenish-yellow color and contains bile salts, bilirubin, cholesterol, lecithin, and the regular electrolytes found in plasma. Not because it contains any enzymes, but rather because of its salts, bile is crucial for the breakdown of fat. Additionally, bile acts as an excretory route for cholesterol and bilirubin, two byproducts of hemoglobin breakdown. With the exception of deoxycholic acid, these are cyclo-pentano-phenanthrene molecules produced from cholesterol. The four main

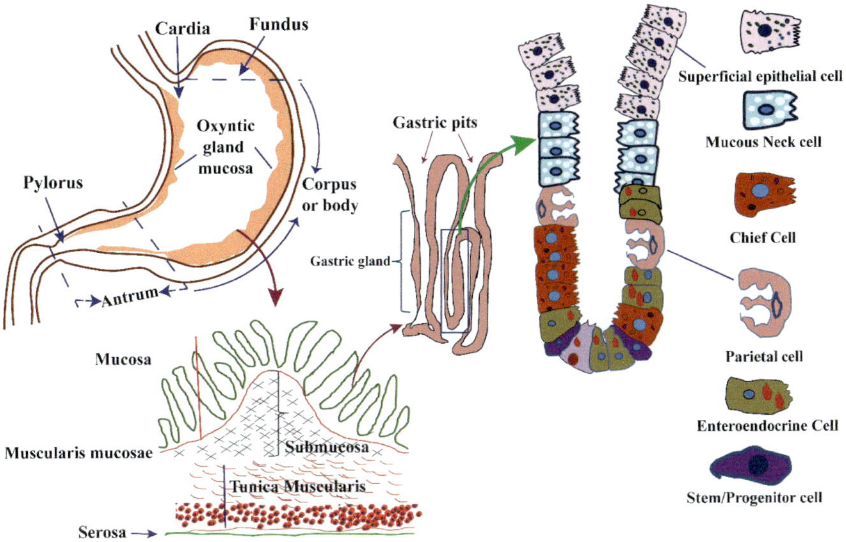

FIGURE 15.1 Schematic diagram of the stomach showing different layers and various cell types in stomach.

bile acids are lithocholic, chenodeoxycholic, deoxycholic, and cholic. They are combined to create taurocholic and glycocholic acids by conjugating either with taurine or glycine. Bile contains the salts of these acids, primarily the sodium salts. Within the gastrointestinal tract, the bile salts serve two crucial purposes: (i) Bile salts' emulsifying or detergent action: This lowers fat's surface tension and enables intestinal movement to break fat globules into tiny pieces. (ii) They facilitate the intestinal tract's absorption of fatty acids, monoglycerides, cholesterol, and other lipids. They accomplish this by joining the lipids to form tiny complexes. The soluble compounds are known as micelles. (Rao, 1998).

15.6.5 Regulation of Bile Secretion

It is regulated by chemicals, hormones, and the nervous system. Although total hepatic denervation does not completely stop bile production, vagal stimulation does. Pancreatic and bile secretions run parallel to one another because secretin and chemicals that stimulate secretin also stimulate bile production. The second part of the bile, which is high in water and electrolytes, is secreted when secretin stimulates the duct cells.

The sphincter of Oddi relaxes, and the gall bladder contracts to release the stored bile, which is caused by cholecystokinin, vagal fibers, and ENS fibers. The most significant choleretics (that induce bile secretion) are bile salts. Although they are fully absorbed in the small intestine and returned to the liver, they are released in the bile. The entero-hepatic circulation is the term for this process of recycling bile salts. Bile salts are typically recycled at least six times a day prior to excretion.

15.6.6 Secretions of the Small Intestine

The first few centimeters of the duodenum include a large number of compound mucus glands known as Brunner's glands. In response to vagal activation, secretin, and tactile or irritating stimuli of the mucosa above, these glands release copious volumes of alkaline mucus. The duodenum wall is shielded from the extremely acidic stomach juice by this mucus. Furthermore, the extra bicarbonate ions in this mucus complement those released in pancreatic juice and bile and balance the hydrochloric acid that enters the duodenum from the stomach. The Brunner's glands are inhibited by sympathetic activation (Carr & Toner, 1984).

Small pits known as Lieberkühn crypts are found throughout the small intestine. The small intestine's interstitial spaces are home to these crypts. In the crypts, a large number of enterocytes secrete large amounts of water and electrolytes (especially bicarbonate and chloride, which are actively secreted followed by sodium and water passively), while the villi's intestinal surface is covered by a moderate number of goblet cells that secrete mucus, which lubricates and protects the small intestinal surface. As the chyme comes into contact with the villi, the fluid that circulates from crypts to villi provides a wet medium for the contents to be absorbed. Enzymes are absent from pure small intestine secretions that do not contain cellular detritus. On the other hand, the enterocytes on the villi do contain enzymes that break down nutrients as they are absorbed. These enterocytes comprise the following enzymes: (i) modest levels of lipases, which divide neutral fat into glycerol and fatty acids; (ii) sucrase, maltase, isomaltase, and lactase, which split disaccharides into monosaccharides; and (iii) peptidases, which split short peptides into amino acids.

15.6.7 Secretions in the Large Intestine

Despite the large intestine having several Lieberkühn crypts as well, the epithelial cell lacks villi and enzymes. Mucus-secreting cells are the only type of cells found in the large intestine crypts. Bicarbonate released by cells other than those that secrete mucus is present in this mucus in rather moderate levels. Mucus secretion is regulated by tactile stimulations and local nerve responses. When the spinal cord stimulates the pelvic nerves, parasympathetic innervation reaches the crypts, resulting in abundant mucus secretion.

In addition to protecting the mucosa, the mucus released in the large intestine serves as an adhesive medium, keeping the feces together. It acts as a barrier to keep the acids produced by bacterial activity from damaging the intestinal wall and shields the wall from the significant amount of bacterial activity that occurs inside the feces. In addition to mucus, there is a strong production of water and electrolytes if there is severe irritation of the large intestine, as occurs in cases of bacterial infections, food poisoning, etc. Diarrhea is caused by the large intestine's excessive motility, which is brought on by the extra water and electrolytes. This has a preventive effect since it removes the irritating substance (Shashikanth et al., 2017).

15.6.8 Digestion

The majority of mammals feed primarily on carbohydrates, proteins, and lipids. They must be broken down into absorbable form because they cannot be absorbed in their original state. The process of moving digested food from the GIT lumen to the circulation (blood/lymph) is called absorption. The absorbed materials are transported by circulation to the tissues for processing, storing, or decomposition.

Limited absorption occurs in mouth and stomach because the digestion is far from complete in these organs. Small intestine is the chief site of absorption both in carnivores and omnivores. In non-ruminant herbivores also small intestinal absorption is of considerable importance. Except for colon, which absorbs water and some electrolytes, large intestines undertake negligible absorption. However, in non-ruminant herbivores like the horse and the pig, caecal microbial fermentation gives rise to large quantities of volatile fatty acids which are absorbed from the large intestine in these animals. The absorptive surfaces of small intestine show many folds called "*folds of Kerckring*" or "*valvulae conniventes*" which increase the surface area by three times. Positioned all over these folds starting from duodenum to illeocecal junction are millions of specified structures known as villi. The villi project into the lumen of the small intestine up to 1-8mm. They further increase the surface area of the small intestinal mucosa by 10 times. On the luminal side of each villus, there are about 1000 microvilli protruding into the intestinal lumen, which increase the surface area, another 20-fold. Thus, a combination of folds of Kerckring, villi and microvilli together increase the surface a 1000-fold. In an adult man the absorptive surface of the small intestine is approx. 250 sq. meters or slightly more than the surface area of a tennis court (Petras, 2013).

The villi (villus) are the most important structural elements involved in absorption in the small intestine. At the core of each villus is a large lymph capillary known as the central lacteal. Each villus has several small arteries, which enter the base and form dense capillary network immediately under its epithelium. Prominent smooth muscle strands are also found in the core of each villus. Villi undergo rhythmic (pumping) contractions including pendular movements and tonic contractions all of which help absorption and transfer of absorbed materials. Contractions of villi are brought about by Meissner's plexus, sympathetic nervous system and a hormone called villi kinin produced by the duodenal mucosa in response to presence of acid chyme.

15.6.9 Mechanisms of Absorption

Absorption through the gastrointestinal mucosa occurs by active transport, diffusion and solvent drag. Transport by solvent drag means that any time a solvent is absorbed because of physical absorptive forces, the flow of solvent shall drag dissolved substances along with the solvent.

15.6.10 Absorption in the Small Intestine

Diffusion is the only mechanism that moves water over the gut barrier. The principles of osmosis apply to this dispersion. Water so absorbs from chyme when it is diluted, but flows into chyme from plasma when it is concentrated. Water is therefore absorbed by isoosmotic absorption through the intestinal mucosa.

Each day, the intestines can absorb up to one-fifth of the body's entire sodium content. Feces lose only 0.5% of the salt that is lost in the gastrointestinal secretions. However, the body can quickly become fatally low in salt when significant amounts of digestive secretions are lost in feces, as occurs in diarrhea.

The renal tubules' method of sodium absorption is similar to that of the intestinal epithelium. As a result, a concentration gradient is formed between the interior of the epithelial cells and the intestinal lumen as sodium is actively transferred from within the cells via the lateral and basal borders. As a result, sodium diffuses more easily over the brush border and into the epithelial cells as it moves down the concentration gradient. In return for potassium and hydrogen ions, some sodium is also absorbed. When released, adrenaline improves the body's ability to absorb salt from the intestine, particularly the colon.

In the upper part of intestine chloride is rapidly absorbed by diffusion due to the electrochemical gradient created between the chyme and the inside of epithelial cells by the absorption of sodium. Thus absorption of chloride follows absorption of sodium.

Pancreatic and bile secretions release large amounts of bicarbonate ions into the intestine. The upper section of the small intestine is where they must be absorbed. Carbon dioxide is absorbed in a way akin to that of the renal tubules when bicarbonate ions are absorbed. As previously stated, in exchange for one hydrogen ion, a certain amount of sodium is absorbed. As a result, the bicarbonate and hydrogen ion released into the chyme mix to generate carbonic acid. After then, the carbonic acid split into carbon dioxide and water. Carbon dioxide readily enters the blood and is exhaled through the lungs, but water stays in chyme.

The duodenum in particular actively absorbs calcium ions from the small intestine. The body precisely meets its daily calcium demands through regulation of calcium absorption. Vitamin D and parathyroid hormone improve the intestines' ability to absorb calcium. From the duodenum, iron ions are actively absorbed. The ferrous state absorbs it more easily than the ferric state. Iron joins forces with transferrin upon intestinal absorption to facilitate circulation. Ninety percent of the iron in the blood is deposited in the liver, where it joins forces with apoferritin to produce ferritin. Because apoferritin and iron are loosely combined, iron can be easily released into the system when needed. Iron absorption from the duodenum is significantly decreased when the whole apoferritin in the liver is saturated with iron.

The intestinal mucosa is also capable of actively transporting a number of other ions, including phosphate, potassium, and magnesium. Monovalent ions are generally readily absorbed in large amounts. Bivalent ions, however, are only slightly absorbed. For instance, calcium ion absorption is only 1/50th of sodium ion absorption.

Monosaccharides are primarily absorbed as carbohydrates. Disaccharides are also absorbed in very minute amounts, but not as bigger polymers. Approximately 80% of the monosaccharides that are absorbed from the intestine are glucose. Galactose and fructose make up the remaining 20% of absorbed carbohydrates. Active transport facilitates the absorption of nearly all monosaccharides. Sodium and glucose travel through the brush border together. Galactose is conveyed in a comparable manner. Nevertheless, fructose is absorbed by enhanced diffusion rather than the sodium co-transport route. Fructose absorbs at half the rate of glucose or galactose because it is not co-transported with salt.

The majority of proteins are absorbed as tri- and di-peptides, as well as some free amino acids. Similar to glucose, these compounds are absorbed through the sodium co-transport process. Some free amino acids are also absorbed through the method of assisted diffusion, similar to fructose. Because different amino acids and peptides have varied binding properties, the intestine has at least five different types of amino acid and peptide transport proteins.

Glycerol, free fatty acids, and monoglycerides are the byproducts of fat digestion. All of these byproducts of fat digestion disintegrate in the bile micelles' inner lipid layer. The micelles' charged exterior and tiny diameter make them soluble in chyme. These micelles travel to the brush border where they get trapped in the nooks and crannies created by the stirring, moving microvilli. Due to their solubility in the lipid membrane of the cells, fatty acids and monoglycerides here both permeate into the epithelial cells right away.

This results in the micelles remaining in the chyme, where they continue to transport progressively more fatty acids and monoglycerides. The fatty acids in the epithelial cells are converted into new triglycerides, which are then carried into the bloodstream by lymph chylomicrons. Because they dissolve in water, small amounts of short and medium chain fatty acids are also absorbed straight into the portal circulation.

15.6.11 Absorption in the Large Intestine

The colon absorbs the majority of the electrolytes and water in chyme. The proximal half of the colon, also known as the absorbent colon, is where the majority of this absorption takes place. The term "storage colon" refers to the function of the distal part of the colon, which stores the undigested food. The large intestine mucosa is highly capable of actively absorbing salt and the resulting chloride. This is especially true in cases where elevated levels of aldosterone are released. Water follows the osmotic gradient that is created throughout the large intestine mucosa by the absorption of sodium and water (Oake et al., 2023).

BIBLIOGRAPHY

Agrawal, S., & Aoun, E. (2014). Physiology of the pancreas. *Practical Gastroenterology*, 38, 48–56.

Assan, D., Mustapha, U. F., Chen, H., Li, Z., Peng, Y., & Li, G. (2021). The roles of neuropeptide Y (Npy) and peptide YY (Pyy) in teleost food intake: A mini review. *Life*, 11(6), 547.

Campbell, J., Berry, J., & Liang, Y. (2019). Anatomy and physiology of the small intestine. In *Shackelford's surgery of the alimentary tract, 2 volume set* C. J. Yeo (ed.), (pp. 817–841). Elsevier.

Carr, K. E., & Toner, P. G. (1984). Morphology of the intestinal mucosa. In *Pharmacology of intestinal permeation I* T. Z. Csáky (ed.), (pp. 1–50). Springer Berlin Heidelberg.

Chandra, R., & Liddle, R. A. (2020). *Regulation of pancreatic secretion*. Pancreapedia: The Exocrine Pancreas Knowledge Base.

Chen, D., Hagen, S. J., Boyce, M., & Zhao, C. M. (2023). Neuroendocrine mechanism of gastric acid secretion: Historical perspectives and recent developments in physiology and pharmacology. *Journal of Neuroendocrinology*, 35(11), e13305.

Chien, H. J., Chiang, T. C., Peng, S. J., Chung, M. H., Chou, Y. H., Lee, C. Y., ... Tang, S. C. (2019). Human pancreatic afferent and efferent nerves: Mapping and 3-D illustration of exocrine, endocrine, and adipose innervation. *American Journal of Physiology-Gastrointestinal and Liver Physiology*, 317(5), G694–G706.

Davis, J. (2018). Hunger, ghrelin and the gut. *Brain Research*, 1693, 154–158.

Davis, R. W., & Davis, R. W. (2019). Feeding and digestion. In *Marine Mammals: Adaptations for an Aquatic Life*, pp. 219–266. First edition, Springer Publisher.

Deloose, E., Verbeure, W., Depoortere, I., & Tack, J. (2019). Motilin: From gastric motility stimulation to hunger signalling. *Nature Reviews Endocrinology*, 15(4), 238–250.

Duan, S., Rico, K., & Merchant, J. L. (2022). Gastrin: From physiology to gastrointestinal malignancies. *Function*, 3(1), zqab062.

Ekstrom, J., Khosravani, N., Castagnola, M., & Messana, I. (2019). Saliva and the control of its secretion. In *Dysphagia: Diagnosis and Treatment*, pp. 21–57. First edition, Springer Publisher (MEDRAD).

Holst, J. J., Gasbjerg, L. S., & Rosenkilde, M. M. (2021). The role of incretins on insulin function and glucose homeostasis. *Endocrinology*, 162(7), bqab065.

Iwasaki, M., Akiba, Y., & Kaunitz, J. D. (2019). Recent advances in vasoactive intestinal peptide physiology and pathophysiology: Focus on the gastrointestinal system. *F1000 Research*, 8.

Jänig, W. (2022). *The integrative action of the autonomic nervous system: Neurobiology of homeostasis*. Cambridge University Press. https://doi.org/10.1017/9781108778411.

Lentle, R. G., & Janssen, P. W. M. (2008). Physical characteristics of digesta and their influence on flow and mixing in the mammalian intestine: A review. *Journal of Comparative Physiology B*, 178, 673–690. https://doi.org/10.1007/s00360-008-0264-x.

Ma, Z. F., & Lee, Y. Y. (2020). Small intestine anatomy and physiology. In *Clinical and basic neurogastroenterology and motility* Z. F. Ma, Y. Y. Lee (eds.), (pp. 101–111). Academic Press.

McQuilken, S. A. (2021). The mouth, stomach and intestines. *Anaesthesia & Intensive Care Medicine*, 22(5), 330–335.

Nightingale, J. M., & Spiller, R. (2023). Normal intestinal anatomy and physiology. In *Intestinal Failure* J. M. D. Nightingale (ed.), (pp. 13-33). Springer International Publishing.

Oake, A., Nesto, N., & Pathak, Y. V. (2023). Gut microbiota and mental health. In *Anxiety, Gut Microbiome, and Nutraceuticals: Recent Trends and Clinical Evidence* Y. V. Pathak, S. Pathak, C. Stough (eds.), 1st Edition, CRC Press.

Paxton, F. (2020). *Foundations of naturopathic nutrition: A comprehensive guide to essential nutrients and nutritional bioactives*. Routledge.

Petras, R. E. (2013). Normal small intestine: Anatomy, specimen dissection and histology relevant to pathological practice. In *Morson and Dawson's Gastrointestinal Pathology*, 5th Edition, pp. 279–292. Wiley Publisher.

Priyanka, S., Moses, J. A., & Anandharamakrishnan, C. (2023). *Food digestion and absorption: Its role in food product development* (No. 42). Royal Society of Chemistry.

Rao, S. B. (1998). *Practical biochemistry for medical students*. Academic Publishers.

Reimann, F., Diakogiannaki, E., Hodge, D., & Gribble, F. M. (2020). Cellular mechanisms governing glucose-dependent insulinotropic polypeptide secretion. *Peptides*, 125, 170206.

Sharkey, K. A., & Mawe, G. M. (2023). The enteric nervous system. *Physiological Reviews*, 103(2), 1487–1564.

Shashikanth, N., Yeruva, S., Ong, M. L. D. M., Odenwald, M. A., Pavlyuk, R., & Turner, J. R. (2017). Epithelial organization: the gut and beyond. *Comprehensive Physiology*, 7(4), 1497–1518.

Stevens, C. E., & Hume, I. D. (2004). *Comparative physiology of the vertebrate digestive system*. Cambridge University Press.

Téllez-Isaías, G. (Ed.). (2023). *Poultry farming: New perspectives and applications*. BoD–Books on Demand.

Wang, H. H., Portincasa, P., & Wang, D. Q. H. (2019). Update on the molecular mechanisms underlying the effect of cholecystokinin and cholecystokinin-1 receptor on the formation of cholesterol gallstones. *Current Medicinal Chemistry*, 26(19), 3407–3423.

Wyatt, B. J. I., & Haugk, B. (2018). Liver, biliary system and pancreas. In *Underwood's Pathology: A Clinical Approach*, 7th Edition, pp. 356–387. Academic Press.

16 Physiology of the Digestive System of Ruminants

Vipin Maurya[1] *and Mukesh Kumar Bharti*[2]

[1]Department of Livestock Production Management, Faculty of Veterinary & Animal Sciences, IAS, RGSC- Banaras Hindu University, Benaras, India.
[2]Department of Veterinary Physiology & Biochemistry, Faculty of Veterinary & Animal Sciences, IAS, RGSC- Banaras Hindu University, Benaras, India.

16.1 INTRODUCTION

Ruminants, cloven-hoofed mammals of the order Artiodactyla, obtain their food by browsing or grazing, subsisting on plant material. Approximately 95% of the population of domesticated ruminants constitutes species: cattle, sheep, and goats, all of them belonging to the Bovidae family. Dairy cows and other animals such as sheep, goats, buffaloes, camels, and giraffes are herbivores because their diets are composed primarily of plant material. Many herbivores also are ruminants. Ruminant animals can be recognized easily because they chew frequently even when they are not eating. This chewing activity, called rumination, is part of the processes that allow a ruminant to obtain energy from plant cell walls also called fiber.

Ruminants have a unique digestive system that allows them to better utilize energy from fibrous plant material than other herbivores. Unlike monogastric animals such as swine and poultry, ruminants have a digestive system designed to ferment feedstuffs and provide precursors for energy for the animal to utilize. By better understanding how the digestive system of the ruminant works, livestock producers can better understand how to care for and feed the ruminant animal.

The ruminant digestive system uniquely qualifies ruminant animals such as cattle to make efficient use of high-roughage feedstuffs such as forages. Anatomy of the ruminant digestive system includes the mouth, tongue, salivary glands, esophagus, 4-compartment stomach (rumen, reticulum, omasum, and abomasum), pancreas, gall bladder, small intestine, and large intestine.

Gastro intestinal tract of a ruminant animal has colonies of microbes residing in it. As the animal grows, the type of microbes gets modified according to the type of diet. There is a complex interaction between the host animal and the microbial population which helps in digestion of different substrates. Ruminant animals do not produce sufficient enzymes to degrade cellulose and other plant polymers; hence they depend on metabolic activities of gut microbes to utilize the fibrous food. Since microbial cellulose digestion is a slow process, they are retained for a longer period in the gut of ruminant animal.

16.2 RUMINANT STOMACH

The ruminant stomach is actually a single stomach modified by marked expansion of the esophageal region into three distinct and voluminous diverticula, the ***rumen***, ***reticulum***, and ***omasum***, collectively known as the ***forestomach***. These are lined with nonglandular stratified squamous epithelium and comprise a series of chambers where food is subjected to digestion by microorganisms before passing through the digestive tract to the smaller glandular portion of the stomach in the ruminant, the ***abomasum*** (Figure 16.1).

16.3 THE RUMEN

Rumen acts as a huge fermentation vat that encompasses the entire left side of the animal's abdominal cavity. The rumen, also known as the "paunch," is lined with papillae for nutrient absorption and divided by muscular pillars into sacs. The rumen is divided into dorsal and ventral sacs by an incomplete partition formed by the cranial and caudal pillars and the right and left longitudinal pillars. The dorsal part is divided into the cranial sac lying between the ruminoreticular fold and the cranial pillar, the dorsal sac, and the caudodorsal blind sac. Approximately 50 to 65 percent of starch and soluble sugar consumed is digested in the rumen. Rumen microorganisms (including bacteria, protozoa, and fungi) digest cellulose from plant cell walls, digest complex starch, synthesize protein from non-protein nitrogen, and synthesize B vitamins and vitamin K. These microorganisms use chemical and physical means of breaking down feed particles into simpler nutrients, thereby producing nutrient forms that both they and the animal need. Rumen pH typically ranges from 6.5 to 6.8. The rumen environment is anaerobic. One major product of carbohydrate fermentation is volatile fatty acids, which provide the animal with most of its energy. The wall of the rumen is

DOI: 10.1201/9781003426851-16

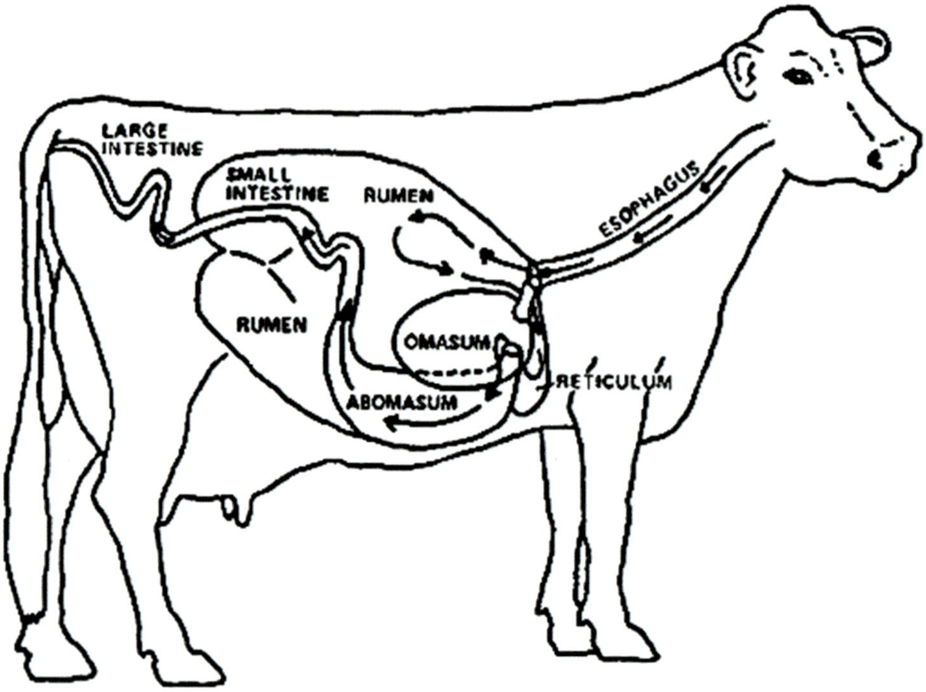

FIGURE 16.1 The figure of the ruminant digestive system describes different stomach parts like the rumen, reticulum, and omasum, as well as small and large intestine.

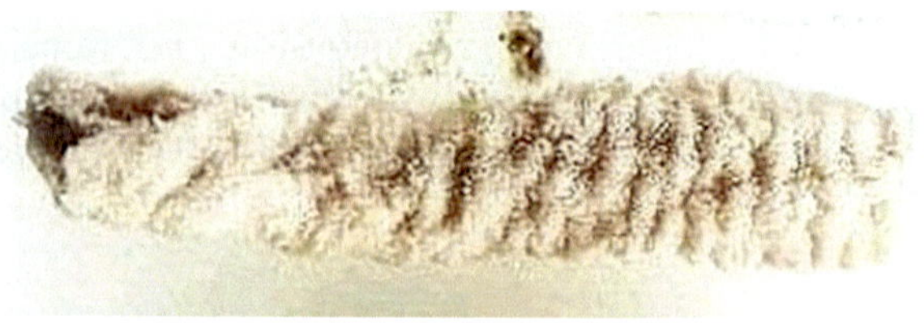

FIGURE 16.2 Rumen/Paunch is lined with papillae for nutrient absorption and divided by muscular pillars into sacs and acts as a fermentation vat.

FIGURE 16.3 Figure of reticulum depicting the "honeycomb" network of low hexagonal ridges.

lined with small, fingerlike projections that are the primary site of absorption for volatile fatty acids. Rumen microorganisms also break down the portion of protein that is susceptible to fermentation and use it as building blocks to synthesize their own amino acids, which are incorporated into the microorganisms. These microorganisms are also capable of synthesizing amino acids from sources of nitrogen that are not true protein, commonly referred to as non-protein nitrogen (Figure 16.2).

In addition to yielding nutrients that are beneficial to the animal, fermentation produces gaseous by-products, such as carbon dioxide, methane, ammonia, and hydrogen sulfide, that must be expelled by the animal. These gases are expelled through a process known as eructation

16.3.1 The Reticulum

The ***reticulum*** is the most cranial compartment of the fore stomach. Its mucosa is thrown up into intersecting ridges that give the reticulum its common name, the "honeycomb" network of low hexagonal ridges. These ridges are papillated in sheep but smooth in cattle (Figure 16.3).

Foreign objects such as wire or nails that are swallowed typically will fall into and remain in the reticulum; contractions of this part of the forestomach may drive sharp objects through the wall of the stomach, leading to traumatic peritonitis or hardware disease. The reticular groove runs ventrally from the cardia to the reticulo-omasal orifice. In young ruminants, the lips of this groove can be brought into contact by contraction of the underlying muscle, so that a tube is formed by which the ingested milk bypasses the ruminoreticulum and directly enters the abomasum for the digestion.

16.3.2 The Omasum

Omasum is dense, kidney-shaped organ whose main function is the absorption of water and water-soluble nutrients.

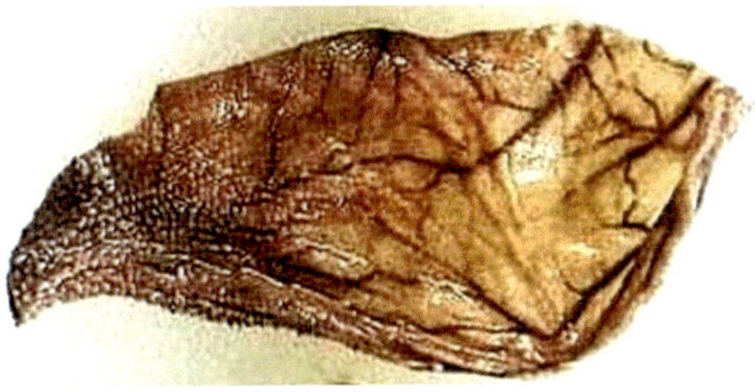

FIGURE 16.4 Omasum is a dense, kidney-shaped organ known as "many plies" due to the presence of many folds or leaves for increasing the surface area, which increases the area that absorbs nutrients from the feed and water.

It is relatively larger in larger ruminants like cattle and buffalo compared to smaller ruminants. It is called the "many plies" due to the presence of many folds or leaves for increasing the surface area, which increases the area that absorbs nutrients from the feed and water. Water absorption occurs in the omasum. It is connected to the reticulum by a short tunnel. After regurgitation of a bolus, it is swallowed and further subjected to ruminal fermentation (Figure 16.4).

16.3.3 The Abomasum

Abomasum also known as the "true stomach," is quite similar to a stomach in a non-ruminant. The abomasum produces hydrochloric acid and digestive enzymes such as pepsin (which breaks down proteins) and receives digestive enzymes secreted from the pancreas, such as pancreatic lipase (which breaks down fats). The digesta undergoes acid and enzymatic digestion and then moves on to the small intestine. These secretions help prepare proteins for absorption in the intestines. The pH in the abomasum generally ranges from 3.5 to 4.0. The abomasum secretes mucus to protect the its wall from acid damage (Figure 16.5).

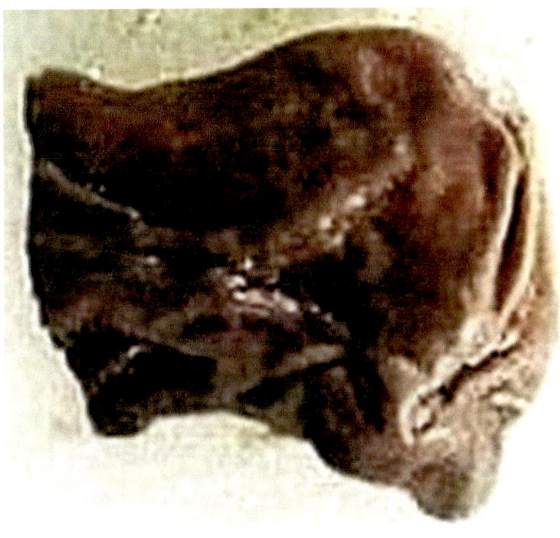

FIGURE 16.5 Abomasum or the "true stomach" is quite similar to a stomach in a non-ruminant dur to presence of digestive glands.

16.3.4 Pregastric Physiology

The organs of prehension vary somewhat in ruminants. In some, as in the cow, the lips, though strong, are not sufficiently movable to be of any use in prehension, while the tongue is so long and the muscles by which it is moved are so powerful that it is the principal organ used in this process. Sheep are grazers; they ramble slowly, eating short plants close to the ground. Goats are browsers and have clefted upper lips; they look for leaves, twigs, vines, and shrubs. A ruminant uses its mouth and tongue to harvest forages during grazing or to consume harvested feedstuffs. Cattle harvest forages during grazing by wrapping their tongue around the plants and then pulling to tear the forage for consumption. They typically spend over one-third of their time grazing, one-third of their time ruminating (cud chewing), and slightly less than one-third of their time idling, where they are neither grazing nor ruminating.

The roof of the ruminant mouth is a hard/soft palate without incisors. The lower jaw incisors work against this hard dental pad. The incisors of grass/roughage selectors are wide with a shovel-shaped crown, whereas those of concentrate selectors are narrower and chisel-shaped. Premolars and molars match between upper and lower jaws. These teeth crush and grind plant material during initial chewing and rumination.

Saliva is the mixed secretion of three pairs of main salivary glands, namely the parotid, submaxillary or submandibular, and sublingual of the mouth. These three glands are the major salivary glands in ruminants. In addition, sheep and cattle have two inferior molar glands and small, numerous buccal and labial glands in the cheeks and lips. Saliva aids in chewing and swallowing, contains enzymes for the breakdown of fat and starch, and is involved in nitrogen recycling to the rumen. When food enters the mouth, a copious secretion of saliva takes place by reflex stimulation of the salivary glands through the buccal receptors and secretory centers. Normal foodstuffs cause salivary secretion rich in mucin and enzymes to facilitate easy swallowing, while dry or noxious materials cause watery secretion with little mucin. The types and volumes of saliva secretion of saliva vary with a diverse range of stimuli. Saliva has a lubricating function that facilitates the mastication and swallowing of ingested food. The mucin content provides adhesion to food to form a bolus for swallowing. Spontaneous secretion of saliva, even during sleep, has an important protective function by keeping the oral mucosa moist and lubricated even in the absence of food. Salivary amylase is absent in horses, cattle, sheep, and goats. Saliva's most important function is to buffer pH levels in the reticulum and rumen. In ruminants, saliva provides a proper medium for bacterial growth and activity in the rumen. The HCO3 and PO4 content of saliva neutralizes the volatile fatty acids produced during microbial fermentation. Urea, a non-protein source, supplies nitrogen for bacterial growth and microbial protein synthesis. A mature cow produces up to 160-200 liters of saliva per day, but this varies depending on the amount of time spent chewing feed, as that stimulates saliva production. Forage and feed mix with saliva containing sodium, potassium, phosphate, bicarbonate, and urea when consumed to form a bolus.

That bolus then moves from the mouth to the reticulum through a tube-like passage called the esophagus. Muscle contractions and pressure differences carry these substances down the esophagus to the reticulum. Ruminants eat rapidly, swallowing much of their feedstuffs without chewing them sufficiently (< 1.5 inches). The esophagus functions bi-directionally in ruminants, allowing them to regurgitate their cud for further chewing, if necessary.

The process of rumination or "chewing the cud" is when forage and other feedstuffs are forced back to the mouth for further chewing and mixing with saliva. This cud is then reswallowed and passed into the reticulum. Then the solid portion slowly moves into the rumen for fermentation while most of the liquid portion rapidly moves from the reticulorumen into the omasum and then the abomasum. The solid portion left behind in the rumen typically remains for up to 48 hours and forms a dense mat in the rumen where microbes can use the fibrous feedstuffs to make precursors for energy.

16.3.5 Fermentative Digestion in Rumen

Ruminal fermentation initially results in the degradation of carbohydrates and proteins to short-term intermediates such as sugars and amino acids. The products of this initial degradation are readily metabolized to microbial mass and carbon dioxide, methane, ammonia, and volatile fatty acids (VFA): primarily acetate, propionate, and butyrate, and to a lesser degree branched chain VFA and occasionally lactate. The rate and extent of fermentation are important parameters that determine protein, vitamins, and short-chain organic acids supply to the animal (Koenig et al., 2003). The host ruminant animal absorbs VFA (mostly through the rumen wall) and digests proteins, lipids, and carbohydrate constituents of microbes and feed residues entering the small intestine to supply its maintenance needs and for the production of meat and milk. Ruminant animals derive about 70% of their metabolic energy from microbial fermentation of feed particles and microbial protein accounts for as much as 90% of the amino acids.

Rumen microorganisms usually adhere to feed particles and form biofilms to degrades plant material. The efficiency of ruminants to utilize of feeds is due to highly diversified rumen microbial ecosystem consisting of bacteria (10^{10}–10^{11} cells/ml, more than 50 genera), ciliate protozoa (10^4–10^6 / ml, 25 genera), anaerobic fungi (10^3– 10^5 zoospores/ml, 5 genera) and bacteriophages (10^8–10^9 /ml). The synergism and antagonism among the different groups of microbes is so diverse and complicated that it is difficult to quantify the role played by any particular group of microbes present in the rumen (Kamra, 2005). Bacterial numbers in the rumen are the highest and bacteria play a dominant role in all facets of ruminal fermentation.

16.3.6 Some Advantages of the Fermentation in the Reticulorumen

1. Allows digestion and then absorption of fermentation products that are of value to the host (e.g., microbial cells, VFAs, and B vitamins) before the acidic abomasum.
2. May change poor quality protein and other dietary nitrogen compounds to good-quality microbial protein.
3. Selective retention of coarse particles extends fermentation time and allows for further mechanical breakdown during rumination (cud chewing).
4. Release of fermentation gas (mostly CO2 & CH4) from the system by eructation.
5. Toxic substances in the diet may be attacked by the microbes before being presented to the small intestine.

16.3.7 Eructation

It is the expulsion of fermentation gases like CO_2 and methane that has accumulated in the rumen. Receptors for eructation reflex are the tension receptors located in the reticulum, cardia, and cranial rumen sac, which are stimulated by accumulation of gas. Centre for eructation reflex are present in Medulla Oblongata. Certain amount of gas is produced in ruminant animal due to fermentation process. The gas layer is moved cranially by the secondary contraction of the dorsal sac. From there it enters into the cardia and then lower oesophageal sphincter will open, and the gas enters into the relaxed oesophagus. By an antiperistaltic wave in the oesophagus and elevation of soft palate the gas is expelled through the mouth. Eructation occurs once in two ruminoreticular contractions. Inhibition of eructation is observed during blokage of ruminoreticular contraction. Blockage of eructation leads to bloating of the rumen, called as tympany which can also takes place when gas production is higher than its elimination. Distension of rumen may cause hindrance of respiratory process as diaphragm did not expand and contract easily. Oils and non absorbable surfactants are used to treat bloat. High grain feeding and feeding of lush alfalfa results in trapping of gases in the rumen called feedlot bloat. This is due to formation of foam in the ruminoreticulum which traps many small gas bubbles. This can be prevented by reducing feeding of grains or lush green pastures.

16.3.8 Ruminal-Reticular Contractions

There are mainly two patterns of motility seen in reticulorumen as primary or mixing contractions called primary cycle and secondary or eructation contractions known as secondary cycle. In the primary contraction, ruminoreticular contractions begin from the reticulum as biphasic contractions of which the first contraction known as mixing contraction. It relaxes for sometime and then again the second contraction with strong force is called as evacuation contraction begin. The function of the secondary contraction is to expel the gases out of the rumen. The reticulorumen contractions occur at a frequency of 1 to 3/ min. It may increase during feeding. Third contraction follows the second strong contraction during rumination. This third contraction is the strongest of the three and is referred to as extra or regurgitation contraction of the reticulum.

16.3.9 Microbial Digestion (Role of Bacteria, Fungi, and Protozoa)

Dietary carbohydrates are the main rumen microbial fermentation substrates. Microbial yields are related primarily to the growth rate that carbohydrate permits. The individual carbohydrates characterized by faster rumen degradation rates result in greater microbial yield (Hall and Herejk, 2001). The enzyme systems produced by microorganisms for carbohydrates hydrolysis are complex; they usually comprise hydrolases from several families, and there may be multiple enzymes hydrolysing each polysaccharide. Nearly all carbohydrate digestion occurs (>90%) within the rumen, but under certain circumstances (e.g., high rate of passage), a significant amount of carbohydrate digestion can occur in the small and large intestine. Carbohydrate digestion in ruminant animals is through microbial fermentation in the rumen. Dietary carbohydrates are degraded (fermented) by rumen microbes (bacteria, fungi, protozoa). The purpose of rumen fermentation is to produce energy as ATP for the bacteria to use for protein synthesis and their own growth. VFAs, also known as short-chain fatty acids, are produced as a product of rumen fermentation and are absorbed through the rumen wall and are utilized by the animal as an energy source. In young ruminants, rumen and the reticulum are not fully developed and are relatively small. The reticular/oesophageal groove reflex, a tube-like fold of tissue, channels milk or water that is sucked from a nipple directly through the omasum to the abomasum. This is a reflex, stimulated by sucking. When the animal is weaned, it normally loses this reflex. Solid food, such as creep feed, passes into the small rumen and fermentation starts. The neonatal ruminant animal has no ruminal bacterial population but from birth, it starts to pick up bacteria from the mother and environment, particularly through contact. Solid food is then fermented forming VFAs, which stimulate the growth and development of the rumen, particularly the growth of the papillae for absorption.

Cellulolytic bacteria tend to produce more acetate, while amylolytic bacteria produce more propionic acid. Typically, three major VFA molar ratios are 65:25:10 with a roughage diet and 50:40:10 with a concentrate-rich diet. Changes in VFA concentration can lead to several disorders of carbohydrate digestion in ruminants. Rumen acidosis occurs when animals are fed high-grain-rich diets or when animals are suddenly changed from pasture- or range-fed to feedlot conditions. Cellulose content is in the range from 35 to 50% of plant dry weight. It is chemically composed of a homogenous polymer of β-1,4-D-glucose linked through β-1,4- glycosidic bonds. *Fibrobacter succinogenes*, *Ruminococcus flavefaciens* and *R. albus* are considered to be the predominant cellulolytic bacteria present in the rumens, these species gain selective advantage in the rumen is by optimizing two catabolic activities: cellulose hydrolysis (depolymerization) and efficient utilization of the hydrolytic products (cellodextrins) (Koike and Kobayashi, 2001). F. succinogenes has a potent ability to solubilize crystalline chains of cellulose (Shinkai and Kobayashi, 2007). F. succinogenes produces primarily succinate (a propionate precursor), and lesser amounts of acetate, R. flavefaciens produces primarily acetate and lesser amounts of succinate converts to propionate by Selenomonas ruminantium (Weimer et al., 1999).

The amylolytic bacteria such as Streptococcus and Ruminobacter are particularly adapted for digesting starch and sugar. Under the strict anaerobic conditions within the rumen these bacteria ferment the starches and sugars to lactic acid, along with some VFAs. The ruminants which are generally high grain fed, the action of the amylolytic bacteria can be a particular concern to those ruminants. The amylolytic bacterial populations can multiply very rapidly in response to the starch in the diet and these bacteria produces large amounts of lactic acid. Lactic acid is nearly 10 times stronger acid than acetate, propionate and butyrate. Due to high amounts of lactic acid and other VFAs in the rumen fluid, pH of rumen fluid fall below 5.7. Because of this pH the cellulolytic bacteria begin to die and also the rumen epithelium can be damaged by the accumulation of large amount of acid. As the bacteria dies, the dying bacteria release endotoxins that enter the blood and can cause shock. This condition is known as rumen acidosis. This type of condition mostly occur in cattle that have not been allowed time to acclimatize to the high grain diet. If the amount of grain in the diet is increased slowly over a period of weeks, this type of condition like ruminal acidosis does not develop because it allows populations of bacteria known as lactate utilizers to populate the rumen. These bacteria metabolize the rumen fluid lactate as an energy source. The lactate utilizers bacteria belong to the Selenomonas and Megasphaera genera. Animals with very high populations of these bacteria in the rumen can be fed very high starch diets with little risk of rumen acidosis because acclimatization already been takes place. Hindgut fermenters like horses, rabbits and others have essentially the same types of cellulolytic bacteria living in their cecum and colon as cows have in their rumen. However, in these species starches and sugars are absorbed by the small intestine before they reach the colon so the risk for very low pH is reduced.

In addition to a wide variety bacteria, the rumen also contains small populations of fungi. Some of these fungal species may help degrade lignin, a woody and indigestible constituent of plant cell walls. Other inhabitants of the rumen are the protozoa. The protozoa feed on ruminal bacteria, plant starch granules and other readily digestible materials including polyunsaturated fatty acids. Most of the protozoa are located in the raft of fibrous digesta in the dorsal ruminal sac and play many important functions like it form a reservoir of microbial protein useful at time intermittent food supply, also helps in prevention of over-population of bacteria, also made available polyunsaturated fatty acids to the lower part of GI tract for their absorption. Protozoa are very sensitive to change in pH of the rumen. The protozoa and primary cellulolytic bacteria require pH 6.2 or greater, whereas the primary amylolytic bacteria require slightly acidic pH of around 5.8.

16.3.10 Microbial Protein

The rumen bacteria used to provide very high quality protein to the animal because its amino acid profile is almost identical to that of muscle and milk, permitting great conversion into meat and milk by the cow. Ruminal bacteria have the ability to form all the essential amino acids from the available resource. When these bacteria die, the proteins within the bacteria can be digested by mammalian proteolytic enzymes and the available amino acids are used by the ruminant animal. Two types of protein are found in ruminant animal, rumen-degradable protein and rumen-nondegradable protein. Dietary protein that the rumen bacteria can break down is referred to as rumen-degradable protein. It is essential to feed high quality proteins to manogastric animals to supply essential amino acids because these animals cannot synthesis high quality protein. In ruminants, if the protein is rumen degradable the essential amino acids will be lost and it will be of no use unless they can be recovered in the form of microbial protein that enters the small intestine. All the dietary protein fed to a cow is not degraded by the rumen bacteria for their use. The protein that bypasses the rumen bacteria, known as rumen-nondegradable protein. These protein can be digested in the small intestine and can be an excellent source of essential amino acids, if it is of high quality.

16.3.11 Formation of Volatile Fatty Acids and Their Absorption

The most important end products of carbohydrates fermentation in ruminants is volatile fatty acids. These volatile fatty acids are the major source of energy for the ruminants. These include acetic acid, butyric acid and propionic acid. Fat and protein contents of milk are also determine by these volatile fatty acids. After their production, these volatile fatty acids are absorbed through the walls of rumen and transported via blood to the liver. The granulosa cells of the forestomach epithelium contain carbonic anhydrase, which promotes the formation of carbonic acids (Figure 16.6).

This carbonic acids are dissociates into bicarbonate ions and hydrogen ions. The hydrogen ions associate with volatile fatty acid anions to form undissociated VFAs that can diffuse more easily across the epithelium, leaving bicarbonate ions in the ruminal fluid. This mechanism also reduces ruminal acidity by exchanging the anions of stronger acids for those of a weaker acid (carbonic acid). Most of the acetate in liver is converted to acetyl CoA via acetyl CoA synthetase. Most of the acetate (approximately 80%) escapes oxidation in liver and is oxidized by citric acid cycle or used for fatty acid acids synthesis. These acetate is the main precursor for lipogenesis in ruminants. Higher activity of acetyl CoA synthetase in ruminant adipose tissue reflects the high rate of acetate conversion to fatty acid. Acetate is also the main precursor for milk fat synthesis. Approximately around 2-5% of propionic acid is converted to lactic acid during its absorption through the rumen ,while the remainder enters in the portal blood as propionate. Most of the propionic acid is oxidized to glucose in liver. Propionyl CoA undergoes CO_2 fixation by the biotin containing enzyme, propionyl CoA carboxylase, to form methylmelonyl CoA and subsequently converted to succinyl CoA. This succinyl CoA then enters the citric acid cycle for the production of ATP as source of energy for the ruminants. Butyrate during its absorption through the ruminal epithelium is converted to ketones and is present in low level in portal blood. More than 80% of the ketones formed occurs in the form of β-hydroxybutyric, whereas the remaining comprised of acetoacetate and acetone. The β-hydroxybutyric acid is mainly used for fatty acid synthesis in adipose and mammary gland tissue, whereas it is oxidized in cardiac and skeletal muscles. Butyrate that is reaching the liver is rapidly metabolized by hepatic tissue.

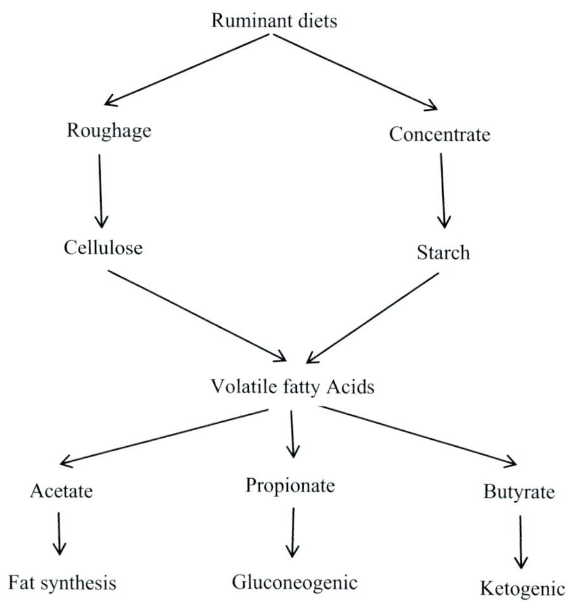

FIGURE 16.6 Formation of volatile fatty acids and their absorption.

16.3.12 Importance of FVAs in Ruminants

VFAs play very important and significant role in ruminant animal. Acetate is the precursor for milk fat synthesis. Propionate is glucogenic acids and provides glucose through gluconeogenesis. Acetate and butyrate contribute the energy needs of the ruminant animal. These fatty acids provide energy by entering Krebs's cycle as different substrate. Butyrate and acetates are mainly ketogenic acids and causes the formation of the ketone bodies like acetone, acetoacetic acid and beta hydroxy butyric acid. The ketone bodies are an important source of energy in tissues like brain and heart.

16.3.13 Other End Products of Fermentation in Ruminants

Lactic acid is a stronger acid than the volatile fatty acids. Certain amylolytic bacteria produce lactic acid along with volatile fatty acid during degradation of starch. Generally lactic acid found in both isomeric form and occurs as a transient fermentation intermediate. With high starch diets, it may accumulate in the forestomach with low pH conditions, initially only slowly absorbed but later largely converted to glucose in the liver. The principal gases produced by ruminant animal after fermentation of substrate is carbon dioxide (60%) methane (30–40%) and variable amount of nitrogen, with trace amount of H_2S, hydrogen and oxygen. Carbon dioxide is produced by decarboxylation reactions of fermentation and from the neutralization of volatile fatty acid hydrogen ions by bicarbonate entering the rumen in saliva. Methane is produced from the reduction of carbon dioxide and formate by methanogenic bacteria. Hydrogen sulfide is produced by the reduction of sulfate and sulfur containing amino acids. This gas is generally toxic even in small quantity. Ruminal environment is generally low in oxygen concentration and sometimes oxygen may enter the rumen through trapped with ingested food and water and also sometimes by diffusion from blood but it is quickly used by the facultative anaerobic bacteria. Ammonia is produced by the deamination of dietary proteins, from non protein nitrogenous substance (NPN) and from urea derived from saliva. Urea feeding is beneficial to the animal and it is always advisable to feed urea as nitrogen supplement (feeding upto 30% total nitrogen as a urea supplement).

16.3.14 Nutrient Reabsorption in Small Intestine

The small intestine is the main site of absorption of nutrients that have by-passed the rumen. Bile and pancreatic juice drain into the small intestine to aid in digestion of certain feed components, such as dietary fat. The liver is a large organ, about 1.5% of an animal's live weight. It is located towards the front of the animal's body cavity, just behind the diaphragm. Newly digested and absorbed nutrients are transported from the absorption sites to the liver for storage and/or further processing. The liver also produces bile that drains into the intestine to aid in the digestion of fats.

REFERENCES

Hall M.B., Herejk C. 2001. Differences in yields of microbial crude protein from in vitro fermentation of carbohydrates. *J. Dairy Sci.* 84:2486–2493

Kamra D.N. 2005. Rumen microbial ekosystem. *Curr. Sci.* 89:124–135.

Koenig K.M., Beauchemin K.A., Rode L.M. 2003. Effect of grain processing and silage on microbial protein synthesis and nutrient digestibility in beef cattle fed barley-based diets. *J. Anim. Sci.* 81:1057–1067.

Koike S., Kobayashi Y. 2001. Development and use of competitive PCR assays for the rumen cellulolytic bacteria: Fibrobacter succinogenes, Ruminococcus albus and Ruminococcus flavefaciens. *FEMS Microbiol. Lett.* 204:361–366.

Shinkai T., Kobayashi Y. 2007. Localization of ruminal cellulolytic bacteria on plant fibrous materials as determined by fluorescence in situ hybridization and real-time PCR. *Appl. Environ. Microbiol.* 73:1646–1652.

Weimer P.J., Waghorn G.C., Odt C.L., Mertens D.R. 1999. Effect of diet on populations of three species of ruminal cellulolytic bacteria in lactating dairy cows. *J. Dairy Sci.* 82:122–134.

17 Physiology of the Respiratory System

Ahmed Hashem El-anwar[1] *and Saadia Abdelfattah Ali*[2]

[1]Department, Faculty of Veterinary Medicine, Beni-Suef University, Al Isma'iliyah, Egypt
[2]Faculty of Veterinary Medicine, Suez Canal University, 4.5 Km Ring Road, Ismailia, Egypt

17.1 INTRODUCTION

Respiration is the gaseous exchange between an individual and its surrounding environment, through which O_2 is transported from the atmosphere to cells and CO_2 and H_2O are transported from cells to the atmosphere.

17.1.1 Functions of Respiration

I) **Primary or main function:**

Through respiration, the living body obtains oxygen, which is used for the oxidation resulting in the liberation of energy and CO_2. The formed CO_2 is eliminated through respiration.

$$C_6H_{12}O_6 + O_2 \rightarrow CO_2 + H_2O + \text{Energy.}$$

II) **Secondary functions:**

1. Elimination of water; control water balance through excretion of water vapour.
2. Regulation of body temperature, especially in panting animals as dogs during summer.
3. Phonation or voice production by the larynx: The movement of air across the vocal cords causes their vibration which initiate the different sounds of speech.
4. Regulation of blood pH.

17.1.2 Classification of Respiration

I) **External or pulmonary respiration:**
In which is the gaseous exchange between the blood in pulmonary capillaries and the surrounding environment.

II) **Internal respiration:**
In which is the gaseous exchange between the blood in the systemic capillaries and the tissues.

Physiological anatomy of the respiratory system:

In animals, the respiratory system consists of two lungs, the air passages leading to them (nasal cavity, pharynx, larynx, trachea, bronchi and bronchioles), the thorax, pleural sac, and respiratory muscles, in addition to respiratory centres and the nerve supply controlling all the previous structures.

17.1.2.1 Nasal cavity

The mucus membrane of the nasal cavity is moist, highly vascular, and containing numerous glands and sensory endings of the olfactory nerve. The caudal part of the nasal cavity contains olfactory (neuro-) epithelium, which is involved in the sense of smell. In common colds and influenza, the act of smelling is decreased due to the blockage of nasal receptors. Additionally, the nasal cavity acts as a buffer, warming excessively cold air, cooling excessively hot air, and moisten dry air. It also removes suspended particles from the inspired air.

17.1.2.2 Pharynx

It is a common chamber for the passage of air and food. During swallowing, the respiratory movements are reflexively inhibited, the soft palate is elevated to close the posterior nares. The laryngeal opening (glottis) is closed by vagal reflex to prevent the escape of food into the trachea and lungs.

17.1.2.3 Larynx

It contains the vocal cords, the vibration of which produces the voice and controls the amount of air passing to the lungs.

17.1.2.4 Trachea

It is always kept open by the presence of an incomplete ring of cartilage in its wall. The mucous membrane is ciliated with the presence of many goblet cells (pseudostratified ciliated columnar epithelium containing many goblet cells). The cilia and the mucous secretion of the glands prevent the entrance of dust and foreign matters into the lungs.

17.1.2.5 Bronchi

Bronchi are similar in structure and function to the trachea. In addition, smooth muscle fibres in the wall of the bronchi show peristaltic movement, which helps in driving foreign materials towards larger air passages. The wall contract or relax to alter the calibre of smaller bronchi and bronchioles, leading to change resistance to air flow (inflow or outflow).

17.1.2.6 Lungs

The lungs are two elastic membranous sacs situated in the thoracic cavity. Each lung has numerous alveoli. The

DOI: 10.1201/9781003426851-17

alveolus is composed of a single layer of respiratory epithelium which is surrounded by blood capillaries that are composed of a single layer of endothelium. Across this layer of cells and the endothelium of capillaries, gaseous exchange between the air in the alveoli and the blood in the adjacent capillaries takes place.

17.1.3 Respiratory Muscles

1. External and internal intercostal muscles.
2. The diaphragm is a musculo-tendinuous organ separating the thoracic cavity from the abdominal cavity. Its thoracic surface is convex, while its abdominal one is concave. It plays an important role in respiration; its contraction increases the length of the thoracic cavity.
3. **The sternomastoid muscle** is active during forced inspiration, i.e., inspiration with effort, leading to elevation of the sternum and ribs.

17.1.4 Protection of Lungs Against Damage

1. Previously mentioned in the functions of the air conducting part.
2. Cells of the reticuloendothelial system are known as "Alveolar phagocytes" or "Dust cells." They engulf dust particles and bacteria entering the alveoli.
3. The presence of a thin layer of fluid between the parietal and visceral layers of the pleura acts as a lubricant, which decrease the friction between the thoracic wall and delicate lung tissue.
4. The tonsils have a protective function against bacteria.

17.1.5 Mechanical Respiration

The mechanism of respiration includes:

a) Mechanism of inspiration.
b) Mechanism of expiration.

17.1.6 Mechanism of Inspiration (Inhalation)

Inspiration means the enlargement of the thorax and lungs and accompanied by an inflow of air. This action is due to:-

a) Contraction of the diaphragm, which is the principal muscle of inspiration. During inspiration, the diaphragm contracts, leading to a reduction in its curvature and an increase the longitudinal diameter of the chest.
b) Movement of the ribs: during inspiration, the external intercostal muscle contractions lead to forward and outward movement of the ribs, resulting in an increase in the transverse diameter of the chest.

Due to a and b:

1) The size of thoracic cavity increases in all dimensions.
2) The intra-thoracic pressure decreases. Pressure between pleural surfaces (intrapleural pressure, already negative; subatmospheric) is reduced from -2 to -6 mmHg (i.e., an increased suction pull is exerted on lung tissue)
3) The air rushes in from the atmosphere through the air passages to fill the lungs (inflow of air).

c) **Sternomastoid muscle** active during forced inspiration. i.e., inspiration with effort → elevation of sternum → rib elevators.

17.1.7 Mechanism of Expiration (Exhalation)

Expiration means a decrease in the size of the thorax and lungs and accompanied by an outflow of air. This action is due to:

a) Relaxation of the diaphragm.
b) Relaxation of the external intercostal muscles.

Due to a and b:-

1) The size of thoracic cavity returns to its normal dimensions (resting position).
2) The intra-thoracic pressure increases. Pressure between pleural surfaces (intrapleural pressure) is reduced from -6 to -2 mmHg (i.e., less pull is exerted on lung tissue).
3) The air rushes in from the lungs to the atmosphere through the air passages (outflow of air).

17.1.8 Respiratory Cycle

At rest, a normal male adult breathes about 16 times (=cycles) per minute. Each cycle consists of:

1. **Inspiration:** Increase in the size of thoracic cavity and lungs leading to the inrush of air into the lungs.
2. **Expiration:** Decrease in the size of thoracic cavity and lungs leading to the outflow of air from the lungs. It is longer than inspiration.
3. **Expiratory pause:** Period of rest following the end of expiration.

N.B.: absent in rapid respiration as in muscular exercise.

17.1.9 Elasticity of the Lungs

When the lung is inflated (stretched), it tends to recoil (collapse) like an elastic rubber band.

(1) Elastic fibres of lungs: Lungs contain a network of elastic fibres responsible for 1/3 of lung recoil.
(2) Surface tension of fluid lining the alveoli: Molecules of fluid lining the alveoli attract each other and try to collapse the alveoli and consequently the whole lung. It is responsible for 2/3 of lung recoil.

17.1.10 Factors Preventing Lung Collapse (Recoil)

1. **Lung surfactant:**
   Nature: Lipoprotein, present in the lumen of alveoli.
   Source: Alveolar epithelium.
   Action: Decrease the attraction between fluid molecules lining the alveoli (decrease surface tension), which prevent collapse of alveoli.
2. **The faster growth of the thoracic cage** compared to that of thoracic contents, including the lungs, causes the lungs to tend to distend to fill the thoracic cavity.
3. **Lymph film in the pleural** cavity pulls the lungs towards the rigid chest wall (hydraulic attraction).
4. **The intrapleural pressure** (sub-atmospheric) is always below the intrapulmonary pressure, helping suction of the lungs towards the rigid chest wall.

17.1.11 Pressure Changes During Normal Respiratory Cycle

1. Intrapleural pressure
2. Intrapulmonary pressure.
3. Intra-abdominal pressure.

1) **Intrapleural pressure (IPP)**

Pressure inside the pleural sac is usually negative (sub-atmospheric; less than atmospheric pressure 760 mm Hg) during normal respiratory cycle.

17.1.12 Pleura

1. Thin, glistening serous membrane; sac (right and left).
2. Each sac encloses the corresponding lung.
3. Consists of: parietal pleura = outer wall lining the inner thoracic wall.
   Visceral pleura = inner wall lining of the lungs.

Pleural cavity: Space between parietal and visceral pleura containing a thin layer of lymph (serous fluid) which acts as lubricant and keeps the lungs always distended.

17.1.13 Values of IPP

Normal inspiration= -6: -10 mm Hg.
Forced expiration = +40 or more.
Normal expiration = -3 mm Hg.
Forced inspiration = -30 mm Hg.

17.1.14 Importance of the Negativity of IPP

1) Help flow of:
   - Venous blood flows from extrathoracic into intrathoracic veins.
   - Lymph flows from extrathoracic into intrathoracic lymph vessels.
   - Blood flows via pulmonary veins.
2) It is a measurement of the elastic recoil of the lung. This recoil is measured by the amount of negative pressure in the intrapleural spaces required to prevent lung collapse.

2) **Intrapulmonary pressure**

It is pressure inside lung alveoli.
Values of intrapulmonary pressure:

17.1.15 During Normal Inspiration

- Beginning = Zero mm Hg (=atmospheric).
- During = -2 mm Hg.(below atmospheric by 2 mm Hg.), as the lung alveoli are distended before air reaches them.
- End = Zero mm Hg again.
- During normal expiration:
- beginning = Zero mm Hg (=atmospheric).
- during = +2 mm Hg (above atmospheric by 2 mm Hg.), as the lung alveoli collapse before air is expelled from them.
- end = Zero mm Hg again.

3) **Intra-abdominal pressure (IAP).**

Pressure inside the abdominal cavity.
Values of intrapulmonary pressure:

17.1.16 During Normal Inspiration

- Beginning = increases due to the descent of the diaphragm.
- End = return to its original level due to decreased tone of muscles of the abdominal wall and pelvic floor.

17.1.17 During Normal Expiration: The Opposite Occurs During

17.1.17.1 Some Respiratory Terms

1. **Respiratory rate:** It is the number of respirations (inspiration and expiration) per minute.
2. **Eupnea:** It is a condition of normal breathing.
3. **Dyspnea:** It is a condition of difficult breathing.
4. Orthopnea = difficult respiration when lying down but not when standing or sitting. It is due to decreased V.C.

5. **Hyperpnea:** It is a condition of breathing in which the rate or depth, or both are increased.
6. **Polypnea:** It is a rapid, shallow, panting type of breathing.
7. **Apnea:** It means temporary cessation (stopping) of breathing.

17.2 LUNG VOLUMES AND CAPACITIES

17.2.1 Lung Volumes

1. **Tidal volume:** Volume of air inspired or expired in each respiratory cycle during rest = 500 ml.
2. **Inspiratory reserve volume**: Maximal volume of air which can be inspired after normal inspiration = 3000 ml
3. **Expiratory reserve volume:** Maximal volume of air that can be expired after normal expiration = 1000 ml.
4. **Residual volume**: Volume of air remaining in the lung after maximal expiration i.e. it cannot be expelled to the atmosphere except after opening the chest = 1000 - 1500 ml.
   a) Significance of residual volume:
   b) Maintains the aeration of blood between breaths.
   c) Diagnosis of respiratory diseases: R.V. is normally below 30% of total lung capacity in adults. It increases and may reach 70% in conditions associated with difficult expiration as:
      1) Bronchoconstriction, e.g. bronchial asthma.
      2) Weak elasticity of lungs, e.g., emphysema.
   d) In these conditions the lungs do not normally collapse during expiration. They remain more distended than normal.

5. **Minimal** air: the volume of air remains in the lungs after opening the thorax and the complete collapse of the lungs.

Significance of minimal air: important in forensic medicine to know whether an infant was born alive then killed or was born dead. If a piece of its lung is:

- Float on water, indicating that minimal air is present. It proves that infant was born alive then killed.
- Not floats on water indicating that minimal air is absent. It proves that infant was born dead.

17.2.1.1 Lung Capacities (Summation of Two or More Lung Volumes Together)

- **Inspiratory capacity** (maximal volume of air which can be inspired after normal inspiration) = Tidal volume (500 ml) + Inspiratory reserve volume (3000 ml).
- ***Functional residual capacity (volume of air remaining in lungs after normal expiration, this volume is known as resting expiratory level) = Expiratory reserve volume (1000 ml) + Residual volume (1200 ml)***
- **Total lung capacity:** (volume of air contained in the lungs after deepest inspiration)
  = All lung volume = 5700 ml
  =Tidal volume (500 ml) + Inspiratory reserve volume (3000 ml) + Expiratory reserve volume (1000 ml) + Residual volume (1200 ml).
- **Vital capacity:** (Volume of air given out by maximal expiration after maximal inspiration).
  = Total lung capacity (5700 ml) - Residual volume (1200 ml) = 4500 ml.
  N.B. It is better measured in relation to body surface area (s.a.). So, it is 2.5 litres/m^2 s.a. in adult males and 2.0 litres/m^2 s.a. in adult females.

Significance of vital capacity (VC): when calculated / mm^2 s.a., VC is an index of pulmonary ventilation efficiency and physical fitness.

17.2.1.2 Factors Affecting V.C.

1. **Conditions of respiratory muscles:**
   a) Strong, e.g., athletes; lung distension is more efficient and increase V.C.
   b) Difficult movement, e.g., spinal cord injuries, ascites, late pregnancy, poliomyelitis. The lung distension is not efficient and decrease V.C
2. **Posture:**
   a) When sitting or standing, the viscera drop down by gravity and the diaphragm so lung distension is more efficient and increase V.C.
   b) In the recumbent position, the viscera press on the diaphragm, resulting in lung distension that is not efficient and decreases V.C.
3. **Decrease lung elasticity:**
   e.g., T.B., emphysema → lung does not distend efficiently → decrease V.C.
4. **Decrease expansion and recoil of the thoracic cage:**
   e.g. deformities of the chest cage, weak respiratory muscles, pneumonia → decrease V.C.
5. **Increase volume of blood (stagnation) in pulmonary circulation:**
   e.g. left-side heart failure → decreased lung distensibility and excess blood takes the space for air → decreased V.C.
6. **Increase resistance to airflow in respiratory passages:**
   e.g. bronchial asthma → decrease V.C

17.2.1.3 Dead Space (D.S.)

It is the space of a conducting passage, e.g., nasal cavity, pharynx, trachea, bronchi, and bronchioles, in which there is no gas exchange.

17.2.1.4 Measurement of D.S. (Using Bohr's Equation)

CO_2 % in alveolar air - CO_2 % in expired air with measurement of tidal volume. It is about 150 ml at the start of expiration.

17.2.1.5 Significance of D.S

1. Protects lungs against damage:
   a) Moisten and warm inspired air.
   b) Filter and clean inspired air from large particles.
2. Responsible for the difference between:
   a) Composition of alveolar and expired air.
   b) Total and true (effective) alveolar ventilation, this is important in conditions of shallow rapid respiration as a large part of tidal volume occupies the D.S → decrease of effective ventilation.
3. One of the pulmonary function tests.

N.B.: Dead Space

Increase	Decrease
• Sympathetic stimulation. • During inspiration due to widening of air passage. • Old age.	• Vagus stimulation. • During expiration due to constriction of air passage.

17.2.1.6 Respiratory Rate of Different Animals

Respiratory rate means the number of respirations per minute. It is affected by several factors.

1. **Animal species:**
   Human: 12-20
   Horse: 9-10
   Cow and buffalo: 12-16
   Camel: 6-8
2. **Body size:** There is an inverse relationship between respiratory rate and animal size.
3. **Exercise**: Muscular exercise increases respiratory rate.
4. **Environmental temperatur**e: High environmental temperature increases respiratory rate.
5. **Excitement:** Fear and fright increase the respiratory rate.
6. **Diseases:** Fever, cardiac, and respiratory diseases increase respiratory rate.

17.2.1.7 Respiratory Sounds (Ralls)

The inlet and outlet of air through the respiratory passages can be heard by the stethoscope over different parts of the chest wall. Diseases of the respiratory system lead to changes in the normal respiratory sound; vesicular-V-sound.

17.3.1 Oxygen Transport

17.3.1.1 In the Lung

If there is a mixture of gases above a liquid, the amount of each gas dissolved in the liquid will depend directly upon the partial pressure of the gas, i.e., upon the pressure exerted by the given gas.

The partial pressure of a gas in a mixture can be calculated if the total pressure and the percentage composition of the mixture are known.

Partial pressure (P.P) of a gas in a mixture of gases = % of a gas in a mixture × Total pressure of the mixture.

17.3.1.2 Inspired Air

Barometric pressure of dry atmospheric air (at sea level) = 760 mm Hg.

	% of gas in inspired air (atmospheric air)	P.P (mm Hg.)
OXYGEN	20.95%	159.22
CARBON DIOXIDE	0.04%	0.304
NITROGEN	79.01%	600.476

17.3.1.3 Alveolar Air

	% of gas in alveolar air	P.P (mm Hg.)
OXYGEN	13.8 %	98.394
CARBON DIOXIDE	5.5 %	39.215
NITROGEN	N2 is an inert gas in respiration. It is not used by the tissues, so it has therefore been omitted.	

N.B.: Total pressure of alveolar air = atmospheric pressure (760 mm Hg) - pressure of water vapour present in the respiratory tract (47 mm Hg) = 713 mm Hg.

17.3.1.4 Venous Blood

	% of gas in venous blood entering lungs	P.P (mm Hg.)
OXYGEN	12.5	40
CARBON DIOXIDE	56	46

- CO_2 pressure is high → CO_2 moves from venous blood to alveolar air (6 volumes % given up by blood).
- O_2 pressure is high → O_2 moves from alveolar air to blood (about 6.5% of volume % taken by blood).

17.3.1.5 Arterial Blood

	% of gas in venous blood entering lungs	P.P (mm Hg.)
OXYGEN	19	96
CARBON DIOXIDE	50	40

17.3.1.6 In the Tissues

	% of gas the tissues	P.P (mm Hg.)
OXYGEN	-------	0-40
CARBON DIOXIDE	-------	50-60

- O_2 pressure is low → O_2 moves from arterial blood to tissues (6.5 volumes% given up by blood).
- CO_2 pressure is high → CO_2 moves from tissues to blood (6 volumes % taken up by blood).

N.B.: Concentrations of the gases and therefore the pressures exerted by them vary in the tissues depending on the metabolic activity of the particular tissue at any one time.

17.3.1.7 Expired Air

	% of gas in expired air	P.P (mm Hg.)
OXYGEN	16.4%	116.932
CARBON DIOXIDE	4.0%	28.52
NITROGEN	79.6%	567.55

Expired air is laden with water vapour exerting a pressure of 47 mm Hg.

In conclusion:

The diffusion of a gas across membranes occurs from the higher tension to the lower tension. Therefore, O_2 passes from alveoli to venous blood and from arterial blood to tissues. On the other hand, CO_2 passes from tissues to venous blood and from venous blood to alveoli, as seen in the following table and figure.

	PP (mm Hg.)	
	O_2	CO_2
Inspired air	159.22	0.304
Alveolar air	98.394	39.215
Venous blood	40	46
Arterial blood	96	40
Tissues	0-40	50-60 ↑↑
Expired air	116.932	28.52

17.3.1.8 Carriage and Transfer of O_2 by Blood

(I) In physical solution: ***(II) In chemical combination:***

17.3.1.8.1 In Physical Solution

- About 3% of O_2 in the lung (alveolar air) is carried this way; 0.3 ml/100 ml blood is dissolved in plasma.

Significance:

- This amount is not enough at all for tissue oxygenation, which is 250 ml/minute at rest.
- O_2 in physical solution is very important in determining the direction and rate of diffusion of O_2 from or to the blood.

17.3.1.8.2 In Chemical Combination with Hemoglobin: ($O_2 + Hb \longleftrightarrow HbO_2$)

- About 97% of O_2 in the lung is carried by this way.
- O_2 combines loosely with hemoglobin to form oxyhemoglobin (HbO2; a reversible compound). This is due to the presence of iron in the ferrous state.

N.B.: Normally each 1 g Hb contains 3.34 mg iron and carries 1.33 ml of oxygen Hbconc.(g.%) = 15 g/100 ml blood.

Oxygen capacity of blood	Oxygen content of blood
Volume of O_2 present in chemical combination with Hb (in 100 ml blood), where Hb is:	
Fully saturated with O_2 e.g. vitro	Not fully saturated with O_2 e.g. vivo.
- Arterial blood (when fully saturated with O_2) contains 1.33 × 15 = 20 ml O_2/ 100 ml blood.	usually, during rest O_2 % saturation: - Arterial blood = 97.5 % (so it contains 19.5 ml O_2/ 100 ml blood i.e. in loose combination with Hb). - Venous blood = 70% (contains 14 ml O_2/ 100 ml blood).

N.B.:* O_2 content of blood varies between zero and 20 ml depending on:

1. Hb content 2. O_2 tension

$$*O_2\ \%\ \text{saturation} = \frac{O_2\ \text{content}}{O_2\ \text{capacity}} \times 100$$

e.g.: In arterial blood:
O_2 content = 19.5 ml/100 ml blood
O_2 capacity = 20.0 ml/100 ml blood

$$*O_2\ \%\ \text{saturation} = \frac{19.5 \times 100}{20} = 97.5\%$$

- * Coefficient of O_2 utilisation = % ratio of the volume taken (utilized) by the tissues from arterial blood to the total O_2 supplied by arterial blood to the tissues.

$$= \frac{O_2 \text{ utilized by tissues}}{O_2 \text{ content of arterial blood}} \times 100$$

$$= \frac{O_2 \text{ content of arterial blood} - O_2 \text{ content of venous blood}}{O_2 \text{ content of arterial blood}} \times 100$$

= 25% at rest and about 75% in muscular exercise.

O_2 utilisation by the tissues depends on:

1) Activity of tissues:	2) Rate of blood flow through tissues:
O_2 utilization is: directly proportional to tissue activity	Inversely proportional to rate of blood flow.

17.3.1.9 O_2 – Dissociation Curve

- It is the relationship between O_2 tension and O_2% saturation of Hb.

The curve takes an S-shape represented by:

- Upper (horizontal) segment where affinity of Hb towards high O_2 tension is high.
- Lower (vertical) segment where affinity of Hb towards low O_2 tension is low

17.3.1.10 Physiological Significance of the S-shape of O_2 – Dissociation Curve

	High O_2 tension (alveoli and arterial blood)			Low O_2 tension (tissues and venous blood)	
O_2 tension (mmHg)	100 eg. at sea level	80 eg., at high altitude	60 eg., lung disease	40 eg., during rest	30:15 eg., muscular exercise
Affinity of Hb toward O_2 (= Hb saturation "%"):	very high (97 %)	++++ (93%)	+++ (90%)	++ (70%)	marked ↓↓
Physiological significance:	The high affinity of Hb toward high O_2 tension → help loading of O_2 on Hb at lung level so:			The low affinity of Hb toward low O_2 tension → help unloading (freedom) of O_2 from Hb at active tissues so,	
	Arterial blood becomes rich with O_2 to ensure good supply of O_2 to tissues.	Persons living at high altitude or with lung disease can get enough O_2.	Tissues take 25% of O_2 of arterial blood which is enough at rest.	PP of CO_2 and H^+ is high, causes Hb to liberate it's O_2 more readily to the active tissues.	

17.3.1.11 Causes of S-shape of O_2 Dissociation Curve of Blood Hb

1. ***Saturation of blood Hb occurs in steps, forming intermediate compounds:***

$$(Hb_4)O_2 \rightarrow (Hb_4)O_4 \rightarrow (Hb_4)O_6 \rightarrow (Hb_4)O_8$$

(N.B.: Hb is made up of 4 subunits)

2. If all Hb is $(Hb_4)O_2$, the curve will be a rectangular hyperbola (like that of myo Hb).

17.3.1.12 Factors Affecting O_2 Dissociation Curve

	Right ← Factors shifting the curve to → Left	
CO_2 tension	increase (e.g. muscular exercise).	decrease (e.g. during sleep)
H^+	increase (e.g. diabetes mellitus)	decrease (e.g. high altitude)
blood temp.	increase (e.g. fever)	decrease (e.g. during sleep)
Hb conc.	increase (e.g. polycythaemia)	decrease (e.g. anaemia)

17.3.1.13 O_2 Dissociation Curve of Myoglobin "(Hb_4) O_2"

Differences between:

Blood Hb	Muscle Hb
It has four iron atoms, each can combine with one O_2 molecule (2 oxygen atoms) → $(Hb_4)O_8$	It has one iron atom which can combine with one O_2 molecule → $(Hb_1)O_2$

Shape of O_2 dissociation curve of myo Hb: Rectangular hyperbola, i.e.

1. Remains horizontal (at high O_2 tension) → myo Hb does not give up its O_2 easily.
2. Then, suddenly descends vertically (at very low O_2 tension, e.g., severe muscular exercise) → myo Hb gives its O_2 easily (as blood Hb has given most of its O_2).

N.B.: After muscular exercise, myoglobin (myo Hb) absorbs O_2 from blood to replenish O_2 store again.

17.3.2 Carriage and Transfer of CO_2 by Blood

17.3.2.1 CO_2 Tension

	CO_2 PP (mm Hg.)
Inspired air	0.304
Alveolar air	39.215
Venous blood	46
Arterial blood	40
Tissues	50-60
Expired air	28.52

17.3.2.2 CO_2 Volume (× cc/ 100 ml Blood)

arterial blood	= 50% (50 cc/ 100 ml blood)
venous blood	= 55% (55 cc/ 100 ml blood)

17.3.2.3 Tidal Volume of CO_2

= Volume of CO_2 added by the tissues to every 100 ml arterial blood.
= CO_2 of venous blood – CO_2 of arterial blood.= 5%.

17.3.2.4 Forms of CO_2 in Arterial Blood

(I) In physical solution (5%)	(II) In chemical combination (95%)	
Free CO_2 gas dissolved in plasma.	(89%)	(6%)
H_2CO_3 or its dissociation products (HCO_3^- and H^+)	$NaHCO_3$ in plasma and $KHCO_3$ in RBCs. , in a ratio of 3:1	In direct combination with hemoglobin forming carbaminohemoglobin and with plasma proteins forming carbaminoproteins.

17.3.2.5 Carriage of Tidal CO_2

1. In physical solution = 5%

2. In chemical combination = 95%

a) **buffered by plasma = 10%:**
   - Proteins: Na proteinate (basic) + H2CO3 → $NaHCO_3$+H Proteinate (Proteinic acid)
   - Phosphates: Na_2HPO_4 (basic) + H_2CO_3 → $NaHCO_3$ + NaH_2PO_4.

b) **buffered by Hb = 85%**

a. 40% of tidal CO_2 is buffered through indirect reaction with K+ released from $KHbO_2$

At the tissue level: CO_2 produced as a result of tissue activity diffuses into plasma with water then into RBCs via its semi-permeable membrane.
   - Bicarbonate formation $CO_2 + H_2O_2 \rightarrow H_2CO_3 \rightarrow HCO_3^- + H^+$.

17.3.2.6 Inside RBCs

1. $CO_2 + H_2O \xleftrightarrow[\text{reduction}]{\text{carbonic anhydrase}} H_2CO_3$, then
2. $KHbO_2$ (K-salt of oxy Hb) → K^+ + Hb (reduced Hb) + O2 (released to tissues)

$$\begin{array}{c} | \\ H_2CO_3 \\ \downarrow \end{array}$$

$KHCO_3 + H^+$, then

Hb (reduced Hb) combines with H^+ → H.Hb (very weak acid)

N.B.:

- Reduced Hb (= Hb) and oxy Hb (=HbO_2) act as weak acids in the alkaline pH of blood. They combine with K^+ to form K salts.
- HbO_2 is a stronger acid (holds more K^+) than Hb. So, on reduction of HbO_2 at the tissue level, Hb cannot hold all K^+ of HbO2; consequently, K^+ combines with the already formed H_2CO_3 inside RBCs to form $KHCO_3$.

a) 25% of tidal CO_2 is buffered by the direct reaction between H_2CO_3 and K salts of Hb acids inside RBCs:

- K. Hb (very weak acid) + H_2CO_3 → H HbO_2 + $KHCO_3$
- $KHbO_2$ (stronger acid than reduced Hb) + H_2CO_3 → HHb + $KHCO_3$

b) 20% of tidal CO_2 combines directly with the free amino group of the Hb molecule to form carbamino Hb. (= Hb. NH.COOH):

Hb. NH_2 + CO_2 → Hb. NH. COOH

Hb's affinity to carry CO_2 as carbamino Hb is double that of HbO_2.

17.3.3 Chloride Shift

17.3.3.1 At Tissues

When arterial blood is delivered by systemic capillaries to the tissues, it is exposed to:

1. Oxygen at very low tension (1-40 mm Hg).
2. Carbon dioxide at high tensions (50-60 mm Hg).

Tidal CO_2 diffuses into plasma with water then into RBCs via its semi-permeable membrane, where it is buffered (previously mentioned).

The continuous accumulation of $KHCO_3$ in red cells upsets the ionic equilibrium, so HCO_3^- diffuses into plasma (K+ cannot accompany HCO_3^-, as it is retained inside the RBCs by negative charges e.g., proteins) where it combines with Na^+ → $NaHCO_3$. On the other hand, Cl^- ions diffuse into red cells forming KCl to restore ionic equilibrium (Chloride shift; Hamburger's phenomenon).

Water Shift: formation of $KHCO_2$ and KCl inside RBCs → increases the osmotic pressure inside RBCs → diffusion of water into RBCs → swelling of RBCs

17.3.3.2 At Lungs

When venous blood flows through the pulmonary capillaries of the lungs it is exposed to:

1. Oxygen at high tension (98 mm Hg).
2. Carbon dioxide at low tensions (40 mm Hg).
   a) O_2 diffuses from alveolar air into plasma (O_2 in solution) then into RBCs, where it combines with reduced Hb (H. Hb; very weak acid) → Oxy Hb (HbO_2) + H^+
   b) HbO_2 (more acidic than Hb and holds more K^+ and less CO_2), so recovers the K^+ that is lost

$$HbO_2 + KCl \rightarrow K\ HbO_2 + Cl^-$$

The continuous accumulation of Cl^- in red cells upsets the ionic equilibrium, so Cl^- diffuses into the plasma where it combines with Na^+ to form NaCl. On the other hand, HCO_3^- ions diffuse into red cells, where they react with H^+ (from reduced Hb).

$$H^+ + HCO_{3^-} \xleftrightarrow{\text{carbonic anhydrase}} H_2CO_3 \xleftrightarrow{\text{carbonic anhydrase}} CO_2 + H_2O$$
$$\downarrow$$

* CO_2 diffuses into plasma, then to alveolar air → atmosphere.
* H_2O shifts from RBCs to plasma.

Net results of chloride shift at:

	Tissues	Lungs
HCO_3^-	increases in both RBCs and plasma	shifts from plasma to RBCs
Cl^- shifts from	plasma to RBCs	***RBCs to plasma***
Na^+ and K^+	do not change	do not change
Water shifts from	plasma to RBCs	RBCs to plasma.
pH of both RBCs and plasma	change very little (7.4 → 7.37)	

Alkali reserve (CO_2 content/ 100 ml arterial plasma)

It is the volume of CO_2 present in chemical combination (e.g., $NaHCO_3$) in 100 ml arterial plasma (not whole blood) = 55-70 ml CO_2/ 100 ml arterial plasma.

17.3.3.3 Measurement

- Venous blood is centrifuged and plasma is introduced into a tonometer.
- Plasma is exposed to CO_2 tension of 40 mm Hg (= CO_2 tension of alveolar air at the last part of deep expiration) for 20 minutes → venous blood is arterialised.
- Plasma is then analysed for CO_2 content per 100 ml.

Why is the measurement of alkali reserve done on plasma but not on whole blood?

- HCO_2^- concentration in plasma is three times greater than that of RBCs.
- Haematocrit value varies under different conditions, even in the same person. For example, in anaemia plasma volume increases while Haematocrit value decreases. If whole blood is used in this case, the alkali reserve will be falsely increased.

17.3.3.4 Importance of Alkali Reserve

1) CO_2 carriage: most of CO_2 in the blood is carried as $NaHCO_3$ in plasma.
2) Acid-base balance: neutralise acid metabolites (e.g., lactic acid) formed by tissues and added to the blood.
3) Gives information about:

Acidosis	Alkalosis
= Alkali reserve:	
Decrease below 55%	Increase above 70%
If accompanied by:	Increase blood pH
Decrease blood pH	↓
↓	Alkalaemia
Acidaemia	

17.3 TRANSPORTATION OF RESPIRATORY GASES

The blood carried O_2 from the alveolar air to the body tissues and CO_2 from the body tissues to the alveoli.

17.4 RESPIRATORY CENTRES

(= Collection of neurones in brain stem)

17.4.1 General Function

1. Responsible for the normal rhythm of respiration.
2. Regulation of respiration (rate and depth) according to body conditions.

17.4.2 Location

Brain stem (within its reticular formation) extending from the upper part of the Pons to lower third of the M.O.

R.C. is composed of 5 bilateral centres

1. Inspiratory centre (I.C.) → M.O. (ventral).
2. Expiratory centre (E.C.) → M.O. (dorsal).
3. Pneumotaxic centre → upper Pons.

4. Apneustic centre → lower $\frac{1}{2}$ Pons.
5. Gasping centre → lower M.O.

17.4.2.1 Inspiratory Centre (I.C.)

Location: Bilaterally situated in the ventral M.O. Nerve cells in the I.C. send out streams of impulses which travel down the M.O. to the AHCs of the spinal cord, which gives:

Phrenic N.: from the 3rd, 4th and 5th cervical segments to supply the diaphragm.

Thoracic Ns. (upper 10 intercostal nerves) supply intercostal muscles. These nerve impulses cause the muscles of inspiration to contract.

1. Stimulation of apneustic centre leads to stimulation of IC.
2. Reciprocal relationship: stimulation of one centre inhibits the other.
3. Stimulation of both centres at the same time → inspiration only.

I.C. is stronger than E.C.

17.4.2.2 Expiratory Centre (E.C)

Location: Bilaterally situated in the dorsal M.O.

Nerve cells in the E.C. send out streams of impulses which inhibit or interrupt the activity of I.C. → the muscles of inspiration relax → expiration follows passively.

1. Stimulation of apneustic centre leads to inhibition of EC.
2. Reciprocal relationship: stimulation of one centre inhibits the other.
3. Stimulation of both centres at the same time → inspiration only.

I.C. is stronger than E.C.

N.B.: I.C. is inherently active, unless inhibited by inhibitory impulses from:

vagal fibres from lungs	pneumotaxic centre
↓	↓
Direct	Indirect
↓	↓
inhibit I.C.	inhibit apneustic centre
↑	↓

17.4.2.3 Apneustic Centre (APC.)

Location: at the lower $\frac{1}{2}$ Pons.

Function: It is called the pacemaker of respiration. It is inherently active, unless inhibited by inhibitory impulses from the pneumotaxic centre and vagal fibres from the lungs.

17.4.2.4 Pneumotaxic Centre (PNC.)

Location: at upper Pons.

Function:

a) During Eupnoea (normal breathing): Functionless! as PNC. is stimulated by I.C. and APC, whose discharges, during eupnoea, are not enough to stimulate PNC.

b) During deep inspiration or cutting vagi: PNC is stimulated, acting similarly to vagi but in a more powerful manner → inhibit APC and I.C. → change their continuous activity into rhythmic (= periodic) one but at a deeper level.

17.4.2.5 Gasping Centre

<Location: at lower M.O.>

Function:

a) During Eupnoea (normal breathing): Functionless (APC. → inhibit G.C.).

b) Just before death: G.C. is responsible for gasping respiration (= deep spasmodic inspiration), which is repeated for a few minutes before death occurs.

N.B.: G-C. is the last centre to fail.

17.4.3 Regulation of Respiratory Centre Activity

1. **Nervous from:**
   a) **Reflexes from respiratory system:**
      1. Hering-Breuer reflex.
      2. Paradoxical reflex.
      3. Sneezing reflex.
      4. Cough reflex.
      5. Afferent impulses from respiratory muscles.

2. b) **Reflexes from the circulatory system:**
      1. From the right atrium (Harrison's reflex).
      2. From the vaso-sensory area.

   c) **Higher centres:**
      1. Cerebral cortex.
      2. Hypothalamus.

3. d) **Other parts of the body:**
   a. **1. Nervous regulation:**
   a. i. Reflexes from respiratory system
   1. **Hering-Breuer reflex:**

*Respiration is regulated by its own function (i.e., change in lung volume)

	Inflation reflex		Deflation reflex
	= Inflation → inhibition of inspiration → expiration start		Excessive Deflation → inhibition of expiration → inspiration start
Stimulus:	Normal inspiration	Deep inspiration	Deep expiration (= excessive deflation of lungs)
Receptors:	stimulation of stretch receptors		
	Inflation (stretch) receptors		Deflation (compression) receptors
	Present in lungs (alveoli, alveolar sac, alveolar ducts), bronchi and bronchioles		
	Send inhibitory impulses		Send excitatory impulses
		"more receptors" "more impulses"	
Afferent:	Vagus (inhibitory fibres)		Vagus (excitatory fibres)
Centers:	Inhibition	More inhibition	Stimulation
APC:	Inhibition		Stimulation
IC:	---	More inhibition	Inhibition
EC:	Functionless		-------
PNC:		Activation Active	
Efferent:	previously mentioned		
Response:	Inspiration is inhibited and exp. follows passively without stimulation of EC.	Insp. is stopped and expiration follows actively.	Expiration is inhibited and inspiration start.

These reflexes stop after cutting the vagus.

17.4.3.1 Mechanism of Deflation During Eupnoea

Deflation (= expiration) → 1) decreases the number of stretch receptors (stimulated inflation receptors) and consequently their discharge as the lung becomes smaller until disappearing by the end of expiration → 2) decreases vagal inhibitory impulses to APC and IC (which recover and restart their spontaneous "inherent" activity).

↓

Inspiration starts, and Expiration is inhibited

N.B.: 1. No excitatory impulses reach respiratory centres in eupnoea.

2. In eupnoea, deflation receptors are not stimulated during expiration (they are stimulated only in deep expiration) because deflation is not sufficient. Therefore, no deflation reflex occurs in eupnoea; only inflation reflex occurs.

17.4.3.2 Importance of Hering-Breuer Reflexes

1. Maintain respiratory rhythm:
   - When the lungs expand, it inhibits inspiration and expiration starts.
   - When the lung empty → it inhibits expiration, and inspiration start.
2. Prevent excessive distension and collapse of lungs:
   - This is important in animals that have no long thoracic cage, e.g., frogs.

N.B. : In some pathological conditions, e.g., pulmonary congestion (=excess blood in the lungs), respiration becomes shallow and rapid.

17.4.4 Para-doxical Reflex

- = Forced inspiration stimulates more inspiration. This is short because receptors of the reflex are rapidly adapting, i.e., they stop function rapidly.

Mechanism: receptors, afferent, centre, efferent, and response → previously mentioned.

17.4.4.1 Importance of Para-doxical Reflex

1. Newborn infant → help to open his lungs.
2. During muscular exercise → helps to fill lungs maximally.
3. During normal respiration, deep inspiration helps to open any collapsed alveoli.

17.4.4.2 Functions of Pulmonary Vagi

1. Carry afferent impulses of Hering-Breuer and paradoxical reflexes.
2. Responsible for shallow, rapid respiration (during eupnea).

N.B.: cutting the vagi → deep and slow respiration (due to the absence of Hering-Breuer reflex). However, in panting animals (without sweat glands), cutting of the vagi does not cause deep and slow respiration, as the PNC is activated by the hypothalamus (heat-regulating centre). This is because panting aims to keep body temperature constant when these animals are exposed to hot weather.

3. Responsible for the rhythmic discharge of APC and IC.

17.4.4.3 Evidence

Isolation of PNC

Intact ⟵ vagic → Cutted

Intact	Cutted
Does not stop the continuous rhythmic discharge from 1C and APC.	Apneustic respiration i.e. deep and slow respiration.

1. **Sneezing reflex**
   Stimulus: Irritation (chemical, mechanical) of muscles of nasal cavity.
   Afferent: Branches of the trigeminal nerve.
   Response: Deep inspiration followed by rapid and strong expiration.
2. **Cough reflex**

Stimulus: Irritation (chemical, mechanical) of mm. of the larynx, trachea, and bronchi.
Afferent: Vagus nerve.
Response: Moderate inspiration followed by strong expiration with closed glottis (N.B. pressure inside the air passage reaches more than 100 mm Hg), then glottis opens suddenly → strong ejection of air and irritants (foreign body).

3. **Afferent impulses from respiratory muscles:**

Produce reflexes similar to Hering-Breuer reflexes, i.e., contraction of inspiratory muscles (diaphragm and intercostal muscles) → affemet impulses (pulmonary vagi) → inhibit IC (which recovers when these muscles relax during expiration).

17.4.4.4 Reflexes from the Circulatory System

1. **From right atrium (Harrison's reflex):**

(= Increased venous returns increase respiratory rate)

Receptors: baroreceptors in the wall of the R. atrium, opening large veins (vena cava and pulmonary veins)
Stimulus: Venous return and distension at the opening of large veins
Afferent: Vagus.
Response: Increase respiratory rate.
Significance: helps oxygenation and removal of CO_2 from excess venous return (that reaches the lung via pulmonary circulation).

2. **From vaso-sensory area:**

17.4.4.5 Carotid Sinus

Site: dilatation at the beginning of internal carotid = after bifurcation of carotid artery into the internal and external carotid.
Receptors: baroceptors present in the tunica adventitia in the wall of carotid artery.
stimulated by increase arterial blood pressure (ABP).
Afferent: Sinus nerve passes to MO as a branch of the glossopharyngeal N. or Herring N.

Function:

(1) When stimulated by increasing ABP than normal it send inhibitory impulses along their afferent nerves to respiratory and cardiovascular centres in the MO → inhibiting cardiac and respiratory centres.
(2) Normal ABP:

Baroceptors continuously send inhibitory impulses to MO centres but with little or no effect on RC.

(3) ↓ ABP than normal (e.g. haemorrhage): ↓number of inhibitory impulses from pressor receptors ↓activity of RC → ↑resp. rate

17.4.4.6 Aortic Arch

- = curve between ascending and descending.

Site: parts of the thoracic aorta.
Receptors: baroceptors present in the tunica adventitia in the wall of carotid artery.
Stimulated by increase arterial blood pressure (ABP).
Afferent: The aortic nerve passes to MO as a branch of the Vagus.
In rabbits, it form a separate bundle = depressor N.:.

Function:

(1) When stimulated by increasing ABP than normal it send inhibitory impulses along their afferent nerves to respiratory and cardiovascular centres in the MO → inhibits cardiac and respiratory centres.
(2) normal ABP:
    Baroceptors continuously send inhibitory impulses to MO centres but with little or no effect on RC.
(3) ↓ABP than normal (e.g. haemorrhage): ↓number of inhibitory impulses from pressoreceptors ↓ activity of RC → ↑respiration rate**.**

17.4.4.7 Carotid Body and Aortic Body

Site: Origin of the carotid artery and aortic arch.
Receptors: Chemoreceptors present in the wall of carotid artery and aortic arch. They are sensitive to CO2, H+, or $O_{2.}$
Afferent: Carotid sinus and aortic arch.

Functions:

1. CO_2 or H^+ or decreased O_2 cause excitatory impulses along their afferent nerves to respiratory and cardiovascular centres in the MO → Stimulate cardiac and respiratory centres.
2. normal CO_2 and O_2 tension in arterial blood: chemoreceptors continuously send excitatory impulses to RC.
3. ↓CO_2 and ↑O_2 :↓resp. rate.

17.4.4.8 Impulses from Higher Centres

1. **Cerebral cortex:**

Voluntary control for a short period: Voluntary hyperventilation: leads to a washout of CO2 from alveolar air and arterial blood → inhibition of respiration (limited duration).

Voluntary apnoea: Accumulation of CO2 in alveoli and arterial blood up to the level which stimulates respiration against the will.

Occurs during: speech, singing, blowing, sucking, micturition, defecation, and parturition. (Can be maintained for 1-3 minutes.)

2. **Hypothalamus:**
   Emotion: e.g., fear, anger, sexual excitement, ... etc.
   a) Accelerate respiration:
      Moderate, mild stimulation of posterior hypothalamic nuclei (which control sympathetic activity).
      Severe, sudden anterior hypothalamic nuclei (which control parasympathetic activity) induce slow respiration.
   b) Body temperature: Increase body temperature → increase resp. rate (to get rid of excess heat via evaporation of water vapour through expiration).
   c) Panting: (occurs in animals that have no sweat glands, e.g., dogs and cat.)
      Respiration rate increases up to 200/min (in hot weather) to increase heat loss via evaporation of H_2O from the lungs and air passage through expiration, leading to the cooling of body temperature back to normal. The increased rate of respiration in panting is due to the stimulation of the panting centre in the hypothalamus, resulting in powerful stimulation of the PNC, which becomes more sensitive than the vagus.

17.4.4.9 Reflexes from Other Part of the Body

1. **Conditioned reflexes:**
   Visual, auditory, or olfactory impulses → may accelerate or inhibit respiration. Centres of these reflexes lie in cerebral cortex. e.g.
   - ↑resp. rate when the swimmer sees the swimming pool.
   - ↑resp. rate when Teaser sees estrous female.
2. **During deglutition:**
   Respiration is reflex inhibited by afferent impulses from the base of the tongue and pharynx along the 5th, 9th, and 10th cranial nerves to the deglutition centre in the MO, which inhibits RC.
3. **Pouring cold water on the skin:**
   Causes apnea followed by deep inspiration. Afferent impulses arise from the cold receptors of the skin and pass along sensory nerves to the RC.
4. **Pain sensation:**
   - Mild pain → accelerates respiration.
   - Severe pain → inhibits or even stops respiration.
5. **Adrenaline apnoea:**
   Injection (i.v.) of large dose of adrenaline → generalised vasoconstriction → ↑ABP → temporary stoppage of respiration (apnoea).

17.4.5 Chemical Regulation of Respiration

Generally, CO_2, O2, and H^+ conc. in arterial blood affect respiration via two mechanisms:

1. **Direct:** via blood supply to RC in MO and pons.
2. **Indirect** (reflex)**:** stimulation of chemoreceptors of carotid and aortic bodies → afferent → centres → efferent → response.

17.4.5.1 Response to Change in CO_2 Tension

CO_2 % (vol/ 100 ml)	Response
Normal:	Normal resp. (as CO_2 act mainly on APC)
Inspired air = 0.04%	
Arterial blood= 48%	
Above normal:	Stimulates RC → increase depth of resp. (and eventually increases rate of resp.)
Up to 7%	
9% or more	Inhibits RC → slow respiration
20-30 %	Comma
30-40 %	Complete anaesthesia
40-50 %	Death
Below normal:	depresses RC → shallow breathing. e.g. voluntary hyperventilation (wash of CO_2 from alveolar air and arterial blood → alkalosis → inhibition of resp. → stoppage of hyperventilation → CO_2 increase again to its normal level and act as a normal stimulus of resp.)

N.B.: The direct mechanism is more powerful (about 10 times) than reflex

evidence:

Cutting the afferent nerves of carotid and aortic bodies → respiration is little affected.

Deep anaesthesia → decreases sensitivity of RC to CO2.

17.4.5.2 Response to Change in O2 Tension

O_2 tension in alveolar air	Response
Normal:* at sea level	O_2 has little role in controlling resp.
O2 **lack** (below 60 mmHg):	Stimulate resp. (reflex mechanism).
• * mild and gradual e.g	Inhibits resp. (direct mechanism)
• high altitude.	
• apnoea following voluntary hyperventilation.	
• * Sever and sudden	
• going suddenly into a room containing pure N_2	

N.B.:

* CO_2 excess is more powerful in stimulating respiration than O_2 lack. Why? Comment:

CO_2 excess	O_2 lack
Stimulates RC.: Directly and reflex	Only reflex (while, directly inhibits)

Evidence:

When a person is allowed to breathe in a bag containing air, after a short time, severe acceleration of respiration occurs due to the accumulation of CO_2. When the experiment is repeated but CO_2 is always removed by a CO_2 absorber, e.g., soda lime, O_2 may drop from 21% (normal) to 14% without acceleration of respiration.

* Giving O_2 to patients under deep anaesthesia may end their life. Why?
Because giving O_2 in this case stops the stimulatory effect of O_2 lack on chemoreceptors, which is the main stimulus to the depressed RC (under deep anaesthesia), a mixture of O_2 and 5% CO_2 is given to keep the stimulation of RC both directly and reflexively.

17.4.5.3 Response to Change in H^+

- Same as CO_2.

17.4.5.3.1 Hypoxia

It means a lack of oxygen lack in tissues.

1. Hypoxic hypoxia.
2. Stagnant hypoxia.
3. Anaemic hypoxia.
4. Histotoxic hypoxia.

1. **Hypoxic hypoxia**
   It is the defect oxygenation of blood in the lung due to:
   a) Decrease air oxygen tension.
   b) Defect gas exchange in lung diseases: pneumonia, emphysema, and pulmonary edema.
   c) Mixing of arterial and venous blood.
2. **Stagnant hypoxia**
   It is a decrease in blood flow through the tissues. There are two types of stagnant hypoxia:
   a) Generalised stagnant hypoxia as in the case of congenital heart failure or polycythemia.
   b) Localised stagnant hypoxia as in the case of thrombosis.
3. **Anaemic hypoxia**
   It is a deficiency of Hb capable carrying O2, as in case of :
   - Lack in Hb.
   - Poisoning by CO (carboxyhemoglobin). CO combines with Hb 210 times more readily than O_2.
   - Formation of met Hb and sulpha Hb.
4. **Histotoxic hypoxia**
   In which the tissues can't utilise the oxygen from the blood as in the case of:
   - Cyanide poisoning, which blocks cytochrome oxidase.
   - Alcohol poisoning that blocks dehydrogenase.

Types of hypoxia	O_2 Tension in blood		O_2 Content		O_2Supply to the tissue
	Arterial	Venous	Arterial	Venous	
Hypoxic	↓	↓	↓	↓	↓
Stagnant	Nil	↓	Nil	↓	↓
Anaemin	Nil	↓	↓	↓	↓
Histotoxic	Nil	Super Nil	Nil	Super Nil	↓

BIBLIOGRAPHY

Ahmed, H. El-Anwar, and Saadia Abdelfattah Ali (2024): *Applied Animal Physiology*, 3rd Edition. Arish University, Egypt.

Anna, Dee Fails, and Christianne, Magee (2018): *Anatomy and Physiology of Farm Animals*, 8th Edition. https://www.wiley .com/en eg /Anatomy and Physiology of Farm Animals, 8th Edition-p-9781119239765.

David, L. (2006): *Cell Physiology*. Lange Medical Books/ McGraw-Hill Medical Publishing Division, New York, Chicago, San Francisco, Lisbon, London, Madrid, Mexico City, Milan, New Delhi, San Juan, Seoul, Singapore, Sydney, and Toronto.

Frandson, R., Wilke, W. L., Fails, A. D. (2003): *Anatomy and Physiology of Farm Animals*. Lippincott Williams and Wilkins, Baltimore.

John, E., Hall, J. E. and Guyton, A. C. (2011): *Guyton and Hall Physiology Review*, 2nd Edition. Saunders, an imprint of Elsevier Inc., Jackson, MS and Philadelphia, PA.

Reddy, P. B. (2014): *Text Book of Animal Physiology*. IMRF Publications, International Multidisciplinary Research Foundation Ratna Prasad Multidisciplinary Research &77 Educational Society (Regd), Andhra Pradesh, India.

18 Physiology of the Excretory System

Mukesh Kumar Bharti, Thulasiraman P. and Manish Kumar

Department of Veterinary Physiology & Biochemistry, Faculty of Veterinary & Animal Sciences, IAS, RGSC- Banaras Hindu University, Barkachha, Uttar Pradesh, India.

*Corresponding Author: Mukesh Kumar Bharti, Assistant Professor, Department of Veterinary Physiology & Biochemistry, Faculty of Veterinary & Animal Sciences, Institute of Agricultural Sciences, RGSC, Banaras Hindu University, Uttar Pradesh. Email: mukeshbharti@bhu.ac.in

18.1 INTRODUCTION

Kidney plays very important role in the excretion of metabolic waste products produced in the body and the maintenance of the body's internal environment. Therefore, the composition of body fluids is determined by what the kidneys retain, not by what the mouth takes in. The kidney has diverse roles in maintaining homeostasis. In mammals, the two kidneys normally receive approximately one-fourth of the cardiac output. The kidneys filter the blood and hence excrete metabolic waste, but they retrieve filtered substances that are needed by the body like water, glucose, electrolytes, and low-molecular-weight proteins. The kidneys alter the rate of reabsorption of water and electrolytes, and in this way, they help in maintaining acid-base balance. These multiple functions are accomplished by an extensive variety of cell types, each with specific responses to direct and indirect signals, arranged in a particular pattern to form the functional unit of the kidney, the nephron. The nephron consists of the glomerulus and its associated renal tubule segments. In the glomerulus, filtration of blood takes place, filtered substances are absorbed from tubular fluid to blood, and plasma components are secreted into the tubule fluid in the tubular segment of the renal tubule. Other than filtration, the kidney has essential functions like the secretion of important hormones and the hydrolysis of small peptides. The hormones participate in the regulation of systemic and renal dynamics, red blood cell production, calcium, phosphorus, and bone metabolism.

18.2 HISTORICAL PERSPECTIVES OF RENAL PHYSIOLOGY

Chronologically Claudius Galen (AD 130–200), a Greek physician who was the first to demonstrate that urine was formed in the kidneys and transmitted from them to the bladder by the ureter. Later on, William of Salicet (1210–1280) put forward his ideas of urine formation and gave his classic account of renal dropsy. Sir William Bowman (1816–1892) discovered that urine is a by-product of blood filtration carried out in the kidney; Bowman's capsule is named after him. Carl Ludwig (1816–1895) proposed that urine formation begins with the separation of a protein-free ultrafiltrate of plasma in the glomerulus. Arthur Cushny (1866–1926) in 1917 proposed the modern theory of urine formation

18.3 FUNCTIONS OF THE KIDNEY

1. Maintenance of homeostasis: The kidney plays a major role in maintaining homeostasis through the following methods:
    i. Excretion of waste products such as urea (end product of amino acid metabolism), uric acid (end product of nucleic acid metabolism), creatinine (end product of metabolism in muscles), bilirubin (end product of hemoglobin degradation). Kidneys also excrete harmful foreign chemical substances such as toxins, drugs, heavy metals, pesticides, etc.
    ii. Maintenance of water balance: Water balance is maintained according to the requirements of the animals
    iii. Maintenance of electrolyte balance: Electrolyte balance is done by the kidney to keep the ratio of important ions in proportion.
    iv. Maintenance of acid-base balance: acid base balance is one of the most important function that kidney is performing in the body.
2. Hemopoietic function: heamatopoietic function is performed by release of certain hormones like erythropoietin and thrombopoietin.
3. Endocrine function: kidney releases different hormones like erythropoietin, thrombopoietin, Renin, 1,25-dihydroxycholecalciferol (calcitriol) and Prostaglandins which having diverse role in maintaining water and electrolyte balance.
4. Regulation of blood pressure is done by regulating the volume of ECF, through renin-angiotensin mechanism
5. Regulation of blood calcium level by activating 1,25-dihydroxycholecalciferol into vitamin D and Vitamin D is necessary for the absorption of calcium from intestine

DOI: 10.1201/9781003426851-18

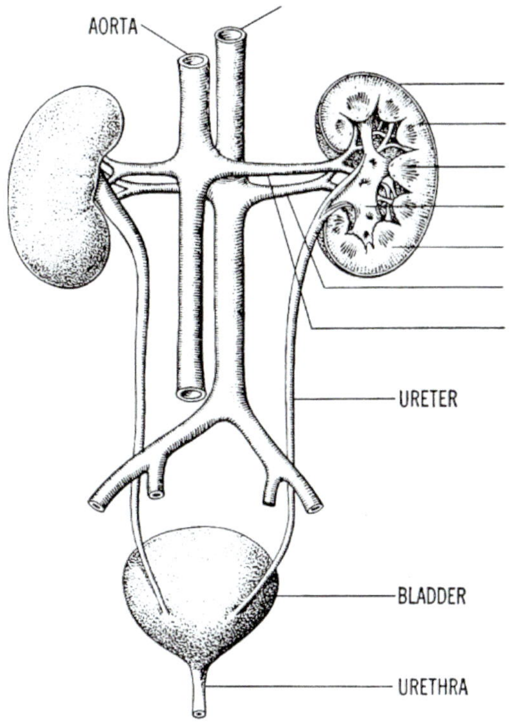

FIGURE 18.1 The renal system in mammals

18.4 THE RENAL SYSTEM

The renal system consists of a pair of kidneys which produces urine, an ureters that transport urine to the urinary bladder, an urinary bladder that stores until voided and it also an urethra through which urine voided out. Renal system mainly involved in removal of waste from the body. Skin, lungs, GIT, salivary glands and the liver are the channels of excretion. Even though various organs are involved in removal of wastes from the body but their excretory capacity is limited. Since urinary system has maximum excretory capacity in excreting substance from body so it plays a major role in homeostasis (Figure 18.1).

18.5 DEVELOPMENT OF THE KIDNEY

In mammals, the final kidney is the metanephric kidney, but development occurs in three stages with the development of three different types of kidney during the embryonic period: pronephros, mesonephros, and metanephros. All three types develop from the intermediate mesoderm sequentially in the cranio-caudal direction (in the direction from the head to the tail of the body). First, the pronephros is formed; in mammals, it is considered rudimentary, meaning it does not function. Then, caudal to the pronephros, the mesonephros develops, which is the functioning kidney of the embryo. Following this, in females, the mesonephros degrades, and in males, it takes part in the development of the reproductive system. The third stage is the formation of a metanephros in the caudal part of the embryo, which is a permanent kidney. In some mammals, kidney organogenesis ends before birth, while in others, it may continue for some time into the postpartum period. When the formation of new nephrons (nephrogenesis) ends, the number of nephrons in the kidney becomes final. Thus, after suffering damage, the kidneys of adult mammals cannot regenerate by forming new nephrons. However, kidneys have other compensatory mechanisms for restoring their function.

18.6 FUNCTIONAL ANATOMY OF THE KIDNEY

The kidneys are described as being retroperitoneal in location, reflecting their position outside the peritoneal cavity. The kidneys of most animals are roughly bean-shaped, with exceptions among domestic animals like the heart-shaped kidney in equines and the distinctively lobulated kidneys present in cattle.

The kidneys are present in the dorsal part of the abdominal cavity on each side of the aorta and caudal vena cava, just ventral to the first few lumbar vertebrae. In most domestic animals, the right kidney is slightly more cranial than the left kidney. The left kidney tends to be more pendulous, especially in ruminants, where the fore-stomach may push the left kidney towards the right, particularly when the rumen is full. The entire kidney is surrounded by a tough connective tissue capsule.

18.7 INTERNAL ANATOMY OF KIDNEYS

The medial aspect of each kidney has a concavity, the hilus, where arteries and nerves enter the kidney, and the ureter, veins, and lymphatic vessels leave. The wide origin of the ureter in the kidney is the renal pelvis, which receives urine from the collecting tubules. The cavity-like structure in the kidney that contains the pelvis is the renal sinus. The bovine kidney does not have a renal pelvis, so the ureter arises directly from the coalescence of individual calyces. The portion of the kidney immediately surrounding the renal pelvis is the renal medulla, which appears striated because of the radially arranged collecting tubules. The medulla also contains some loops of Henle (descending and ascending loops), and it is surrounded peripherally by the renal cortex in which the renal corpuscles are present (Figure 18.2).

The cortex has a granular appearance because of the large number of renal corpuscles, as well as proximal and distal convoluted tubules and other segments of loops of Henle. The medulla and cortex are arranged in units called lobes, which is a cone-shaped aggregates of renal tissue. The medullary portion of each lobe constitutes a renal pyramid, whose apex, i.e., renal papilla, is directed toward the renal pelvis. In the bovine kidney, each pyramid is associated with one of the grossly obvious lobes, each of which communicates with a minor calyx. The renal cortex of the equine, ovine, and caprine kidneys lacks visible divisions from outside into individual lobes, so the kidneys in these species appear smooth. In these species, the tips of the renal

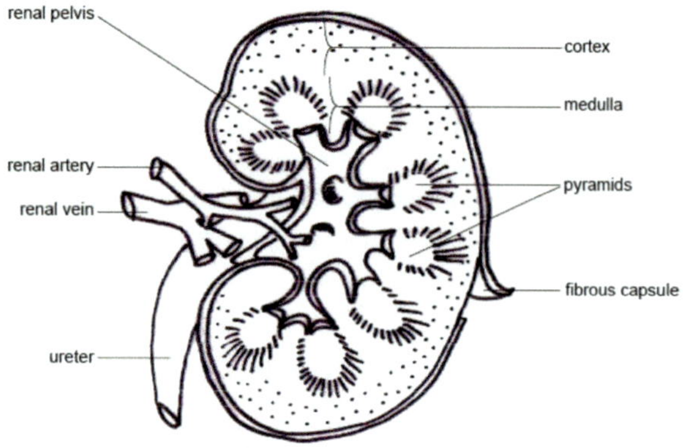

FIGURE 18.2 Internal structure of kidney

pyramids fuse with each other to form a common renal papilla, called the renal crest. The porcine kidney lacks external divisions into lobes, but the renal papillae of the medulla distinguish each lobe internally. In the kidney of the ox and pig, individual pyramids project into minor calyces (cup-like diverticula of the common collecting space within the renal hilus). These, in turn, empty into major calyces in these species empty directly into the ureter.

18.8 URETERS, URINARY BLADDER, AND URETHRA

The ureter is a muscular tube that transports urine to the urinary bladder. The smooth muscle of the ureter helps in this transport process as it undergoes peristaltic waves of contraction that stimulate the flow of urine to the urinary bladder. Each ureter originates at the renal pelvis (or the major calices of the bovine kidney) and empties into the urinary bladder near its neck at the trigone. The backflow of urine to the kidney is prevented by a type of valve in the ureter. The urinary bladder is a muscular organ that varies in size with the amount of urine it contains. The pelvis, ureter, bladder, and urethra are all lined with transitional epithelium. This epithelial lining is useful in these areas, since considerable distension of the lumen may occur.

18.9 BLOOD AND NERVE SUPPLY TO THE KIDNEY

The blood supply to the kidney is extensive because of its important role in adjusting the composition of extracellular fluid (including plasma) (Figure 18.3).

The two renal arteries may receive as much as one-fourth of the total cardiac output. The origin of renal arteries is directly from the abdominal aorta. While passing through the renal sinus, the renal artery divides into many segmental arteries. Each segmental artery subdivides into interlobar arteries. The interlobar artery passes between the medullary pyramids. At the base of the renal pyramid, the interlobar artery forms the arcuate artery. Each arcuate artery gives rise to interlobular arteries. From each interlobular artery, numerous afferent arterioles arise. The afferent arteriole enters the Bowman capsule and forms the glomerular capillary tuft.

In the glomerulus, the capillary loops unite and form the efferent arteriole, which leaves the Bowman capsule. After leaving the Bowman capsule, efferent arterioles form a second capillary network called peritubular capillaries, which surround the tubular portions of the nephrons. In this way, the renal circulation forms a portal system by the presence of two sets of capillary networks, namely glomerular capillaries and peritubular capillaries.

18.9.1 Venous System

Venous system runs opposite to the arterial system. It starts with peritubular venules and continues as interlobular veins, arcuate veins, interlobar veins, segmental veins, and finally the renal vein. The renal vein leaves the kidney through the hilus.

18.9.2 Innervation

The innervation to the kidney is mainly by efferent sympathetic nerve fibers. There is no authentic proof for the innervation of the kidney by parasympathetic nerves. Normal physiological stimulation of the efferent sympathetic nerves is involved in maintaining the water and sodium balance in the body. Efferent renal sympathetic nerve activity (ERSNA) produces significant changes in renal hemodynamics, tubular ion and water transport, and renin secretion. The renal nerves are the communication link between the central nervous system and the kidneys. Reno-renal reflexes are defined as responses occurring in one kidney as a result of intervention on the same or the opposite kidney that are mediated by neurohumoral mechanisms.

18.10 THE NEPHRON

Nephron is the structural and functional unit of the kidney. Kidney has millions of nephrons, and each nephron is made up of two parts: a renal corpuscle or Malpighian corpuscle, which is a blind end, and a tubular portion called the renal tubule. The renal corpuscle is situated in the cortex of the kidney, either near the periphery or near the medulla. The renal corpuscle is formed by two portions: 1. Glomerulus 2. Bowman capsule.

18.10.1 Glomerulus

Glomerulus is a bunch of capillaries surrounded by the Bowman capsule. It consists of glomerular capillaries interposed between the afferent arteriole and the efferent arteriole. Thus, the vascular system in the glomerulus is purely arterial. Glomerular capillaries are composed of a single layer of endothelial cells. The endothelium has many pores called fenestrae. The presence of fenestrae is proof of the filtration function of the glomerulus (Figure 18.4).

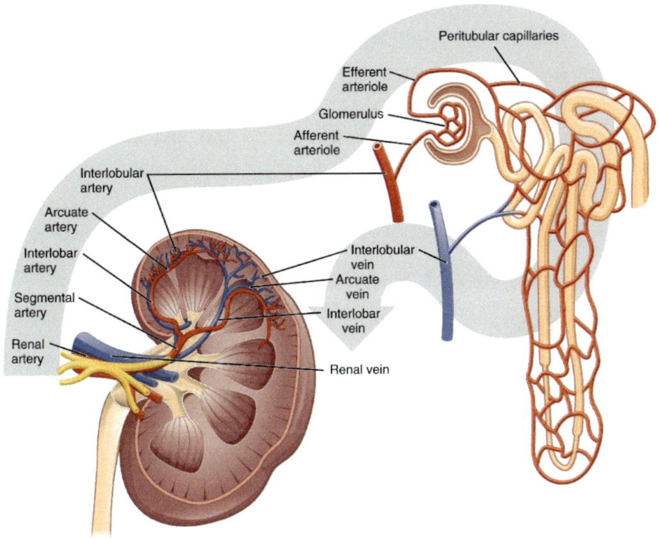

FIGURE 18.3 Blood flow to the kidney

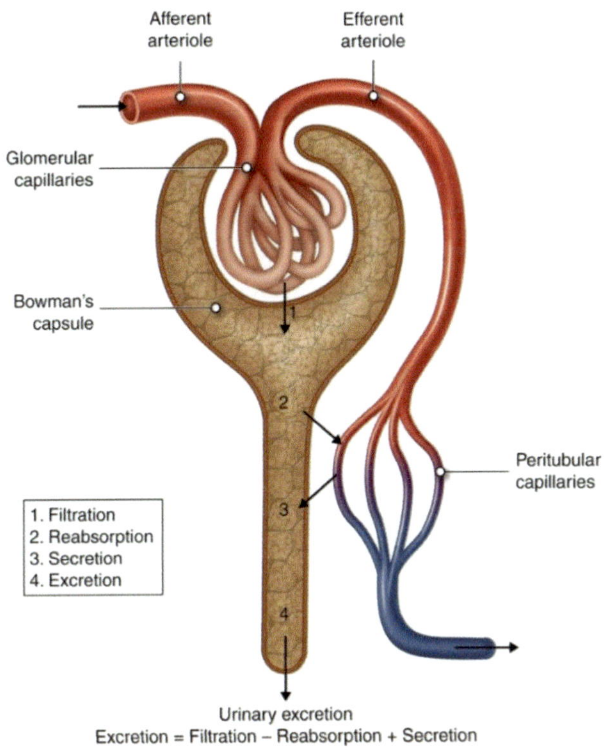

FIGURE 18.4 Bowman capsule and renal corpuscles

18.10.2 Bowman Capsule

Bowman capsule is a capsular structure that encloses the glomerulus. It is formed by two layers; the inner layer is the visceral layer, and the outer one is the parietal layer. The visceral layer covers the glomerular capillaries, and the parietal layer is continuous with the wall of the tubular portion of the nephron. Bowman capsule resembles a funnel with filter paper. Epithelial cells of the visceral layer fuse with the basement membrane, but the fusion is not complete, leaving small cleft like spaces in between, called slit pore.

18.10.3 Tubular Portion of the Nephron

Tubular portion starts with the proximal convoluted tubules (PCT), which lie in the cortex of the kidney. The proximal tubule is continued by the loop of Henle, which dips into the medulla. The loop of Henle is composed of three segments: the thin descending limb, thin ascending limb, and thick ascending limb. The descending thin segment of each loop is continuous with the ascending thin segment at the hairpin curve. The ascending limb has a thickened segment in the cortex known as the macula densa. Beyond the macula densa, the tubule is known as the distal tubule, which resides entirely in the renal cortex. This is followed by the cortical collecting tubule, which leads to the cortical collecting duct. The cortical collecting ducts join to form a single medullary collecting duct and finally a larger collecting duct

18.10.4 Approximate Number of Nephrons in Each Kidney

Approximate number of nephrons in each kidney of cattle and dogs is around 400,000. Nephron numbers vary considerably among species, while within a species, nephron numbers are relatively constant.

18.11 TYPES OF NEPHRONS

The mammalian kidney has two types of nephrons, based on (i) the location of their glomeruli and (ii) the depth of penetration of the loops of Henle into the medulla.

Cortical nephrons: glomeruli in the outer and middle cortices. They are associated with a short loop of Henle that extends to the junction of the cortex and medulla or into the outer zone of the medulla.2. Juxtamedullary Nephrons: Glomeruli in the cortex close to the medulla. Juxtamedullary nephrons have long loops of Henle that extend deeper into the medulla. However, the tubular fluid of all nephrons (cortical and juxtamedullary) is emptied into shared collecting ducts that proceed through the medulla to the renal pelvis.

18.12 THE JUXTAGLOMERULAR APPARATUS

Juxtaglomerular apparatus is a specialized organ situated near the glomerulus of each nephron. The JGA is made up of three different structures: 1. Macula densa 2. Extraglomerular mesangial cells 3. Juxtaglomerular granular cells. The JGA is involved in feedback mechanisms that help in the regulation of renal blood flow and glomerular filtration rate.

Macula densa is the end portion of the thick ascending segment before it opens into the distal convoluted tubule. It is situated between the afferent and efferent arterioles of the same nephron and is very close to the afferent arteriole. The macula densa is formed by tightly packed cuboidal epithelial cells. It helps in regulating the glomerular filtration rate, renal blood flow, and renin release. Macula densa cells detect intratubular salt concentrations to control renal blood flow through afferent arteriole constriction and dilation (Figure 18.5).

Extraglomerular mesangial cells are situated in the triangular region surrounded by the afferent and efferent arterioles and macula densa. These cells are also called agranular cells or Lacis cells. They resemble smooth muscle cells and play a role in renal autoregulation of blood flow to the kidney and the regulation of systemic blood pressure through RAS.

1. Juxtaglomerular granular cells: The smooth muscle cells of the afferent arteriole that make contact with the macula densa are specialized smooth muscle cells and are called juxtaglomerular (JG) granular cells. Juxtaglomerular cells are also called granular cells because of the presence of secretory granules in their cytoplasm that contain renin, a proteolytic enzyme.

2. Glomerular mesangial cells: The space between the macula densa and the afferent and efferent arterioles, as well as the space between the glomerular capillaries, is known as the mesangial region and consists of mesangial cells. Mesangial cells secrete the matrix, glomerular basement membrane, provide structural support, have phagocytic activity, secrete prostaglandins, and also exhibit contractile activity and can influence blood flow through glomerular capillaries.

The secretion of hormones is the primary function of the juxtaglomerular apparatus. It also regulates the glomerular blood flow and glomerular filtration rate. The juxtaglomerular apparatus secretes two hormones, i.e., renin and prostaglandin. Renin plays an important role in the maintenance of blood pressure. Secretion of renin is stimulated by four factors: i. A fall in arterial blood pressure ii. A reduction in the ECF volume iii. Increased sympathetic activity iv. A decreased load of sodium and chloride in the macula densa. Once renin is secreted, it converts angiotensinogen to angiotensin I. Angiotensin I is converted to angiotensin II by another enzyme called angiotensin-converting enzyme (ACE). Angiotensin II acts as a potent vasoconstrictor and helps in the restoration of decreased blood volume.

18.13 REGULATION OF RENAL BLOOD FLOW

Renal blood flow is regulated mainly by autoregulation. Autoregulation is the intrinsic ability of an organ to regulate its own blood flow. Some vital organs in the body, such as brain, heart, and kidneys, have autoregulation mechanisms. It is highly significant and more efficient in the kidneys. Renal autoregulation is important in maintaining the glomerular filtration rate (GFR). Blood flow to the kidneys remains normal even when the mean arterial blood pressure varies widely. This helps maintain normal GFR.

Two mechanisms are involved in renal autoregulation: Myogenic response: In the afferent arteriole, increased

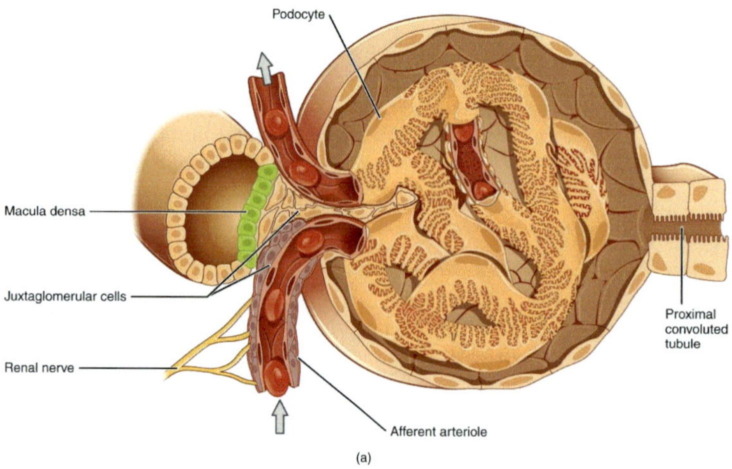

FIGURE 18.5 Juxtaglomerular apparatus

blood pressure would increase the stretch, and the arteriole would respond by contracting. In this way, renal blood flow would be decreased. A reduction in blood pressure would cause less tension, and the blood vessel would dilate, hence increasing renal blood flow. Tubuloglomerular feedback: The macula densa plays an important role in tubuloglomerular feedback, which controls the renal blood flow and GFR. Macula densa cells sense changes in volume delivery as well as NaCl concentration to the distal tubules.

18.13.1 Urine Formation

Urine formation consists of three processes: glomerular filtration, tubular reabsorption from the renal tubules into the blood, and tubular secretion from the blood into the renal tubules. Filtration is the function of the glomerulus, whereas reabsorption and secretion are the functions of the tubular portion of the nephron.

18.13.2 Glomerular Filtration

Most substances must be cleared from the blood, especially the end products of metabolism such as urea, creatinine, uric acid, and urates. These substances are poorly reabsorbed and therefore excreted in large amounts in the urine. Certain foreign substances and drugs are also poorly reabsorbed, so their excretion rates are high. Electrolytes like sodium ions, chloride ions, and bicarbonate ions are highly reabsorbed from renal tubule, so that only very slight amounts appear in the urine. Substances such as amino acids and glucose are completely reabsorbed from the renal tubules and hence do not appear in the urine.

The formation of urine begins when an ultrafiltrate of plasma passes through the fenestrated capillary endothelium, the basement membrane, and the epithelium of Bowman's capsule. The glomerular capillaries are impermeable to proteins, so the glomerular filtrate is protein-free and devoid of cellular elements, including RBCs. Other constituents of the glomerular filtrate are similar to the concentrations in the blood plasma

Size of solute is an important parameter for their filtration. Filtration of solutes is inversely proportional to their size. The barrier present in glomerular membrane is selective in determining the molecules to filter, based on their size and electrical charge. Negatively charged large molecules are filtered less easily than positively charged molecules of equal molecular size. Albumin is restricted from filtration, because of its negative charge and the size. In certain kidney diseases, the negative charges on the basement membrane are lost and as a result of this loss, some of the lower-molecular-weight proteins, especially albumin, are filtered and appear in the urine.

18.14 GLOMERULAR FILTRATION RATE (GFR)

Glomerular filtration rate (GFR) is defined as the total quantity of filtrate formed by the nephrons in the given time period. Normal GFR is 125 ml/minute or about 180 L/day in human being. GFR is determined by the balance of hydrostatic and colloid osmotic forces acting across the capillary membrane and the capillary filtration coefficient (K_f). The glomerular capillaries have a much higher rate of filtration because of a high glomerular hydrostatic pressure and a large filtration coefficient. Glomerular hydrostatic pressure is the pressure exerted by the blood in glomerular capillaries. This pressure favors glomerular filtration (60 mmHg). Colloidal osmotic Pressure is the pressure applied by plasma proteins in the glomeruli. It opposes glomerular filtration (32 mmHg). Hydrostatic pressure in Bowman capsule is the pressure exerted by the filtrate in Bowman capsule. It is also called capsular pressure and it also opposes glomerular filtration (18 mmHg)

Net Filtration pressure = $HP_{glomerular}$ - (COP + $HP_{capsular}$) = 60 - (32+18) = 10mm Hg.

Factors increasing GFR by vasodilatation: i. Atrial natriuretic peptide (ANP), ii. Brain natriuretic peptide (BNP), iii. Bradykinin, iv. Endothelial derived nitric oxide, v. Prostaglandin (PGE2 & PGI2)

Factors decreasing GFR by vasoconstriction: i. Angiotensin II, ii. Endothelins, iii. Noradrenaline, iv. Platelet derived growth factor (PDGF), v. Prostaglandin (PGF2).

Filtration fraction: The ratio of GFR to RPF is referred to as the filtration fraction (FF) i.e. basically the fraction of the renal plasma flow that is filtered. The FF is the fraction (or percentage) of plasma flowing through the glomerulus that becomes glomerular filtrate. Renal plasma flow (RPF) is that part of the RBF that is plasma. Filtration fraction = GFR / Renal plasma flow x 100. It is expressed in percentage.

18.14.1 Tubular Reabsorption

Tubular reabsorption is the process through which water and other important substances are transported from renal tubules to the blood. Tubular reabsorption is known as selective reabsorption because the tubular cells will reabsorb only those substances that are necessary for the body. Essential substances such as glucose, amino acids and vitamins are completely reabsorbed from renal tubule whereas the metabolic waste products are not reabsorbed and excreted through urine.

18.14.2 Routes of Reabsorption

Water and solutes can be transported either through Transcellular route: In this route the substances move through the cell (through apical or luminal surface of the cell membrane) or Paracellular route: In this route, the substances move through the intercellular space (through the tight junction between the cells) (Figure 18.6).

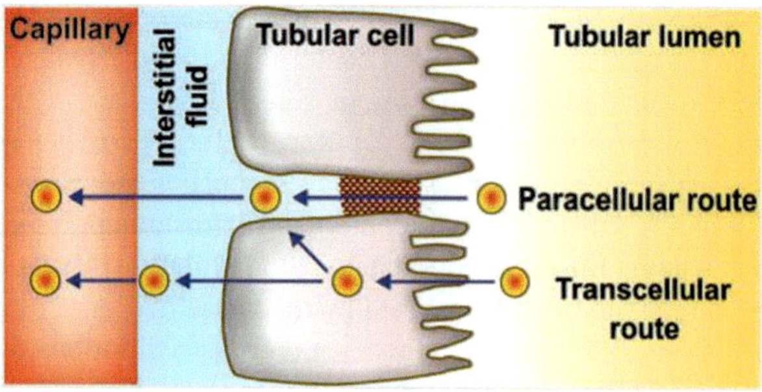

FIGURE 18.6 Route of reabsorption.

18.14.3 Mechanism of Reabsorption

Tubular reabsorption is of two types:

1. **Active transport:** Active transport is the movement of molecules against the electrochemical (uphill) gradient. This process needs energy which is derived from ATP. Certain substances are reabsorbed actively from the renal tubule like sodium, potassium, calcium, phosphates, bicarbonates, glucose, amino acids, ascorbic acid, uric acid, and ketone bodies.

Active transport is of two types:

Primary active transport: Transport that is coupled directly to an energy source, such as the hydrolysis of ATP. The best example of this type of transport is the sodium-potassium ATPase pump that functions throughout most parts of the renal tubule.

Secondary active transport: Transport that is coupled indirectly to an energy source, such as that due to an ion gradient, is referred to as secondary active transport. Glucose reabsorption by the renal tubule is through secondary active transport. In secondary active transport, two or more substances interact with a specific membrane protein (carrier molecule) and are transported together across the membrane. As one of the substances diffuses down its electrochemical gradient, the energy released is used to bring another substance against its electrochemical gradient. In this way, secondary active transport does not require energy directly from ATP.

2. **Passive reabsorption:** Passive reabsorption is the movement of molecules along the electrochemical gradient. This process does not need energy. Substances reabsorbed passively are chloride, urea, and water.

18.14.4 Co-Transport and Counter-Transport of Solute

Simultaneous transport of two or more compounds on the same carrier in the same direction (e.g., Na+ plus glucose, or Na+ plus an amino acid) is known as co-transport.

Counter-transport refers to the movement of a compound in one direction, driven by the movement of a second compound in the opposite direction (e.g., Na^+–H^+ countertransport).

18.14.5 Reabsorption of Important Substances

18.14.5.1 Reabsorption of Sodium

65% of sodium is reabsorbed in the PCT, the remaining 35% continuing beyond the proximal tubules. The energy requirement is derived from the Na+/K+-ATPase located in the basal and lateral borders of the proximal tubule epithelial cells. The direction of transport is from the nephron proximal tubule to the peritubular capillaries. The reabsorption of Na+ must be accompanied by the reabsorption of anions to maintain electrical neutrality.

18.14.6 Reabsorption of Glucose

Glucose is completely reabsorbed in the proximal convoluted tubule. It is transported through the secondary active transport mechanism. Glucose and sodium bind to a common carrier protein in the luminal membrane of the tubular epithelium and enter the cell. Here the carrier protein is called sodium-dependent glucose cotransporter 2 (SGLT2) which actually helps in the transport of glucose. From the tubular cell glucose is transported into the medullary interstitium by another carrier protein called glucose transporter 2 (GLUT2).

18.14.7 Reabsorption of Amino Acids

Complete reabsorption of amino acids takes place in the proximal convoluted tubule. It is reabsorbed actively by the secondary active transport mechanism along with sodium.

18.14.8 Reabsorption of Bicarbonates

Bicarbonate is reabsorbed actively, mostly in the proximal tubule. Its reabsorption occurs in the form of carbon dioxide. In the filtrate, bicarbonate is mostly occurs in the form of sodium bicarbonate. Sodium bicarbonate dissociates into sodium and bicarbonate ions in the tubular lumen. Sodium diffuses into the tubular cell in exchange for hydrogen. Bicarbonate combines with hydrogen to form carbonic acid. Bicarbonate from the tubular fluid enters the interstitium where it combines with sodium to form sodium bicarbonate.

18.14.9 Reabsorption of Proteins and Peptides

Proteins with a molecular weight of less than 69,000 can become part of the glomerular filtrate. Because they are important nutrients, most of these proteins are reabsorbed in the proximal tubule and are not lost in the urine. However, a small quantity of protein is present in normal urine. Proteins (and polypeptides) are reabsorbed by endocytosis and subsequently degraded by cellular lysosomes to their constituent amino acids. The amino acids are expected to move from inside of the cell to the peritubular space by facilitated diffusion.

18.14.10 Transport Maximum (TM)

Tubular transport maximum or TM is defined as the rate at which the maximum amount of a substance is reabsorbed from the renal tubule. When the TM for the substance is exceeded in a nephron, the substance will appear in the urine. In diabetes mellitus, the movement of glucose from plasma into body cells is impaired because of the lack of insulin. Therefore, glucose concentration in the plasma increases, and hence the plasma and tubular loads of glucose increase. In absence of a carrier molecule glucose reabsorption is hampered and the excess glucose continues its flow through the tubules into the urine. Because of the presence of glucose within the tubules, the effective osmotic pressure of the tubular fluid increases, and water remains in the tubular fluid. Thus, in diabetes mellitus, glucose can be detected in the urine and a greater volume of urine is formed. The affected animal drinks more water to compensate for the loss of urine. Not all of the hundreds of thousands of nephrons have same TM. First appearance of glucose in the urine does not represent the Tm for the kidney but rather the renal threshold. Renal threshold for glucose is defined as the plasma glucose concentration at which glucose first begins to appear in urine. The TM for the kidney is reached when all nephrons are reabsorbing to their maximum ability. In Human renal tubular filtered load is 125 mg/min whereas Tm is 375 mg/min. However renal threshold is 180 mg/dl.

18.14.11 Tubular Secretion

It is the process by which the substances are secreted from blood into renal tubules.

18.14.12 Substances Secreted in Different Segments of Renal Tubules

Potassium (K+) is secreted actively by sodium potassium pump in proximal and distal convoluted tubules and collecting ducts. Renal K+ transport is unique in that it is reabsorbed in some parts of the tubule and secreted in others. When dietary potassium intake is extremely low there is greater reabsorption of K+ in the distal nephron, and when dietary potassium intake is high there is greater K+ secretion. Ammonia is secreted in the proximal convoluted tubule. H+ secretion is primarily in distal nephron which is an active process and occurs in the intercalated cells of the collecting duct. Urea is secreted in loop of Henle.

18.14.13 Glomerulotubular Balance

This property of the proximal tubule to reabsorb a consistent fractional amount of glomerular filtrate is known as glomerulotubular balance. The amount of filtrate reabsorbed by the proximal tubule is consistently a certain percentage of the filtrate (about 65% for water and NaCl) rather than a constant amount for each unit of time. If the GFR is low, only a fractional amount of filtrate is reabsorbed in the proximal tubule, and the remaining fraction continues to the distal nephron. If the GFR is high, the additional amount of filtrate does not continue to the distal nephron but, rather, only about one-third of it.

18.14.14 Concentration of Urine

The mechanism for concentration of the tubular fluid depends on the existence of a very high osmolarity in the interstitial fluid of the renal medulla. The osmolarity increases with the distance from the cortex, reaching a maximum in the innermost aspect of the medulla The magnitude of this value varies with species, as in human it is about 1200 mOsm/L water, Dog (2400 mOsm/L water). Plasma osmolarity is about 300 mOsm/L water. The high osmolarity exists because of the countercurrent mechanism which is established by the activity of the loops of Henle and is maintained by the vasa recta

18.14.15 Countercurrent Mechanism

A countercurrent system is a system of 'U-shaped tubules where the inflow of fluid runs parallel to, counter (opposite) to, and in close proximity to the outflow. These characteristics are common to the anatomical arrangements of the loops of Henle and the vasa recta. Thus, the countercurrent mechanism in the kidney comprises two countercurrent systems: the countercurrent multiplier (loops of Henle) and the countercurrent exchanger (vasa recta).

18.14.16 Countercurrent Multiplier System

Countercurrent multiplier system is an active process occurring in the loops of Henle, which is responsible for the production of concentrated urine. The loop of Henle uses the countercurrent multiplier system to increase the concentration of solute and ions within the medullary interstitium. These solutes accumulate in the medullary interstitium and increase the osmolarity. Because of increased osmolarity nephrons reabsorb more water and concentrate the urine. Tubular fluid entering the descending limb of loop of Henle from proximal tubule has an osmolarity of 300 mOsm/L, 280 contributed from NaCl and 20 from urea (280,20).

From the descending thin limb water diffuses outward (osmosis) and solute remain within. Tubular fluid at the hairpin turn has an osmolarity of 1200 mOsm (1120, 80). In the ascending thin limb water retained while NaCl diffuses outward and urea diffuses inward. Osmolarity decreases to 500 mOsm (400, 100) when reaches the thick segment. In the thick ascending limb and above, due to continuous outflow of NaCl outward, osmolarity decreases to about 200 mOsm (100, 100) which is hypotonic to plasma. At the end of the distal tubule, and before the fluid enters the cortical collecting duct, the osmolarity is about 150 mOsm (50, 100) (Figure 18.7).

In the cortical collecting duct, outer medullary collecting duct, and the inner medullary duct, Na+ reabsorption is stimulated by aldosterone and water & urea reabsorption is stimulated by ADH. So, at the end of the cortical collecting duct, the osmolarity is 300 mOsm (125, 175), at the end of the outer medulla it is 600 mOsm (200, 400), and before it empties into the renal pelvis it is 1200 mOsm (600, 600).

Here, the sodium and chlorine ions are repeatedly recirculated between the descending limb and ascending limb of Henle's loop through medullary interstitial fluid. In addition to that, there is regular addition of more new sodium and chlorine ions into the descending limb through constant filtration. Because of all these phenomena, there is increase or multiplication of the osmolarity of the medullary interstitial fluid and medullary gradient. Hence, it is called countercurrent multiplier.

18.14.17 Countercurrent Exchanger System

The vasa recta act as countercurrent exchangers. Vasa recta are permeable to solutes and water throughout their length. In the descending limbs, water is drawn out by osmosis from the vasa recta to the hyperosmotic peritubular fluid, and the solutes diffuse in reverse. In the ascending limbs, solutes diffuse back into the peritubular fluid, and water is drawn by osmosis back into the vasa recta. The net result is that the solutes responsible for the vertical medullary gradient are mostly retained in the peritubular fluid of the medulla.

18.14.18 Medullary Washout

An increased rate of medullary blood flow would reduce the time for diffusion of solute from the ascending limb back to the peritubular fluid; therefore, the blood leaving the ascending limb would have a higher concentration of solute. This may result in a gradual loss or washout of the medullary gradient, called medullary washout. To prevent medullary loss of solutes and to maintain the medullary gradient, the blood flow to the vasa recta is reduced and it is often sluggish. If countercurrent blood flow in the vasa recta did not exist, the solute of the renal medulla would be quickly removed instead of being retained.

18.15 WATER BALANCE

18.15.1 Water Gain

Water gains occur by ingestion of water along with food and also from metabolic water. Metabolic water is derived from the chemical reactions of cellular metabolism in the mitochondria. The metabolism of proteins, carbohydrates, and fats requires different amounts of cofactors, with the greatest amounts required for fats. Accordingly, the yield

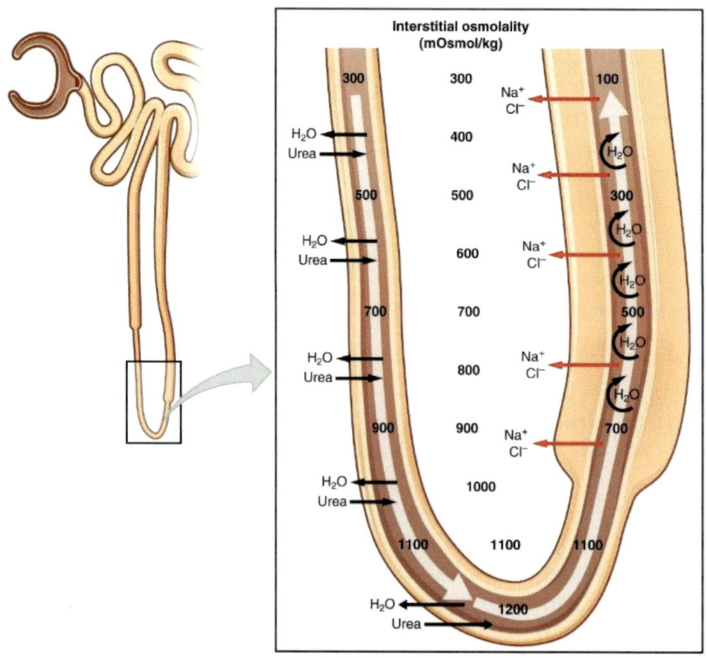

FIGURE 18.7 Countercurrent multiplier in loop of Henle.

of metabolic water is greater for a certain amount of fat than for an equal amount of protein or carbohydrate. For example, the metabolic water yield from each of 100 g of carbohydrate, protein, and fat is 60 ml, 40 ml, and 110 ml respectively.

18.15.2 Water Loss

Water loss from the body is classified as either an insensible loss or a sensible loss. Insensible losses are associated with vapor losses and occur constantly by evaporation from the skin and by loss of water vapor in exhaled air. Inhaled air becomes saturated with water vapor in the respiratory passages and lungs, but there is no body mechanism to remove moisture from the respiratory gases before exhalation. Sensible losses are the visible losses in the form of urine, feces, and body secretions that leave the body. Sensible losses can become excessive in certain conditions, such as diarrhea, and threaten body stores of water.

18.15.3 Regulation of Acid-Base Balance

The regulation of acid-base balance actually means regulation of $[H^+]$ in the body fluids. The $[H^+]$ is usually expressed in terms of pH, which is the negative log of the hydrogen ion concentration. pH = log 1 /$[H^+]$ = -log$[H^+]$. The pH of the extracellular fluid (ECF) is generally 7.4. Relatively small pH changes are associated with correspondingly large changes in H+ concentration. Enzymatic reactions and metabolic processes in the cells of the body operate optimally within a very narrow range of pH (7.35 to 7.45). Therefore, if any deviations occur from this pH value, mechanisms are present in the body to make corrections for the deviations. Most of the acids and bases in the extracellular fluid that are involved in normal acid-base regulation are weak acids and weak bases. The normal range of blood pH may be assumed to be 7.35 to 7.45. To resist acid-base disturbances, the body uses three basic mechanisms

1. 1. Chemical buffering – immediately combining with acid or base to prevent excessive changes in H^+ concentration.
2. Respiratory adjustment of blood CO2 concentration – regulates the removal of CO_2 (and therefore, H_2CO_3) from the extracellular fluid.
3. Excretion of H^+ or HCO_3^- ions by the Kidneys – can excrete either acid or alkaline urine, thereby readjusting the extracellular fluid H^+ concentration toward normal during acidosis or alkalosis.

A chemical buffer system consists of a mixture of a weak acid and its conjugate base, example is a solution of carbonic acid and bicarbonate ion. The Henderson-Hasselbalch equation describes the relationship between pH and the mixture of a weak acid and its conjugate base as pH= pKa log [Base]/[Acid]. Ka is the dissociation constant of the weak acid and Pka is the negative logarithm of Ka. pKa is defined as the pH at which the acid and base of the buffer system are in equal concentration. Thus, pH = p*K*a when base and acid are equal.

When an acid is added to the buffer system, a shift of pH is resisted, but the ratio of base to acid decreases and pH is somewhat depressed. If further acid will be added, buffers system will be exhausted and acidemia will result. The restoration requires the formation and secretion of H+ ions by the kidney tubules. If instead strong base were added to the body fluids, it would react with the acid of the buffer and thus form more buffer base. This ultimately must be corrected by renal excretion of base and retention of hydrogen ions. The principal buffer systems of the blood are the bicarbonate, protein, phosphate and Haemoglobin.

18.15.4 Bicarbonate Buffer System

The bicarbonate buffer system is the principal buffer of the blood and the ECF. The weak acid component of the bicarbonate buffer system is H2CO3 and the conjugate base is $HCO3^-$. pH = pK + log $[HCO_3^-]$/ $[H_2CO_3]$. Bicarbonate is formed in the body by the reaction of carbon dioxide with water.

When a strong acid (Hcl) is added to the bicarbonate buffer solution, the increased H^+ from the acid is buffered by HCO3-. As a result, more H2CO3 is formed, causing higher production of CO_2 & H_2O. The excess CO_2 greatly stimulate respiration which eliminate the CO_2 from the ECF. Opposite reaction takes place when strong base, NaOH is added to bicarbonate system. Here concentration of H_2CO_3 decreases, and net result is decreased concentration of blood CO_2 level. This inhibits respiration and decreases the rate of CO_2 expiration. For bicarbonate buffer system, pK is 6.1

18.15.5 Phosphate Buffer System

It plays a major role in buffering renal tubular fluid and intracellular fluids (ICF). The main component is H_2PO_4 and HPO4- so when strong acid (Hcl) is added.

$$HCl + Na_2HPO_4 \rightarrow NaH_2PO_4 \text{ (weak acid)} + NaCl$$

Here strong acid, Hcl is replaced by weak acid NaH_2PO_4 and the decrease in pH is minimized. When a strong base, NaOH is added, the OH- is buffered by the H2PO4 to form additional amount of HPO4- & H2O. Phosphate buffer system has pK of 6.8 which is very near to the normal pH of 7.4. The phosphate buffer is especially important in tubular fluid of kidney for two reasons. Phosphate usually becomes highly concentrated in the tubules, hence increasing the buffer power of the phosphate system. Tubular fluid usually has lower pH than the ECF, hence it brings the operating range of buffer closer to the pK (6.8) of the system. The phosphate buffer system is important in buffering ICF for two reasons: The concentration of phosphate is many times higher in ICF than ECF. Also, the pH of ICF is lower than that of ECF and therefore it is closer to the pK of the phosphate buffer system compared with the ECF.

18.15.6 Protein Buffer System

Proteins are among the most plentiful buffers in the body because of their high concentration intracellularly. Approximately 60-70% of the total chemical buffering of the body fluid is inside the cells, and most of this results from the intracellular proteins. In addition to the high concentration of proteins in the cells, another factor that contributes to their buffering power is the fact that the pK of many of the proteins system are very close to intracellular pH.

18.15.7 Excretion and Osmoregulation

The maintenance of fluid balance is critical for maintaining many bodily functions, including metabolic and biochemical reactions, transport of nutrients and maintenance of temperature. The average adult animal has roughly 60% fluid volume. These body fluids may again be subcategorized into intracellular and extracellular fluid compartments.

Intracellular fluid is all the fluid contained within cells. The intracellular fluid makes up roughly two-thirds of the total body volume.

The remaining one-third of fluid volume in the body is constituted by extracellular fluid and may further divide into its interstitial, intravascular and transcellular compartments.

The maintenance of fluid homeostasis in each of these compartments is mainly dependent on the concentration of electrolytes and this electrolyte is responsible for generation of osmotic pressure. This process regulation of osmotic pressure is known as osmoregulation.

To maintain homeostasis, it is essential that the excretion of water and electrolytes match an individual's intake. The kidneys play a very essential role in osmoregulation by controlling the quantity of fluid reabsorbed from the glomerular filtrate. This fluid is reabsorbed in the renal tubes and may be modulated by the hormones like antidiuretic hormone (ADH), aldosterone, and angiotensin II. The kidney has large capacity to alter fluid excretion, as well as electrolyte excretion.

The glomerular filtration rate value in an average human being is 180L/day. The advantage of this high filtration rate in terms of osmoregulation is that it allows the kidneys to rapidly and precisely regulate the volume and composition of body fluids.

At the hypothalamus level, osmoreceptor response to extracellular fluid hypertonicity, and facilitate ADH release from axons down to the posterior pituitary into the circulation. ADH serves a primary function to increase solute-free water reabsorption in the nephrons to bring down body fluid hypertonicity.

18.15.8 Pathophysiology of Electrolyte Imbalance

Due to excessive release of ADH one important disease occurs named as Syndrome of inappropriate ADH secretion (SIADH). This excessive ADH release may results from hypothalamic hyperactivity, or other sources like small-cell carcinoma. Increased ADH promotes free-water reabsorption from the filtrate, leading to an inappropriately elevated urine osmolality compared to blood plasma and by consequence, hyponatremia may develop.

In both, chronic or acute kidney disease glomerular function affected so heavily, thereby reducing the production of the filtrate. Typically, this leads to increased water retention, increased potassium retention and dilution of plasma sodium concentrations due to reduced water excretion.

Oedema is one of the consequences of kidney dysfunction. Oedema is the presence of excess fluid in either the intracellular or the extracellular fluid compartment. Extracellular oedema, may occur clinically due to acute or chronic kidney failure, mineralocorticoid excess, decreased plasma proteins or decreased hepatic synthesis of proteins.

BIBLIOGRAPHY

Berliner RW, Levinsky NG, Davidson DG, Eden M. Dilution and concentration of the urine and the action of antidiuretic hormone. *Am J Med*. 1958 May;24(5):730–44.

Cooke CR, Turin MD, Walker WG. The syndrome of inappropriate antidiuretic hormone secretion (SIADH): pathophysiologic mechanisms in solute and volume regulation. *Medicine* (Baltimore). 1979 May;58(3):240–51.

Farnsworth EB. Electrolyte partition in patients with edema of various origins; sodium and chloride. *Am J Med*. 1948 Mar;4(3):338–42.

Hammel HT, Schlegel WM. Osmosis and solute-solvent drag: fluid transport and fluid exchange in animals and plants. *Cell Biochem Biophys*. 2005;42(3):277–345.

Lameire N, Van Biesen W, Vanholder R. Electrolyte disturbances and acute kidney injury in patients with cancer. *Semin Nephrol*. 2010 Nov;30(6):534–47.

Levey AS, Inker LA, Coresh J. GFR estimation: from physiology to public health. *Am J Kidney Dis*. 2014 May;63(5):820–34.

Seifter JL, Chang HY. Extracellular acid-base balance and ion transport between body fluid compartments. *Physiology* (Bethesda). 2017 Sep;32(5):367–379.

Singh RG. The physiology and emerging roles of antidiuretic hormone. *Int J Clin Pract*. 2002 Dec;56(10):777–82.

19 Physiology of the Endocrine System

Keyvan Sobhani
Department of Animal Science, University of Kurdistan, Sanandaj, Iran

19.1 INTRODUCTION

Many chemical messengers have been recognized for their role in facilitating communication between cells to achieve a specific homeostatic or homeorhetic response. The first system functions as a rapid channel, while the second one operates more gradually to produce a lasting reaction. The endocrine system consists of various ductless endocrine glands that produce hormones, a specific type of chemical messengers.

19.1.1 Glands and Classification

Glands are a collection of cells that are both structurally and functionally arranged to collaborate in producing a substance, such as an enzyme, sweat, saliva, milk, or hormone, and releasing it into either a duct or the bloodstream. Whether or not a duct is present is commonly used to categorize various glands into three main classes:

1. ***Endocrine glands:*** Secretions from ductless glands are frequently released directly into the blood. For example: T3/T4, Somatostatin, GH
2. ***Exocrine glands:*** Comprised of secretory acini that release their substances into a specific duct. For example: Bile, Saliva and etc.
3. ***Mixed glands:*** Glands that carry out both exocrine and endocrine roles. For example: Pancreas (*Exo*: enzymes, *Endo*: insulin), Testes (*Exo*: spermatozoa, *Endo*: testosterone)

19.1.2 Hormone Properties

Hormones are substances produced by specific endocrine glands or cells in small amounts, released into the bloodstream or interstitial fluid to reach target cells. This definition includes all hormonal chemical messengers, regardless of where they come from or how they work. Important traits of hormones include:

1. *Specificity of receptors:* Every hormone interacts with a particular high-affinity receptor, resulting in the formation of a hormone-receptor complex (HRC) that initiates a response in the target. The concentration of receptors differs between target cells and can be adjusted in response to hormone levels, known as upregulation or downregulation. This active communication functions as enhancers and filters of hormone signaling.
2. *Slow onset of action:* Hormones generally take a long time to start showing biological effects, usually requiring hours to days, as they affect genetic processes to kickstart the transcription of effector genes.
3. *Absence of enzyme function:* Hormones do not have inherent enzyme function and are unable to directly accelerate intracellular enzymatic reactions.
4. *Signal transduction:* Hormones trigger intracellular cascades of reactions by acting as "first messengers" when they bind to target cell receptors. This frequently results in the formation of intracellular "second messengers."
5. *Metabolic clearance rate:* Once hormones have produced effects in the body, they are either deactivated by the tissues they target or removed from the bloodstream through the excretory functions of the liver and kidneys. The elimination rate is opposite to their binding strength to carrier plasma proteins.
6. *Feedback mechanisms:* Feedback mechanisms involve close regulation of hormone secretion and concentration, mainly through negative feedback mechanisms, maintaining a narrow range.

19.1.3 Hormone Classification

Different hormones can be classified based on criteria such as where they come from, what chemicals they are made of, what they do in the body, how they work, and how easily they dissolve.

1. ***Their secretion source:***
   Classified as pituitary and non-pituitary hormones based on their site of production.
   a. *Pituitary hormones:* Synthesized and secreted from different lobes of the pituitary gland. For example: GH, prolactin (PRL), LH, FSH, ACTH, MSH, ADH, and oxytocin.
   b. *Non-pituitary hormones:* Include hormones produced by other endocrine glands except for the pituitary gland. For example: Somatocrinin (from the hypothalamus), insulin (from the pancreas), aldosterone (from the adrenal

DOI: 10.1201/9781003426851-19

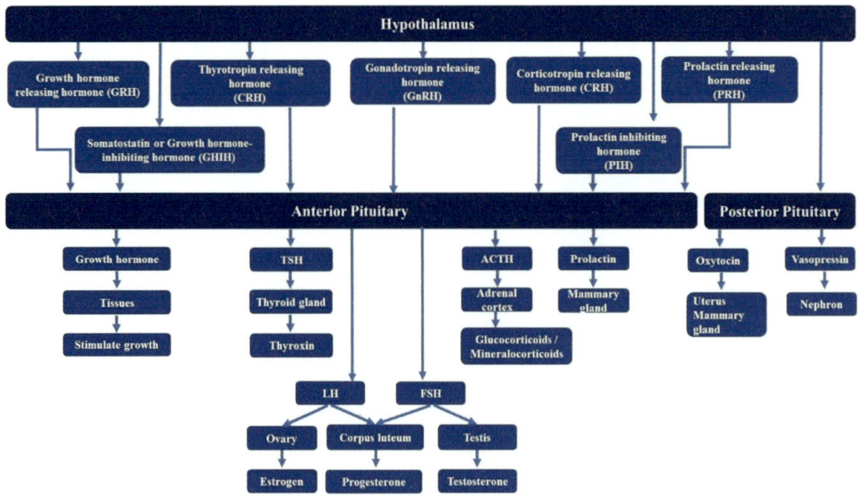

FIGURE 19.1 Classification of hormones in body.

cortex), T3/T4 (from the thyroid gland), etc (see Figure 19.1).

2. ***Chemical nature:***
   Different types of hormones can be classified based on their biochemical structure.
   a. Protein hormones: These are composed of more than 50 amino acids. Protein hormones can be further categorized into:
      i. Simple protein hormones (GH, Insulin, PRL, etc).
      ii. Glycoprotein hormones (LH, FSH, TSH, and chorionic gonadotropins (eCG/hCG))
   b. Polypeptide hormones: These consist of fewer than 50 amino acid residues. Examples include Calcitonin, TRH, Glucagon, etc.
   c. Amine hormones: These are derived from amino acids through decarboxylation.
   d. Steroid hormones: These are derived from cholesterol. They are primarily produced by gonads (e.g., estrogen, testosterone) and the adrenal cortex (e.g., Aldosterone, Cortisol).
   e. Fatty acid derivatives: These hormones are synthesized from fatty acids. An example is the prostaglandin family (e.g., PGF2α, PGE2) derived from arachidonic acid (see Figure 19.1).
3. ***Their physiological action:***
   The integration of endocrine and nervous systems is crucial in controlling most basic physiological events such as growth, reproduction, intermediary metabolism, stress response (fight or flight response, emotional/physical stress, environmental stress), feeding responses, and water and electrolyte balance. Hence, the appropriate functioning of the endocrine system is vital for animal survival, optimum reproduction, production, and adaptation to various environmental conditions (see Figure 19.2).
4. ***Their degree of solubility:***
   a. Water-soluble/hydrophilic hormones: The polypeptide/glycoprotein hormones are soluble in water and hence do not require carrier proteins for their transport in blood. In addition, they cannot freely pass through the lipid bilayer membrane and require membrane-bound receptors or transporter molecules. E.g. LH, catecholamines, GH, etc.
   b. Water-insoluble/lipophilic hormones: Hormones derived from cholesterol, fatty acids, and the thyroid gland are insoluble in water and require carrier proteins for their transport in the systemic circulation. These hormones possess intracellular (cytosolic/nuclear) receptors as they can freely cross the cell membrane. E.g. T3/T4, testosterone, aldosterone, etc.

19.1.4 Synthesis of Hormones

The regulation of hormone synthesis relies heavily on fluctuations in circulating hormone or metabolite levels, as well as sensory input. Furthermore, hormones released by the hypothalamus can either stimulate or inhibit the synthesis of other hormones. When prompted by an appropriate stimulus, genes responsible for encoding specific hormones undergo increased transcription and translation. However, the precise processes of biosynthesis and post-translational modifications differ across various types of hormones, including peptides/proteins and steroids.

1. ***Protein or peptide hormones:***
   During translation, the rough endoplasmic reticulum (RER) serves as the primary site for synthesizing large precursor proteins known as "pre-prohormones" for various peptide and

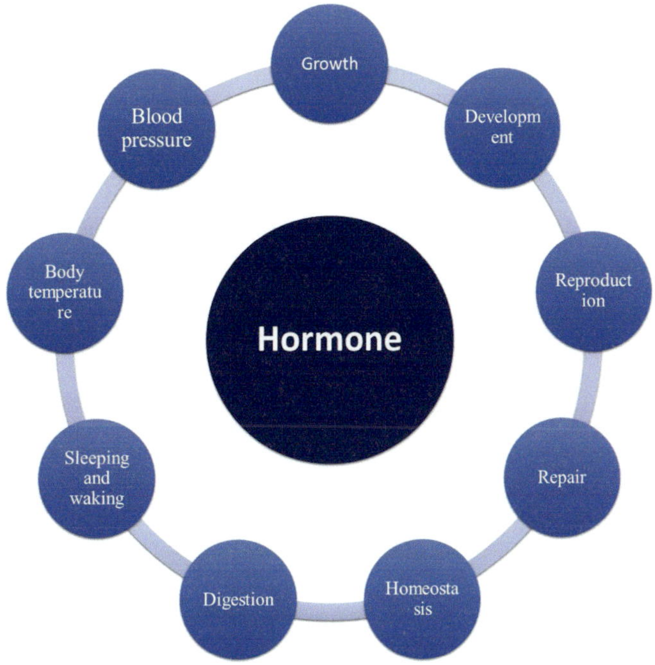

FIGURE 19.2 Functions of hormones in body.

protein hormones. Within the RER, signal peptidases cleave the signal peptide from pre-prohormones, forming prohormones. These prohormones undergo further processing by site-specific endopeptidases, or prohormone convertases, within the Golgi apparatus (GA), leading to the production of mature hormones. These mature hormones are subsequently stored within cytoplasmic vesicles. Additionally, enzymatic cleavage, facilitated by carboxypeptidase and aminopeptidase, occurs within secretory vesicles, alongside post-translational modifications, enhancing the biological activity of specific hormones. In response to specific stimuli, such as increased intracellular calcium (Ca^{+2}) and cAMP concentration, storage vesicles undergo fusion with the cell membrane via exocytosis, releasing hormones into the extracellular fluid.

2. ***Steroid hormones:***
Cholesterol, either obtained from the circulation or synthesized in the smooth endoplasmic reticulum (SER) by condensing acetyl-CoA with the assistance of the rate-limiting enzyme HMG-CoA-reductase, serves as the precursor for all steroid hormones. Cells with the capacity to release steroid hormones contain a lot of smooth endoplasmic reticulum and, as a result, lipid droplets in the cytosol. Frequently, trophic peptide hormones enhance steroid hormone-producing cells by boosting cholesterol uptake, new synthesis, and activated enzymatic machinery. A suitable stimulus activates StAR and causes a quick steroidogenic response by moving cholesterol swiftly from the OMM to the IMM. Steroidogenesis begins with cholesterol being converted into pregnenolone by an enzyme called cholesterol side chain cleavage enzyme, also known as P450scc or CYP11A1. The previous step is seen as the step that limits the rate for all types of steroid hormones. The transformation of pregnenolone into various steroid hormones relies solely on the specific cell's mitochondrial enzymes and SER.
3. ***General mechanism of action:***
Hormones create a hormone-receptor complex (HRC) by attaching to their specific receptor on target cells or tissues. The receptors' cellular location changes depending on the type of hormone. Peptide hormones bind to receptors on the cell membrane, while steroid hormones have receptors in both the cytosol and nucleus. Thyroid hormones have a distinctive quality in that they attach to nuclear receptors to produce their biological impacts. The HRC starts the cellular signaling pathways downstream, which ultimately regulate the metabolic and transcriptional activities of the cell either directly or indirectly.

19.1.4.1 Functioning Process of Peptide Hormones

The fact that peptide and protein hormones are attracted to water requires the existence of receptors bound to the membrane. They interact with target cells by attaching to either G-protein-coupled receptors (GPCRs) or receptor tyrosine kinases (RTK).

1. **Signaling mechanism of peptide hormones:**

   Polypeptide and protein hormones predominantly initiate signal transduction by binding to membrane-bound G protein-coupled receptors (GPCRs). These receptors consist of three distinct regions: an extracellular domain (N-terminus), a transmembrane domain, and an intracellular domain (C-terminus). Inactive GPCRs are associated with a GDP molecule bound to the α subunit of the G-protein. Hormone-receptor complex (HRC) formation activates GPCRs, leading to the phosphorylation of GDP to GTP and the subsequent dissociation of the GTP-bound α subunit (Gα-GTP). Gα-GTP can then activate or inhibit membrane-bound enzymes such as adenylate cyclase (AC), guanylate cyclase (GC), phospholipase C (PLC), or ion channels. G-proteins that activate specific membrane-bound enzymes or ion channels are termed stimulatory G-proteins (Gs, Gq), while those with inhibitory effects are known as inhibitory G-proteins (Gi). Receptor tyrosine kinases (RTK) are transmembrane receptors that includes tyrosine kinase receptors and tyrosine kinase-associated receptors.

Both tyrosine kinase receptors and tyrosine kinase-associated receptors play crucial roles in cellular signaling. Tyrosine kinase receptors, upon hormone binding, undergo dimerization and activate their intracellular kinase domain, leading to phosphorylation of tyrosine residues and subsequent activation of downstream signaling pathways such as Ras-MAPK and Ras/PI3K/AKT. Examples include insulin acting via tyrosine kinase receptors. Tyrosine kinase-associated receptors lack intrinsic kinase domains but are coupled with protein tyrosine kinases (PTKs), notably from the Janus kinase (JAK) family. Hormone binding induces receptor dimerization, activating associated JAKs, which phosphorylate tyrosine residues on the receptor. This trigger signaling cascades such as the JAK-STAT pathway, influencing gene transcription and cellular functions. Notably, these pathways are crucial for various physiological processes, including metabolism, proliferation, and differentiation.

19.1.4.2 Functioning Process of Steroid Hormones

The majority of steroid hormone receptors are located in the cytoplasm, although some, like estrogen (E2), have receptors localized in the nucleus. These intracellular receptors feature three primary domains: the ligand-binding domain (LBD) at the C-terminus, the DNA-binding domain (DBD), and the amino-terminal domain (NTD). The LBD, housing the activation function 2 (AF2) region, serves as the binding site for steroid hormones. The DBD allows them to bind to specific regions of the genome known as hormone response elements (HRE) or steroid response elements (SRE), directly modulating the transcriptional rate of specific genes. The NTD contains the activation function 1 (AF1) region, determining optimal transcriptional activity. In their inactive state, cytosolic steroid receptors are bound with heat shock proteins (HSP). Upon binding to a steroid hormone, HSP removal and receptor phosphorylation occur, facilitating the migration and binding of hormone-receptor complexes (HRC) to SRE/HRE regions in the DNA, thus contributing to the classic steroid hormone signaling pathway. Alternatively, AF2 can indirectly regulate cellular transcription by binding to DNA-bound transcription factors like AP1 or SP1, instead of directly binding to DNA.

19.1.5 Hypothalamus

The hypothalamus is a central regulator of various bodily functions such as temperature, appetite, and emotional states, and it secretes hormones that integrate the nervous and endocrine systems. These hormones are synthesized by specialized neuroendocrine cells organized in distinct nuclei, including magnocellular and parvocellular neurons. The hypothalamic-hypophyseal portal system facilitates communication between the hypothalamus and the anterior pituitary gland, allowing for the regulation of hormone release. Hypothalamic hormones act on target cells in the pituitary gland via G protein-coupled receptors, influencing the synthesis and secretion of pituitary hormones. Pituitary tropic hormones and target hormones, originating from systemic circulation, can enter the brain and modulate hypothalamic hormone secretion directly or indirectly.

19.1.6 Pituitary

The pituitary gland, often termed the "master gland," originates from the ectoderm and is regulated by hypothalamic stimuli to secrete hormones. Comprising the anterior, intermediate, and posterior lobes, it plays a pivotal role in regulating growth, metabolism, and reproduction through the secretion of tropic hormones. The anterior lobe, housing neuroendocrine cells responsible for synthesizing major tropic hormones, can be broadly classified into chromophils and chromophobes based on histological staining properties. Chromophils, further categorized as acidophils or basophils, actively secrete hormones, while chromophobes consist of reserve cells and undifferentiated stem cells.

19.1.7 Growth Hormone (GH)

Growth Hormone (GH), also known as Somatotropin, plays a crucial role in animal growth, spanning pre-natal and post-natal stages and varying across physiological states such as pre-pubertal, pubertal, postpubertal, and senescence. Synthesized by somatotropes, GH is initially produced as a prohormone with 217 amino acids, eventually maturing into a 191-amino acid single-chain protein with two intramolecular disulfide bridges. Binding to specific membrane-bound growth hormone receptors (GHRs), GH initiates signal transduction primarily via the JAK-STAT pathway, influencing cellular genetic machinery, and activating other pathways like Ras/MAPK and PI3K.

GH regulates body metabolism and affects various organs, including the liver, bones, skeletal muscle, adipose tissue, and gonads. A key effect of GH is stimulating hepatocytes to produce somatomedins (IGFs), particularly IGF-1 and IGF-2, which act as vital mediators of GH actions outside the liver, influencing metabolic and functional changes in target organs. GH also enhances hepatic glucose production, triglyceride secretion, and lipogenesis, while decreasing glucose utilization by skeletal muscle and adipose tissues, leading to insulin resistance and increased circulating glucose levels.

Additionally, GH stimulates insulin synthesis and secretion by pancreatic β-cells to counteract insulin resistance. Furthermore, GH promotes transcription, translation, and amino acid uptake, favoring protein anabolism and suppressing amino acid catabolism, particularly in skeletal muscles, thereby contributing to growth and development.

GH stimulates hepatocytes and adipocytes, increasing FFA, TG, and cholesterol levels for energy use, spares carbs and amino acids. It promotes protein accumulation, muscle hypertrophy via GH-IGF1 axis, and bone growth via chondrogenesis. GH induces cardiomyocyte hypertrophy, vascular expansion, and regulates testicular function. It triggers lipolysis, depleting fat reserves. Hypothalamic hormones regulate GH secretion, while IGF-1 inhibits it. Various factors like exercise and stress can stimulate GH secretion.

19.1.8 Prolactin (PRL)

Prolactin (PRL) is a multifunctional hormone with diverse roles beyond lactation. It acts through its receptors, PRLR, initiating various signaling pathways like JAK-STAT and Ras/Raf/MAPK. While primarily known for its role in lactation, PRL also influences behaviors such as maternal care and reproductive functions in animals. In mammals, PRL regulates mammary gland development, promoting lobuloalveolar growth during gestation and lactation. It stimulates the transcription and translation of casein genes, essential for milk production, and activates lactose synthesis through α-lactalbumin. This cascade of events facilitates lactogenesis and galactopoietic, ensuring adequate milk production to nourish offspring. Moreover, PRL impacts maternal behaviors, such as nest-building, nursing, and cleaning of young ones, which are crucial for offspring survival. Elevated PRL levels post-parturition contributes to the suppression of the estrous cycle, allowing the mother to focus on nurturing her offspring. Additionally, PRL secretion is influenced by estrogen levels during the estrous cycle, affecting sexual receptivity in female animals. In avian species, PRL regulates plumage patterns and influences broodiness in hens. It also stimulates the secretion of crop milk, a specialized form of nourishment provided by some bird species to their young. The regulation of PRL secretion in birds involves both dopaminergic inhibition and positive regulation by vasoactive intestinal polypeptide (VIP). Regular milking practices can also affect PRL levels, highlighting its dynamic regulation in response to environmental and physiological cues. Overall, PRL plays a pivotal role in coordinating various physiological processes essential for reproductive success and offspring care across a wide range of animal species.

19.1.9 Adrenocorticotropic Hormone (ACTH)

Adrenocorticotropic hormone (ACTH) orchestrates the synthesis of corticosteroids, vital for managing stress responses in animals. Its action is diverse, primarily focusing on stimulating the secretion of glucocorticoids to mitigate the impacts of various stressors.

Hypothalamic corticotropin-releasing hormone (CRH) triggers the synthesis of pro-opiomelanocortin (POMC) by corticotropes in the anterior lobe and neuroendocrine cells in the intermediate lobe. Cleavage of POMC in these lobes yields ACTH, a critical peptide hormone consisting of 39 amino acids. ACTH binds to the melanocortin 2 receptor (MC2-R) on adrenal cortical cells, initiating a cascade of events involving cyclic adenosine monophosphate (cAMP) production and subsequent activation of protein kinase A (PKA). This activation sets off downstream cellular pathways crucial for corticosteroid synthesis.

Upon PKA activation, there is a robust steroidogenic response marked by enhanced cholesterol synthesis, facilitated by the activation of hormone-sensitive lipase (HSL), steroidogenic acute regulatory protein (StAR), and cytochrome P450 side-chain cleavage enzyme (CYP11A1 or p450scc). This collaborative action leads to escalated conversion of cholesterol into pregnenolone, a pivotal step in corticosteroid biosynthesis. Additionally, zone-specific hydroxylases are stimulated, further diversifying corticosteroid classes.

While CRH positively regulates ACTH levels, corticosteroids, notably cortisol, exert a negative feedback loop on the CRH-ACTH axis, maintaining hormonal equilibrium essential for homeostasis. This intricate regulatory network ensures precise control over stress response mechanisms in animals.

19.1.10 Thyroid Stimulating Hormone (TSH)

Thyroid-stimulating hormone (TSH) orchestrates a finely tuned process within the thyroid gland, ensuring the synthesis and secretion of thyroid hormones align with the body's metabolic demands. As a heterodimeric glycoprotein, TSH's α and β chains bind specifically to thyroid-stimulating hormone receptors (TSHR) on the surface of thyroid follicular cells. This interaction triggers a cascade of intracellular events, including the activation of adenylyl cyclase and subsequent elevation of cyclic adenosine monophosphate (cAMP) levels. The rise in cAMP initiates downstream signaling pathways, prominently involving protein kinase A (PKA), which in turn modulates gene transcription and translation pivotal for thyroid hormone synthesis.

TSH's influence spans the entirety of the thyroid hormone production process. It promotes the synthesis of essential proteins such as sodium-iodide symporter protein (NIS), thyroid peroxidase, and thyroglobulin, facilitating the uptake of iodide ions and their incorporation into thyroglobulin molecules. Moreover, TSH enhances the endocytosis of stored colloid, thereby increasing the availability of iodinated thyroglobulin for subsequent hormone release.

Crucially, the regulation of TSH secretion is tightly controlled by feedback mechanisms. Thyrotropin-releasing hormone (TRH) and growth hormone (GH) act as primary stimulators, promoting TSH release from the anterior pituitary gland. However, the system is delicately balanced, as elevated levels of circulating thyroid hormones, particularly T3 and T4, exert negative feedback, inhibiting further TSH secretion. This intricate feedback loop ensures that thyroid hormone levels remain within a narrow physiological range, vital for metabolic homeostasis (see Figure 19.3).

Overall, TSH serves as a master regulator of thyroid function, modulating hormone synthesis and secretion to meet the dynamic metabolic needs of the organism. Its finely tuned control mechanisms underscore its critical role in maintaining metabolic balance and overall physiological health.

The thyroid gland, a focal point in physiological psychology discussions, is a roughly 30-gram organ distinguished by its vivid red hue and encased in connective tissue. Comprising two lobes joined by an isthmus, it originates from a pharynx floor invagination during fetal development's second month, gradually releasing hormones with profound physiological impacts. This article explores the thyroid gland's pivotal role and the significance of its hormones.

Measuring around 5 centimeters in length and 2 centimeters in width and thickness, the thyroid lobes are relatively large endocrine glands nestled alongside the trachea, larynx, and carotid artery sheath. The isthmus, connecting the lobes anteriorly to the second and third tracheal rings, is crucial for structural integrity. Blood supply to the thyroid is facilitated by the superior and inferior thyroid arteries.

The thyroid tissue comprises over one million follicles or closed acini, each with a diameter of 150 to 300 microns. These follicles' walls consist of a single layer of cuboidal epithelial cells, enclosing colloid – a substance predominantly composed of thyroglobulin, a protein synthesized by follicular cells. Parafollicular cells interspersed among the follicles produce calcitonin, a secondary hormone involved in calcium metabolism. The thyroid's extensive vascular branching ensures efficient blood supply and hormone entry into circulation, with both sympathetic and parasympathetic nerves innervating the gland.

The thyroid gland secretes two main iodine-containing hormones – thyroxine (T4) and triiodothyronine (T3) – synthesized by follicular and parafollicular cells. Thyroxine, or T4, features four iodine atoms, while triiodothyronine, or T3, contains three. Follicular cells synthesize thyroglobulin from blood-borne tyrosine, storing it within follicles' lumen. These follicles, grouped into lobules, actively participate in hormone synthesis.

Approximately 93% of thyroid hormone secretion comprises thyroxine, with the remaining 7% being triiodothyronine. However, tissues predominantly utilize triiodothyronine, derived from thyroxine conversion. Thyroglobulin iodination, facilitated by follicular cells, progresses from monoiodotyrosine to diiodotyrosine, ultimately forming triiodothyronine and thyroxine. Upon

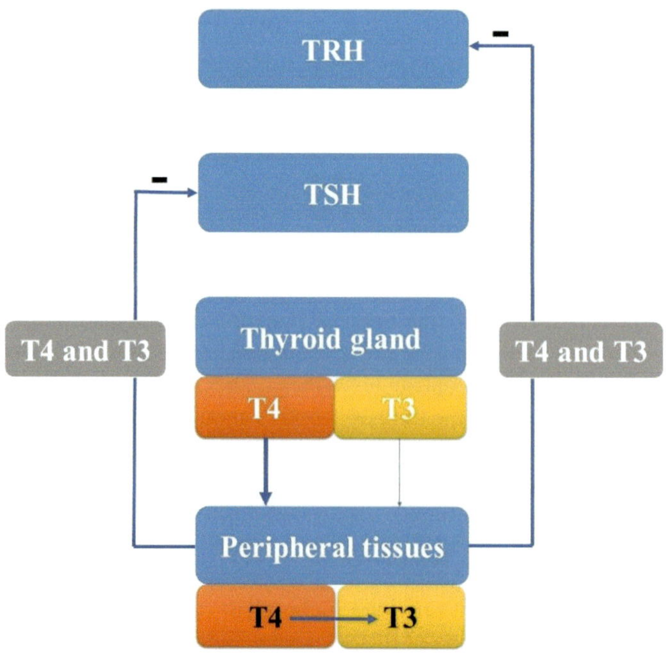

FIGURE 19.3 Thyroid hormone: function and mechanism of action.

synthesis, these iodine-containing hormones bind to plasma proteins for transportation to target tissues, where they modulate gene transcription, enhance metabolic activity, and influence growth and development.

Thyroid hormones play diverse physiological roles, including stimulating carbohydrate and fat metabolism, regulating plasma lipid levels, increasing basal metabolism, and influencing cardiac function, nervous system activity, and muscle function. Moreover, thyroid hormone imbalances can affect sleep patterns, muscle tremors, and the secretion of other hormones.

19.1.11 Pituitary Gonadotropins (LH and FSH)

Pituitary gonadotropins play a central role in regulating reproductive functions, ensuring the production of steroid hormones and the maturation of gametes, particularly in animals after reaching sexual maturity, which is essential for maintaining optimal reproduction in farm animals.

Comprising α and β subunits, gonadotropins are glycoprotein heterodimers, with the unique composition of the β subunit determining their specific activities. The glycosylation of specific residues within these subunits is crucial, influencing their effects and duration within the body. Upon binding to their respective receptors, such as the luteinizing hormone receptor (LHR) and follicle-stimulating hormone receptor (FSHR), gonadotropins initiate intricate signaling cascades. These cascades lead to an increase in cyclic adenosine monophosphate (cAMP) levels and activate protein kinase A (PKA), which subsequently regulates various cellular processes essential for steroid hormone production and cell cycle control.

Luteinizing hormone (LH) acts on thecal cells within ovarian follicles, stimulating the production of testosterone, a precursor for estrogen synthesis and a key factor in ovulation in females. In males, LH targets Leydig cells, promoting testosterone production, which is vital for male fertility. Furthermore, LH facilitates the transformation of granulosa cells into luteal cells and plays a pivotal role in regulating male and female reproductive processes.

Follicle-stimulating hormone (FSH) predominantly affects granulosa cells in females, fostering estrogen synthesis and supporting spermatogenesis and libido maintenance in males by acting on Sertoli cells. FSH sensitivity is highest in the dominant follicle in females and plays a critical role in regulating essential processes in males.

Gonadotropin-releasing hormone (GnRH) serves as the primary regulator of gonadotropin secretion, orchestrating the intricate hormonal balance necessary for reproductive health. Additionally, factors such as estrogen, activin, and inhibin modulate FSH and LH secretion, contributing to the intricate regulation of reproductive processes to ensure optimal fertility in animals.

19.1.12 Posterior Pituitary

The posterior pituitary, originating from neural ectoderm, comprises axons stemming from magnocellular (MC) neurons found in the paraventricular nucleus (PVN) and supraoptic nucleus (SON) of the hypothalamus. Upon receiving appropriate stimulation, these nerve endings in the posterior pituitary release oxytocin (OT) and antidiuretic hormone (ADH). Additionally, the neurohypophysis contains glial cells known as pituicytes.

Both oxytocin and ADH are nonapeptides, differing slightly in their amino acid composition. They are synthesized within MC neurons, accompanied by their specific precursor transport protein called neurophysin, and stored as secretory granules. However, during axonal transport, the transport protein component is cleaved to yield the active hormone. Action potentials generated in response to neural stimuli on MC neurons trigger the release of these hormones from their nerve terminals (see Figure 19.4).

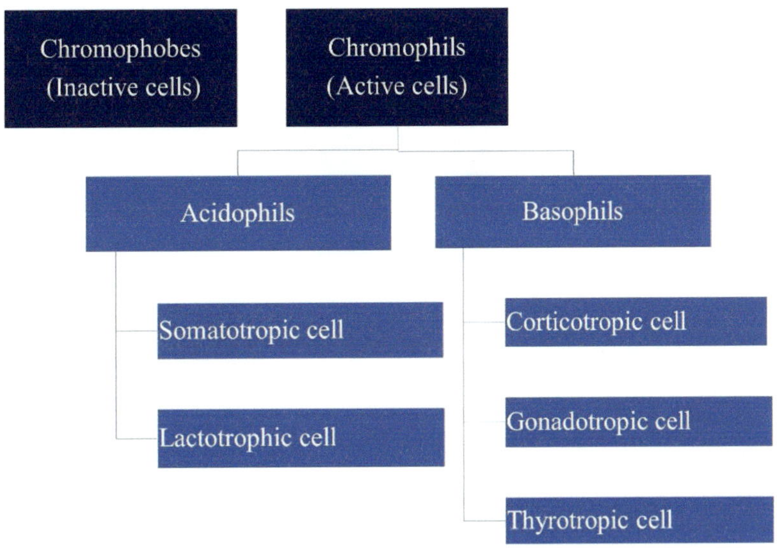

FIGURE 19.4 Cell types in the pituitary gland.

19.1.13 Oxytocin

Oxytocin, well-known for its crucial role in childbirth, also affects milk ejection, sperm transport, social bonding, and ovulation in animals. It binds to oxytocin receptors (OTR), specific G protein-coupled receptors, triggering the phospholipase C (PLC) system to generate secondary messengers such as diacylglycerol (DAG) and inositol triphosphate (IP3). These secondary messengers activate protein kinase C (PKC) and release calcium ions (Ca+2) from the endoplasmic reticulum. Elevated Ca+2 levels activate myosin-light chain kinase (MLCK), while PKC inhibition of myosin-light chain phosphatase (MLCP) increases actin-myosin bridge formation, initiating smooth muscle contraction.

The afferent neural stimuli resulting from fetal entry into the cervical region induce oxytocin release, triggering Ferguson's reflex, a neuroendocrine reflex. This reflex stimulates smooth muscle cells in the myometrium to initiate uterine contractions, facilitating fetal expulsion, hence its moniker, the "birth hormone." Oxytocin is also released during and after mating in both male and female animals, aiding spermatozoa transport in the reproductive tracts of both sexes.

Contraction of myoepithelial cells surrounding the alveoli prompts milk ejection from the mammary gland. In dairy animals, tactile stimulation of the udder, along with visual or auditory stimuli, induces oxytocin secretion, facilitating milk ejection.

Oxytocin additionally stimulates prostaglandin F2α (PGF2α) synthesis, which plays a crucial role in follicular membrane rupture, leading to ovulation.

19.1.14 Anti-Diuretic Hormone

ADH, crucial for water conservation during dehydration, combats urine water loss and prevents ECF osmolality rise.

It binds to vasopressin receptors (VR), influencing various tissues depending on receptor type: V1, V2, and V3.

Binding to V2 receptors in tubular epithelial cells triggers cAMP elevation, prompting aquaporin gene activity. This leads to aquaporin insertion, enhancing water reabsorption and displaying ADH's anti-diuretic effect.

V1 receptor activation in arteriolar smooth muscle cells stimulates PLC, generating DAG and IP3. This leads to calcium influx, inducing muscle contraction, hence its alias, vasopressin. This constriction reduces GFR, conserving water and maintaining blood pressure during dehydration.

Osmoreceptors in the SON detect ECF osmolality shifts, stimulating ADH release. Heart stretch receptors also prompt ADH secretion in response to blood volume reductions (see Figure 19.5).

19.1.15 Pineal Gland

Often termed the "biological clock," the pineal gland regulates mammalian circadian rhythms by producing melatonin and comprises glial cells and pinealocytes.

Melatonin, derived from tryptophan, undergoes hydroxylation and carboxylation to form serotonin and then N-acetyl serotonin, respectively, through enzymatic processes involving AANAT and ASMT.

While the pineal gland is essential for circadian rhythm control, the suprachiasmatic nucleus (SCN) of the hypothalamus houses the underlying cellular and molecular mechanisms. SCN's sympathetic innervation activates pineal gland melatonin synthesis during darkness by stimulating AANAT and ASMT gene expression. Additionally, rhythmic clock gene expression, particularly Perl, contributes to melatonin synthesis, peaking at night. Melatonin then influences clock gene activity in the hypothalamus and peripheral tissues.

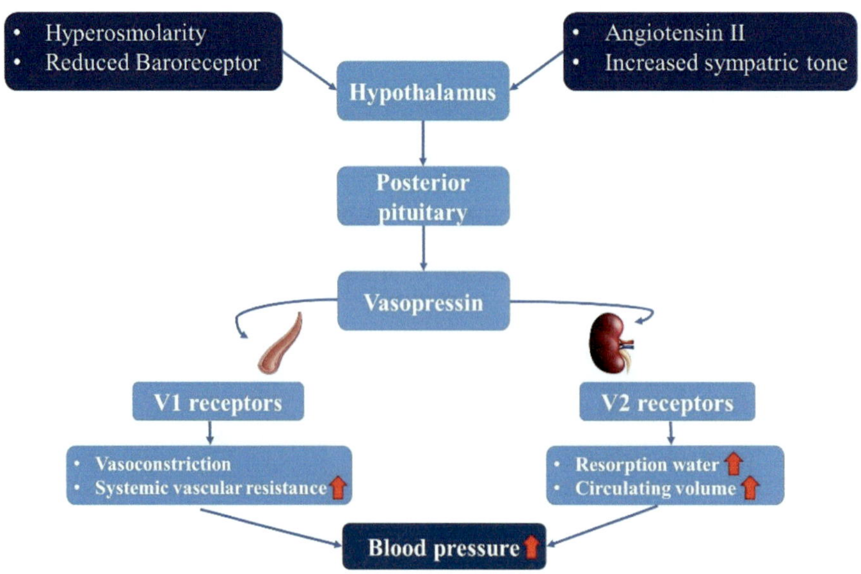

FIGURE 19.5 Mechanism of secretion and action of ADH.

TABLE 19.1
Thymic Hormones and their Function Concerning the Immune System

Thymic Hormone	Chemical Nature	Function in the Immune System
Thymosin α1	Peptide	Stimulates T cell development, enhances T cell function, modulates immune responses
Thymosin β4	Peptide	Promotes wound healing, regulates immune cell migration and differentiation
Thymulin	Peptide	Stimulates T cell maturation and differentiation, enhances T cell function, modulates immune responses
Thymopoietin	Peptide	Supports T cell development and maturation, regulates immune cell proliferation

Melatonin receptors, including MT1 and MT2, inhibit adenylyl cyclase activity, reducing cAMP levels, and affecting various signaling pathways.

While known for its circadian rhythm, sleep, and reproductive regulation, melatonin also combats oxidative stress and modulates the cell cycle and apoptosis. Its anti-apoptotic effects are crucial for follicle growth, oocyte maturation, and luteinization, ensuring fertility. In males, melatonin's antioxidative properties protect spermatozoa, supporting fertility by impacting steroidogenesis and mitigating oxidative stress on gametes.

Melatonin levels fluctuate with seasons, influencing breeding cycles by signaling optimal reproductive timing. In short-day breeders like sheep, increased melatonin levels with shorter daylight durations trigger the estrous cycle onset, while in long-day breeders, it aligns with longer daylight periods. Melatonin's effects on seasonal breeding alter pituitary gonadotropin and thyrotropin secretion patterns.

19.1.16 Gastrointestinal Hormones

Considered as the only site for nutrient assimilation, the gastrointestinal tract (GIT) also produces hormones that are responsible for regulating the secretions of exocrine glands, gastrointestinal motility, cellular proliferation, and differentiation in the GIT. The first hormone to be discovered by Bayliss and Starling (1902) is secretin, which is a gut hormone. Considered the largest endocrine organ in the animal's body, the GIT produces more than 20 peptide hormones.

The hormone-producing cells are also known as enteroendocrine cells (EECs), interspersed between gut mucosal cells in the GIT. They are derived from the pluripotent intestinal stem cells in the intestinal crypts. Different types of EECs are present in the GIT, with each cell type having the ability to synthesize and secrete at least one kind of hormone. The hormones thus produced can also act as autocrine or paracrine or neurocrine factors on the nearby target cells. Based on structural homologies, the gut hormones are further classified into different families: the secretin family (secretin, glucagon-like peptides, VIP), the gastrin family (gastrin, CCK), and the PP-fold family (Neuropeptide Y (NPY), PYY, pancreatic polypeptide).

The gut hormones have the unique property of activating at least one specific type of GPCR present on the target tissues (Table 19.1). The signal transduction mechanisms include the activation of adenylyl cyclases, protein kinases, and membrane-bound ion channels. With overlapping target organs, multiple gut hormones are needed to produce the desired biological effects. The major biological effects of gut hormones include the regulation of ingestion, digestion, and the metabolism of nutrients (see Table 19.2).

19.1.17 Pancreas

The pancreas, often termed the "abdominal gland," is an indispensable component of the digestive system, orchestrating a symphony of functions crucial for metabolic equilibrium. Positioned anteriorly to the lumbar vertebrae, its name, rooted in Greek, underscores its multifaceted role. With its distinct pinkish-gray hue and robust texture, spanning about 15 to 20 centimeters in length, 2.5 to 4 centimeters in width, and weighing approximately 90 grams, the pancreas commands attention.

Within the realm of digestion, the pancreas plays a pivotal role, acting as a hub for the production of digestive juices and enzymes crucial for breaking down complex nutrients such as carbohydrates, fats, and starches. Simultaneously, it operates as an endocrine powerhouse, secreting hormones that meticulously regulate blood sugar levels and stomach acid secretion. Deep within its tissue lie specialized cell clusters known as pancreatic islets, each contributing unique hormones: beta cells produce insulin, alpha cells yield glucagon, delta cells release somatostatin, and PP cells generate pancreatic polypeptides.

Functioning as an exocrine gland, the pancreas diligently synthesizes an array of digestive enzymes, including trypsin, chymotrypsin, amylase, and lipase, essential for efficient nutrient breakdown. These enzymes embark on their digestive journey through the pancreatic duct, converging with the common bile duct to form the ampulla of Vater, culminating in their discharge into the duodenum. Meanwhile, the endocrine facet of the pancreas, embodied by the Islets of Langerhans, dispatches hormones such as insulin and glucagon directly into the bloodstream, meticulously overseeing blood sugar levels. In the grand scheme of metabolic regulation, insulin and glucagon emerge as

TABLE 19.2
Gastrointestinal-Secreted Hormones: Source, Chemical Structure, Receptors and Effects on Feed Intake

Hormone	Source	Chemical Structure	Receptors	Effects on feed intake
Ghrelin	Stomach, Small Intestine	Peptide	Growth hormone secretagogue receptor (GHS-R)	Stimulates appetite, increases feed intake
Cholecystokinin (CCK)	Duodenum, Jejunum	Peptide	CCK-A receptors	Suppresses appetite, reduces feed intake
Glucagon-like peptide 1 (GLP-1)	L-cells of the intestine	Peptide	GLP-1 receptors	Suppresses appetite, reduces feed intake
Peptide YY (PYY)	L-cells of the ileum and colon	Peptide	Y2 receptors	Suppresses appetite, reduces feed intake
Leptin	Adipose tissue	Protein	Leptin receptors in the hypothalamus	Suppresses appetite, reduces feed intake
Insulin	Pancreas	Peptide	Insulin receptors	Modulates appetite, can reduce feed intake
Neuropeptide Y (NPY)	Hypothalamus, Enteric nervous system	Peptide	NPY receptors	Stimulates appetite, increases feed intake
Gastrin	Stomach, Duodenum	Peptide	Cholecystokinin B receptors	Stimulates gastric acid secretion, may influence appetite
Secretin	Duodenum	Peptide	Secretin receptors	Regulates bicarbonate and water secretion, may indirectly affect appetite
Motilin	Duodenum, Jejunum	Peptide	Motilin receptors	Stimulates gastrointestinal motility, may influence feeding behavior
Neurotensin	Small Intestine	Peptide	Neurotensin receptors	Regulates intestinal motility and secretion, may have an effect on appetite

pivotal players, engaged in a delicate dance to maintain blood sugar homeostasis (see Table 19.3).

19.1.17.1 Insulin

Function: Acting as a key regulator of glucose metabolism, insulin is released by beta cells in response to elevated blood sugar levels. It facilitates glucose uptake by cells, promoting glycogen storage in the liver and muscle cells, and converting excess glucose into fat.

Storage: Insulin promotes the storage of excess glucose in the liver and muscle cells in the form of glycogen, facilitating long-term energy storage.

Protein Synthesis: Insulin facilitates protein synthesis by promoting the uptake of amino acids into cells.

Lipid Metabolism: Inhibiting fat breakdown (lipolysis) and promoting fat storage, insulin contributes to overall lipid metabolism.

Anabolic Effects: Insulin fosters growth, energy storage, and cellular repair (see Table 19.3).

19.1.17.2 Glucagon

Function: Glucagon serves as a counter-regulatory hormone to insulin, released by alpha cells when blood sugar levels drop. It stimulates glycogen breakdown (glycogenolysis) in the liver, promoting glucose release into the bloodstream.

Glycogenolysis: Activating liver enzymes, glucagon breaks down glycogen into glucose, ensuring adequate blood sugar levels.

TABLE 19.3
Pancreas Hormones: Chemical Nature, Effects, and Plasma Half-Life

Hormone	Chemical Nature	Effects	Plasma Half-life (minutes)
Insulin	Peptide	Facilitates glucose uptake into cells, promotes glycogen synthesis, inhibits gluconeogenesis and lipolysis	~5
Glucagon	Peptide	Stimulates glycogenolysis, gluconeogenesis, and lipolysis, raises blood glucose levels	~5–10
Somatostatin	Peptide	Inhibits the release of insulin and glucagon, slows down gastrointestinal motility and nutrient absorption	~2–3
Pancreatic Polypeptide (PP)	Peptide	Inhibits exocrine pancreatic secretion, regulates gastrointestinal function	~30

Gluconeogenesis: Glucagon prompts the liver to produce glucose from non-carbohydrate sources, maintaining a steady glucose supply.

Lipolysis: Glucagon promotes fat breakdown into fatty acids and glycerol, serving as an alternative energy source.

Opposing Actions: Glucagon's actions oppose those of insulin, preventing hypoglycemia and maintaining blood glucose levels within a narrow range (see Table 19.3).

19.1.18 Reproduction Hormones

19.1.18.1 Placenta

The placenta acts as the site of attachment of the fetus, delivers nutrients and gases derived from maternal circulation, prevents the fetal allograft from the maternal immune system, and eliminates fetal metabolic-waste products. In addition, it acts as a transient endocrine organ by producing hormones that are vital for the maintenance of gestation, fetal growth, and parturition (see Table 19.4).

19.1.18.2 Progesterone

Progesterone stands as an absolute requirement for the maintenance of gestation in all mammals. Although corpus luteum is the primary source of progesterone in domestic animals, placental-derived progesterone also helps in the maintenance of gestation. The shift of progesterone production from the corpus luteum to the placenta is regarded as an essential phenomenon seen in the case of equines, sheep, and primates. This luteo-placental shift in progesterone synthesis has minor relevance in goats and pigs. Luteal-progesterone plays a prominent role in myometrial quiescence, endometrial growth and differentiation, immunosuppression to prevent fetal rejection, and cervical closure. However, the progesterone derived from the placenta is speculated to be useful for species with longer gestation (see Table 19.4).

19.1.18.3 Estrogen

Unlike follicle that produces 17β-estradiol, placenta in ungulate animals primarily secretes estrone in sulfate form. In primates, placental estrogen is implicated in trophoblast differentiation, mammary gland development, and utero-placental blood flow. However, in cattle, maternal estrogen levels rise along with a concomitant decrease in progesterone levels at the end of gestation period due to an increased conversion of progesterone to estrogen by CYP17A1 (17-α-hydroxylase) in the placenta. Thus, the increased circulatory levels of estrogen lead to the ductular development of the mammary gland, remove the progesterone-mediated negative feedback on lactation, excite the myometrial tissue, and prepare the birth canal for parturition.

19.1.18.4 Chorionic Gonadotropins

As of now, only two chorionic gonadotropins: equine chorionic gonadotropin (eCG) and human chorionic gonadotropin (hCG) are reported in equines and primates, respectively. Like the pituitary gonadotropins (LH & FSH), the chorionic gonadotropins are made of a common α-glycoprotein chain. However, β-chains of chorionic gonadotropins (βCG) are similar to the βLH chain, except that they are heavily glycosylated. The hCG released by blastocyst in primates is responsible for the maternal recognition of pregnancy and maintenance of corpus luteum by binding to LH receptors. Whereas, distinct areas of the placenta known as endometrial cups secrete eCG to promote the formation of accessory corpora lutea, thereby helping in the maintenance of gestation.

19.1.18.5 Placental Lactogen (Chorionic Somatomammotropin)

The binucleate cells of trophectoderm in ruminants, primates, and rodents secrete placental lactogen, a hormone that has both somatotropic and lactogenic functions. Mature placental lactogen (PL) is a single polypeptide chain with 200 amino acids. In bovines, PL shares a structural homology of 50% and 23% with PRL and GH, respectively. The circulatory levels of PL start raising from Day 30 onwards, reaching a peak during the last trimester, and begin to fall when the animal approaches parturition. In ruminants, PL has luteotrophic actions and augments progesterone secretion from the corpus luteum. It plays a major role in the partitioning of maternal nutrients to support fetal growth by stimulating the uptake of maternal nutrients, and glycogenesis in fetal tissues. In addition, it stimulates lobuloalveolar growth in the mammary gland and exerts galactopoietic effects by stimulating dry matter intake.

19.1.18.6 Relaxins

They are peptide hormones belonging to the insulin family, produced primarily by the corpus luteum and placenta. Relaxin1 (RLN 1) plays a major role in reproductive functions in mammals. It is a polypeptide heterodimer hormone with 53 amino acids. The production of RLN1 rises at the end of gestation to promote cervical ripening, dilatation of pubic symphysis, and relaxation of the sacrosciatic ligament to aid the process of parturition.

19.1.19 Ovary and Testes

Gonads refer to the testes and ovaries present in male and female animals, respectively. They are responsible for the production of gametes and hormones, which are crucial elements governing sexual reproduction in animals (Table 19.4). The hypothalamic–pituitary axis secretes GnRH, LH, and FSH, which are essential for regulating the endocrine and gametogenic activities of gonads. In turn, the gonadal hormones produced exert negative inhibition on the release of hypothalamic GnRH and pituitary gonadotropins (see Table 19.4).

TABLE 19.4

Reproductive Hormones: Source, Chemical Structure, Function, and Characteristics

Hormone	Source	Chemical Structure	Function	Characteristics
Testosterone	Testes (males); ovaries and adrenal glands (females)	Steroid hormone	Regulates the development of male secondary sexual characteristics, such as facial hair, deep voice, and muscle mass. It also plays a role in sperm production, libido, bone density, and overall energy levels.	- Most potent androgen in males - Produced in larger amounts in males than females - Converted to estrogen in adipose tissue via aromatase
Estrogen	Ovaries (females); testes and adrenal glands (males, in smaller amounts)	Steroid hormone	Regulates the development of female secondary sexual characteristics, such as breast development and the menstrual cycle. It also plays a role in bone health, reproductive function, and cardiovascular health.	- Primary female sex hormone - Includes estradiol, estrone, and estriol - Levels fluctuate during the menstrual cycle - Promotes endometrial thickening during the menstrual cycle
Progesterone	Ovaries (females); testes and adrenal glands (males, in smaller amounts)	Steroid hormone	Regulates the menstrual cycle and supports pregnancy by preparing the uterus for implantation and maintaining the uterine lining. It also plays a role in maintaining pregnancy and modulating mood.	- Produced primarily in the ovaries (corpus luteum) - Helps maintain pregnancy by preventing uterine contractions - Levels rise during the luteal phase of the menstrual cycle
Follicle-stimulating hormone (FSH)	Anterior pituitary gland	Glycoprotein	Stimulates the growth and development of ovarian follicles in females and spermatogenesis in males. In females, it also helps regulate the menstrual cycle and fertility.	- Secreted by gonadotroph cells in the anterior pituitary - Regulated by gonadotropin-releasing hormone (GnRH) - Essential for pubertal development and fertility
Luteinizing hormone (LH)	Anterior pituitary gland	Glycoprotein	Stimulates ovulation and the formation of the corpus luteum in females. In males, it stimulates Leydig cells in the testes to produce testosterone. It also plays a role in regulating the menstrual cycle and fertility in females.	- Secreted by gonadotroph cells in the anterior pituitary - Regulated by gonadotropin-releasing hormone (GnRH) - Surge in LH triggers ovulation in females - Stimulates the production of androgens in males
AMH	Sertoli cells in males; granulosa cells in females	Glycoprotein	In males, inhibits the development of Müllerian ducts. In females, serves as a marker of ovarian reserve and inhibits the development of ovarian follicles.	- Also known as Müllerian inhibiting substance (MIS) - Levels decline with age in females - Used clinically to assess ovarian function and fertility
Activin	Various tissues including gonads, pituitary gland, and placenta	Dimeric glycoprotein	Stimulates FSH secretion from the anterior pituitary gland, regulates cell growth and differentiation, and plays a role in reproductive processes.	- Part of the TGF-β superfamily - Acts as a local regulator of cell growth and differentiation - Plays a role in embryonic development and adult tissue homeostasis
Inhibin	Granulosa cells in females; Sertoli cells in males	Dimeric glycoprotein	Inhibits FSH secretion from the anterior pituitary gland by negative feedback, regulates follicular development in females and spermatogenesis in males.	- Consists of two subunits: α and β - Inhibin A (αβA) and Inhibin B (αβB) subtypes exist - Plays a key role in modulating FSH secretion to regulate ovarian function and spermatogenesis
Follistatin	Various tissues including gonads, pituitary gland, and placenta	Glycoprotein	Binds and inhibits the actions of activin, regulating activin-mediated signaling pathways. It also plays a role in cell growth, differentiation, and embryogenesis.	- Binds to and neutralizes activin, preventing its actions - Regulates cell proliferation, differentiation, and apoptosis in various tissues - Implicated in various physiological processes, including fertility and muscle growth

19.1.20 Other

19.1.20.1 The Heart as an Endocrine Organ

The endocrine activity of the heart comprises the release of various natriuretic peptide hormones from the cardiomyocytes. The atrial natriuretic peptide family includes three structurally similar peptides, i.e., atrial natriuretic peptide (ANP), brain or B-type natriuretic peptide (BNP), and C-type natriuretic peptide (CNP). The secretion of these hormones enables the heart to participate in regulating the circulatory volume and systemic blood pressure in animals.

The target effects of natriuretic peptides are mediated by cell membrane-bound receptors coupled with guanylyl cyclase. Three different types of receptors were identified: NPR-A, NPR-B, and NPR-C. ANP and BNP exert their biological effects by binding to NPR-A. The binding of ANP or BNP to NPR-A results in a conformational change with consequent activation of the intracellular guanylyl cyclase domain. The cGMP formed from GTP in the presence of guanylyl cyclase is regarded as the secondary messenger in the signal transduction mechanism. The elevated cGMP levels affect various cGMP-dependent protein kinases (PKGs), cGMP gated ion channels, and cGMP-regulated phosphodiesterases to bring about specific target effects. Kidneys, blood vessels, and the adrenal cortex are regarded as the principal target tissues of natriuretic peptides.

The natriuretic peptides (except CNP) are responsible for lowering the blood volume by enhancing the renal excretion of water (diuresis) and sodium (natriuresis). Apart from kidneys, they also cause the relaxation of vascular smooth muscles thereby bringing down the arterial pressure. They act on the pituitary gland, adrenal cortex, and small intestine to inhibit ADH, aldosterone secretion, and sodium absorption, respectively.

19.1.20.2 The Kidney and the Liver as Endocrine Organs

Erythropoietin (EPO) is a glycoprotein hormone (167 amino acids) released from the kidneys in response to hypoxia. The fetal liver and kidneys are major sites for the production of EPO. The interstitial peritubular cells in proximal tubules synthesize and secrete EPO during hypoxia and anemic stress.

The activation of erythropoietin receptors (EPOR) when bound with EPO results in their dimerization and phosphorylation by Janus kinase 2 (JAK2). The transphosphorylation of EPOR initiates the phosphorylation of signal transducer and activator of transcription 5 (STAT5), PI3K/AKT, and SHC/MAPK pathways to regulate cellular proliferation and survival. The EPOR is distributed majorly in erythrocyte progenitor cells residing in the bone marrow, thus making them the primary target tissue.

EPO is essential for the survivability of erythrocyte progenitor cells and their subsequent differentiation to colonyforming unit-erythrocyte (CFU-E) cells. It results in the swift passage of erythroblasts through different stages in erythropoiesis, thereby increasing the rate of erythropoiesis (Figure 19.2). Apart from the erythrocyte progenitors, EPO is found to have angiogenic effects on the endothelial cells and neurotrophic effects in the brain and regulate bone mass and formation of skeletal muscle fibers.

BIBLIOGRAPHY

Bayliss W. M., & Starling E. H. (1902). The mechanism of pancreatic secretion. *The Journal of Physiology*, 28(5), 325–353

Bie, P., Secher, N. H., Astrup, A., & Warberg, J. (1986). Cardiovascular and endocrine responses to head-up tilt and vasopressin infusion in humans. **American Journal of Physiology*, 251(4 Pt 2), R735–R741.

Bousfield, G. R. (2019). *Biosynthesis and posttranslational processing of peptide hormones.* https://doi.org/10.1016/B978-0-12-801238-3.92822-8

Brent, G. A. (2012). Mechanisms of thyroid hormone action. *Journal of Clinical Investigation*, 122(9), 3035–3043.

Casteel, C. O., & Singh, G. (2023). *StatPearls: Physiology, gonadotropin-releasing hormone.* StatPearls Publishing. Retrieved from https://www.ncbi.nlm.nih.gov/books/NBK556130/

Clarke, I. J. (2011). Hypothalamus as an endocrine organ. **Comprehensive Physiology*, 5(1), 217–253. https://doi.org/10.1002/cphy.c140019

Conn, P. M., & Crowley, W. F. (1991). Gonadotropin-releasing hormone and its analogues. *New England Journal of Medicine*, 324(2), 93–103.

Crowley, W. F., Filicori, M., Spratt, D. I., & Santoro, N. F. (1985). The physiology of gonadotropin-releasing hormone (GnRH) secretion in men and women. *Recent Progress in Hormone Research*, 41, 473–531.

Devesa, J., Almengló, C., & Devesa, P. (2016). Multiple effects of growth hormone in the body: Is it really the hormone for growth? *Clinical Medicine Insights: Endocrinology and Diabetes*, 9, 47–71. https://doi.org/10.4137/CMED.S38201

Giustina, A., & Veldhuis, J. D. (1998). Pathophysiology of the neuroregulation of growth hormone secretion in experimental animals and humans. *Endocrine Reviews*, 19(6), 717–797.

Hu, J., Zhang, Z., Shen, W. J., & Azhar, S. (2010). Cellular cholesterol delivery, intracellular processing and utilization for biosynthesis of steroid hormones. **Nutrition & Metabolism*, 7(47), 1–25. https://doi.org/10.1186/1743-7075-7-47

Jackson, I. M. (1982). Thyrotropin-releasing hormone. *New England Journal of Medicine*, 306(3), 145–155.

Johnstone, C., Hendry, C., Farley, A., & McLafferty, E. (2014). Endocrine system: Part 1. *Nursing Standard*, 28(38), 42–49.

Magner, J. A. (1990). Thyroid-stimulating hormone: Biosynthesis, cell biology, and bioactivity. *Endocrine Reviews*, 11(2), 354–385.

Melmed, S., Polonsky, K. S., Larsen, P. R., & Kronenberg, H. M. (2017). *Williams textbook of endocrinology* (13th ed.). Elsevier Inc. https://doi.org/10.1016/C2013-0-15980-6

Mendoza, A., & Hollenberg, A. N. (2017). New insights into thyroid hormone action. *Pharmacology & Therapeutics*, 173, 135–145.

Miller, W. L. (1988). Molecular biology of steroid hormone synthesis. *Endocrine Reviews*, 9(3), 295–318.

Nussey, S. S., & Whitehead, S. A. (2001). *Endocrinology: An integrated approach*. Retrieved from https://www.ncbi.nlm.nih.gov/books/NBK22/

Schwanzel-Fukuda, M., Bick, D., & Pfaff, D. W. (1989). Luteinizing hormone-releasing hormone (LHRH)-expressing cells do not migrate normally in an inherited hypogonadal (Kallmann) syndrome. *Brain Research Molecular Brain Research*, 6(4), 311–326.

Van Herle, A. J., Vassart, G., & Dumont, J. E. (1979). Control of thyroglobulin synthesis and secretion. (First of two parts). *New England Journal of Medicine*, 301(5), 239–249.

Wu, P., Lechan, R. M., & Jackson, I. M. (1987). Identification and characterization of thyrotropin-releasing hormone precursor peptides in rat brain. *Endocrinology*, 121(1), 108–115.

20 Physiology of the Male Reproductive System

M. M. Pathan[1] *and Tanmoy Rana*[2]

[1]Department of Veterinary Physiology and Biochemistry, College of Veterinary Science and Animal Husbandry, Anand, Gujarat, India
[2]Department of Veterinary Clinical Complex, West Bengal University of Animal & Fishery Sciences, Kolkata, India

20.1 INTRODUCTION

The male animal reproductive system is a vital component of an organism's biology, essential for the propagation and survival of species. It encompasses a series of organs and structures designed to produce, mature, and deliver spermatozoa, facilitating successful fertilization. The primary organs include the testes, which are responsible for the generation of sperm and the secretion of androgens, particularly testosterone, which play a crucial role in male reproductive physiology and secondary sexual characteristics (Johnson, 2014). Spermatogenesis, the process of sperm production, occurs within the seminiferous tubules of the testes and is regulated by a complex interplay of hormones, including gonadotropin-releasing hormone (GnRH) from the hypothalamus and luteinizing hormone (LH) and follicle-stimulating hormone (FSH) from the pituitary gland (Hess & Renato de Franca, 2008). Post-spermatogenesis, spermatozoa are transported to the epididymis, where they undergo further maturation and are stored until ejaculation (Cooper, 2011).During ejaculation, sperm travel through the vas deferens and mix with seminal fluid secreted by accessory glands such as the seminal vesicles, prostate gland, and bulbourethral glands. This seminal fluid provides a medium that enhances sperm viability and motility, critical for the journey through the female reproductive tract (Poole & Sansone, 2012). The coordinated function of these structures ensures the efficient delivery of sperm, a key determinant of reproductive success. Variations in the anatomy and function of the male reproductive system across different animal species illustrate the diverse strategies evolved to maximize reproductive efficiency under varying ecological pressures. This introduction aims to provide an overview of the critical components and functions of the male animal reproductive system, setting the stage for a more detailed examination of its anatomy, physiology, and relevance across species.

20.2 TESTES

Spermatozoa are produced by the testes. They have a similar structure, even if they differ slightly in size, shape, and position across species. The majority of each testicle is occupied by the convoluted seminiferous tubules. Within them, spermatozoa are created. The tunica albuginea is a capsule made of connective tissue that encloses the testicle. The tunica albuginea's connective tissue extensions, referred to as septa or trabeculae, support the seminiferous tubules inside the testis. The link between the seminiferous tubules and their connective tissue support (interstitial tissue) is revealed in a cross-section of the testicles (Figure 20.1).

20.2.1 Sertoli Cell (Sustentacular Cell) and the Leydig cell (Interstitial Cell)

To help developing spermatozoa, the Sertoli cell acts as a "nurse." Sertoli cell processes encircle spermatids and spermatocytes, establishing close contact with them during the entire spermatozoa generation process; for this reason, they are referred to as sustentacular, or supporting, cells. The Sertoli cells extend towards the center of the seminiferous tubules from their base at their periphery. The blood-testis barrier is created by the basal junction, also known as the tight junction, forming with nearby Sertoli cells. This barrier allows control over the tubule's environment and prevents spermatozoa from entering the interstitium.

The seminiferous tubules are divided into two compartments by the Sertoli cells: the adluminal compartment, which is the space between Sertoli cells and which communicates centrally with the tubule lumen, and the basal compartment, which provides space to germinal epithelial cells and interacts with interstitial fluid. A replacement cell and another cell that needs to pass through the Sertoli cell junction in order to access the adluminal compartment are produced when a germinal epithelial cell (spermatogonium) divides in the basal compartment. Here, more divisions take place, and the final spermatozoa are produced. A fluid that is secreted into the adluminal compartment by the Sertoli cells is advantageous to the growing spermatozoa. **The Leydig cells** are found in the connective tissue surrounding the seminiferous tubules and are responsible for testosterone production (Figure 20.2).

DOI: 10.1201/9781003426851-20

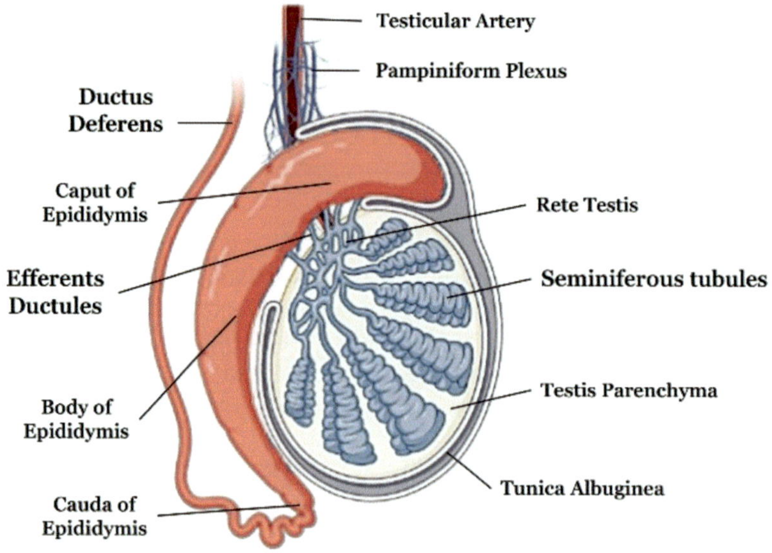

FIGURE 20.1 Cross-section of the testicles

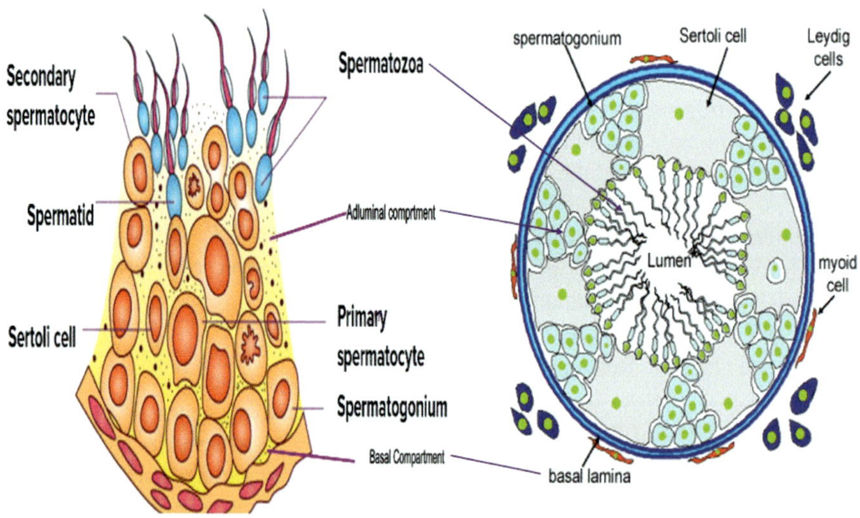

FIGURE 20.2 Illustrative diagram showing sertoli cells and Leydig cells

20.2.2 Epididymis

The testis's collecting and storing tubule is called the epididymis. The head of the epididymis, where blood vessels and nerves enter at the pole of the testis, is where it starts. The body of the epididymis, which eventually ends and turns upward to become the tail, extends from the head along one side of the testis (Figure 20.1).Sperm and adluminal fluid from the rete testis are delivered to the head of the epididymis via efferent ducts. Rete testis is the intratesticular network of straight tubules that receives content from the convoluted seminiferous tubules. Fluid from the adluminal gaps enters the lumen of the seminiferous tubules and transports spermatozoa to the epididymis. The spermatozoa mature and become motile while being stored in the epididymis.

A large portion of the seminiferous tubular fluid is reabsorbed in the epididymis head.

20.2.3 Ductus Deferens

The ductus deferens, also known as the vas deferens, is the duct system that extends from the pelvic urethra to the tail of the epididymis. Within the visceral layer of the vaginal tunic, the ductus deferens, testicular artery, vein, nerve, and lymphatic vessels are encapsulated as they leave the testis and move towards the abdomen. The spermatic cord is the name given to this group of components. The visceral layer of the vaginal tunic also envelops the testis and epididymis. It originates from the embryonic abdominal peritoneum that was formed when the testes passed through the inguinal canal and reached the scrotum. The inguinal canal, which runs from the interior to the exterior inguinal rings, is an oblique passageway that connects the abdominal cavity to the outside of the body. The tendinous attachments of the two flat abdominal muscles to the pelvis include gaps called inguinal rings.

The ductus deferens splits from the spermatic cord to reach the pelvic urethra after the spermatic cord travels through the inguinal rings. The ampulla of the ductus deferens, which is lacking in the dog and boar, is an expanded, glandular region at the end of the ductus deferens that varies in size among species.

20.2.4 Scrotum

The testes are located in a cutaneous bag called the scrotum. The subcutaneous layer of smooth muscle fibers in the scrotum, known as the tunica dartos, contracts during the colder months to draw the testes closer to the abdominal wall. The parietal layer of the vaginal tunic, which extends the parietal peritoneum into the scrotum, lines the inside of the scrotum. During embryonic development, the testes are intra-abdominal but outside the peritoneum. They have not yet entered the scrotum, but each has a fibrous connection to the scrotum known as the gubernaculum testis. As development and growth progress, the gubernaculum testis "pulls" the testes through the inguinal canal into the scrotum, which creates a double-walled tube of peritoneum. The testis, epididymis, ductus deferens, and testicular vessels, nerves, and lymphatics are enveloped by the inner tube of the peritoneum, known as the visceral vaginal tunic.

20.3 ACCESSORY SEX GLANDS AND SEMEN

Secretions from the accessory sex glands discharge into the pelvic urethra in close proximity to where they originate. They vary in size and shape among species and can be absent in some.

The accessory sex glands are: 1) the ampulla of the ductus deferens, 2) the vesicular glands (or the seminal vesicles), 3) the prostate gland, and 4) the bulbourethral glands (or the Cowperglands).

The ampulla is the enlargement of the terminal part of the ductus deferens, and its secretion empties into the lumens of the ductus deferens. The ampulla is absent in the boar and dog. The vesicular glands (absent in the dog) are paired glands that empty into the pelvic urethra along with the ductus deferens. The prostate gland is present in all domestic animals. It is prominent in the dog, encircling the urethra. The paired bulbourethral glands (absent in dogs) are the most caudal of the accessory glands.

The ejaculating semen is a mixture of the accessory sex gland secretions, seminal plasma, sperm, and fluid of the epididymis. The seminal plasma provides an environment for the survival of sperm within the female reproductive tract. It is rich in electrolytes, fructose, ascorbic acid, and other vitamins. The greatest fertilization potential is

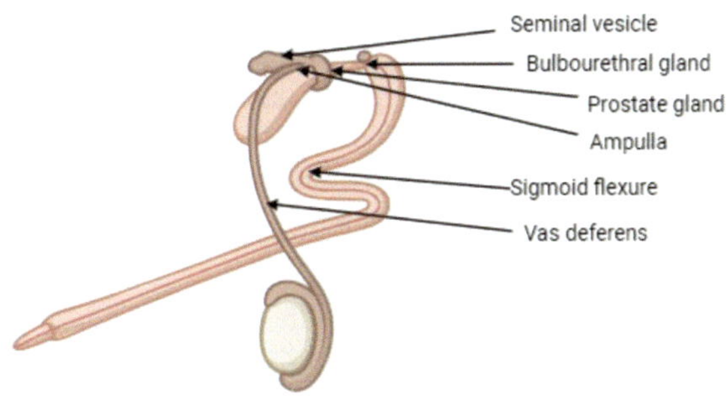

FIGURE 20.3 Illustrative diagram of accessory sex glands (Created with BioRender.com)

achieved with seminal plasma. Prostaglandins of seminal plasma help during fertilization by reacting with cervical mucus and making it more receptive to sperm and by smooth muscle contraction, so it is believed that reverse peristalsis is initiated in the uterus and oviducts to facilitate transport of sperm toward the ovaries (Figure 20.3).

20.4 PENIS

The penis is the male copulatory organ that allows semen and urine to go through the penile urethra. The penis is made up of three compartments: 1) The roots (crura) of the penis begin at the caudal border of the pelvic ischial arch, 2) The body is a forward extension from the roots, and 3) the glans is a free extremity. The corpus cavernosum, also called erectile tissue, is the cavernous tissue that makes up the majority of the interior structure and fills with blood during an erection. Surrounding the corpus cavernosum is a dense fibroelastic coating called the tunica albuginea, which also contributes to the sheets of connective tissue that are present within. Depending on the amount of connective tissue found in the corpus cavernosum, the penis is classified as musculocavernosus, which contains more cavernous space and less connective tissue, or fibroelastic having less cavernous space and more connective tissue. The musculocavernous penis is characteristic of the horse and dogs, and the fibroelastic penis **is** found in ruminants (cattle, sheep, goats) and pigs. During erection, the musculocavernous penis greatly increases in both length and girth as the cavernous spaces fill with blood, while in the fibroelastic penis, the increase in length is largely brought about by a straightening of the sigmoid flexure. The urethra is on the ventral aspect of the body of the penis and is surrounded by the corpus spongiosum. An evident urethral process is found in the ram and buck**.** The purpose of the urethral process in rams and bucks is thought to be to spray semen during ejaculation into the cervical region. The dog has a bulbus glandis at the caudal part of the glans. The enlargement of the bulbus glandis is responsible for prolonged retention of the penis during coitus, commonly known as the tie. The penis of bulls, rams, and boars has an S-shaped sigmoid flexure that gets extended during erection.

The prepuce is an invaginated fold of skin that surrounds the free extremity of the penis. The stallion has a double-folded pouch. Waxy accumulations, known as beans, sometimes form in the outer fold and must be removed manually. The boar has a preputial diverticulum (pouch) on the dorsal wall, which often contains decomposing urine and macerated epithelium.

20.5 MALE GENITAL MUSCLES

The cremaster muscle pulls the testis up against the superficial vaginal ring, particularly in cold weather. The cremaster muscles are responsible for the testes being drawn into the abdominal cavity of the elephant, deer, and rabbit during times other than the breeding season. A skeletal muscle, the urethralis, is the pelvic continuation of the smooth muscle wall of the urinary bladder. The peristaltic action of this muscle assists in the transport of urine or semen through the pelvic urethra. The bulbospongiosus muscle is a striated muscle continuation of the urethralis. It continues throughout the length of the penis in the horse, but in other animals, it is only present up to a short distance along the penile urethra. The ischiocavernosus muscles pull the penis upward against the floor of the pelvis. The retractor penis muscles pull the flaccid penis back into the prepuce.

20.6 BLOOD AND NERVE SUPPORT

Blood to the testicles is supplied by the testicular arteries and veins. The pampiniform plexus, a network of testicular veins in close association with the convoluted part of the testicular artery, helps in the cooling of blood entering the testis. In addition to autonomic nerve fibers to the testes, penis, and accessory sex glands, the penis is supplied by a spinal nerve, the pudendal nerve.

20.7 SPERMATOGENESIS

The complete process by which germinal epithelial cells develop into spermatozoa is referred to as spermatogenesis. It is divided into two phases: spermatocytogenesis and spermiogenesis.

Spermatocytogenesis is the proliferative phase whereby spermatogonial cells multiply by a series of mitotic divisions, followed by the meiotic divisions that produce the haploid (n) number of chromosomes. The spermatogonia are located in the basal compartment of the seminiferous tubules. A single cell replaces the recently divided cell in a spermatogonium during mitotic division, and it remains in the basal compartment. The other cell becomes a type A spermatogonium, which migrates through the Sertoli cell barrier to the adluminal compartment. Type A spermatogonia undergo mitotic division until large numbers of type B spermatogonia have been produced. Type B spermatogonia undergo the last of the mitotic divisions, which results in the formation of primary spermatocytes with 2n chromosome numbers. Primary spermatocytes undergo meiotic

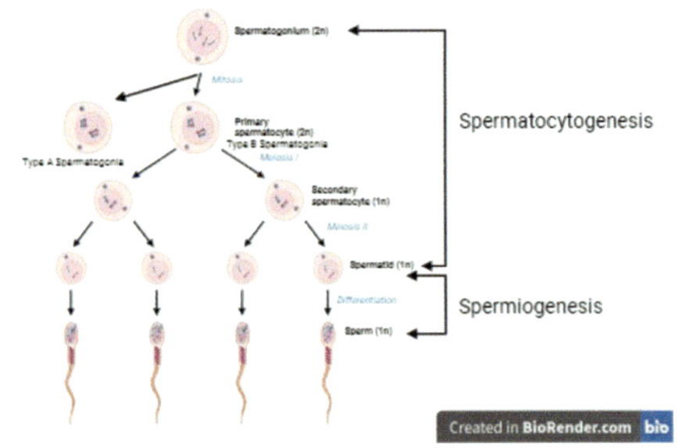

FIGURE 20.4 Spermatogenesis

division to form secondary spermatocytes, which in turn undergo meiotic division to form spermatidswith n chromosome numbers.In the bull, 64 spermatids are formed from one type A spermatogonium.

Spermiogenesis involves the maturation of the spermatids while they are still in the adluminal compartment. Spermiogenesis comprises a series of nuclear and cytoplasmic changes and transformations from a nonmotile cell to a potentially motile cell in which a tail has formed. The mature spermatids produced during the final phase of spermiogenesis are released into the lumen of the seminiferous tubules as spermatozoa. The release of mature spermatids into the lumen of the seminiferous tubules is known as spermiation (Figure 20.4).

20.8 SPERMATOGENIC WAVE

Because the adluminal compartment needs roughly 64 days in the bull to complete, spermatozoa production would not continue if all segments of the seminiferous tubules were engaged in the same activity at the same time. While this development is proceeding, a new type A spermatogonium migrates through the Sertoli cell barrier into the adluminal compartment to begin its development behind the developing type A spermatogonium that preceded it. This happens once every 14 days in the bull. Given that spermatozoa require 64 days to form, there will be 4.6 cycles (64/14) of initial development prior to the first cycle from a specific region of the seminiferous epithelium beginning to reach the rete testis.

This sequential change in the stage of the cycle along the length of the tubule is known as the spermatogenic wave.

20.9 TRANSPORT OF SPERMATOZOA

The newly formed spermatozoa are less motile, so the transportation of sperms to the epididymis occurs by the fluid of the seminiferous tubules and rete testis, as well as by the activity of contractile elements in the testis that direct fluid flow to the head of the epididymis.

The fertilizing ability of an animal is attained progressively during the transit of spermatozoa through the epididymis. Changes include the development of unidirectional motility, changes in nuclear chromatin (DNA-protein complex), and changes in the nature of the surface of the plasma membrane. The major site of sperm storage within the male reproductive tract is the tail of the epididymis. About 70% of the total number of spermatozoa are found in this region.

20.10 MALE REPRODUCTIVE HORMONES AND THEIR CONTROL

The hormones are secreted by Leydig and Sertoli cells in the testes. **Leydig cells** produce testosterone, which is controlled by luteinizing hormone (LH) or interstitial cell-stimulating hormone (ICSH). Low levels of testosterone cause an increase in LH secretion by the anterior pituitary. The increase in LH secretion causes the Leydig cells in the testes to secrete testosterone. Increased testosterone inhibits the further secretion of LH, and testosterone levels are thus stabilized and the cycle is repeated; this is known as a negative feedback system. Testosterone must diffuse into the seminiferous tubules from the interstitial tissues in order to affect spermatogenesis. It appears that testosterone supports the meiotic process within the seminiferous tubules, hence maintaining spermatogenesis.

Sertoli cells are stimulated by follicle-stimulating hormone (FSH) and produce androgen-binding protein. The androgen-binding protein is secreted into the lumen of the seminiferous tubules and binds with testosterone and other androgens to stabilize their concentrations and ensure appropriate amounts for spermatogenesis. It is also believed that FSH stimulates the secretion of estrogens by the Sertoli cells. The actual secretion of estrogen might arise from the intracellular conversion of testosterone (originating from Leydig cells) by the Sertoli cells. The Sertoli cells also secrete a hormone known as **inhibin,** which inhibits the secretion of FSH by the anterior pituitary. FSH is not necessary for the maintenance of spermatogenesis once it has begun, but LH is constantly needed for spermatogenesis (testosterone-supported meiosis). FSH is essential for the initiation of spermatogenesis throughout puberty and following physiological or pathologic disruptions.

20.11 SECONDARY FUNCTIONS OF TESTOSTERONE

The growth and maintenance of libido, the secretory activity of the accessory sex glands, and the general masculine physical characteristics are among the additional functions of testosterone. The structural development and physiologic functioning (production of secretions) of the accessory sex glands are influenced by testosterone. Secondary sexual characteristics are influenced by testosterone. These features include increased bone growth (heavier bones), greater muscle, thicker skin, and a deeper voice (in the bull). During fetal growth, testosterone directs the descent of the testes. The presence or absence of testosterone determines the respective development of the penis and scrotum, or the clitoris and vagina. With normal male hormonal stimulation, the Wolffian (mesonephric) ducts become tubular portions of the male reproductive system, and the Müllerian (paramesonephric) ducts regress. In the female, the Müllerian ducts become tubular portions of the reproductive system, and the Wolffian ducts regress.

BIBLIOGRAPHY

Cooper, T. G. (2011). *The Epididymis, Sperm Maturation and Fertilisation*. Springer Science & Business Media.

Hess, R. A., & Renato de Franca, L. (2008). *Spermatogenesis and Cycle of the Seminiferous Epithelium*. Springer.

Johnson, M. H. (2014). *Essential Reproduction*. John Wiley & Sons.

Poole, D. H., & Sansone, S. A. (2012). The effects of nutrition on reproduction in beef cattle. *Veterinary Clinics of North America: Food Animal Practice*, 28(3), 497–514.

Sjastad, O. V., Knut Hove, K., & Sand, O. (2005). *Physiology of Domestic Animals*. International Book Distributing Co.

William, O. R., & Eric, W. R. (2017). *Functional Anatomy and Physiology of Domestic Animals*. Wiley Blackwell.

21 Physiology of the Female Reproductive System

M. P. S. Tomar[1], *S. K. I. Vasantha*[1], *S. S. R. Kona*[1] *and Devendra Pathak*[2]
[1]Sri Venkateswara Veterinary University, Tirupati, India
[2]Guru Angad Dev Veterinary & Animal Sciences University, Ludhiana, India

21.1 INTRODUCTION

In males with fertile ejaculates, the processes of steroidogenesis and gametogenesis continue even during the nonbreeding seasons. After semen is deposited in the female reproductive tract during mating, the male's role simply ends, and the female's reproductive participation begins. During their reproductive lifespan, females will only release fewer gametes during ovulation, coinciding with the behavioral changes that attract males for insemination. In order to successfully complete a gametic encounter, the fertilization, embryonic development, implantation, gestation, and delivery, she must also create a synchronous and sufficient oviducal and uterine environment. The mother's primary responsibility is to supply the energy needed for the baby's birth, upbringing, and protection until it is able to sustain itself. Therefore, compared to males, females devote a greater portion of their physical energies to reproduction.

The duct system, external genitalia, and two ovaries make up the female reproductive system (Figure 21.1). The entire tubular portion is composed of the following: submucosa of variable thickness, muscularis (comprising outer longitudinal and inner circular layers of muscle), mucosa (whose lumen is lined by mucosal epithelium), and outer serosa (continuation of the peritoneum). The mucosal epithelium is altered in accordance with the origin and necessity. The broad ligament supports the uterus, oviduct, and ovaries inside the abdominal cavity. The mesovarium is the portion of the broad ligament that supports the ovaries, while the mesosalpinx and mesometrium, respectively, support the oviduct and uterus.

21.2 THE OVARY

The paired structures known as mammalian ovaries are located within the abdominal cavity (Figure 21.2). The animal's age, species, and reproductive state all affect the size and form of its ovaries. However, the ovaries of monotocous animals are ovoid, while those of polytocous animals typically resemble a cluster of grapes. The ovulation fossa, a prominent depression in the kidney-shaped ovary that is the site of every ovulation, is present in mares. The cells known as surface (germinal) epithelium, which encircle the ovary and pierce through the tunica albuginea, provide granulosa cells – not eggs – to the developing ovarian follicles. The ovarian cortex, which displays corpora lutea and ovarian

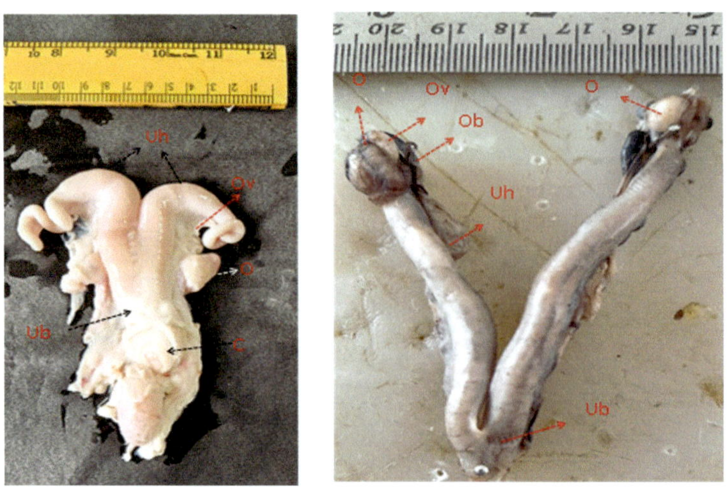

FIGURE 21.1 Female reproductive organs in goat (left) and dog (right): Ovary (O), Oviduct (Ov), Ovarian bursa (Ob), Uterine horn (Uh), Uterine body (Ub), and Cervix (C).

DOI: 10.1201/9781003426851-21

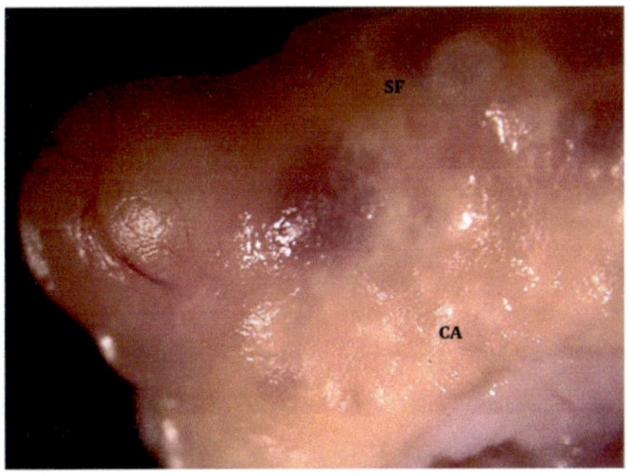

FIGURE 21.2 Gross morphology of ovary in cats: Surface follicle (SF) and Corpus albicans (CA).

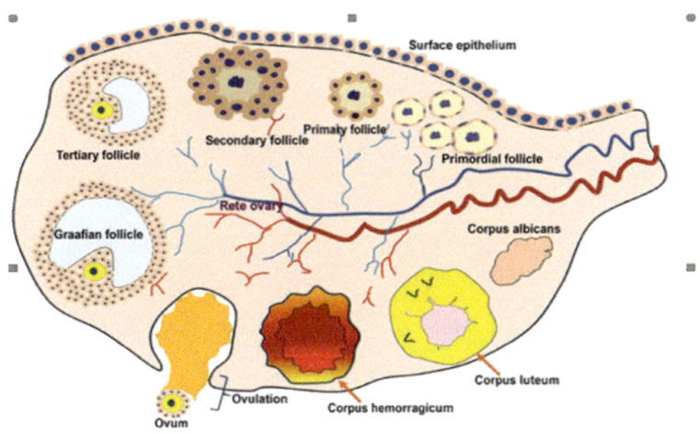

FIGURE 21.3 Schematic diagram of the ovary showing various stages of follicles and corpus luteum.

follicles at various developmental stages, is located next to the tunica (Figure 21.3). The ovarian medulla has a vast nerve and vascular supply as well as irregular fibroblastic tissue. Although they are found in the medulla as well as the cortex, smooth muscle cells are more prevalent in the perifollicular areas of the cortex. These adrenergic receptors are found in smooth muscles. The ovary's supply of adrenaline is thought to play a role in the ovarian follicles' selection for maturation and/or ovulation. The luteal phase is when arterial blood flow to the ovary is greatest, and ovulation is when it is lowest. Estrogen is transported from the ipsilateral ovary to the lymph vessels that drain the uterus.

21.3 THE OVARIAN FOLLICLES

The ovary serves two major purposes: (i) gametogenesis, or the production of eggs, and (ii) steroidogenesis, or the production of female sex hormones. Both are the duties of the ovarian follicle. It is possible to categorize ovarian follicles into five different groups: primordial, primary, secondary, tertiary, and Graafian (Figure 21.3).

Mesonephric cells enter the undifferentiated gonad during fetal development, undergo numerous mitotic divisions, and ultimately give rise to primordial follicles and oocytes. In postnatal life, these primordial follicles containing the primary oocytes do not go through mitotic division, in contrast to the spermatogonia in males. The thick layer of primordial follicles, which are spherical or circular in shape, lies beneath the tunica albuginea (Figure 21.4). They display oocytes with a single layer of flattened or cuboidal follicular cells surrounding them, devoid of the zona pellucida and vitelline membrane. The basement membrane is present in these primordial follicles in its entirety. There is no distinct vasculature present in these unilaminar follicles. Primordial follicles are created at birth and, as an animal ages, their quantity decreases rather than increases. A heifer calf's ovary may contain up to 150,000 of these primordial follicles. All mammals experience this same phenomenon, though the exact number varies by species. Cats have characteristic clustering of primordial follicles, which are called Multi-oocyte follicles (Figure 21.4).

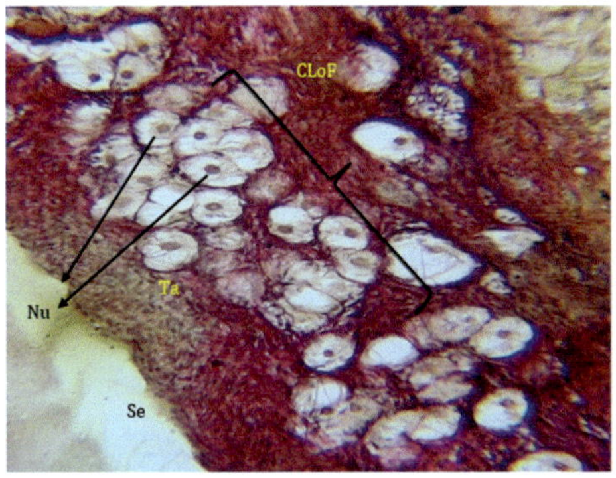

FIGURE 21.4 Photo micrograph showing clusters of primordial follicles in cats: Surface epithelium (Se), Tunica albuginea (Ta), Nuclei (Nu) & Clusters of follicles (CLoF).

A tiny portion of all follicles in a species develop into Graafian follicles and eventually ovulate. These are known as primordial follicles. Before reaching ovulatory size, a number of them begin to grow but then begin to regress at different times. Others, however, never start the process of growth! A follicle that is growing is one that has begun to grow, has more than one layer of follicular cells, a clear zona pellucida, an oocyte with a greater diameter, and other characteristics. The follicular cells in the secondary follicles multiply and layer more, taking on polyhedral forms. *These cells are now known as granulosa cells because their cytoplasm starts to show granular inclusions. Within these developing/secondary follicles, spindle-shaped cells arrange themselves outside the basement membrane to form the theca. The follicle itself enlarges, becomes more oval, and migrates from the cortex in the direction of the medulla. Follicle growth is not dependent on pituitary gonadotropins at this point. Gonadotropins cause the granulosa cells to secrete a fluid that builds up in the spaces between cells. The follicle gradually develops the antrum, a fluid-filled space, as the fluid continues to be secreted and accumulate. The follicle is now referred to as a vesicular, antral, or tertiary follicle.* The corona radiata, a solid mass of cells, has now encircled the oocyte. Theca externa and interna, two layers of thecal cells, protrude from the

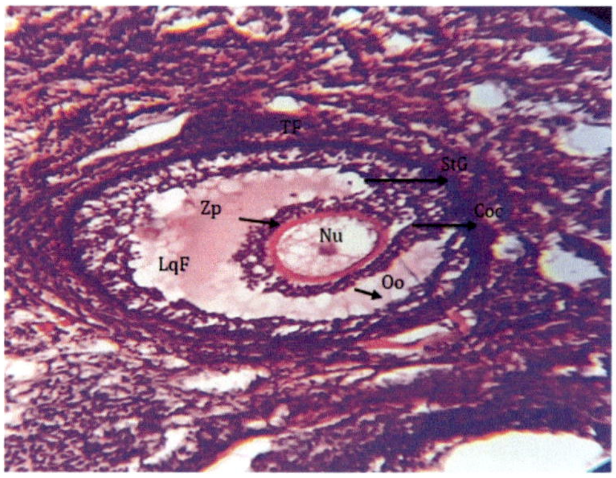

FIGURE 21.5 Photo micrograph showing Graafian follicle and cumulus oophorous oocyte complex in bitch: Stratum granulosum (StG), Cumulus oophorous oocyte complex (Coc), Tertiary follicle (TF), Oocyte (Oo), Nucleus (Nu), Zona pellucida (Zp) & Liquor folliculi (LqF).

surface of the ovary in Graafian follicles, and the oocyte's diameter increases (varying from 80 to 120 µm in different species). The follicle wall is composed of numerous granulosa cell layers and thecal layers (Figure 21.5). Myoid cells form the follicle's outermost wall, or theca externa, and they have smooth muscle-like properties. Fibroblasts containing a high concentration of organelles and cytoplasmic granules form theca interna. Granulosa cells, which are isolated from thecal layers by the basement membrane, make up the innermost layer of the follicular wall. During ovulation, this membrane becomes discontinuous. Now, a solid mass of cells known as the corona radiata envelops the oocyte. The ovary's surface is adorned with Graafian follicles, which have two layers of thecal cells called theca externa and interna. The diameter of the oocyte also grows, ranging from 80 to 120 µm depending on the species. The wall of the follicle is made up of many granulosa cell layers and thecal layers. Myoid cells, which have properties similar to smooth muscles, form the follicle's outermost wall, or theca externa. Fibrocytes that are abundant in organelles and cytoplasmic granules form theca interna. Granulosa cells form the innermost layer of the follicular wall, which is divided from thecal layers by the basement membrane. At the time of ovulation, this membrane becomes irregular.

Intrafollicular fluid pressure causes the follicle to become extremely turgid and causes the volume of the follicular fluid to increase dramatically. The granulosa cells are forced up against the follicular wall by this increase in follicular fluid, and as a result, they become known as *membrana granulosa*. The ovum separates from its attachment to the follicular wall and floats freely in the liquor folliculi as a result of additional follicular fluid accumulation.

21.4 FOLLICULAR FLUID

The transudate of plasma across the follicular wall is the initial source of follicular fluid, which builds up in the antrum, created by the fusion of tiny fluid-filled pockets. The metabolic activity of the cells within the follicle changes the composition of the follicular fluid. The follicular fluid contains proteins, amino acids, carbohydrates, enzymes, and hormones (sex steroids, prostaglandins, and gonadotropins). The follicular fluid is essential for the maturation of oocytes, ovulation, granulosa cell regulation, and egg transport. The oviduct receives large volumes of follicular fluid, which is released during ovulation and contributes to the conditions that allow for sperm capacitation, fertilization, and the early development of embryos. The follicle and the oocyte grow at the same time when certain unique components in the follicular fluid appear. The purpose of the immunoglobulins (IgA and IgG) present in the follicular fluid is unknown.

21.5 FOLLICULOGENESIS

Graafian follicles develop from a pool of primordial follicles that are at rest through a process known as folliculogenesis. Throughout a female's entire reproductive life, this process is ongoing. Three physiologically separate stages can be distinguished in the development of a primordial follicle into a Graafian follicle: the stages from primordial to preantral, preantral to antral, and antral to Graafian follicle.

(i) **Primordial to preantral follicle development:** The typical number of primordial follicles in a species, most likely controlled by an unidentified intraovarian factor, begins to increase daily. An initial developing primordial follicle persists until ovulation or degenerates. It is said that a primordial follicle needs anywhere from 20 days to 6 months to mature to the ovulatory stage. Every day, a number of follicles reach the preantral stage, regardless of the time it takes for them to grow from the primordial to the preantral stage or the hormonal and reproductive status of the animal. To the degree that the growth of primordial follicles is an ongoing process, gametogenesis in females is thus similar to spermatogenesis in males. Gonadotropins have no effect on the preantral stage of primordial follicle development. After completing this stage of hormone-independent growth, primordial follicles still need to produce receptors for LH in the theca, FSH, and estrogen in the granulosa cells, and other hormones to move into the hormone-dependent phase of development and reach the Graafian follicle stage.

(ii) **Preantral to Antral follicle development:** Each species' characteristic number of preantral follicles is attracted by the FSH surge, which in all farm animals takes place 20–30 hours after the preovulatory LH surge. Preantral follicles undergo antrum formation in response to this FSH surge, which happens before the CL starts secreting progesterone. Some of these follicles go on to ovulate in a matter of cycles. The number of antral follicles on the ovary about 17 days later and the ovulation rate are related to the magnitude of this FSH surge in farm animals. Either 17 or 34 days pass from the start of antrum formation to the time needed for ovulation. The granulosa cells produce their own receptors as a result of FSH. It could be this capacity of FSH that gives rise to the term "follicle recruitment by FSH." Increased FSH also causes granulosa cell LH receptors to grow. LH stimulates theca cells, which then release androgens. Granulosa cells use FSH to change these androgens into estrogens. In granulosa cells, estrogen stimulates FSH receptors even more. Additionally, the amount of estrogen secreted dictates which follicle will acquire the LH receptors required for luteinization and ovulation. In granulosa cells, FSH and estrogen stimulate mitosis. It is also in charge of gap junction formation within and between granulosa cells as well as between them. Under the effect of FSH, granulosa cells produce and secrete substances that lead to granulosa cell separation and the development of an antrum.

(iii) **Antral to Graafian follicle development:** Few follicles that were encouraged to form an antrum reach the ovulatory stage; the rest experience atresia, most likely as a result of a drop in FSH levels that takes place during the luteal phase. It is true that injecting FSH or hormones similar to FSH during the luteal phase of an animal's cycle can cause super ovulation in domesticated animals. During ovulation, multiple ovum are released due to the administration of FSH exogenously. Domestic animals' luteal phases last comparatively long because progesterone from the CL can cause follicular growth to decline for up to 17 days. It's interesting to note that when progesterone levels are lowered through hormonal or manual removal of the corpus luteum, follicular growth accelerates and the next estrus soon ensues. This is the physiological underpinning of the widely recognized practice of synchronizing estrus following prostaglandin-induced luteolysis. Progesterone therapy can be used to extend the life of CL and lengthen the cycle. The transformation of preantral follicles into Graafian follicles is largely dependent on FSH. Every estrous cycle, a crop of developing follicles is stimulated to continue growing and maturing by the pituitary gland's adequate release of FSH. Throughout the early stages of the estrous cycle, many follicles grow, but only a small number mature before the ovum. On the granulosa, FSH increases LH receptors while LH decreases them, particularly during the preovulatory LH surge. The transition of granulosa from estrogen secretion (during follicular phase) to progesterone secretion (during luteal phase) is dependent on these receptor alterations.

21.6 OOGENESIS

The final generation of oogonia, which are created by the mitotic proliferation of primordial germ cells, continue to grow until they reach the prophase of the first meiotic division. As primary oocytes at this stage, they go into a protracted resting phase of development. This resting phase, known as the "dictyate stage," is specific to female gametes. These dictyate oocytes require the preovulatory LH surge to resume meiosis. Oogonia in domestic animals mature into primary oocytes either prior to or soon after birth. Oocytes begin to grow along with the primordial follicles, and by the time the follicles reach the preantral stage, the oocytes have finished growing and formed the zona pellucida and vitelline membrane. The glycoprotein structure known as zona is secreted by the inner granulosa cells. The material secreted by follicular cells and the oocyte are incorporated into the zona, which thickens as the follicle grows. The main components of a primary oocyte are the germinal vesicle, a large spherical nucleus, and the cytoplasm, or vitellus. The oocyte can resume meiosis when the preovulatory surge of LH occurs because it eliminates the inhibition caused by the OMI. It finishes the first meiotic division, extrudes the first polar body, moves on to the second meiotic division's metaphase, and then stops once more. It is referred to as the secondary oocyte at this point.

Similar to spermatogenesis, the first meiotic division – known as reduction division – does not establish the haploid state because chromosome replication occurs before this division. Only after being activated by the sperm during fertilization does the secondary oocyte resume and finish the second meiotic division and extrude the second polar body. Momentarily, the oocyte is an Ootid. The haploidy is established by the second meiotic division, and the oocyte's cytoplasmic maturation happens concurrently with nuclear maturation. Since the granulosa and other cells surrounding the oocyte begin to loosen at this point, it is likely that the follicular cells' release of an inhibitory influence is what causes this cytoplasmic maturation. The first meiotic division in the majority of domestic animals is finished a few hours before ovulation. However, in bitch, the first meiotic division is finished two to three days following ovulation. This could be the cause of the oocytes in bitch oviducts having a longer fertile life span – up to seven days. The first meiotic division in a mare can be completed either before or after ovulation.

21.7 THE OVUM OR THE EGG

Ovulated eggs are larger than sperm in all species, despite the fact that they differ in size and appearance among many species. The existence of a substance in the yolk called duetoplasm accounts for the ovum's large size. In goat and rabbit eggs, the amount of yolk material in the cytoplasm is greater, but in mare eggs, it is less. Cows have the smallest eggs among domestic animals, while rabbits have the largest. The layers of cells known as the cumulus oophorus and corona radiata envelop freshly ovulated eggs. These cells have cytoplasmic extensions that interdigitate with the microvilli on the egg's vitelline surface. The purpose of these extensions is to carry nutrients into the egg. However, the connections are broken and the cells are separated from the egg after ovulation as a result of ciliary movement and fibrinolytic enzymes found in the oviduct. Sperm enzymes may also cause these cells to be removed during fertilization. The vitelline membrane and the zona pellucida are the two membrane structures found in the ovum. The ovum contains the nucleus (in the metaphase II stage), the endoplasmic reticulum, multivesicular bodies, and all other common organelles. Furthermore, cortical granules are located directly below the vitelline membrane. The eggs of various domestic animals exhibit a number of anomalies, such as giant, small, oval, or flat eggs, ruptured zona pellucida, and absence of cortical granules.

21.8 THE DUCT SYSTEM OF THE FEMALE

The female duct system is made up of the uterus, cervix, internal vagina, and two oviducts, also known as fallopian tubes or uterine tubes. The duct system performs the following tasks (Table 21.1).

21.9 THE OVIDUCT OR THE FALLOPIAN TUBE

The ovary and uterus are joined by an oviduct. Its four functional divisions are the isthmus, ampulla, infundibulum, and fimbriae (Figure 21.6). The oviduct's ovarian end enlarges to form fimbriae, which have finger-like projections or fringed edges. The ovarian bursa is made up of fibriae and a thin meso-salpinx peritoneal fold. In farm animals, the ovarian bursa is typically open, but in pigs, it covers the entire ovary. Only the ovulation fossa is covered by the bursa in mares. During ovulation, free-moving fimbriae sweep over the ovarian surface. The second portion of the oviduct, known as the infundibulum, is shaped like a funnel and presents the ostium abdominale, the oviduct's abdominal opening. The animal's age and size affect the infundibulum's dimensions. Ampulla joins the infundibulum to the isthmus and makes up more than half of the oviduct's length. At the utero-tubal junction (UTJ), the isthmus and uterus are directly connected.

Both inner circular and outer longitudinal smooth muscles are present in the oviduct walls. From the ovarian to

TABLE 21.1

The Functions of Various Parts of Duct System

Oviduct	Ideal conditions for preimplantation growth and fertilization
Uterus	Early embryogenesis, attachment of the conceptus, and fetus expulsion through fetal membranes.
Cervix	Barrier; during estrus, secretes mucus; during pregnancy, the cervical seal is produced.

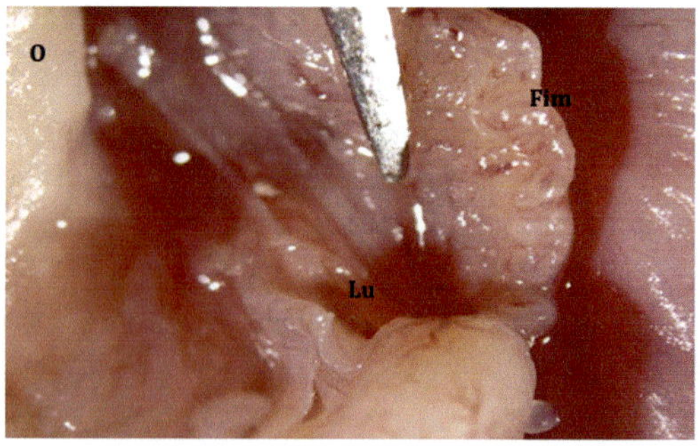

FIGURE 21.6 Gross morphology of the oviduct showing fimbriated end and lumen in cats: Ovary (O), Fimbriae (Fim), & Lumen of oviduct (Lu).

the uterine end, the muscle thickness increases. The oviduct gets shorter when muscles contract longitudinally, and annular constriction occurs when muscles contract circularly. Three different types of contractions are seen in oviducal musculature, viz., the localized peristalsis and segmental contractions.

Due to the possibility of simultaneous contractions or specific patterns, the oviduct's motility is complex. Different oviduct segments also have different patterns and amplitudes of contractions. During ovulation, oviductal muscle contractions peak in intensity and complexity before gradually declining. The primary, secondary, and tertiary folds comprise the oviducal mucosa.

The mucosa in the ampulla is arranged into high-branched folds that become low ridges at the UTJ and get shorter as they approach the isthmus. At the ovarian end of the oviduct, the lumen is entirely filled with these folds.

One layer of columnar cells, both ciliated and non-ciliated in type, makes up the mucosa (Figures 21.7 & 21.8). Hormonal activity affects the ciliary beat. During ovulation, the movement of fimbriae and this ciliary beat are synchronized. These cilia beat at their fastest during ovulation and then gradually slow down afterward. Normally, these cilia beat in the direction of the uterus. Along the oviduct, the concentration of ciliated and non-ciliated cells varies, most likely in relation to its function. For instance, the ciliated cells are most abundant at the fimbriated end, where their primary role is to pick up the ovum. On the other hand, the non-ciliated secretory cells (Figure 21.7) are most abundant in the ampulla and other areas where fluids are required to support gamete interaction and embryonic nourishment.

The nature of non-ciliated cells is secretory. They have secretory granules with many microvilli covering their apical surface. The oviductal lumen receives the release of secretory granules that accumulate in the epithelial cells during the follicular phase of the estrous cycle, thereby reducing their height. Oviductal fluid flows in the direction of the ovary. For two to three days following estrus, the majority of it is lost into the peritoneal cavity. However, it then descends the tube into the uterus, where it carries the

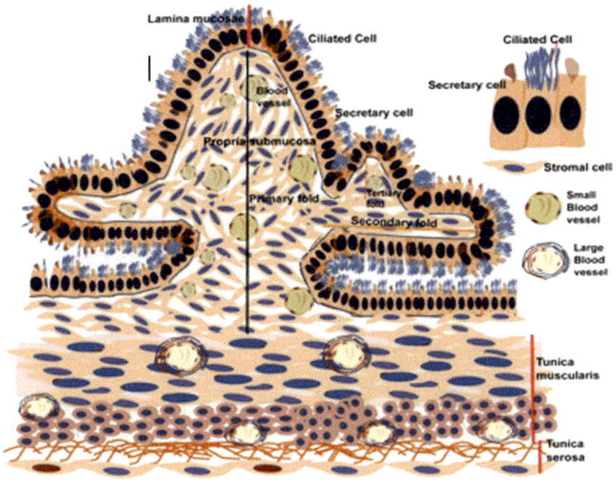

FIGURE 21.7 Schematic diagram of various histologic layers of oviduct.

embryos for implantation. The functional activity of ciliated and secretory cells in the oviduct is reliant on ovarian estrogens. The oviduct is responsible for transporting spermatozoa and eggs in opposing directions at roughly the same time. These disparate activities never fail because of how well the tubes' structure and function complement one another. The ovum is transferred from the ovarian surface into the infundibulum by the fimbriae. The delicate balance between the ciliary beating towards the ostium and the cumulus-oocyte complex's adhesion to the infundibulum is what causes this ovum pickup. The extracellular matrix that is present between the cumulus cells that are attached to the ciliary tips of the fimbriae in the infundibulum facilitates the adhesion of the COC. It is possible for further fimbriae to absorb eggs shed from the contralateral ovary. Oviductal secretions serve several purposes: (i) they capacitate sperm; (ii) they create an environment that is favorable for fertilization; (iii) they nourish the developing embryo; (iv) they protect oocytes, sperm, and embryos from the mother's immune system as foreign antigens; and (v) they cause zygote cleavage by secreting a particular protein that causes cleavage. The ciliary movement, in conjunction with fluid movement and muscular contractions, aids in the following processes: (i) oviductal contents mixing, which increases the likelihood of male and female gametes encountering each other; (ii) denudation of follicular cells adhering to the ovum during ovulation; (iii) zygote transport into the uterus; and (iv) prevention of tubal implantation, or ectopic pregnancy. Oviductal peristalsis does not speed up but rather delays the entry of early embryos into the uterus.

21.10 THE UTERUS

All domestic animals have a uterus, which is made up of a body and two horns. The uterus is characterized as bipartite in cows, ewes, and mares, with the horns fully divided by a septum with a separate uterine body. This is substantial, particularly in the mare. The sow's uterus is bicornuate, with a short body connected to extremely long (4-5 feet) folded horns. Anatomically speaking, mammals that bear children have long uterine horns. Other mammal uteri types include simplex (found in primates), duplex (found in rabbits and guinea pigs), and diadelphous (found in platypuses). In animals with diadelphous uteri, the male has a penis that is forked and enters both reproductive openings at the same time. It's interesting to note that farm animals occasionally have primitive uterine structures like duplex or diadelphous, but these do not affect fertility.

The uterus is made up of three distinct layers: the inner endometrium, which is made up of epithelium, glands, and connective tissue; the outer serous membrane, which is an extension of the peritoneum; and the middle myometrium, which is made up of outer longitudinal, inner circular smooth muscles and a vascular layer in between (Figures 21.9 & 21.10).

All domestic animals have simple columnar non-ciliated luminal epithelium in their uteri. During estrus, these cells are tall and actively secretory; however, over the next two days, they become low cuboidal. Estrogens therefore stimulate the endometrial epithelium. The uterine lumen is where the coiled and branching tubular structures known as endometrial glands open. At estrus, they are straight, but as their secretory activity rises during the luteal phase of the cycle, they become more coiled and complex. The endometrial

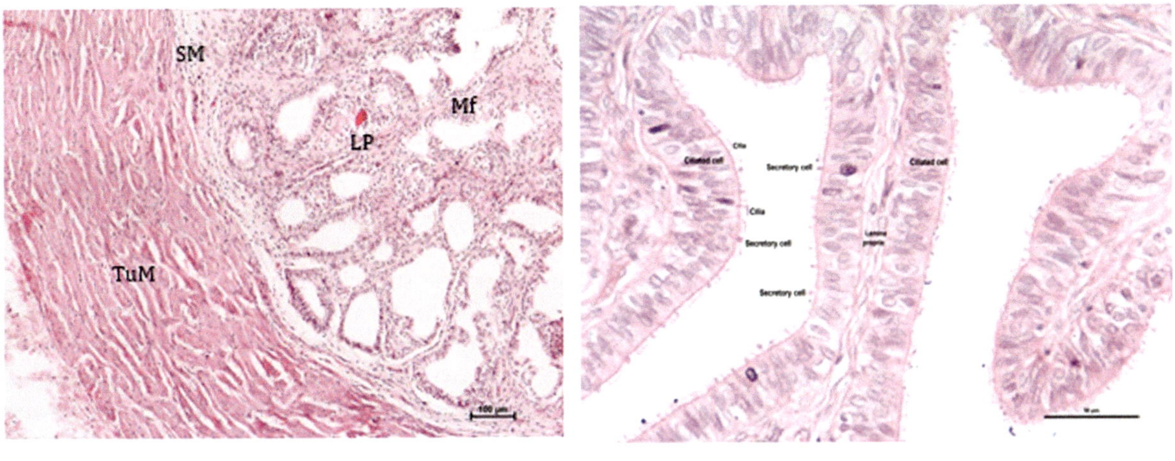

FIGURE 21.8 Photomicrograph showing various histologic layers of the oviduct (a) and the different cell types in the epithelium (b): Tunica muscularis (TuM), Submucosa (SM), Lamina propria (LP), and Mucosal folds (MF).

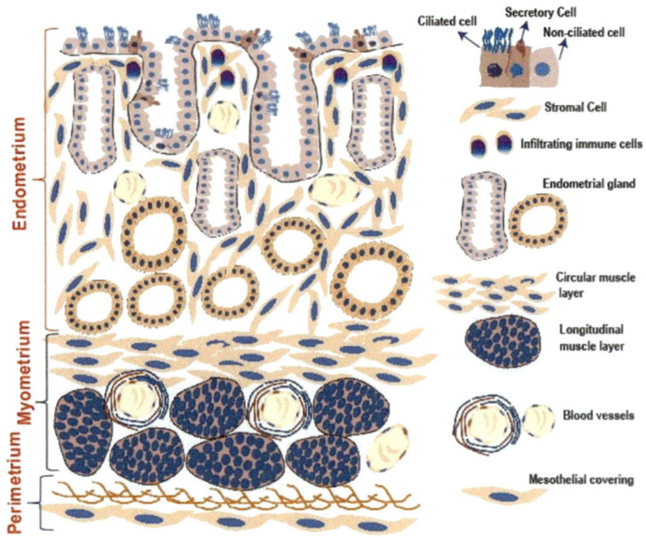

FIGURE 21.9 Schematic diagram of various histologic layers of the uterus.

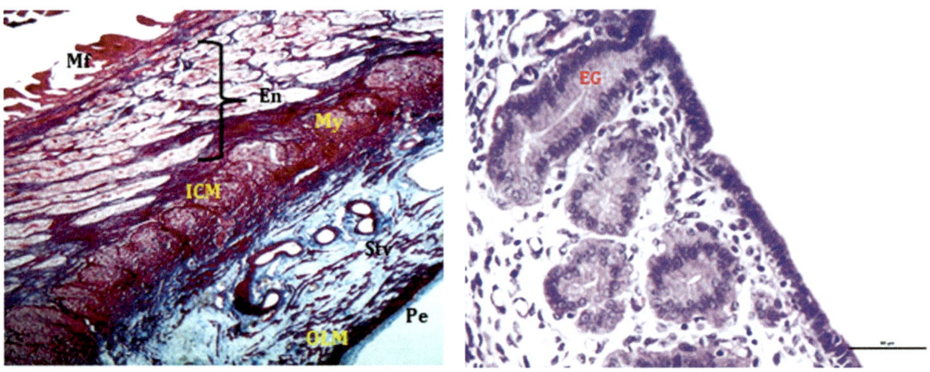

FIGURE 21.10 Photo micrograph showing various histologic layers of uterus (a) and uterine glands (b): Mucosal folds (Mf), Endometrium (En), Myometrium (My), Inner circular muscle layer (ICM), Outer longitudinal muscle layer (OLM), Stratum vasculosum (Stv), Perimetrium (Pe) & Endometrial glands (EG).

secretions' volume and composition change according to the stage of the reproductive cycle. Various critical proteins, including uteroferrin, glyceryl phosphoryl choline diesterase (GPC), blastokinin, and implantation promoting factor, are detected in endometrial secretions during distinct stages of the estrous cycle. The ruminant endometrium has specific places where the fetal membranes can attach. "Cotyledons" are the elevated structures that make up these sites. At the cotyledon, endometrial glands are absent.

The regular motions of the oviduct are timed with the contractions of the myometrium. Generally, during estrus, the contractions travel towards the ovary; later, they move towards the cervix. During various stages of the estrous cycle, uterine contractility is altered by ovarian steroid hormones. Progesterone has the opposite effect of estradiol, which causes uterine contractions to occur more frequently.

Sperm must be transported from the site of ejaculation to the site of fertilization by means of uterine contractions that take place during mating. Therefore, it is thought that rubbing the uterus gently at AI will increase fertility. Following estrus, the uterine contractions move the zygote into the uterus for implantation as they are directed towards the cervix. Elevated progesterone levels prevent uterine contractions during the course of pregnancy. This progesterone-induced inhibition disappears as parturition gets closer, and the strong uterine contractions that ensue are essential for the fetus's expulsion. During estrus, endometrial secretions cause sperm to become capacitated, which is a necessary stage of maturation for fertilization. They also speed up the metabolism of sperm. In this regard, GPC plays a crucial role in releasing glycerol from phosphoglycerol so that sperm can use it. Preimplantation and early post-implantation embryos receive nourishment from endometrial secretions during the post-estrous/luteal phase. In farm animals, endometrial secretions appear to play a significant role in the nutrition of early post-implantation embryos. For instance, cow embryos do not firmly adhere to the endometrium; instead, they live freely and multiply widely for up to 30 days following the estrus. The sensitivity of the myometrium and the proliferative response of the endometrial epithelium and glands to ovarian steroids are modulated by endometrial protein secretions, particularly those of the estrogen and progesterone receptors. A particular protein called blastokinin is found in the uterine secretions of rabbits, and it helps morula stage embryos form blastocysts. Uterine secretions in mice contain an implantation-promoting factor. It is also thought that uteroferrin, which is present in ewes' endometrial secretions, promotes embryonic development and implantation. Only the uterus could serve as a home for pregnancy. Ectopic pregnancy, or placement of the fetus elsewhere, is dangerous for the mother as well as the fetus. The uterus is uniquely adapted to experience profound alterations in dimensions, composition, and blood flow that are necessary to support the preimplantation embryo from the beginning of gestation to the end of parturition. The uterus is crucial to the upkeep or decline of CL. The uterus notifies the CL of the existence or lack of a conceptus by sending the proper signals. A functional CL lives longer after a hysterectomy, but a partial hysterectomy causes it to regress. In a similar vein, depending on when they are implanted, intrauterine devices may also lengthen or shorten CL's life. The utero-ovarian veno-arterial pathway is a local pathway that carries these signals.

21.11 THE CERVIX

The uterus and vagina are separated by the cervix, a structure with thick walls that resembles a sphincter. The cervix primarily consists of connective tissue, but its walls also contain some smooth muscles. The cervix's lumen is small and has annular rings. In cows and ewes, these annular rings are highly noticeable and form interlocking ridges. The corkscrew-shaped rings in the sow are broken by the spiral twisting in the boar's penis tip. Even though the vaginal end forms a well-constricted os uteri, the cervical rings are less noticeable in mares. Tall columnar cells and goblet cells that secrete mucus line the cervix's lumen.The cervical mucosa has a huge secretory surface because of its intricate folding. Glycoproteins make up cervical mucus, with 25% amino acids and 75% carbohydrates. Certain distinct biophysical characteristics of cervical mucus include ferning, elasticity, viscosity, and stickiness (tack), cervical mucus's physical and chemical characteristics, as well as how much of it changes throughout the estrous cycle. For instance, the macromolecules in cervical mucous are oriented to increase intermolecular space to 2.5–5 μm when estrogens are present, but progesterone during the luteal phase causes it to decrease. The cervix performs a number of vital tasks, including: (i) sperm transport; (ii) sperm reservoir; (iii) sperm selection center; and (iv) shielding the upper reproductive tract from harmful substances. In most animals, the sperm are oriented towards the cervix as they are ejaculated into the vagina close to it. Enzymes that hydrolyze backbone proteins and cross-links in mucin increase the sperm penetrability of the cervical mucous as they beat and travel towards the upper reproductive tract. Sperm transit channels between the cervical mucous and semen soon form. Many sperm lodge in the cervical crypts during their journey through the cervix, and they may then be released phasically. As a result, the cervix may serve as a sperm reservoir, preventing a large number of sperm from entering the uterus at once. Additionally, sperm may become more or less continuously available at the site of fertilization with such periodic release (over a period of about 24 hours). Contrarily, sperm that enter cervical crypts may never be released, so the cervix restricts entry into the upper reproductive tract to only the strongest sperm. During the luteal phase, the cervical mucous that allows sperm to enter the uterus during estrus transforms into a thick, viscous secretion that seals the cervix. By doing this, the upper reproductive tract is shielded from toxins and other infections. This seal becomes liquefied at parturition and offers protection during the entire gestation period. When parturition approaches, the cervix's connective tissue undergoes significant biochemical alterations that cause it to greatly expand in size, allowing the fetus to pass through.

21.12 THE VAGINA

The vagina is divided in to two parts i.e., vestibular, also known as the outer vagina, and posterior vagina (from the urethral opening to the cervix). Surface epithelium, muscular coat, and serosa make up the vaginal wall; mucosa is absent. There is a thick layer of circular fibers and a thin layer of longitudinal fibers in the less developed muscular coat of the vagina. There are plenty of blood vessels, nerves, and both dense and loose connective tissue within the muscular coat. In addition to the posterior sphincter present in all animals, cows also have an anterior sphincter in their vagina. With the exception of a few mucous-secreting cells in cows, the vaginal epithelium is squamous and lacks glands. Changes in the vaginal epithelium occur in synchrony with the estrous cycle. During the luteal phase, it is low cuboidal and stratified squamous during estrus. Only the rat and bitch could have their various estrous cycle phases accurately detected using a vaginal smear; other animals could not. IgA and IgG antibodies are secreted by plasma cells situated beneath the epithelium, along with antisperm antibodies. These cells proliferate during the luteal phase to offer defense against invasive organisms. The female copulatory organ called the vagina is where semen is deposited. It provides spermatozoa to the cervical reservoir by acting as a semen reservoir. Vaginal distension is made possible by the muscles in the walls and the tiny ridges on the vaginal surface, which aid in parturition and mating. Sperm transport and psychosexual satiety are significantly influenced by the muscular contractions of the vaginal wall. The vagina absorbs seminal plasma. When absorbed in the vagina, certain components of seminal plasma affect the activity of other reproductive tract organs. The cervix and uterus' excretory duct, is the vagina. The majority of the mucus secreted through the vagina – erroneously referred to as vaginal mucous – is generated within the cervix. The vagina receives fluids from the vascular supply even though it lacks secretory glands. These fluids, in conjunction with seminal plasma and cervical mucous, form a buffer system that adjusts the vagina's normally unfavorable pH to promote sperm survival until they are transported. Insufficient cervical mucous, little vaginal fluid, or low ejaculate volume all result in insufficient buffering, which negatively impacts sperm survival. Bulls and dogs can detect the distinct scent of vaginal secretions in estrous cows. Sperm could be produced by the vagina, at least in rabbits. During parturition, the fetus passes through this portion of the birth canal.

21.13 EXTERNAL GENITALIA

External genitalia include the vestibular portion of the outer vagina, the vestibular glands, the labia majora, minora, and clitoris. A structure known as the hymen may occasionally be seen where the outer and posterior vaginas meet. When an animal begins to cycle, the hymen normally dissolves on its own, but in rare circumstances, it might need to be surgically removed to allow for copulation. The embryological counterpart of the penis, the clitoris, has some erectile tissue. It is deeply buried in the vestibular mucosa of the cow, but it is highly visible in the mare. It is crucial for psychosexual satiety and contains a lot of nerve endings. Although the labia are similar to the scrotum, animals' labia minora are not as well developed. Sebaceous and tubular glands abound in the integument of the labia majora. They have a thin layer of smooth muscles, elastic tissue, fat deposits, and some hair on the outside. During estrus, labia majora become enlarged, clogged, and edematous. Breeding management could benefit from these morphological changes. Similar to male bulbo-urethral glands, sebaceous and Bartholin glands are located in the vestibule and secrete a lubricating mucus to enhance the copulatory process.

BIBLIOGRAPHY

Broi, M. G., V. S. I. Giorgi, F. Wang, D. L. Keefe, D. Albertini, and P. A. Navarro. 2018. Influence of follicular fluid and cumulus cells on oocyte quality: Clinical implications. *Journal of Assisted Reproduction and Genetics* 35 (5): 735–751.

Erickson, B. 1966. Development and radio-response of the prenatal bovine ovary. *Journal of Reproduction and Fertility* 10: 97–105.

Gordon, I. 2003. *Laboratory production of cattle embryos*. CABI Publishing, pp. 112–157. https://www.cabidigitallibrary.org/doi/book/10.1079/9780851996660.0000

Klisch, K., and A. Mess. 2007. Evolutionary differentiation of Cetartiodactyl placentae in the light of the viviparity-driven conflict hypothesis. *Placenta* 28 (4): 353–360. doi: 10.1016/j.placenta.2006.03.014.

Morel, M. 2015. *Equine reproductive physiology, breeding and stud management*, 4th edition. CAB International.

Senger, P. L. 2012. *Pathways to pregnancy and parturition*, 3rd edition. Current Conceptions.

Suarez, S. S., and M. F. Wolfner. 2021. Cilia take the egg on a magic carpet ride. *Proceedings of the National Academy of Sciences (PNAS) USA* 118, no. 27 (July 6): e2108887118. doi: 10.1073/pnas.2108887118. PMID: 34162738; PMCID: PMC8271566.

Walt, M. L., G. H. Stabenfeldt, J. P. Hughes, D. P. Neely, and R. Bradbury. 1979. Development of the equine ovary and ovulation fossa. *Journal of Reproduction and Fertility* 27: 471–477.

22 Physiology of the Lactation System

Mohamed Hasif G[1], *Manju G Preedaa*[2], *Nikhil Kumar Tej*[3], *Manish Kumar*[4], *Mukesh Kumar Bharti*[4], *Harivarman Segar*[5] *and Thulasiraman Parkunan*[*4]

[1]Department of Veterinary Physiology, Institute of Veterinary Science and Animal Husbandry, Siksha O Anushandan University, Bhubaneshwar, Odissa, India.
[2]Department of Veterinary Physiology and Biochemistry, Veterinary College and Research Institute, Theni, Tamil Nadu Veterinary and Animal Sciences University (TANUVAS), Tamil Nadu, India
[3]Centre for Translational Research, Rajiv Gandhi Institute of Veterinary Education and Research, Kurumbapet, Puducherry, India
[4]Department of Veterinary Physiology and Biochemistry, Faculty of Veterinary and Animal Sciences, Institute of Agricultural Sciences, Rajiv Gandhi South Campus, Barkachha, Banaras Hindu University, Uttar Pradesh, India
[5]Department of Gynaecology and Obstetrics, Veterinary College and Research Institute, Orathnadu, Tamil Nadu Veterinary and Animal Sciences University (TANUVAS), Tamil Nadu, India

22.1 EXTERNAL ANATOMY OF THE MAMMARY GLAND

The mammary glands, or mammae, are specialized organs derived from sweat glands that play a crucial role in nurturing offspring through the production of milk and colostrum. These glands are present in both sexes, but their functionality is primarily evident in females, where hormonal influences during pregnancy and lactation stimulate the development and activity of the glandular structures. Structurally, the mammary gland comprises the udder, teat or nipple, and an intricate network of ducts and alveoli lined with epithelial cells responsible for milk secretion. The gland is encapsulated within fibroelastic tissue that provides support and is interspersed with adipose tissue.

Hormonal regulation is key to the gland's function; it dictates the balance between the secretory tissue and connective tissue, with the former expanding significantly during lactation. Post-lactation, the gland undergoes involution, where the secretory cells diminish, and the connective tissue becomes more prominent. Evolutionarily, the distribution and number of mammary glands vary across species, often correlating with litter size. For instance, carnivores and sows, which typically have multiple offspring, possess a more extensive range of mammary glands, while ungulates like horses and cows have fewer glands, reflecting their tendency to bear single offspring (Table 22.1). This variation underscores the mammary gland's adaptability and its integral role in mammalian reproductive strategies.

The mammary gland anatomy in ruminants and horses exhibits a complex yet systematic structure where the individual glands are closely associated, often collectively referred to as an udder. This association is evident through the presence of a single teat per gland, which facilitates milk discharge via a distinct duct system. The development of these mammary glands is an intricate embryological process, originating from bilateral thickenings known as mammary ridges on the ventrolateral ectoderm of the embryo. These ridges invaginate to form the duct system that serves individual glands, each associated with a teat.

Interestingly, during the developmental phase, multiple mammary buds may form, with the surplus typically regressing quickly. However, some may persist, leading to the formation of supernumerary teats that are generally non-functional and can hinder milking practices; thus, they are often removed in domesticated animals. The dairy cow's udder has undergone significant selective breeding, enhancing its capacity to produce volumes of milk well beyond the basic nutritional needs of its offspring, a testament to both natural and artificial selection pressures in agricultural practices.

ARE YOU AWARE?

Marine mammals have developed remarkable adaptations to thrive in the aquatic environment, particularly in the way they nurse their young. The Carnivora order, which includes seals and sea lions, features retractable nipples that allow for efficient feeding without compromising the streamlined body necessary for swimming. Cetaceans, such as dolphins and whales, possess mammary glands situated in the inguinal region, concealed by mammary slits to maintain hydrodynamic form. It is hypothesized that calves may utilize their tongue's dexterity or that the mother's specialized muscles contract to facilitate

DOI: 10.1201/9781003426851-22

TABLE 22.1
Locations of Mammary Glands and Number of Openings in Teats Across Different Species

Species	Location of Mammary Gland	Number of openings in teats/teat pores
Goat, Sheep	Inguinal	2 glands & 2 teats (1 opening/teat)
Mare	Inguinal	2 glands & 2 teats (2 opening/teat)
Cow	Inguinal	4 glands & 4 teats (1 opening/teat)
Buffalo	Inguinal	4 glands & 4 teats (1 opening/teat)
Sow	Thorax, abdomen & inguinal	12 to 14 teats (2 Inguinal pairs, 3 abdominal pairs, 2 thoracic pairs) – 2/3 opening/teat
Cat	Thorax, & abdomen	8 teats (1 inguinal pair, 1 abdominal, 2 thoracic pairs) – 3–7 teat openings
Bitch	Thorax, abdomen & inguinal	10 teats (1 inguinal pair, 2 abdominal pairs, 2 thoracic pairs) & 8–14 teat opening
Elephants	Thoracic	2 glands & 2 teats (10–11 teat opening)
Red Kangaroo	Abdominal	4 glands & 15 openings/teat
Opossum	Abdominal	13 glands & 8 openings/teat
Rat & Mouse	Thoracic, Abdominal & Inguinal	10 glands (4 thoracic, 2 abdominal, 4 inguinal) & 1 opening/teat
Rabbit	Thoracic, Abdominal & Inguinal	10 glands (4 thoracic, 4 abdominal, 2 inguinal) & 6–10 opening/teat

Modified and inspired from Akers RM (2002), Nickerson and Akers (2011) & Mukherjee et al. (2023); Singh and Roy (2017).

milk transfer directly into the offspring's mouth. The mammary glands of these creatures are substantial, weighing approximately 113 kilograms and extending nearly 5 feet, with the capacity to produce about 200 liters of milk daily, rich in fat content ranging from 35-50%. This high-fat milk is crucial for rapid growth and insulation in the cold marine environment. These evolutionary innovations underscore the diverse strategies marine mammals employ to nurture their young while navigating the challenges of their aquatic habitats (Mukherjee et al., 2023).

22.2 MAMMARY GLAND

The udder of the cow comprises four individual glands, referred to as quarters. The skin of the udder is covered with fine hair; however, the teat is completely hairless. The right and left halves of the udder each consist of a cranial (front) quarter and a caudal (hind) quarter. Each side of the udder operates with a degree of autonomy due to its distinct blood supply, nerve supply, and suspensory apparatus. Ventrally, the two halves of the udder are demarcated by a longitudinal furrow, the intermammary groove, which corresponds to a median septum of connective tissue dividing left and right halves. Because of the relative isolation of each side, half of the udder can be removed surgically without damaging the other half, as might be done to treat a tumor. The two quarters in each half are separate from one another as far as the gland tissue and duct system are concerned. Thus, all the milk from one teat is produced by the glandular tissue of that quarter. The vasculature, nerve supply, and lymphatic drainage, however, are common to both quarters of a given half. The parenchyma of the lactating mammary gland consists of secretory tissue and the ducts of the gland. The secretory units, the alveoli, are lined by a simple epithelium that varies from columnar to cuboidal in height. The alveoli are the chief structures for actual milk production, although the initial portion of the associated duct is also lined with secretory epithelium.

The various small initial ducts converge to form larger ducts, and these converge to form yet larger ones, all of which eventually terminate in a large single basin, the lactiferous sinus. The lactiferous sinus is sometimes described as being divided into a large cavity within the quarter itself, the pars glandularis (gland cistern), and a smaller cavity within the associated teat called the pars papillaris (teat cistern). The demarcation between the gland cistern and teat cistern is frequently marked by a circular ridge (annulus) that contains a vein and some smooth muscle fibers. The wall of the empty cistern contains numerous overlapping longitudinal and circular folds that are obliterated through expansion of the wall when it is full of milk. There may also be diverticula (pockets) in the wall of the gland cistern. The teat cistern is continuous with the exterior of the teat through a narrow opening at the end of the teat, the papillary duct (commonly called streak canal or teat canal), which opens at the ostium papillae. The bovine streak canal is about 8.5 mm long, and its lumen is normally closed by epithelial folds that project inward from the wall of the streak canal, leaving only a star-shaped potential opening. A sphincter of smooth muscle fibers surrounds the streak canal at the distal end of the teat.

22.3 SUSPENSORY APPARATUS

The udder of a lactating dairy cow can weigh up to 60kg, and it is supported by a dense system of fibroelastic ligaments known as the suspensory apparatus. This apparatus is primarily comprised of two medial laminae, originating from the abdominal wall's linea alba and the pelvis's symphysis. These laminae pass between the two halves of the udder, with one layer covering each half closely. While proximal regions are thicker, they thin out near the intermammary groove as they diverge into the udder parenchyma. The lateral laminae, composed mostly of dense white fibrous connective tissue, originate from the body wall's aponeurotic tissues and meet the medial laminae at both ends of each udder half. They also thin out ventrally, releasing connective tissue sheets into the gland's substance.

The udder of a lactating dairy cow can weigh as much as 60kg, so the organ is supported by a dense system of fibroelastic ligaments called the suspensory apparatus. The primary supportive elements of the suspensory apparatus are its two medial laminae, which take their origin together from the linea alba of the abdominal wall and the symphysis of the pelvis. Each medial lamina passes ventrad between the two halves of the udder so that one layer intimately covers the medial side of each half. The two medial laminae can be readily separated, as they are united only by a small amount of loose areolar connective tissue; practically no vessels or nerves pass through the medial ligament from one half of the udder to the other. Proximally (close to the body wall), the laminae are thickest. As they descend, they give off sheets of connective tissue that diverge from the midline and interdigitate into the parenchyma of the udder so that the two medial laminae are thinnest near the intermammary groove.

The lateral laminae of the suspensory apparatus are composed largely of dense white fibrous connective tissue, making them less elastic than the medial laminae, which are mostly elastic connective tissue. The cranial part of the lateral lamina derives from the aponeurotic tissues of the body wall near the external inguinal ring and more caudally from the regions of the pelvic symphysis and prepubic tendon (the tendon of insertion of the m. rectus abdominis). From its origins, the lateral lamina passes ventrad and around the lateral side of each half of the mammary gland, meeting the medial lamina at the cranial and caudal aspects of each half. Like the medial lamina, the lateral lamina is thick close to the body wall and thin progressively ventrally as it gives off sheets of connective tissue into the substance of the gland.

22.4 BLOOD SUPPLY

The vascular system of the udder is a complex network that ensures the delivery of nutrients and the removal of waste products, vital for the maintenance and production of milk. A modestly performing dairy cow is expected to generate at least 500 L of blood through its udder for each liter of milk. This corresponds to around 10% of cardiac output during nursing. The caudal aorta provides the blood supply for the cow's udder. It then splits into the right and left iliac arteries and flows bilaterally to supply the udder's left and right parts. The iliac arteries bifurcate into external and internal iliac arteries. The perineal artery, a branch of the internal iliac artery, supplies blood to the caudal udder.

The majority of blood supply to the mammary gland comes from the external iliac artery, which transforms into the prepubic artery until it passes through the internal inguinal ring, where it becomes the external pudendal artery and enters the mammary gland as the mammary artery. The external pudendal artery, branching from the pudendoepigastric trunk, plays a crucial role in supplying blood to the udder (Figure 22.1). Its division into cranial and caudal branches allows for distributed blood flow to both the anterior and posterior parts of the udder on the same side of the artery. The mammary artery further divides into cranial and caudal mammary arteries, which supply different areas of the udder. The right and left mammary arteries interconnect behind the medial suspensory ligament, and collateral circulation potential exists via the perineal artery connecting with the caudal mammary artery. Branches of the mammary arteries form a plexus above the teat, supplying the teat walls.

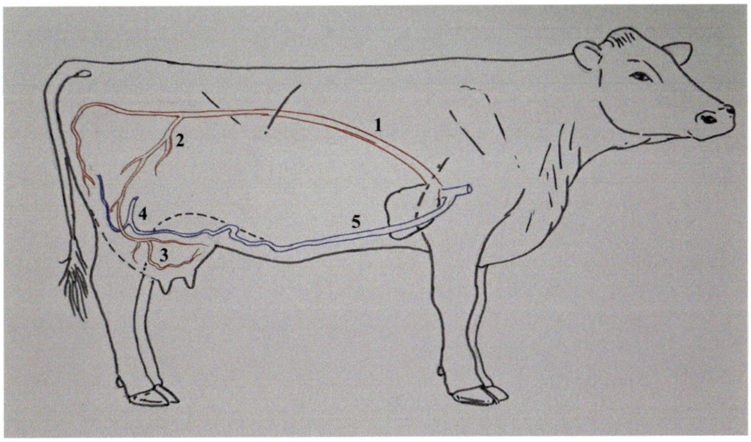

FIGURE 22.1 Vascular supply to and from the mammary Gland. 1-caudal artery; 2-external iliac & pudic artery; 3-cranial & caudal mammary artery; 4-external iliac & pudic vein; 5-subcutaneous abdominal vein.

The venous drainage from the udder is largely by way of a venous circle at the base of the udder, where it attaches to the abdominal wall. This venous circle is formed from the main veins that drain the udder. The external pudendal vein of each side receives blood from both the cranial and caudal quarters of the same side. Cranially, each external pudendal vein is continuous with the caudal superficial epigastric vein and caudally with the perineal vein. An anastomosis between the two caudal superficial epigastric veins just at or in front of the udder completes the venous circle.

The caudal superficial epigastric vein passes forward in a sagittal plane lateral to the midline on the ventral abdominal wall and joins with the cranial superficial epigastric vein, which ultimately drains to the internal thoracic veins and then to the cranial vena cava. Before the heifer comes into milk, the connection between cranial and caudal superficial epigastrics is poorly developed. During the first pregnancy, when the udder undergoes a marked increase in size and, consequently, blood supply, the two veins develop a functional anastomosis, after which they collectively constitute the subcutaneous abdominal vein or the milk vein. In high-producing dairy cows, sheep, and goats, the subcutaneous abdominal vein is large and tortuous. It passes through a foramen in the rectus abdominis muscle (the milk well), joins the internal thoracic vein, and ultimately drains into the cranial vena cava.

22.5 NERVE INNERVATION

The mammary gland is innervated by both sensory and motor nerves. Sensory nerves, while not directly connected to alveolar cells or milk secretion, play a crucial role in milk ejection. The sensory nerve supply to the udder originates from ventral branches of the first through fourth lumbar nerves. These nerves unite to form the inguinal nerve, which divides into cranial and caudal branches. The cranial branch innervates the cranial part of the front quarter, including the teat, while the caudal branch supplies the glandular tissue and rear quarter. The skin of the teat is rich in sensory receptors, particularly near the teat sphincter, stimulating sensory nerves essential for milk letdown via oxytocin release.

Motor nerve supply primarily consists of fibers from the sympathetic system, regulating vasoconstriction and muscular constriction of the teat sphincter. Despite the cutaneous origin of the mammary gland, there is no parasympathetic nerve innervation. This complex innervation pattern complicates local anesthesia but simplifies teat surgery due to the vertical orientation of teat nerves within the teat wall.

22.6 LYMPHATIC VESSELS

During lactation, the udder experiences a significant increase in blood flow, resulting in the production of a substantial volume of lymph, approximately 1.6 times the volume of milk produced. This lymph travels through a complex network of vessels to the supramammary lymph nodes, located in the mammary fat pad above the caudal glands (Figure 22.2). While some lymph may pass through the prefemoral lymph nodes, the primary drainage route is through the supramammary lymph nodes. Efferent vessels exit these nodes, travel through the inguinal ring into the abdomen, and may pass through the deep inguinal, external iliac, or pre-temporal lymph nodes before entering systemic circulation. The lymphatic vessels draining the udder are visible superficially just under the skin, especially in high-producing cattle.

In cases of udder edema, which can occur in prepartum cattle or those with mastitis, the occlusion of lymph vessels due to increased pressure or loosening of tight junctions between epithelial cells can lead to the accumulation of lymph and milk components in the interstitial space, reducing lymph clearance. In severe cases, the deep inguinal node may be palpable per rectum in cows.

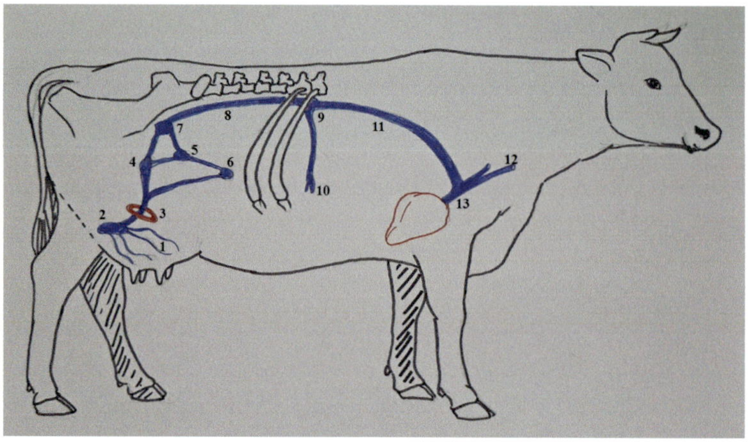

FIGURE 22.2 Diagram showing lymphatic vasculature of the mammary gland. Lymphatic vessels of the mammary gland (1); supramammary lymph gland (2); inguinal ring (3); deep inguinal lymph gland (4); external iliac lymph gland (5); prefemoral lymph gland (6); internal iliac lymph gland (7); lumbar lymph trunk (8); cisterna chyli (9); intestinal lacteals (10); thoracic lymph duct (11); jugular vein (12); anterior vena cava (13).

22.7 THE MICROSCOPIC STRUCTURE OF THE MAMMARY GLAND

The mammary gland is classified as a compound tubuloalveolar gland. It consists of connective tissue interstitium, parenchyma (secretory epithelium), ducts, vessels, and nerves. In the dry (non-lactating) state, the gland has proportionately more stroma than parenchyma; during lactation, the parenchyma undergoes marked growth and constitutes the bulk of the gland. The surface of the bovine and porcine teat is covered with glabrous (hairless), aglandular stratified squamous epithelium. The teats of other domestic species have more typical haired and glandular skin. This is continuous with the stratified squamous epithelium that lines the papillary duct (streak canal). The duct is encircled by smooth muscle fibers that act as a sphincter. At the junction of the papillary duct and the teat cistern, the epithelial lining changes abruptly from stratified squamous to stratified columnar epithelium that is usually two cells thick. This stratified columnar epithelium lines the teat and gland cisterns and the larger lactiferous ducts. As the ducts branch and become smaller, the epithelial lining changes first to simple columnar and then to secretory epithelium in the alveoli. The height of the alveolar epithelium varies considerably with the level of activity of the gland.

The mammary gland differs from most other exocrine glands in that the secretory portion is not limited to the terminations of the smallest ducts; milk-secreting structures also empty directly into the larger ducts and directly into the gland cistern and the teat cistern. A group of alveoli surrounded by a connective tissue septum forms a more or less distinct unit called a lobule (Figure 22.3). A group of lobules within a connective tissue compartment forms a lobe. Correspondingly, the ducts are classified as intralobular, interlobular, intralobar, and interlobar as they increase in size. The alveoli making up the lobule empty into small ducts within the lobule, the intralobular ducts. These intralobular ducts drain into a central collecting space from which the interlobular ducts emerge.

Within the lobe, the interlobular ducts unite to form a single intralobar duct, which is called an interlobar duct as soon as it emerges from the lobe. The interlobar duct may enter the gland cistern directly, or it may join one or more other interlobar ducts before entering the gland cistern. Many of the ducts have numerous dilations that act, in addition to the lactiferous sinus, as collecting spaces for milk. The alveoli and ducts are surrounded by contractile myoepithelial cells, which are also called basket cells. These cells contract to eject milk (called milk letdown) in response to oxytocin release.

In addition to the epithelial parenchyma and the myoepithelial cells, the mammary gland is made up of an interstitium of white fibrous connective tissue and yellow elastic connective tissue. Blood vessels, lymph vessels, and nerves ramify throughout the interstitium to reach the epithelial structures. The veins of the mammary gland are valveless and form a rich network throughout the gland and within the wall of the teat. The vascular layer of the teat is called a corpus cavernosum because of its resemblance to the erectile tissue of the penis in the male. Lymphatic plexuses are found throughout the udder just deep to the skin and scattered throughout the parenchyma of the gland. Nerves appear to be primarily sensory and vasomotor.

22.8 INTERNAL ANATOMY

The tissue that makes up the udder is pink and white. White tissues, or non-parenchymal cells, are composed of connective, fibrous, and fatty tissue, whereas pink tissue is known as parenchyma or epithelial tissue. The parts of the udder's internal structure are as follows:

Teat Pore: During milking, the milk is drained out of the teat through this small orifice. The muscle known as the teat sphincter keeps it firmly covered. Between milkings, the muscles of the teat sphincter prevent milk from leaking.

The teat contains a portion of the canal called the "fat streak" that opens up in the teat cistern. There are unique folds at the intersection of the teat cistern and teat canal that stop milk flow against gravity. These folds are called

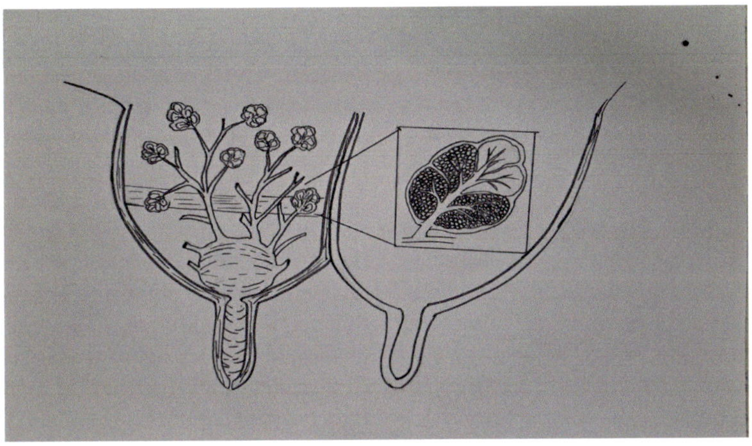

FIGURE 22.3 Diagram showing the internal structure of the mammary gland.

Furstenberg's Rosettes. One of the most significant defenses against microorganisms entering the udder is the teat canal. The teat canal ranges from 8 to 12 mm in length, and depending on the teat's size, it is capable of holding 15 to 40 ml of milk. Slow milkers are cows with narrow teat canals or tight sphincter muscles, which make them take longer to milk. Large teat canals or weak sphincter muscle animals leak milk between milkings and are more susceptible to pathogens that cause mastitis. Additionally, the smooth muscles in the milk ducts and the vascular system are related to the bladder by nerves. Every teat has its own cistern for retaining milk, holding around 400-500 ml between milkings.

The gland cistern is partly divided by annular folds of tissue and is situated directly above the teat cistern. Milk is carried into the gland cistern via the giant milk ducts, which are opened by the small ducts. Each alveolus produces milk, which is transported by the tiny ducts. The udder's ductal system is made up of several varying-sized ducts arranged in the following order: lobar ducts, intralobar duct, interlobular duct, and intralobular ducts. The main function of the duct system is to collect milk from the secretory tissue, store a part of the milk between milkings, and transport the milk into the gland cistern. Constrictions are placed at the tip of both large and tiny ducts to stop milk from dripping and leaking into the teat.

Lobes and Lobules: Each mammary gland comprises several lobes, and each lobe consists of numerous lobules. Layers of connective tissue divide lobes and lobules. A lobule can contain between 200 and 250 alveoli. Each alveolus is composed of a single layer of epithelial cells, responsible for producing milk, which is then discharged into the alveolar lumen.

The spherical alveolus of the udder is the functional unit responsible for milk production. It is lined by a single layer of epithelial cells, and a small duct called the alveolar duct drains the milk.

The outermost layer of the alveolus also contains muscle-like cells called myoepithelial cells, which contain receptors that bind to the hormone oxytocin, necessary for milk secretion. When oxytocin binds to these receptors, the myoepithelial cells contract, releasing milk into the alveolar lumen. The contraction of myoepithelial cells can increase intramural pressure to 60 mmHg or more, practically stopping milk secretion. Each quarter's primary lobe is formed by a lobule, which consists of several alveoli. There are approximately 5 x 10^12 secretory cells in the udder's alveolar tissue of dairy cows.

The mammary gland of a cow or buffalo is composed of two components: connective tissue, which is non-parenchymal, and pink secretory parenchymal tissue. The amount of milk produced by animals is determined by the number of secretory tissues or secreting cells. Thus, the rate-limiting component in the udder's ability to produce milk is parenchyma. Contrary to popular assumption, a large udder does not necessarily indicate a great ability for milk production. There might not be as much space to accommodate milk-secreting tissue in an udder with a large volume of connective and adipose tissue. After milking, some udders shrink significantly and become flabby due to having less connective tissue. Milk is generated by the secretory cells, sometimes referred to as epithelial cells, which are organized in a spherical shape known as alveoli, forming a single layer atop a basal membrane.

22.9 MAMMARY GLAND IMMUNITY

The secretions of the mammary gland create an environment that can support bacterial growth if bacteria manage to enter the gland. The immune function in the mammary gland differs from other parts of the body and may be compromised at various stages of lactation. Most new infections in the mammary gland occur through the streak canal, so the primary defense of the immune system in the mammary gland relies on anatomical and physiological barriers. The smooth muscle of the teat sphincter closes the entrance to the streak canal, preventing bacterial entry. However, there are instances, such as immediately after machine milking or during decreased muscular activity due to periparturient paresis, when sphincter activity decreases, allowing easier access for microorganisms.

Similar to other body parts, the skin lining the teats and udder, as well as the streak canal, act as physical and chemical barriers against bacterial invasion. These areas have a natural bacterial flora that inhibits the colonization of bacterial pathogens. Additionally, the surface of the streak canal is lined with keratin, which contains antibacterial lipid and protein compounds. Keratin binds bacteria to its surface and is shed during milk secretion, carrying away bacteria. Furthermore, keratin desiccation can lead to the formation of a plug in the streak canal during non-lactating periods, blocking bacterial entry. Many of these primary anatomical and physiological defense mechanisms are genetically determined, including udder and teat characteristics such as size, shape, length, end shape, placement, udder depth, teat sphincter function, and keratin production.

22.10 CELLULAR DEFENSE

The bovine udder possesses a complex and efficient immune system geared towards safeguarding against bacterial invaders and maintaining udder health. Serving as the frontline defense, the teat of the udder acts as a physical barrier, preventing microbial entry. Within the mammary gland, mammary epithelial cells (MECs) play a crucial role in the synthesis of antibacterial proteins, bolstering the innate immune response with immune cells like macrophages, dendritic cells, and lymphocytes. Complementing this defense, leukocytes, including neutrophils and macrophages, are mobilized to produce inflammation mediators, phagocytize bacteria, and recognize antigenic structures, thereby enhancing the immune defense within the mammary gland. Humoral immunity is further fortified by the presence of key immunoglobulins, such as IgG1, IgG2, IgM, and IgA, which combat bacterial pathogens. Additionally, the complement system, comprised of various proteins, reinforces the inflammatory process, fortifying the immune response against microbial invaders. Further, different pathogens trigger distinct responses in MECs based on the activation of various pattern recognition receptors (PRRs), leading to cytokine production, antimicrobial petites expression, and modulation of host's immune response. Chemical antibacterial agents and humoral defense components like lactoferrin, lysozyme, and antimicrobial peptides actively contribute to shielding the bovine mammary gland from infection. Understanding these intricate immune mechanisms is crucial for devising effective strategies to uphold udder health and prevent infections in bovine populations.

22.11 GENESIS OF MAMMARY GLAND (MAMMOGENESIS)

Successful lactation is vital for the completion of the reproductive cycle in many mammalian species, with the processes of reproduction and lactation intricately linked. The journey begins with coordinated growth processes that commence in utero and extend through the delivery of the first offspring and into early lactation. This growth and development of the mammary gland, known as **mammogenesis**, lays the foundation for milk production. Lactogenesis marks the onset of lactation, during which alveolar cells undergo differentiation into milk-producing and -secreting cells. **Galactopoiesis** encompasses the maintenance and potential enhancement of established lactation. Following the cessation of milking, involution occurs, involving anatomical and physiological changes in the mammary gland, both immediate and long-term. These lactation processes, from **mammogenesis** to **galactopoiesis** and involution, are intricately regulated by hormonal interactions with the mammary gland and exchanges with other body tissues. Key endocrine glands involved in successful lactation include the hypothalamus, pituitary, adrenal glands, ovaries, and placenta, while other endocrine glands also exert influence.

22.11.1 Pre-natal Development

The initial formation of the mammary band is observed, followed by the appearance of a streak within the band. The transformation of this streak into the mammary line is noted. After a certain stage, the line progresses into a mammary crest, which eventually develops into two or three layers. Subsequently, the mammary crest forms a hillock. The hillock then gives rise to a small bud, which emerges in the middle. A solid core of epithelial cells forms the primary sprout. The secondary sprout is characterized by the canalization of the teat area, as depicted in the figure. The teat pores are formed during the tertiary sprout stage. The proliferation and eversion of epithelial cells demonstrate notable differentiation from the bud stage to the secondary sprouts, as illustrated in the Figure 22.4.

22.11.2 Birth to Puberty

During the initial stage of development, from birth to puberty, young animals possess well-formed teats, although their glandular tissue remains rudimentary. Throughout the prepubertal period, which extends from birth until sexual maturity, mammary growth follows a pattern closely aligned with overall body growth. This phase is primarily characterized by the elongation of ducts and the maturation of stromal tissue, pivotal for future mammary function. Notably, the nutritional intake during this phase emerges as a critical determinant, as excessive feeding may lead to undesirable fat accumulation, potentially impeding the development of mammary tissue.

22.11.3 Puberty to Conception

As animals transition from puberty to conception, mammary growth undergoes a notable shift towards a more accelerated trajectory, marked by a rate surpassing overall body growth. This phase witnesses significant advancements in parenchymal growth, chiefly characterized by the elongation and branching of ducts, accompanied by concurrent stromal enhancement, particularly in blood vessel

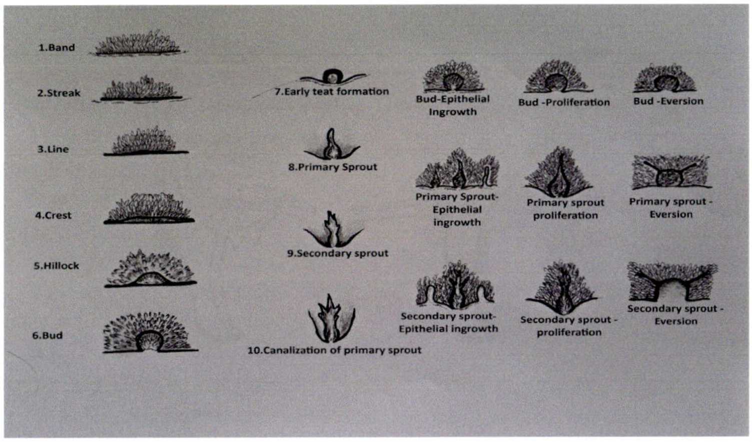

FIGURE 22.4 Diagram showing different stage of mammary gland development.

formation. Estrogen assumes a pivotal role during this period, fostering heightened mammary growth, particularly in females. The intricate interplay between estrogen and other regulatory hormones influences the branching patterns of mammary ducts, laying the groundwork for future milk production.

22.11.4 Conception to Parturition

The period spanning from conception to parturition signifies a crucial phase in mammary development, where factors associated with the maternal-conceptus axis exert profound influences on growth. This period witnesses exponential mammary expansion, with a substantial portion occurring during the latter stages of pregnancy. Hormones such as estrogen, progesterone, prolactin, growth hormone, and placental lactogen orchestrate mammary development, facilitating the formation of extensive lobuloalveolar structures. Alveolar cells undergo proliferation and differentiation under the influence of these hormones, priming the mammary glands for milk secretion postpartum. The orchestrated hormonal profile, coupled with adequate nutritional support, is imperative for optimal mammary growth and subsequent lactation.

22.11.5 Hormonal Influence of Mammary Gland Development

Role of different hormones and their function in mammary gland is enlisted in Table 22.2. Follicle Stimulating Hormone (FSH) and Luteinizing Hormone (LH), produced by the adenohypophysis, regulate ovarian hormones such as estrogen and progesterone, influencing their effects on mammary gland development. Additionally, placental lactogen serves as a mammotrophic hormone.

The adenohypophysis exerts direct influence on the mammary gland through prolactin and somatotropic hormone, and indirectly through its control over thyroid and adrenal cortex hormones.

22.11.6 Growth Factors of Mammary Gland Development

Various unidentified growth factors have been proposed for both normal and cancerous mammary cells based on activities observed in tissue extracts and conditioned medium. These factors originate from non-mammary tissues, such as the pituitary and liver (endocrine), as well as from the mammary gland itself (paracrine or autocrine), associated with mammary mesenchyme or epithelium respectively. Until these factors are fully characterized, whether they function as cell nutrients or polypeptide growth factors, activating DNA synthesis in dormant cells via high-affinity receptors, remains unclear. Examples of tissue-derived mitogens identified as cell nutrients include phospho-ethanolamine from bovine pituitary and transferrin from pig pituitaries. Studies have demonstrated proliferation and improved epithelial morphology in mouse mammary epithelium cultures grown in or on collagen when exposed to adipose tissue-conditioned medium, suggesting a potential role for unsaturated fatty acids. In addition, Insulin-like Growth Factors (IGF) and Epidermal Growth Factor (EGF) play significant roles. .

22.11.7 Fat Pad and Mammary Gland Significance

The fat pad in the mammary gland serves several important functions in mammary physiology and development:

- **Structural Support:** The fat pad provides structural support to the mammary gland. It helps maintain the shape and integrity of the glandular tissue, ensuring proper positioning and function of the mammary epithelial cells.

TABLE 22.2
Function of Different Endocrine Secretions in Mammary Gland

Endocrine component (organ)	Function in mammary gland
FSH (Anterior pituitary)	Estrogen secretion from ovarian follicles
Prolactin (Anterior pituitary)	Mammary growth; initiation and maintenance of lactation
GH (Anterior pituitary)	Stimulates milk production
Oxytocin (Posterior pituitary)	Milk ejection
T3, T4 (Thyroid)	Stimulates O2 consumption, protein synthesis and milk yield
Calcitonin (Thyroid)	Calcium and phosphorus metabolism
Glucocorticoids (Adrenal cortex)	Initiation and maintenance of lactation
Epinephrine, norepinephrine (Adrenal medulla)	Inhibition of milk ejection
Estradiol (ovary)	Mammary duct growth
Progesterone (ovary)	Mammary alveolar – lobule growth, inhibition of lactogenesis
Estradiol (Placenta)	Mammary duct growth
Progesterone (Placenta)	Mammary alveolar – lobule growth, inhibition of lactogenesis
Placental lactogen (Placenta)	Mammary growth

Modified and inspired from Singh and Roy (2017).

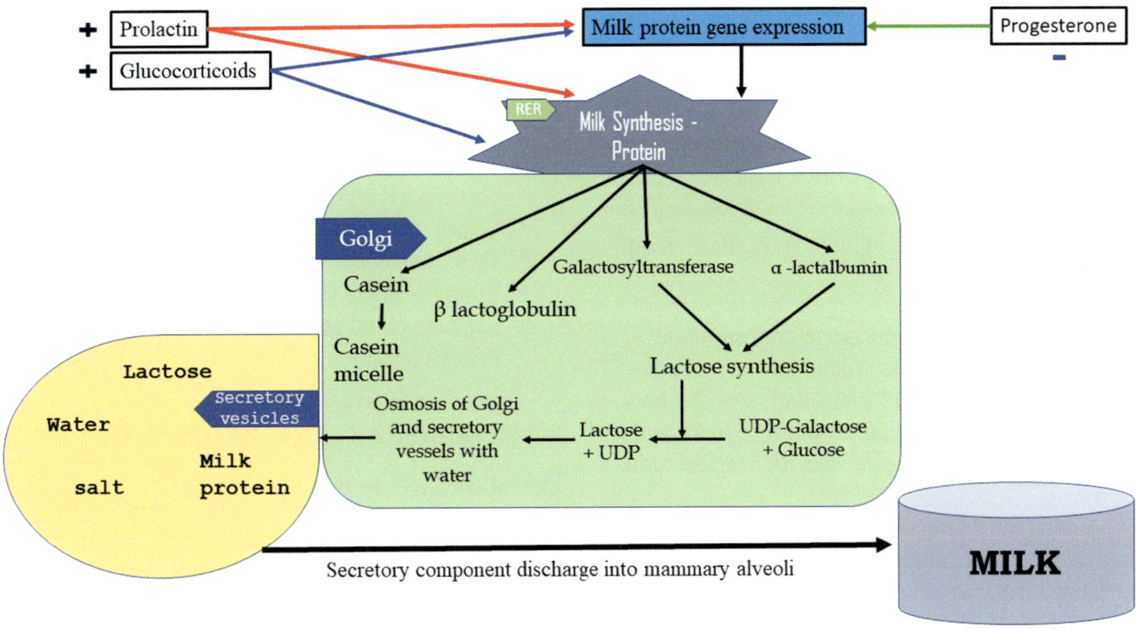

FIGURE 22.5 Flow diagram showing the overview of the lactogenesis process.

- **Energy Storage:** Adipose tissue within the fat pad serves as a reservoir for energy storage. During lactation, when there is a high demand for energy to support milk production, adipocytes in the fat pad can release stored triglycerides as fatty acids to provide energy for mammary epithelial cells.
- **Insulation:** The fat pad provides thermal insulation to the mammary gland, helping to maintain a stable internal environment. This is particularly important during lactation when the mammary gland is actively producing and secreting milk.
- **Hormonal Regulation:** Adipose tissue is an endocrine organ that secretes hormones and adipokines, which can influence mammary gland development and function. For example, adipose-derived hormones such as leptin and adiponectin may regulate aspects of mammary gland physiology, including lactation.
- **Mechanical Protection:** The fat pad can also provide mechanical protection to the mammary gland, cushioning it against physical trauma or injury.

Overall, the fat pad plays a crucial role in supporting mammary gland function and ensuring optimal lactation and reproductive success in mammals.

22.12 LACTOGENESIS

For successful lactation, three key events must transpire: (i) the prepartum proliferation of alveolar epithelial cells, (ii) the biochemical and structural differentiation of these cells, and (iii) the synthesis and secretion of milk components. Lactogenesis, the initiation of lactation, entails a process of differentiation whereby mammary alveolar cells gain the capability to secrete milk. This process is delineated by a two-stage mechanism. The initial stage involves partial enzymatic and cytological differentiation of alveolar cells, accompanied by limited secretion. Subsequently, the second stage commences with a copious secretion of all milk components shortly before parturition, extending into the early postpartum period across most species. Elevated levels of progesterone during late pregnancy impede lactogenesis. However, a decline in progesterone levels near parturition facilitates the initiation of cell differentiation and lactation by the lactogenic complex. Estrogen exerts a stimulatory effect on prolactin (PRL) secretion. Glucocorticoids, upon becoming less bound to globulin proteins, form a complex with PRL, initiating differentiation and early enzymatic activities crucial for milk production and secretion. Additionally, hormones such as growth hormone (GH), insulin-like growth factor (IGF), insulin, and thyroid hormones play roles in this process. At the onset of lactation, alveolar cells undergo significant maturation of organelles like the rough endoplasmic reticulum, smooth endoplasmic reticulum, and Golgi apparatus under the influence of PRL and glucocorticoids. This maturation enhances the cellular capacity to synthesize and secrete proteins, fats, and lactose (Figure 22.5).

22.13 GALACTOPOIESIS

Lactation initiation is a continuous process post-parturition, persisting until peak yield and lactation establishment. Galactopoiesis denotes the maintenance of lactation or milk secretion. Key factors responsible for lactation maintenance include milk synthesis and secretion, milk removal, and adequate release and action of galactopoietic hormones. Milk synthesis and secretion are closely interlinked, with impaired synthesis if milk isn't removed from the udder during milking. Growth hormone, prolactin, thyroid hormone, insulin, and adrenal corticoids play vital roles in milk secretion, albeit with species-specific differences. Galactopoiesis terminology often indicates lactation enhancement in dairy animals due to galactopoietic hormones. Evaluating substances or hormones as galactopoietic agents involves two methods:

1. Suppression of plasma hormone concentration to observe effects on milk production and composition. A decline in milk yield indicates the hormone's necessity for lactation maintenance, with effects varying by species, lactation stage, and parity.
2. Exogenous hormone administration during lactation to enhance milk production, indicating the hormone's role as a rate-limiting factor in milk secretion. From different studies, even some non-hormonal preparations and herbal products like shatavari, guduchi, protected choline chloride, protected amino acids, or prill fat augment milk production in cows and buffaloes, acting as galactagogue substances.

The mammary gland comprises numerous epithelial cells and alveoli in varying functional states during lactation stages. Dynamic changes occur in mammary glands during lactation, extensively studied in goats. Milk yield typically increases after parturition, peaking around 45 days in cows and buffaloes. Persistence varies among individuals, with some reaching peak yield early and others later. Factors influencing milk yield include nutrient availability, metabolic state, management, and environment, particularly pronounced during lactation phases. Late pregnancy's programmed cell death, apoptosis, leads to decreased milk secretion. Augmenting milk production involves maintaining alveolar cell number, synthetic activity per mammary cell, milk ejection reflex efficacy, using galactagogues substances, and balancing cow feeding based on milk yield. Galactopoietic hormones effectively influence milk yield when milk ejection is complete and frequent milk removal occurs. Inefficient removal despite adequate hormonal status leads to discontinued milk synthesis. Maintenance of lactation necessitates both milk secretion and removal. Qualitative changes in milk yield, mammary secretory cell count, and milk yield per epithelial cell are extensively studied in goats. Mammary DNA increases during early lactation in goats due to continued lactogenic enzyme activity and increased secretory cell differentiation. In dairy cattle, mammary DNA increases during early lactation, but precise postpartum mammary growth timing and extent remain undefined. Mammary cell count changes and milk yield per cell are regulated by galactopoietic hormones and local mammary factors.

22.13.1 Hormonal Regulation of Galactopoiesis

Hormones released from various sources regulate the maintenance of lactation in both ruminants and non-ruminants.

- Adrenal gland (glucocroticoids and mineralocorticoids)
- Ovary (estrogen)
- Oxytocin (essential for let down of milk)
- Pituitary gland (prolactin, GH, ACTH, TSH)
- Thyroid gland (I, & T) – Mammary glands local growth factors

22.14 MAMMARY INVOLUTION

Involution refers to the process that unfolds after milking cessation, leading to both structural and physiological alterations within the mammary gland. The decrease in secretory activity is a consequence of heightened pressure and elevated levels of the lactation feedback inhibitor hormone (FIL). In rodents, the loss of alveolar structure occurs swiftly (within days), with only ducts evident in the fat pad due to diminished prolactin (PRL) levels. Conversely, dairy cows undergo a gradual decline in secretory activity, retaining their alveolar structures for weeks to months. Throughout involution, extensive differentiation of alveolar cells occurs to facilitate milk synthesis and secretion. Moreover, metabolic shifts transpire, encompassing heightened appetite (for both food and water), augmented gastrointestinal tract dimensions and function, alterations in nutrient metabolism, and weight loss, all of which are orchestrated to meet the demands of lactation (Figure 22.6).

22.15 MILK SYNTHESIS AND SECRETION AND ITS COMPOSITION

22.15.1 Secretion and its Constituents

Milk secretion is the process in which the milk constituents synthesized by epithelial cells are released into the alveoli's lumen. Most of these constituents originate from the alveolus itself. While milk fat, protein, and lactose are produced within the alveolar epithelial cells, certain elements like vitamins, minerals, and specific proteins are not manufactured in the mammary gland but are instead sourced from the bloodstream. These non-synthesized components enter the cells from the blood and are then discharged into the milk. Lactose, fat, and the majority of proteins are created in the epithelial cells using precursors supplied by the bloodstream.

The secretion of milk constituents, including fat and protein, involves distinct processes within the epithelial cells. Milk fat forms as small droplets within these cells, which then merge as they move towards the cell's apical part. These fat droplets are subsequently released from the epithelial cell through a pinching-off mechanism (Figure 13). During this process, the fine apical membrane of the epithelial cell encases the milk fat globule's outer layer.

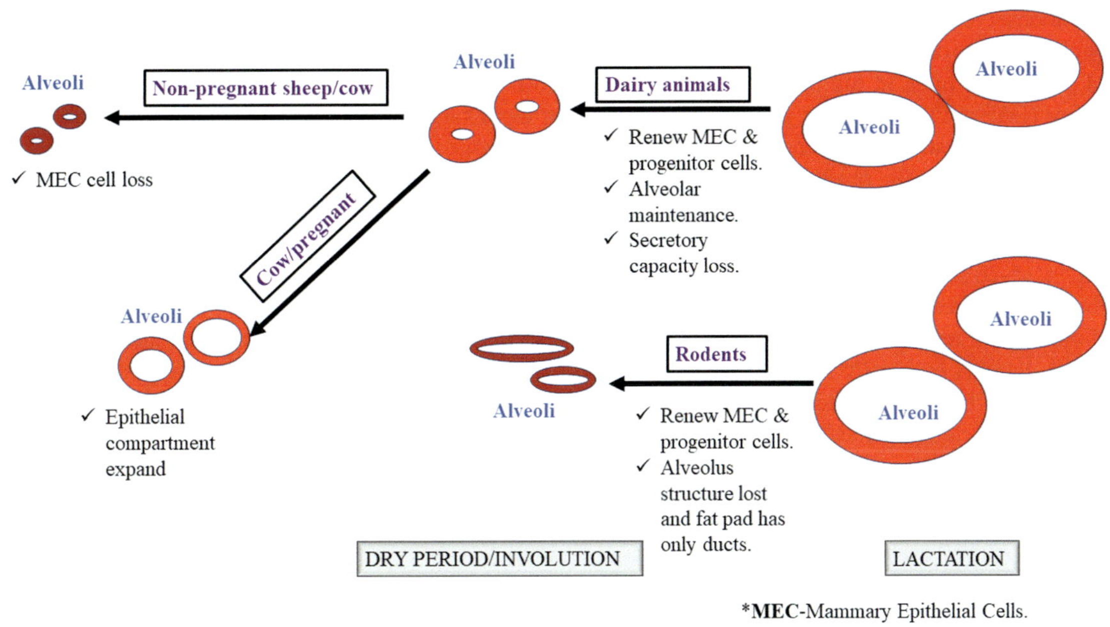

FIGURE 22.6 Diagram showing involution of the mammary gland in bovine and rodents.

Casein and lactose molecules are released as secretory vesicles surrounded by a membrane. These vesicles migrate towards the apical part of the cell before being discharged into the alveolar lumen. Before release, the secretory vesicle's wall fuses with the cell membrane, and its contents are discharged into the alveolar cavity, repairing the membrane loss caused by the release of fat droplets. This secretion process is continuous, with each epithelial cell contributing to the overall secretion.

The molecules of proteins, milk sugar, vitamins, and minerals are evenly distributed within the milk. However, milk fat globules float in water upon entering the alveolar lumen. These fat droplets aggregate into small clusters that cannot easily pass through the ducts draining the alveoli. Meanwhile, water flows freely along the ducts, gradually filling the cisterns of quarters. Some fat remains within the secretory cell and is only released when udder pressure decreases during milking. Consequently, the milk initially drawn from the udder contains only 1 to 2% fat, with subsequent milk possibly containing more fat.

22.15.2 Rate of Secretion of Milk

Epithelial cells display a heightened milk synthesis rate within the first 10 hours post-milking, which gradually diminishes afterward. When a cow or buffalo goes unmilked, milk synthesis and secretion typically cease within a specific timeframe following the last milking session. The cessation of milk secretion is due to increased pressure within the alveolus, which constricts blood vessels, hindering the flow of blood and the supply of milk precursors to epithelial cells. Low-producing cows are less affected by extended milk intervals due to having less secretory tissue in the udder. Consequently, younger and low-producing cows experience a greater increase in udder pressure per unit milk, underscoring the importance of short milking intervals for such animals.

Leaving a significant amount of milk (2kg or more) in the udder after incomplete milking can lead to a permanent reduction in milk production. Milk secretion is influenced by various factors including parity, lactation stage, milking interval, frequency of milking, residual milk, and milking management practices. The rate of milk secretion plays a crucial role in determining the additional milk yield achievable with more frequent milking. For instance, milking high-producing cows (>12–15kg/day) three times daily instead of twice daily can result in a 20% increase in milk yield. Further increases in milk yield (5 to 10%) can be achieved by milking four times per day. However, considering feeding and labor costs, milking four times per day may not be economically feasible or advisable. Economically, three times daily milking becomes favorable for cows producing over 15kg/day. Sleeping one milking session per week may lead to a reduction in total milk yield by 5 to 10%. Maintaining regular milking intervals of 12 hours for cows and buffaloes with moderate production potential is crucial in dairy farming, particularly in rural settings. Agalactia refers to complete lactation failure, while hypogalactia denotes partial lactation failure.

22.15.3 Role of Nerves in Milk Secretion Regulation

The regulation of milk secretion by the nervous system has not been elucidated extensively. Many studies indicate that frequent milking of dairy animals stimulates milk secretion. Mammary glands respond to the following neural factors for milk synthesis.

It has been observed that the production capacity of the mammary gland in rats can be modulated by the nutritional demand of the pups or litter. In ruminants, increased milking frequency (IMF) of 2-3 times or intense suckling by the calf results in increased milk synthesis. Thus, IMF can be utilized to augment milk production and production efficiency.

In dairy animals, such as cows, milking three times a day results in 15-20% more milk production. Similarly, milking four times is preferable to three times, leading to an additional increase in milk production. This indicates that the mammary gland is especially sensitive to the demand of offspring in early lactation, influencing the shape of the lactation curve and lactation persistency.

Increased milking frequency (IMF) in ruminants and rodents leads to:

1. An increase in mammary cell number
2. An increase in mammary cell activity
3. Additionally, the effect of suckling intensity on mammary gland development has also been reported.

In rodents, mammary development with increased suckling is accompanied by:

1. An increase in DNA
2. An increase in RNA
3. An increase in the RNA:DNA ratio

22.15.4 Regarding the Impact of IMF and Suckling Intensity on Hormones

In rodents, the act of suckling prompts the release of both prolactin and oxytocin, wherein prolactin serves to stimulate mammary cell differentiation. An increased litter size is associated with elevated prolactin levels within the pituitary gland. This elevation in prolactin levels is linked to increased mammary cell activity. When oxytocin is injected into rats, it prolongs lactation by facilitating the complete evacuation of the mammary gland, although it does not impact milk yield. During late lactation in rodents, cortisol is recognized as the limiting factor; however, administering cortisol does not consistently result in a significant increase in milk yield. This suggests that local factors play a crucial role in regulating the mammary response to systemic hormones.

22.16 LACTATION CYCLE

The lactation cycle in farm animals encompasses the process of milk production for nourishing offspring, starting from birth until weaning. It comprises several distinct phases. Initially, after giving birth, the female animal secretes colostrum, a vital fluid rich in antibodies and nutrients crucial for the newborn's immunity and gut health. Subsequently, the lactation shifts into a phase of increased milk production, known as Lactogenesis I, characterized by a rapid surge in milk volume and composition alterations. This stage transitions into Lactogenesis II, where milk production stabilizes at its peak, providing consistent nutrition for the growing offspring. As the young animals mature and consume less milk, a gradual decline in milk production occurs. Ultimately, the lactation cycle concludes with the weaning of the offspring, prompting the cessation of milk production. Post-weaning, the female undergoes involution, whereby the mammary glands return to their non-lactating state. Various factors, including genetics, nutrition, environment, and management practices, influence the lactation cycle. Breeds selectively bred for dairy production, such as Holstein and Jersey cows, may exhibit extended lactation periods compared to meat breeds.

22.17 DISEASE CONDITIONS ASSOCIATED WITH THE MAMMARY GLAND

22.17.1 Mastitis

Mastitis poses a significant challenge for dairy farmers globally, impacting milk production, animal well-being, and farm profitability. Let's delve into the dynamics of mastitis in cows:

1. Causes: Mastitis in cows can arise from a multitude of factors, ranging from bacterial infections and subpar milking hygiene to unfavorable environmental conditions like unclean bedding or damp pastures. It can also stem from inadequate management of udder health and the spread of contagious pathogens through cow-to-cow contact or milking equipment.
2. Types: Within the bovine realm, mastitis manifests in two distinct forms: clinical and subclinical. Clinical mastitis presents visible signs of udder inflammation, including swelling, warmth, redness, and the presence of abnormal milk textures or discoloration. On the other hand, subclinical mastitis lurks beneath the surface, devoid of overt clinical indicators but detectable through laboratory assessments like somatic cell count analysis.
3. Symptoms: Recognizable symptoms of clinical mastitis in cows encompass observable changes in the udder, such as swelling, warmth, or redness, alongside anomalies in milk consistency or appearance. Affected cows may also exhibit discomfort during milking, reduced milk output, or systemic signs of illness like fever or lethargy.
4. Diagnosis: Diagnosis of mastitis in cows involves a blend of astute clinical observation, meticulous udder examination, and precise laboratory scrutiny of milk samples. Tests like the California Mastitis Test (CMT) or somatic cell count analysis

serve as pivotal diagnostic tools, particularly for uncovering subclinical cases.

5. Treatment: Managing mastitis in cows typically entails a tailored approach centered on antibiotic therapy targeted at the specific causative bacteria. In severe instances, supplementary measures such as anti-inflammatory medications or intramammary infusions may be warranted to alleviate discomfort and expedite recovery.
6. Prevention: Proactive measures to thwart mastitis in cows hinge on fostering optimal udder health and implementing sound management protocols. This entails upholding meticulous milking hygiene, vigilantly monitoring udder health, and promptly identifying mastitis cases. Furthermore, minimizing exposure to environmental pathogens through pristine housing and pastures, alongside strategic vaccination and genetic selection, are integral components of a comprehensive prevention strategy.

22.17.2 Neoplasia

Neoplasia, characterized by abnormal cell growth, was a concerning condition associated with the mammary gland in cows. Here's an overview of mammary gland neoplasia in cows:

1. Types: Mammary gland neoplasia in cows could manifest as either benign or malignant tumors. Benign tumors, like adenomas or fibroadenomas, typically grew slowly and remain localized. In contrast, malignant tumors, including adenocarcinomas or sarcomas, were more aggressive and could metastasize to other parts of the body.
2. Causes: The exact causes of mammary gland neoplasia in cows were not fully understood, but factors such as hormonal imbalances, genetic predisposition, and environmental factors might have played a role. Chronic inflammation and exposure to carcinogens in the environment or feed might also have contributed to the development of mammary tumors in cows.
3. Symptoms: Symptoms of mammary gland neoplasia in cows could vary depending on the type and stage of the tumor. Common signs included:
   - Palpable masses or lumps in the udder or mammary gland area
   - Changes in udder size, shape, or texture
   - Abnormalities in milk production, such as changes in milk yield or quality
   - Pain or discomfort in the udder
   - Ulceration or discharge from the udder
4. Diagnosis: Diagnosis of mammary gland neoplasia in cows typically involved a combination of physical examination, imaging techniques (such as ultrasound or mammography), and biopsy of the affected tissue. Laboratory analysis of tissue samples could help determine the nature of the tumor (benign or malignant) and guide treatment decisions.
5. Treatment: Treatment options for mammary gland neoplasia in cows might have varied depending on the type, size, and stage of the tumor, as well as the overall health of the cow. Treatment might have included surgical removal of the tumor, chemotherapy, radiation therapy, or supportive care to manage symptoms and improve quality of life.
6. Prognosis: The prognosis for cows with mammary gland neoplasia depended on factors such as the type and stage of the tumor, response to treatment, and overall health of the cow. Early detection and prompt intervention could have improved the chances of successful treatment and long-term survival.

22.17.3 Hypocalcemia

Hypocalcemia, commonly referred to as milk fever, was a significant disease condition associated with the mammary gland in cows. Here's an overview of cow hypocalcemia:

1. Definition: Hypocalcemia referred to low levels of calcium in the blood, which could lead to various health issues in cows, including those related to the mammary gland.
2. Causes: The primary cause of hypocalcemia in cows was the sudden demand for calcium during the onset of lactation. This demand often exceeded the cow's ability to mobilize calcium from its bones or absorb it from the diet, leading to a deficiency. Factors such as inadequate dietary calcium intake, improper feeding management, breed predisposition (e.g., high-producing dairy breeds), and hormonal imbalances could contribute to the development of hypocalcemia.
3. Symptoms: Symptoms of hypocalcemia in cows could include: muscle weakness or tremors, stiff gait or difficulty walking, reduced milk production, loss of appetite, lethargy or depression, sternal or lateral recumbency (in severe cases)
4. Diagnosis: Diagnosis of hypocalcemia in cows was based on clinical signs and could be confirmed through blood tests to measure serum calcium levels. Additionally, veterinarians could evaluate other blood parameters such as magnesium and phosphorus levels to rule out concurrent metabolic disorders.
5. Treatment: Treatment of hypocalcemia in cows typically involved intravenous administration of calcium solutions to rapidly increase blood calcium levels. Additionally, oral calcium supplements might have been provided to maintain calcium levels in the long term. In severe cases,

supportive care such as fluid therapy and management of secondary complications might have been necessary.

6. Prevention: Prevention of hypocalcemia in cows focused on optimizing dietary calcium intake, particularly during the transition period from late pregnancy to early lactation when calcium demands were highest. Strategies included feeding high-calcium diets, providing access to mineral supplements, and implementing proper management practices to reduce stress and maintain cow health during this critical period.

22.17.4 Puerperal Tetany or Eclampsia

Puerperal tetany, also referred to as eclampsia or hypocalcemia, a critical condition associated with the mammary gland in cows, particularly during the postpartum period. Here's an overview:

1. Definition: Puerperal tetany, or eclampsia, a metabolic disorder characterized by low blood calcium levels, primarily occurring in cows shortly after calving or during early lactation. This condition could have severe consequences for both the cow and her offspring if left untreated.
2. Causes: The main cause of puerperal tetany in cows was the sudden increase in calcium demand during the onset of lactation, which exceeded the cow's ability to mobilize calcium reserves or absorb calcium from the diet. This imbalance could lead to hypocalcemia and subsequent clinical signs of tetany.
3. Symptoms: Clinical signs of puerperal tetany in cows could include: muscle tremors or spasms, stiff gait or difficulty walking, hyperesthesia (increased sensitivity to touch), fever, restlessness or agitation, and in severe cases, convulsions or seizures might have occurred.
4. Diagnosis: Diagnosis of puerperal tetany in cows was typically based on clinical signs and could be confirmed through blood tests to measure serum calcium levels. Additionally, veterinarians might have evaluated other blood parameters to rule out concurrent metabolic disorders.
5. Treatment: Treatment of puerperal tetany in cows involved immediate correction of calcium levels. This often required intravenous administration of calcium solutions to rapidly increase blood calcium levels. Additionally, oral calcium supplements might have been provided to maintain calcium levels in the long term. Supportive care, such as fluid therapy and management of secondary complications, might also have been necessary.
6. Prevention: Prevention of puerperal tetany in cows focused on optimizing dietary calcium intake during late pregnancy and early lactation. This included feeding high-calcium diets and providing access to mineral supplements. Proper management practices, such as minimizing stress and ensuring adequate nutrition during the transition period, were also crucial in preventing this condition.

Puerperal tetany was a serious condition that required prompt recognition and intervention to prevent complications and ensure the health and productivity of the cow and her offspring.

22.18 RECENT ADVANCES IN LACTATION PHYSIOLOGY

Recent advances in lactation physiology have seen significant progress in understanding the molecular mechanisms underlying milk synthesis, regulation of milk composition, and mammary gland development. Here are some key areas of advancement are,

1. **Heat stress and its effect on dairy buffaloes:** Various innovative measures like combination of fan and sprinklers together or wallowing the animal completely under the water/pools found to have increased production of milk as well as reproductive ability of animals. By combined way of using fan shade and sprinklers along with the supplementary like minerals, vitamins and antioxidants was also led to increased milk production.
2. **Thermotolerance and impact of heat stress in dairy cattle:** Climate change exacerbates environmental challenges for livestock, particularly dairy cattle, by intensifying heat stress through more frequent hot days and heatwaves. Dairy cows, with their high metabolic heat load, are especially vulnerable. Heat stress disrupts various biological processes, leading to significant economic losses. To combat heat stress, cows employ physiological and cellular mechanisms that divert energy away from other functions, impacting milk production, reproduction, and health. Selecting thermotolerant dairy cattle becomes imperative. Strategies include reducing milk production, crossbreeding with heat-tolerant breeds, selecting based on physiological traits, and enhancing immune responses. Each approach has its merits and drawbacks, all detailed in the literature. This review explores the multifaceted issues surrounding heat stress in dairy cattle and evaluates proposed selection strategies for thermotolerance (Cartwright, Shannon L. et al. 2023).
3. **Contribution on dairy production in one health perspective:** Taking enough care and precision on the dairy production by providing proper nutrients and environment to the animal concerned to reduce the stressful environment such as heat

stress during climate change. Also, by reducing the occurrence of the disease condition using the help of veterinarians to avoid cross contamination of the milk with milk borne diseases. All these cross contamination is reduced by the use of standard measuring ways using somatic cell count by tracking the WBC status of the cow and has the control over the infections of the udder (Nguyen et al., 2023).

4. **Microbiomes of milk and udder in correlation:** Many study highlighted the variability in results regarding the characterization of the microbiota in bovine milk. While other studies had characterized the microbiota of teat skin, teat canal, and milk samples, there was still no consensus on the existence and function of a core commensal microbiota in bovine milk. Additionally, the idea of dysbiosis (microbial imbalance) as a causal factor in udder health issues was still speculative. One consistent finding by Ruegg (2022) was that milk contained a low biomass of microbial DNA. However, there was a pressing need for standardization of methods and replication of experiments to enhance understanding in this emerging field.
5. **Microbiome-mammary gland interactions:** McGuire, M.K., et al. (2017) found that the microbiome of the mammary gland and its influence on lactation performance and milk composition are areas of growing research interest, highlighting the importance of the gut-mammary axis in lactation physiology.
6. **Epigenetic regulation of lactation:** The interaction between the microbiome of the mammary gland and lactation performance, as well as milk composition, has been an increasingly studied area, emphasizing the significance of the gut-mammary axis in lactation physiology as observed by Lemay, D.G., et al. (2013).
7. **Nutrigenomics and milk composition:** Bach, A. (2018) mentioned that progress in nutrigenomics has provided insights into how maternal nutrition during lactation impacts the composition of milk and the health of offspring, guiding the development of optimal dietary strategies to enhance milk quality.
8. **Role of hormones and signaling pathways:** Ongoing research into hormonal regulation and intracellular signaling pathways involved in mammary gland development and lactation has revealed potential targets for improving both the quantity and quality of milk production.
9. **Painless or non-ivasive method of tracking the thermal condition of the animals for production:** Sejian, V et al. (2022) summarized various non-invasive methods used to measure heat stress response in farm animals, focusing on behavioral and physiological changes. Technologies for measuring cortisol from hair, feces, urine, saliva, and milk offer valuable insights into heat stress levels. Infrared thermography facilities can accurately gauge animal surface temperatures, aiding in assessing climate resilience. GPS and GNSS applications help understand behavioral changes in grazing animals during heat stress. Emerging techniques like sensors and machine learning show promise but require further development for broader use. These advancements could inform policymakers in developing strategies to sustain livestock production in a changing climate.
10. **Genomic selection for lactation traits:** Advances in genomic technologies and bioinformatics have facilitated the discovery of genetic markers associated with lactation performance, enabling more precise selection methods to enhance milk yield and quality in dairy animals Van Raden (2008).
11. **Mammary stem cells and regeneration:** Research has made strides in understanding mammary stem cell populations and their role in mammary gland regeneration and lactation persistence. (Wang, D., et al. 2020).
12. **Role of exosomes in milk:** Chen, X., et al. (2021) said that exosomes, small extracellular vesicles secreted in milk, have emerged as important mediators of intercellular communication in the mammary gland and are implicated in milk quality and neonatal development.
13. **Maternal stress and milk composition:** Studies have investigated the impact of maternal stressors on milk composition and its implications for offspring health and development, highlighting the complex interplay between maternal physiology and lactation outcomes reported by Prosser, C. G., and Davis, S. R. (2022).
14. **Sustaining longer lactation cycle:** Throughout history, farmers have been advised to maintain a one-year calving interval to optimize milk yield peaks, yet this led to increased risks of diseases and disorders due to frequent drying-off, calving, and lactation cycles. Extending the voluntary waiting period (VWP) between inseminations was identified as a solution to mitigate these challenges, reducing the frequency of calving events and benefiting cow health, fertility, and reducing labor and surplus calf numbers. Concerns arose about prolonged lactation periods, such as decreased milk yield in late lactation and potential cow fattening, impacting herd economics. Limited knowledge existed about subsequent lactation performance and calf outcomes with extended lactations. Responses of dairy cows to extended VWP varied depending on individual characteristics like parity, milk yield, and body condition. Customized strategies based on individual cow

traits emerged as a potential future approach to select high-producing cows with consistent lactation curves, thus reducing risks of fattening and milk yield reduction while minimizing challenging calving eventsvan Knegsel et al. (2022).

15. **Non-coding RNAs in milk:** Non-coding RNAs, including microRNAs and long non-coding RNAs, present in milk have been shown to regulate gene expression in the neonate and may have implications for milk functionality and offspring health.
16. **Environmental exposures and milk quality**: Ly, A., et al. (2023) explored the effects of environmental contaminants and pollutants on milk quality, including the presence of toxins and their impact on human health through breastfeeding.
17. **Extending life of dairy cows as producer:** The sustainability of milk production on farms is closely tied to the longevity of dairy cows, referring to their productive lifespan. Longevity is a complex aspect influenced by various factors, lacking a standardized definition or measurement. Its assessment is crucial as it impacts farm profitability, environmental footprint, and animal welfare. This paper aimed to review metrics for measuring longevity and assess its status in major milk-producing nations. Enhancing dairy cow longevity involves achieving early age at first calving and a profitable, extended productive life. Evaluating longevity comprehensively involves considering age at first calving, productive lifespan, and overall profitability. The study revealed a concerning trend of decreasing dairy cow longevity in many top milk-producing countries, highlighting industry and stakeholder concerns. Extending cow lifespan could lower health expenses, boost profitability, and enhance both animal welfare and industry sustainability Dallago et al. (2021).
18. **Mastitis and lacatting animals:** Mastitis is the most common disease in dairy cows, impacting their welfare and farm profits. Since modern dairy farming began, efforts to minimize mastitis have been ongoing. This paper reviews advances in mastitis detection, management, and prevention since 1917. Initially, research focused on understanding bacteria causing intramammary infections, particularly Streptococcus agalactiae and Staphylococcus aureus. Successful control programs emerged by identifying infection mechanisms, clinical states, screening tests, and pathogen-specific traits. Control strategies shifted to preventing new infections and reducing infection duration, incorporating teat dipping, hygienic milking, and strategic antibiotic therapy. Researchers now address mastitis from environmental pathogens and define antimicrobial usage for animal welfare. While significant progress has been made, evolving herd structures and stricter standards ensure ongoing mastitis research remains crucial.

These recent advances provide valuable insights into various aspects of lactation physiology, contributing to our understanding of milk production, composition, and the health outcomes for both the cow and offspring.

22.19 CONFLICT OF INTEREST

None Declared.

BIBLIOGRAPHY

Admyre, C., Johansson, S. M., Qazi, K. R., Filén, J.-J., Lahesmaa, R., Norman, M., Neve, E., Scheynius, A., & Gabrielsson, S. (2007). Exosomes with immune modulatory features are present in human breast milk. *The Journal of Immunology, 179*(3), 1969–1978.

Akers, R. M. (2002). *Lactation and the Mammary Gland*. January 2002, Iowa State Press. SBN: 0-8138-2992-5..

Allen, A. J., Barrington, G. M., & Parish, S. M. (2008). Physiologic mastectomy via flank laparotomy. *Veterinary Clinics of North America: Food Animal Practice, 24*(3), 511–516.

Bach, A. (2018). Recent advances in amino acid nutrition during early lactation. *Advances in Dairy Technology, 30*(1), 43–54.

Bach, A. (2019, January). Decomposing efficiency of milk production and maximizing profit. In *Journal of Dairy Science* (Vol. 102, pp. 304–304). STE 800, 230 Park Ave, New York, NY 10169: Elsevier Science Inc.

Brito, A. F., & Silva, L. H. P. (2020). Symposium review: comparisons of feed and milk nitrogen efficiency and carbon emissions in organic versus conventional dairy production systems. *Journal of Dairy Science, 103*(6), 5726–5739.

Cartwright, S. L., Schmied, J., Karrow, N., & Mallard, B. A. (2023). Impact of heat stress on dairy cattle and selection strategies for thermotolerance: a review. *Frontiers in Veterinary Science, 10*, 1198697.

Chen, X., et al. (2021). Mammary epithelial autophagy and developmental gene regulation revealed by single-cell RNA sequencing. *Cell Reports, 35*(4), 109103.

Cohick, W. S. (2022). The role of the IGF system in mammary physiology of ruminants. *Domestic Animal Endocrinology, 79*, 106709.

Comerford, K. B., Miller, G. D., Boileau, A. C., Masiello Schuette, S. N., Giddens, J. C., & Brown, K. A. (2021). Global review of dairy recommendations in food-based dietary guidelines. *Frontiers in Nutrition, 8*, 671999.

Dallago, G. M., Wade, K. M., Cue, R. I., McClure, J. T., Lacroix, R., Pellerin, D., & Vasseur, E. (2021). Keeping dairy cows for longer: A critical literature review on dairy cow longevity in high milk-producing countries. *Animals, 11*(3), 808.

Goff, J. P. (2008). The monitoring, prevention, and treatment of milk fever and subclinical hypocalcemia in dairy cows. *The Veterinary Journal, 176*(1), 50–57.

Grace, D., Wu, F., &Havelaar, A. H. (2020). MILK Symposium review: Foodborne diseases from milk and milk products in developing countries—Review of causes and health and economic implications. *Journal of Dairy Science*, *103*(11), 9715–9729.

Harrison, M. T., Cullen, B. R., Mayberry, D. E., Cowie, A. L., Bilotto, F., Badgery, W. B., ... Eckard, R. J. (2021). Carbon myopia: The urgent need for integrated social, economic and environmental action in the livestock sector. *Global Change Biology*, *27*(22), 5726–5761.

Hoffmann, G., Herbut, P., Pinto, S., Heinicke, J., Kuhla, B., & Amon, T. (2020). Animal-related, non-invasive indicators for determining heat stress in dairy cows. *Biosystems Engineering*, *199*, 83–96.

Hogeveen, H., Huijps, K., & Lam, T. J. G. M. (2011). Economic aspects of mastitis: new developments. *New Zealand Veterinary Journal*, *59*(1), 16–23.

Ji, B., Banhazi, T., Perano, K., Ghahramani, A., Bowtell, L., Wang, C., & Li, B. (2020). A review of measuring, assessing and mitigating heat stress in dairy cattle. *Biosystems Engineering*, *199*, 4–26.

Katsafadou, A. I., Politis, A. P., Mavrogianni, V. S., Barbagianni, M. S., Vasileiou, N. G., Fthenakis, G. C., & Fragkou, I. A. (2019). Mammary defences and immunity against mastitis in sheep. *Animals*, *9*(10), 726.

Lemay, D. G., et al. (2013). RNA sequencing of the human milk fat layer transcriptome reveals distinct gene expression profiles at three stages of lactation. *PLoS One*, *8*(7), e67531.

Ly, A., et al. (2023). Maternal exposure to environmental pollutants during lactation and its effects on breast milk composition and infant health: a systematic review. *Environmental Research*, *204*, 112084.

McGuire, M. K., et al. (2017). What's normal? Oligosaccharide concentrations and profiles in milk produced by healthy women vary geographically. *The American Journal of Clinical Nutrition*, *105*(5), 1086–1100.

Merewood, A., Mehta, S. D., Chamberlain, L. B., Philipp, B. L., & Bauchner, H. (2006). Breastfeeding rates in US Baby-Friendly hospitals: results of a national survey. *Pediatrics*, *118*(2), e70–e75.

Middleton, J. R., Luby, C. D., & Adams, D. S. (2019). Mammary tumors in cattle. *The Veterinary Clinics of North America. Food Animal Practice*, *35*(2), 283–296.

Mukherjee, J., Das, P. K., & Banerjee, D. (2023). Lactation physiology. In *Textbook of Veterinary Physiology* J. Mukherjee, P. K. Das, D. Banerjee (eds.), (pp. 639–674). Springer Nature Singapore.

Neethirajan, S. (2024). Net zero dairy farming—advancing climate goals with big data and artificial intelligence. *Climate*, *12*(2), 15.

Neville, Margaret. (2009). Classic studies of mammary development and milk secretion: 1945–1980. *Journal of Mammary Gland Biology and Neoplasia*, *14*, 193–197. 10.1007/s10911-009-9151-7.

Nguyen, B. T., Briggs, K. R., & Nydam, D. V. (2023). Dairy production sustainability through a one-health lens. *Journal of the American Veterinary Medical Association*, *261*(1), 12–16.

Nickerson, S. C. (2011). *Nutritional Programming of the Dairy Calf View Project*. Agricultural and Food Sciences.

Nickerson, S. C, & Akers, R. M. (2011). Mammary gland anatomy. In: *Encyclopedia of Dairy Sciences* J. W. Fuquay (ed.), pp. 328–337. San Diego: Elsevier Ltd.

Nickerson, S. C., & Akers, R. M. (2022). *Mammary Gland: Anatomy* In *Encyclopedia of Dairy Sciences* P. L. H. McSweeney and J. P. McNamara (eds.). Academic Press.

Pakrashi, A., Ryan, C., Guéret, C., Berry, D. P., Corcoran, M., Keane, M. T., & Mac Namee, B. (2023). Early detection of subclinical mastitis in lactating dairy cows using cow-level features. *Journal of Dairy Science*, *106*(7), 4978–4990.

Petrocchi Jasinski, F., Evangelista, C., Basiricò, L., & Bernabucci, U. (2023). Responses of dairy Buffalo to heat stress conditions and mitigation strategies: a review. *Animals*, *13*(7), 1260.

Pinto, S., Hoffmann, G., Ammon, C., & Amon, T. (2020). Critical THI thresholds based on the physiological parameters of lactating dairy cows. *Journal of Thermal Biology*, *88*, 102523.

Prosser, C. G., & Davis, S. R. (2022). Impact of maternal stress on lactation performance and milk composition: a review. *Journal of Dairy Science*, *105*(1), 1–15.

Radostits, O. M., Gay, C. C., Hinchcliff, K. W., & Constable, P. D. (2007). *Veterinary Medicine: A Textbook of the Diseases of Cattle, Horses, Sheep, Pigs, and Goats* (10th ed.). Saunders Ltd.

Ruegg, P. L. (2017). A 100-year review: Mastitis detection, management, and prevention. *Journal of Dairy Science*, *100*(12), 10381–10397.

Ruegg, P. L. (2022). The bovine milk microbiome–an evolving Science. *Domestic Animal Endocrinology*, *79*, 106708.

Sejian, V., Shashank, C. G., Silpa, M. V., Madhusoodan, A. P., Devaraj, C., & Koenig, S. (2022). Non-invasive methods of quantifying heat stress response in farm animals with special reference to dairy cattle. *Atmosphere*, *13*(10), 1642.

Singh, M., & Roy, A. K. (2017) *Textbook on Physiology of lactation*. Satish Serial Publishing House.

Soteriades, A. D., Foskolos, A., Styles, D., & Gibbons, J. M. (2020). Maintaining production while reducing local and global environmental emissions in dairy farming. *Journal of Environmental Management*, *272*, 111054.

Tao, S., Rivas, R. M. O., Marins, T. N., Chen, Y. C., Gao, J., & Bernard, J. K. (2020). Impact of heat stress on lactational performance of dairy cows. *Theriogenology*, *150*, 437–444.

Tricarico, J. M., Kebreab, E., & Wattiaux, M. A. (2020). MILK Symposium review: sustainability of dairy production and consumption in low-income countries with emphasis on productivity and environmental impact. *Journal of Dairy Science*, *103*(11), 9791–9802.

van Knegsel, A. T., Burgers, E. E., Ma, J., Goselink, R. M., & Kok, A. (2022). Extending lactation length: consequences for cow, calf, and farmer. *Journal of Animal Science*, 100(10), skac220.

VanRaden, P. M. (2008). Efficient methods to compute genomic predictions. *Journal of Dairy Science*, 91(11), 4414–4423.

Vlasova, A. N., & Saif, L. J. (2021). Bovine immunology: implications for dairy cattle. *Frontiers in Immunology*, *12*, 643206.

Wang, D., et al. (2020). Single-cell transcriptomic atlas of primate ovarian aging. *Cell*, *180*(3), 585–600.

Wankar, A. K., Rindhe, S. N., & Doijad, N. S. (2021). Heat stress in dairy animals and current milk production trends, economics, and future perspectives: the global scenario. *Tropical Animal Health and Production*, *53*(1), 70.

Wellnitz, O., & Bruckmaier, R. M. (2021). The role of the blood-milk barrier and its manipulation for the efficacy of the mammary immune response and milk production.

23 Physiology of Growth

Rachana Sharma, Gurpreet Singh, Tanmay Mondal and Pallavi Khajuria
College of Veterinary Science, Rampura Phul, GADVASU-Ludhiana

23.1 INTRODUCTION

Growth is a biological process which occurs through a series of cellular proliferation, differentiation, and morphogenesis resulting in a continuously increase in the size and weight of an animal. Growth takes place through the processes of increase of cell numbers and size. Growth and development of cells, tissues, and organs occur during prenatal and postnatal life of an organism. Various factors that regulate growth during prenatal periods include heredity, endocrine factors, maternal age and nutrition, sex of the fetus, litter size, and environment (Figure 23.1).

Several factors which influence postnatal growth include genetic factors, nutritional, climatic, managemental, and health status of animal. The factors and biological process that influences the growth of meat-producing animals and poultry species is very much essential to maximize the growth for meat and egg production efficiency, and quality of the product yield by the animals. However, the growth process may be manipulated by using various growth promoter, feed additives, and supplementation of growth factors to improve the growth of animals.

Growth is defined as an increase in cell number and an increase in cell size, resulting in a permanent change in the form/shape of an animal is called as growth.

Growth is also defined as a function of heredity of a specific subject and varies with the fact that animals kept on maintenance or even sub-maintenance diets may increase in size due to bone growth, although weight may remain constant or may diminish.

Broody defined growth as "Relatively irreversible time change in the measured dimensions"

Maynard distinguished true growth from the deposition of fat in adipose tissues. True growth is an increase in the weight of organs and of structural tissues like muscles and bones, while no such changes occur during growth due to the deposition of fat in adipose tissues. Growth is characterized by increase in protein, mineral and water.
Increase in cell number (Hyperplasia), e.g., blood corpuscles, hair follicle cells, skin cells, etc., always grow by an increase in cell number. Hyperplasia occurs as a consequence of DNA replication and protein synthesis.

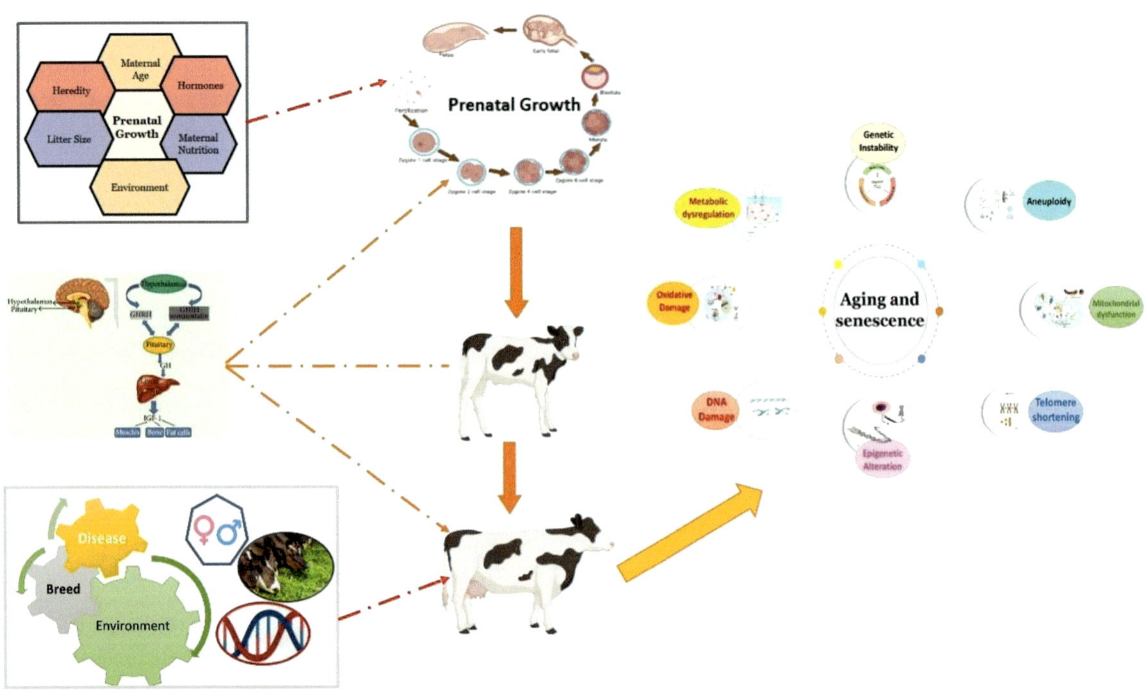

FIGURE 23.1 Several factors influence the growth of animals.

 DOI: 10.1201/9781003426851-23

Increase in cell size (Hypertrophy), e.g., nerve tissues and skeletal muscle tissue, grows by increase in protoplasm of cells.

23.2 TYPES OF BODY CELLS ON THE BASIS OF GROWTH

According to growth pattern, cells can be categorized as:

Renewing type which continuously lost and replaced by new cells, e.g., epidermis, endometrium, skeletal tissue (partly).

Expanding type which expand continuously until the adult size of the organ is reached, e.g., liver, kidney, glands (endocrine and exocrine), skeletal tissue (partly).

Static type which is a part of the expanding system in which cell division is restricted to early stages of development. Later, only hypertrophy may occur, e.g., neurons, retinal rods and cones, muscle fibers.

23.2.1 Types of Growth

Compensatory Growth: If one organ or a part of an organ is removed, then remaining part grows faster to compensate the function of the other organ or part of the organ. This is known as compensatory growth. For example, if one kidney is removed, the other kidney increases the number and size of nephrons to normalize body function to normal. If a part of the liver is cut, it grows faster to compensate for liver functioning.

Differential Growth: When the growth in different organs of the body takes place during different stages of life. For e.g. Nervous system grows up to the age of 5 years, while reproductive organs grow up to puberty.

Negative Growth: It is a natural process and usually occurs during old age when cells are gradually lost and are not replaced. More catabolism than anabolism, Decrease in body weight, Irreversible change

Growth Retardation: This is the failure of DNA replication and protein synthesis due to exogenous processes. Factors causing growth retardation include hormonal imbalance (e.g., GH, T_3, T_4, Chromosomal abnormality, Radiations, Toxins, Hypoxia, Drugs, Infections.

Sympathetic nervous system also influences the rate of body growth. In case of stress, increased sympathetic activity leads to growth retardation as epinephrine released during stress may alter the secretion of growth hormones.

Growth retardation accompanies metabolic diseases in which the body is unable to synthesize or use protein, there is an abnormal ratio of amino acids in the extracellular fluid, and impaired transport of amino acids from the GIT to plasma or into body cells. Reduced protein synthesis is due to low intake of protein or increased loss of protein (nephrosis).

Catch up growth The growth of an individual which does not take place when normal condition is not provided. When suitable environment is provided, normal growth resumes. For example, in poultry used in case of poultry for meat purposes, this needs to be with great precaution. If used for longer periods, it causes permanent damage to animal.

23.3 GROWTH PATTERN

Isometric Growth: When growth rate of organs and the body is the same. Growth of young animals follows an isometric pattern.

Allometric Growth: When growth rate of organs and the body is different. Growth of the mammary gland during puberty follow allometric pattern.

23.3.1 Growth of Cells, Tissues, and Organs

Any increase in size inevitably involves a change in form (relative dimensions) e.g. a sphere.

If a sphere is doubled in weight and volume, its surface area is increased by 60% and diameter by 26%.

Now, if the sphere happens to be a **cell**. The surface area of the cell in relation to its mass may be very important for exchange between cells and between a cell and external environment i.e. incorporation of nutrients and removal of waste products through cell surface to the extra-cellular fluid.

There is a limit to the size to which an individual cell can grow.

Some organs have characteristically smaller cells than others and these have particularly higher rate of metabolism e.g. liver and kidney have smaller cells than the heart.

Differential growth: Proportional dimensions of the body organs may change during prenatal or postnatal period of development. The different growth centers of the body become active at different times and exhibit different rates of activity. Activity of these centers is coordinated to form pre-determined proportion.

Huxley gave an equation to pre determine the size of an $\boxed{Y = bxk}$ organ.

Where Y = is the size of organ, x = is the size of body, b = co efficient (which is function of body size that organ represents), k = 1; when the growth is isometric (K can be +ve or –ve)

Body organs and tissues grow at characteristic rate determined by the genetic set up. Some grow rapidly during early stage of life and others grow later. The maximum rate of growth occurs in a definite sequence; e.g. CNS has maximum growth rate first, followed by bones, muscle and adipose tissue respectively. The growth gradients run from cranium backwards and from extremities of limbs upwards to meet in lumbar region. Loin region grows maximum followed by pelvis, thorax, neck, head and limbs grow least. The composition of the body varies according to growth, breed, sex, age, nutritional status and physical activity.

23.4 MEASURE OF GROWTH

The mostly growth measure by live weight. The live weight recording is most widely used technique in experimental field and day to day record to determine growth rate in animals. The advantage of this technique – it is easy, less time consuming and less expensive. But care must be taken in respect of – precision of weighing machine and human error.

The growth rate may be expressed as absolute gain in weight in unite time, by formula- w_2-w_1 / t_2-t_1, where w_1 and w_2 are initial and final body weight and t_1 and t_2 are initial and final time.

The relative growth calculated by w_2-w_1 / w_1, where w_1 and w_2 are initial and final body weight. The relative growth rate is also very much misleading when final weight is more than the initial weight. Brody has suggested alternative formula W-Ae Kt.

Where, w = the natural logarithm of weight of the animal at time t

e = natural logarithm of W when

t = 0 and K is the constant relative growth rate. This value when multiplied by 100 gives the percentage growth rate.

23.5 BODY MEASUREMENTS

Body measurement, exhibits highest correlation with body weight is heart girth. This measurement mostly used in case of cattle and horses and to a limited used in case of sheep and goat very limited extent on pigs.

In small animals like goat accurate estimation of body weight from body measurement has been suggested (Bhattacharya et al., 1984). The best equation for body weight suggested are

(i) $Y = -24.5382 + 0.4878X_1 + 0.2325X_3$ ($R^2 = 0.8329$) and

(ii) $Y = -23.8391 + 0.5244X_1 + 0.2665X_4$ ($R^2 = 0.8338$)

Where Y = body weight in kg and X_1, X_3 and X_4 respectively are heart girth, length and circumference of neck in cm.

23.6 PHASES OF GROWTH

Growth in higher animals start even before birth, the day zygote is formed and continue till aging/senesence. Thereafter, negative growth starts. The rate of growth is not uniform throughout life. It varies during different periods of life. According to Broadly, growth is divided into two phases: Prenatal Phase and Postnatal Phase.

23.7 PRENATAL GROWTH

Prenatal growth starts when zygote is formed and continue till the birth of animal. This includes a series of orderly differential processes that transform a single celled zygote into multicellular organism (Figure 23.2).

23.7.1 Three Stages of Prenatal Growth

23.7.1.1 Cleavage

Division and re-division of cells without an increase in the volume of cytoplasm (Figure 23.3).

The cleaving cells are known as blastomeres. Cleavage is characterized by a progressive increase in cell number but a decrease in cell size. Nuclear multiplication involves the synthesis of significant amounts of DNA, which is one of the main constituents of chromosomes and chromatin. The early cleavages are synchronous and occur at specific times after fertilization. During later cleavages, cell division is asynchronous since different parts of the embryo cleave at different rates. When the dividing cells reach a certain number characteristic of the species, the cells start pushing each other due to the accumulation of fluid. The embryo, then called a blastocyst, which is a spherical or flattened hollow ball. The cavity developed inside is called

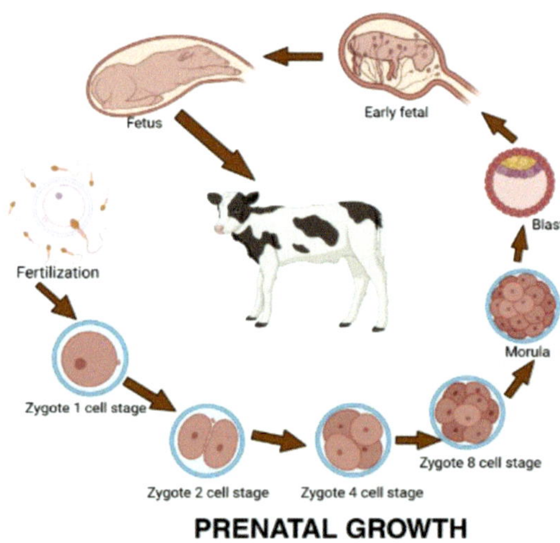

FIGURE 23.2 Prenatal growth

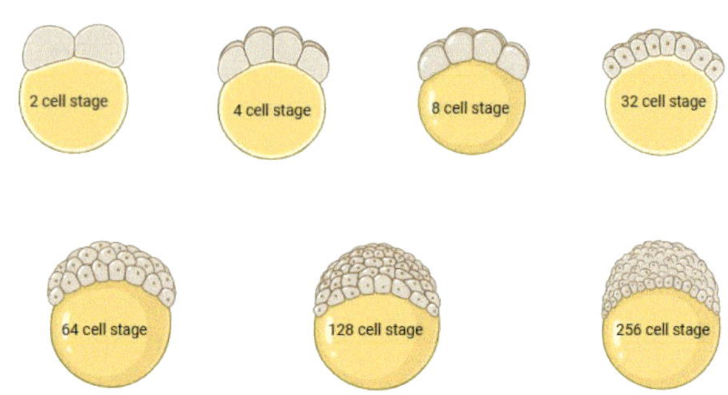

FIGURE 23.3 Cleavage of cells

blastocoelic cavity. The blastomeres vary in size, yolk content, and cytoplasmic organization, but until this stage, no organ formation starts.

23.7.1.2 Differentiation

Progressive specialization of cells both structurally and functionally. During the 1st week of gestation, the blastocyst differentiates into two types of cells: a single layer of outermost cells known as the trophoectoderm/trophoblast and a small aggregation of cells beneath the trophoectoderm known as inner cell mass (ICM)/embryonic disc. The trophoectoderm leads to the establishment of the placenta, whereas inner cell mass develops into the embryo proper.

The blastocyst then further differentiates into gastrula, which is a three-layered structure

- Outermost layer: Ectoderm
- Middle layer: Mesoderm
- Innermost layer: Endoderm

These three layers are also known as germinal layers/germ layers. These three germinal layers give rise to the organs of the embryo.

23.7.1.3 Organogenesis (Organ Formation)

Organogenesis starts when the three germinal layers are formed at the gastrula stage. It occurs by:

- Invagination (in folding)
- Envagination (out folding)
- Budding
- Hollowing out

Cells from each of the three germ layers aggregate into primordial cell masses from which special organs will be formed. The outermost layer or ectoderm forms a mid-dorsal ridge along the antero-posterior axis of the blastocyst quite early in development. This elongated ridge of neural ectoderm subsequently forms the brain, spinal cord, and other derivatives of the nervous system. Ectodermal cells lateral to the neural ectoderm develop into the skin and its derivatives such as mammary glands, sweat glands, sebaceous glands, hair, nails, and hoofs.

Innermost layer (endoderm) forms:

- Inner lining of the gut and its glands,
- Liver and pancreas
- Lining of trachea, bronchi and lungs
- Thyroid, parathyroid and thymus gland
- Urinary bladder

Middle layer (mesoderm) gives rise to:

- Bones and muscles
- Reproductive and excretory system
- Blood and blood vessels
- Inner layer of skin/dermis

Organogenesis completed in 35 days in cows, 40 days in horses, and 60 days in humans.

Most organs are not functional when they first appear in the embryo. The embryonic structures start functioning only after further differentiation. The differentiation and development of embryonic and fetal structures are called **developmental horizons** and are useful criteria for determining the age of embryos and fetuses. For example, in cattle, the earliest formation of most organs and body parts occurs during the 2nd – 6th week of gestation. The digestive system, lungs, liver, and pancreas develop from the primitive gut during this period.

23.7.1.4 Fetal Growth

Once the organogenesis is completed, the embryo is now called a fetus. After organogenesis, there is a period of **rapid increase in the organ size** of fetus (Figure 23.4).

This increase in dimensions of the fetus is a result of **cell proliferation (Hyperplasia) and an increase in cell size (Hypertrophy)**. This is called fetal growth. The phases of cellular differentiation in which the fetal tissues develop their definite characteristics are known as **histogenesis** (e.g., cells which become muscle, develop contractile fibrils). The **rate of growth** in early stages of fetal development depends upon:

Genetic make up which is variable among species. In general, the growth rate is faster in smaller species than in larger ones.

Supply of adequate nutrition. Nutrition has a major influence on growth. It should be adequate in the early stage of fetal development so that the fetus becomes larger and gains considerable weight per day.

Maternal blood supply. The mother's blood supplies the fetus with all its nutrients. The quantity of blood reaching the fetus determines its rate of growth.

Variation in Prenatal Growth. There are many variations in the prenatal growth of different species that might be due to their different genetic make up, developmental process and gestation period

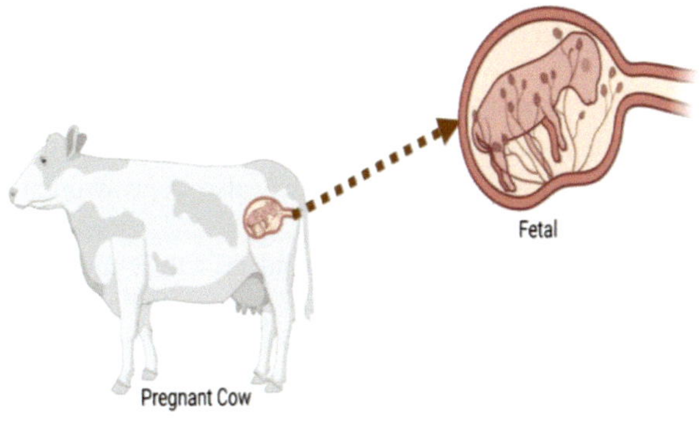

FIGURE 23.4 Fetal growth

In **sheep**, a lamb is born with about the same weight as a human infant but the gestation period is only half as long (150 days vs. 280 days). The reason being is that the fetal lamb grows more rapidly than the human fetus during the first stage of development, and this difference is maintained throughout its development in utero.

In **rats**, during first 3 weeks of growth, the increase in the number of cells goes from single cell to 2-3 million cells. By the time when it leaves the uterus, it weighs about 5 g, while in humans, in the same time period from conception, a human fetus attains only about one-hundredth of this weight and number of cells.

Small rodents (mouse, rat, rabbit, etc.) grow extremely rapidly, while cat, dog, pig, goat, sheep, cattle, horses, etc., grow less rapidly in the early stages. But these species grow for a longer time in the uterus, so they grow larger in size than rat.

23.7.2 Growth of Fetus Occurs Differentially

The birth weight of the fetus = approx. 60% of the total weight of the conceptus. The fetal membranes (amnion, chorioallantois, and chorionic villi) and placental fluids (amniotic and allantoic) make up the remaining 40%. Different parts of the conceptus grow at different rates during gestation. For example, during the **early stages of gestation**, the fetal membranes grow rapidly, and their absolute weight reaches a maximum during mid-pregnancy. The increase in the volume of placental fluids corresponds more or less to that of the fetus. Toward the **end of gestation**, the placental fluids tend to decrease in volume, although the fetus is still growing.

In the **early stage** of embryonic development, the cephalic region grows rapidly, so that the head of the fetus is disproportionately large. During the **later stages** of fetal life, when growth in the cephalic region is less rapid, the proportions of the neonate are established. The growth gradients of the fetus follow a definite order (e.g., the head, limbs, and the forequarter are relatively more developed than other parts of the body at the time of birth). Since the various organs of the fetus grow at vastly different rates, the conformation of the fetus is continuously changing.

23.7.3 Factors Affecting Prenatal Growth

Heredity/genetics: Genes influence the growth of an individual animal. The maternal contribution is more important than the paternal, which is directly affects litter size and fetal size. Further variations between species, strains, etc., also depend upon genetic set up.

Size of dam: The faster prenatal growth is directly related to larger maternal size. Walton and Hammond conducted experiments on reciprocal crosses between two different breeds of horses: -
Large Female X Small Male (Shire X Shetland) resulted in foals bigger in size
(Shire) (Shetland)
Larger Male X Small Female resulted in foals smaller in size
This difference in size was maintained even up to 4 years of age. The smaller maternal environment restricts the growth of the fetus, making parturition easier. A small placenta restricts its growth.

Age of dam: A young mother that is not fully developed during pregnancy competes with the fetus for available nutrients, thus not let the fetus to grow to its full potential.

Parity of dam: Young ones born to aged females are also smaller, because there is a lot of fat deposition around the uterus, which restricts further growth of the young one. (Less space availability).

Maternal nutrition: The fetus continues to grow up to near normal with poor nutrition during early pregnancy. However, young ones born to poorly fed dams during late pregnancy show stunted growth and these mothers may give birth to premature young ones.
In **sheep/cattle**: if the animal during early pregnancy but properly during late pregnancy, there is no effect on fetal growth, the birth weight is also normal. But if poorly fed during late pregnancy, lambs have stunted growth.
During last stage of pregnancy the variation in fetal weight reflects the differences in genetic factor, litter size, **nutritional status**, and health of the dam. When the mother is adequately fed, the birth wt. is towards the upper limit (male fetus grow more than the female). Poor nutrition of mother affects glycogen content of the fetal muscle and liver that may be the cause of neonatal mortalities. Fetal glycogen uses as a source of energy immediately after birth and normally builds up during later stage of pregnancy. Restriction of diet of mother produces differences in growth of different organs of fetus body. The nervous system, skeletal system and heart are least affected, whereas kidneys, lung and muscles are moderately affected. The skin, thymus, spleen and liver are highly affected organ of fetus.

Litter size: In polytocous species, increase in litter size retards the prenatal growth. Similarly, in monotocous species, multiple births usually retard prenatal growth but the combined weight of two or more fetus is more than the single fetus. This is due to variation in function of placenta. The placenta determines the prenatal growth. Smaller growth with increase in litter size is mainly because lesser nutrients reach each fetus. Thus, a small fetus could be due to Inherent low growth potential and Poor supply of blood, nutrients from mother to fetus.

Placenta size: Smaller size of placenta retards the size of fetus as the nutrients pass to the fetus through the placenta. The placenta acts as a blood reservoir. So bigger the placenta more blood (nutrients) reach the fetus. When there is large no. of placenta in one uterine horn, it decreases the flow of blood and oxygen supply to the distally positioned fetus/ placenta which leads to intra uterine transfusion syndrome which markedly affects the prenatal growth.

Ambient temperature: Stunted prenatal growth is observed when pregnant animals exposed to high temperature. Degree of stunted growth of fetus in uterus depends on the length of exposure to the environmental temperature.

Fetal hormones: Fetal hormones regulate functioning of the fetus. When the levels of these hormones are not sufficient they may affect the fetal growth, e.g. Growth hormone, thyroid hormones, adrenal hormones etc. GH regulates metabolic processes in the growing fetus while thyroxin and androgens are required for morphogenesis. In man and rat, the brain fails to develop normally when thyroxin is deficient. Morphogenesis of external genitalia (e.g. scrotum and penis) is androgen dependent. In absence of fetal androgens, morphogenesis of external genitalia fails to occur and external genitalia will not develop properly.

23.8 POSTNATAL GROWTH

The period of postnatal growth extends from birth until death of animal. Visible growth stops long before death and there is restricted growth that occurs in certain tissues only till death. Postnatal growth has 4 phases: 1. Prepubertal Phase 2. Pubertal Phase 3. Reproductive Phase 4. Senescence Phase

1. **Pre-pubertal phase:** This phase extends from birth of animal up to the period of acquisition of puberty of animal. During this phase, animal grows proportionately i.e. when growth rate of organs and the body is the same (Isometric pattern). There are 3 main tissues which develop very fast during this phase: Bones, Muscles and Adipose tissue.

   Bones grow before birth as well as after birth. A bone grow in length through the ossification i.e. deposition of calcium at both the ends. Due to ossification, when epiphysis gets closer, the bone growth stops However, bones have the capability that they can repair themselves if broken. Bone elongation stops but bone tissue is constantly being deposited and resorbed.

   Muscle fibers are formed during prenatal growth and there is no further increase in the number of muscle fibers after birth. Postnatal growth of muscle fiber is characterized by increase in length of muscles and diameter of muscles i.e. increase in the size of muscles (hypertrophy). As muscle fibers are mainly made up of proteins i.e. actin and myosin, therefore, the growth of muscles is decided by the rate of protein synthesis.

   Growth of **adipose tissue** is totally depends upon the nutritional status of the animal. There are two types of fat tissue

   a) **White fat** which stores energy.
   b) **Brown fat** which maintain constant body temperature especially in young animals.

   Fat is deposited in four different areas of the body

   a) **Intra abdominal fat** i.e. Around kidneys and pelvic area in abdominal cavity
   b) **Subcutaneous fat** i.e. under the skin
   c) **Intramuscular fat** i.e. between muscles

The level of intramuscular fat deposition is called the degree of marbling. It has a importance in meat industry as marbling increases the palatability of meat. Bones, muscles and fat tissue growth occur differently throughout the different phases of postnatal growth of animals. Majority of the muscle tissue grow during the pre pubertal periods and at peak level during the puberty. Thereafter, the growth of muscles slower down. Bone growth stops when the animal reaches a mature body size i.e. once puberty attain, bone growth stops but bone tissue deposition and resorption continue till death. Growth in adipose tissue depends upon the diet of animal.

2. **Pubertal phase:** Phase of growth when the animal has attained puberty i.e. growth of reproductive organs and mammary tissue (in females) starts.

   Pubertal age of different animals:
   Rabbit = 3–4 months
   Sheep = 6–7 months
   Goat = 6–7 months
   Pig = 6–7 months
   Dogs = 6–20 months
   Cats = 5–12 months
   Cattle = 12–18 months
   Buffalo = 18–24 months
   Horse = 15–24 months

   Pubertal Phase: Phase of growth of reproductive organs

   In females: Growth of ovary, uterus, mammary glands and other secondary sexual characters.

   In males: Growth of testes, musculature and other secondary sexual characters. Prior to this phase, reproductive organs grow in proportionate to whole of the body (Isometric Growth) but, at the time of puberty, anatomical and physiological changes occur in hypothalamus which causes release of gonadotropins and thus extensive reproductive growth takes place (Allometric Growth).

After puberty, bone elongation stops due to closure of epiphysis. Therefore, skeletal size does not increase, but, live body weight continuously increased due to Increase in deposition of proteins and fats.

3. **Reproductive Phase:** During this phase, reproductive organs become functional. Females start showing annual or rhythmic cycle of estrus. If there is successful mating in this period, there will be successful pregnancy and after full term of pregnancy, there will be parturition of the young one and then there will be phase of lactation.
4. **Senescence/Aging Phase:** It is a natural process and usually occurs during old age when cells are gradually lost and are not replaced. More catabolism than anabolism, Decrease in body weight and Irreversible change and ultimately death of animals

    Three types of changes occur during Aging: Anatomical changes, Physiological changes and Degenerative changes.

    Anatomical changes: Decrease in no. of muscle fibers, Decrease in muscle performance, Decrease in protein deposition, Decrease in fat deposition, Loss in CNS activity and Change in organs & glands.

    Physiological changes: The metabolism and thermoregulatory activity is reduced with increase in age, Decrease in Hb, RBC concentration, WBC concentration, Blood supply and Bone marrow efficiency, Decrease in ACTH, Estrogen, Progesterone, testosterone, FSH & LH secretion and Increase in cholesterol & triglycerides.

    Degenerative Changes: Degenerative changes start from the periphery of the organs i.e. the cells at the periphery degenerate first and then the cells in the middle portion. Irreversible loss of cells, Gradually functional loss and finally Death of animal.

    Senescence: The process of aging or beginning of old age is known as senescence.

    Longevity: It means how long an animal lives. It can be measured by median life span or how long an animal remains active.

Life span of different animals:

Man	90 years
Elephant	70 years
Horse	50 years
Donkey	50 years
Monkey	30 years
Goat	20 years
Dog	20 years
Cow	20 years
Fox	15 years
Rabbit	15 years
Guinea pig	8 years
Rat	5 years

23.8.1 Measurement of Postnatal Growth

Growth in postnatal animals can be measured by height, length, girth/circumference and body weight.

Height of an animal is the vertical distance from the ground to the highest point of withers. It is measured in case of horses in hands (a hand is equal to 4 inches) and in inches in other animals.

Length of animal is measured in inches from the point of shoulder to the point of buttocks in case of horses and from the point of shoulder to pin bone in case of cattle, sheep and goat.

23.8.2 Factors Affecting Postnatal Growth

Genetics: Genetic composition of the animal decides the potential for growth and development of animal. All animals have set genotype that determine the growth. However, phenotype is affected by external factors like nutrition, climate, disease status of animal etc.

Nutrition: Balanced supply of calories, amino acids, lipids, vitamins, minerals and water is necessary for proper growth of animal as these nutrients furnishes energy and raw material for growth. Nutrition is important aspect to determine the postnatal growth. Restriction of caloric intake and nitrogen retention greatly reduced intracellular DNA synthesis, although the cells may continue to grow. If the dietary protein is restricted but caloric intake and nitrogen retention is normal, DNA synthesis continues but the cell size is reduced. Thus both protein and total caloric intake are important. Vitamins and minerals are also essential for normal growth of the organism. Severe deficiencies or excess may cause striking changes in metabolism and thus, growth and development of an animal. e.g., Iodine deficiency causes hypothyroidism and also inhibits brain development during early phases of postnatal growth.

Climate: There are direct and indirect effects of climate on animal growth as the living organisms respond to changing weather conditions i.e. temperature, humidity, rain etc. in that particular area.

Change in Body Weight: Environment affect the food intake, energy utilization and protein metabolism in animals. During cold environmental conditions, plenty of food is available and moreover, food intake is increased, that results in gain in body weight of animals. However, feed availability is reduced as well as food intake is reduced due to hot environmental condition which results in loss of body weight.

Development: *Change in body form*: Animals living in the **cold environment**, typically appear **short having thick rough coats**, while those living in the **warm conditions** are **long, lean and sleek coated**.

Body Composition: Deposition of protein is less affected by environmental stress and it is the fat which is more affected. Animals will be fattest in a thermoneutral environment and those outside this zone will be leaner.

Weight of organs: Visceral organs are enlarged as a result of cold exposure. Hypertrophy of lungs, heart spleen, liver and kidneys occurs.

During cold→ ↑ in food intake → ↑ metabolism → means more work for organs like liver/ Kidney → Hypertrophy results

Pathological status of animal Various disease conditions may cause change in body weight of animal (may increase or decrease) which adversely affect the growth of animal.

Hormones:

Hormones directly or indirectly influence the growth by altering the biochemical reactions. Hormones which have a direct effect on whole body growth include Somatotropin/Growth Hormone

Thyroxine

Androgens

Estrogens

Corticosteroids

Somatotropin (Growth Hormone): GH causes a general increase in body mass. GH affects skeletal and protein metabolism. GH stimulates growth of epiphyseal plates of long bones in growing animals. GH administration promotes nitrogen retention, decrease in plasma amino acid levels and urea excretion (due to increased transport of amino acids to skeletal muscles, heart, kidney and liver). GH also stimulates protein synthesis by directly increasing the ability of ribosomes to incorporate amino acids into protein. GH can act alone to promote growth and nitrogen retention, but its full effects are observed only in the presence of insulin and thyroxine. GH also increases lipid mobilization from adipose tissue and reduces lipid synthesis.

Thyroxine: Essential for the maintenance of basal metabolic rate (BMR). Hypothyroidism decreases the BMR, the ability to respond to GH secretion, appetite, food intake and thus slows down the growth. Thyroid hormones and goitrogens such as thiouracil or its derivatives given to swine, sheep, poultry, beef and dairy cattle showed a positive response on growth, milk/egg production wool production.

Corticosteroids: Excessive corticosteroids have adverse effects on bone growth and results in growth retardation in young animals. Excessive amounts of corticoids, especially mineral corticoids, are catabolic to bone and produce a negative calcium balance. Glucocorticoids cause an increase in urinary nitrogen excretion and carbohydrate synthesis from amino-acids (gluconeogenesis). Glucocorticoids increase amino acid transport to the liver and decrease the same to skeletal muscles. As a result, there is decrease in protein synthesis in the skeletal muscles. Thus affect the growth of animals.

Specific tissue growth factors affect proliferation and development of specific type of cells. Important tissue growth factors are

- Epidermal growth factor (EGF)
- Nerve growth factor (NGF)
- Erythropoietin (EP)
- Retine and promine
- Chalone

23.9 EPIDERMAL GROWTH FACTOR (EGF)

Present in sub-maxillary and parotid glands and play important role in cell proliferation and differentiation. Injection of salivary gland extract from the male mice to the newborn mice causes gross anatomical changes like

- Opening of eyelids occurs as early as 7 days instead of 14 days.
- Eruption of incisors occurs as early as 6–7 days instead of 8–10 days.

Nerve growth factor (NGF): NGF is synthesized in neurons and enhances the growth and differentiation of sympathetic ganglia. NGF stimulates incorporation of amino acids and increased RNA synthesis. In newborn (<9 days old) mice, the sympathetic ganglia respond to NGF by hyperplasia and hypertrophy. In adult (>9 days old) mice, these ganglia respond by hypertrophy.

Erythropoietin (EP): Mainly produced by kidney in all species of animals. It facilitates the entry of Fe into the bone marrow before synthesis of hemoglobin and increases the release of erythroid cells into the circulation.

- **Retine and promine**
  Retine – Growth inhibitors.
  Promine – Growth promoters.

Present in thymus, muscle, tendon, urine, large blood vessels. Both are present together in the tissue. They stimulate or inhibit cell growth depending upon their ratio and concentration.

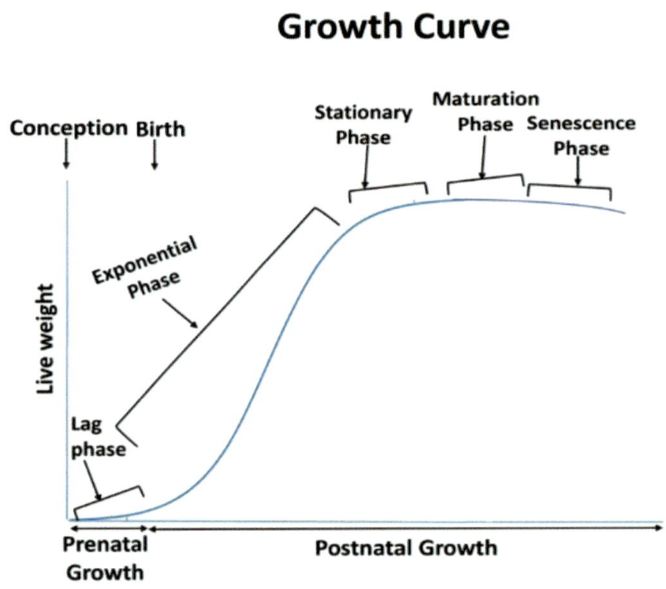

FIGURE 23.5 Growth Curve

Chalone Substances inhibit the mitotic activity of specific tissue. Isolated from liver, kidney, epidermis and granulocytes.

23.9.1 Growth Curve

The growth rate of an individual at different phase of life can be represented by plotting the live weight of individual at different Phase of life on a graph paper is called growth curve. The growth curve of animal is "sigmoid or S" shaped. The difference between the initial and final weight or initial and final size of an individual for any period of time is the absolute growth (Figure 23.5).

23.9.2 Sigmoid Growth Curve Has Three Phases

Lag Phase: It is the first period during the growth phase, where the curve rises gradually. In this phase, the organism is getting prepared for growth by synthesizing enzymes, cells, and accumulating substances to metabolize protoplasmic components.

Log/Exponential Phase: During this period, growth begins slowly at first and becomes rapid later on. Hence, the curve rises steeply.

Stationary Phase: The exponential growth does not continue indefinitely. The curve, therefore, rises slowly and becomes horizontal, signifying the limit of growth. During this phase, the rate of anabolism is exactly equal to catabolism. Thereafter, **negative growth** starts where catabolism exceeds anabolism, which finally leads to the **death** of animal.

Growth is a biological process which occurs through a series of cellular proliferation, differentiation, and morphogenesis resulting in a continuously increase in the size and weight of an animal. Growth takes place through the processes of increase of cell numbers and size. Growth and development of cells, tissues, and organs occur during prenatal and postnatal life of an organism. Various factors that regulate growth during prenatal periods include heredity, endocrine factors, maternal age and nutrition, sex of the fetus, litter size, and environment. Several factors which influence postnatal growth include genetic factors, nutritional, climatic, managemental, and health status of animal. The factors and biological process that influences the growth of meat-producing animals and poultry species is very much essential to maximize the growth for meat and egg production efficiency, and quality of the product yield by the animals. However, the growth process may be manipulated by using various growth promoter, feed additives, and supplementation of growth factors to improve the growth of animals.

BIBLIOGRAPHY

Bhattacharya, B., Ghosh, T.K., Duttagupta, R., and Maitra, D.N. (1984) Estimation of body weight in Black Bengal goats from body measurements. *Indian Vet J* 61:406–408.

Cunningham, J.G. and Klein, B.G. (2007) *Textbook of Veterinary Physiology*. 4th Ed. Saunders Elsevier, St. Louis, Mo.

Duke's Physiology of Domestic Animals, 12th. Edited by M. Swenson and W. Reece, Cornell University Press, 2004 [A thorough and comprehensive presentation of domestic animal physiology for advance students]

Frandson, R.D., Wilke, W.L. and Fails, A.D. (2003) *Anatomy and Physiology of Farm Animals*. 6th Ed. Lippincott, Williams and Wilkins, Philadelphia, 450P. [introduction to physiology of farm animals.

Reece, W.O. (1997) *Physiology of Domestic Animals*. Williams and Wilkins, Baltimore. 464P [A concise presentation of introductory systems physiology in animals.

24 Environmental Physiology

Vishwa Ranjan Upadhyay, Swagatika Priyadarsini and Aruna Kuniyal
ICAR-National Research Centre on Camel, Bikaner, Rajasthan, India

24.1 INTRODUCTION

The organisation and development of molecular and organismal variations in nature, observed on a global, regional, and local level, follow structured patterns and are not arbitrary. These patterns exhibit consistency across different forms of life and are positively linked to, and to some extent, can be anticipated by both the biotic and abiotic environmental diversity and challenges (Nevo 2001). As a branch of physiology, environmental physiology scrutinises the physiological responses of organisms to the diverse physical, chemical, and biological components within their surroundings. This field encompasses a wide range of topics to understand how animals function and respond to their natural environments at all stages of their life cycles, as well as how their bodies' physiological processes deal with the effects of wide-ranging temperature, relative humidity (RH), radiant energy, precipitation, air circulation, atmospheric pressure, altitude, radiation, and pollution. Individuals or species demonstrate distinct responses to stimuli through unique behavioural patterns, thereby significantly impacting their ability to thrive in their natural surroundings. When animals are confronted with changes in their environment, they normally display one of three categories of response: avoidance, conformity, or regulation. Regulation, the major physiological aspect, can be defined as the maintenance of a constant internal environment under varying external environments, typically known as homeostasis. However, there are no perfect regulators and depending on conditions, some animals may behave as ionoconformers, osmoconformers, thermoconformers, or oxyconformers. The formal scientific exploration of the behavioural responses, reflecting how animals interact with their native environment and react to stimuli, is termed ethology.

Bioclimatology, a subset of animal ecology, deals with the inter-relationships between components of the environment, viz., climate, soil, plants, and animals. A satisfactory environment is one that satisfies various criteria, like physical comfort without any manipulation in the design of the fixtures and fittings within the shelter, disease control without any stringent biosecurity measures, thermal comfort without any ameliorative actions, and lastly, behavioural satisfaction and physiological comfort. When animals are continuously exposed to major environmental challenges, they slowly develop certain morphological and functional mechanisms that make them fit for those environmental conditions. Broadly, these coping mechanisms for extreme climatic conditions are categorised into habituation, acclimation, acclimatisation, genetic or biological adaptation, and phenotypic or physiological adaptation (Gaughan 2012). Acclimation refers to the synchronised phenotypic adjustments exhibited by the animal in response to a specific environmental stressor. In contrast, acclimatisation denotes a coordinated response to multiple concurrent stressors, encompassing factors like temperature, humidity, wind speed, and solar radiation. Both acclimation and acclimatisation mechanisms contribute to enhancing the adaptability of animals to dynamic climatic alterations. Correspondingly, adaptation involves genetic modifications that occur gradually over extended periods, aiming to mitigate the adverse effects of environmental variables on an animal's biological systems as these stressors persistently impact the organism. In instances of extreme climatic conditions, animals display peculiar adaptive responses such as decreased metabolic rates or states of dormancy, which are manifested through mechanisms like sleep, torpor, hibernation, and estivation.

24.2 ENVIRONMENTAL STRESS

Stress within the microenvironment of an animal arises due to the interplay among crucial interacting biotic and abiotic factors, stress level, the extent and duration of variations, and the energy or resources accessible to manage stressors. The interconnections among these elements significantly influence the array and variety of animals present in that environment, as well as the specific type of evolutionary selection that takes effect (Wilmer et al. 2009). Abiotic environmental stress, defined as the negative impacts of non-living factors on living organisms under a specific environment, also affects the pasture and forage availability to livestock (Dawson et al. 2014). Despite the neutralising tendency of the vegetation, the introduction of an animal into an area also alters local conditions by adding excretory by-products (such as CO_2), depleting oxygen, and modifying humidity and temperature levels. Consequently, an environment hosting an animal undergoes distinct changes compared to its state before the animal's presence. Environmental stress such as drought, high/low temperature, ozone, elevated CO_2, soil water logging, and salinity reduces the productivity and health of livestock in all phases of production resulting in significant economic

DOI: 10.1201/9781003426851-24

losses. Under field conditions, most of the time, the multiple stresses, i.e., heat stress, nutritional stress, water stress, and walking stress, occur in combinations and simultaneously cause a cascade of stressful conditions and ultimately culminating in serious health anomalies. The outcomes of multiple stress include decreased growth, body weight, production potential, reproductive performance and increased susceptibility to diseases.

Environmental stressors induced by climate change encompass diminished availability of feed and water, alterations in temperature, and an escalation in extreme weather events. Anticipated global climate change is expected to impact temperature, precipitation, atmospheric CO_2 concentrations, and water availability which will likely influence the productivity of both crop and livestock systems (Hatfield et al. 2008). Instabilities in the atmospheric concentrations of various greenhouse gases, namely carbon dioxide (CO_2), methane (CH_4), nitrous oxide (N_2O), hydrofluorocarbons (HFCs), perfluorocarbons (PFCs), and sulphur hexafluoride (SF_6), impact the earth's capacity to absorb radiation. These greenhouse gases trap infrared radiation by reflecting it from the surface, functioning akin to a thermal insulating layer that sustains warmth within the planet's atmosphere. Over the past 250 years, the concentrations of these greenhouse gases have markedly increased, primarily attributed to the amplified utilisation of fossil fuels which potentially induces indirect alterations in the hydrologic cycle. Environmental stressors have the potential to disrupt an organism's internal equilibrium, resulting in physiological alterations that can either be advantageous or disadvantageous. The individual response to these stress-inducing factors is influenced by various determinants, including species, breed, prior exposure to the stressor, health condition, performance levels, body condition, metabolic status (such as pregnancy or lactation), mental state, and age (Figure 24.1). Adding complexity, stressors may manifest as acute (sudden changes), typically short-term alterations spanning hours to days or chronic (prolonged exposure), spanning weeks to months. Animals exhibit distinct responses to acute and chronic stressors by employing behavioural strategies as an initial defence, resorting to intricate and resource-demanding physiological adaptations only when behavioural strategies prove ineffective. Although acute responses play a crucial role in adaptation and aiding survival, chronic stressors prompt supra-physiological reactions that potentially contribute to health hazards (Sejian et al. 2010). At higher limits, stress triggers diverse neuroendocrine responses in animals, initiating the activation of various hormonal pathways and the release of regulatory hormones. Changes in the hormonal balance primarily involve a decrease in anabolic hormones and an increase in catabolic hormones. These alterations can serve as potential indicators of physiological changes within an animal's body (Bernabucci et al. 2010).

24.3 ADAPTATION AND ADAPTABILITY

Adaptation is commonly referred to as the observable characteristics or traits found in animals, which are outcomes of the process of selection. However, a more precise definition views adaptation as a process whereby natural selection adjusts the traits impacting an organism's fitness and its offspring's survival in successive generations. The adaptation process typically unfolds slowly over generations and is often irreversible, which is discernible by various adaptive traits under different environmental conditions (Table 24.1). Conversely, acclimatisation denotes a swifter phenomenon where an individual animal undergoes physiological or biochemical changes due to exposure to new environmental conditions. Acclimation refers to akin processes occurring in laboratory settings in response to experimentally induced alterations in conditions. Both acclimatisation and acclimation can be described as immediate compensatory alterations in response to environmental disruptions and may be reversible. Additionally, short-term changes due to acclimation or acclimatisation lead to phenotypic plasticity that aligns with local conditions (Wilmer et al. 2009). Livestock initially responds to adverse environmental conditions through coordinated and interconnected behavioural mechanisms, and later by mutual interactions among the immune system, central nervous system, and endocrine system. This behavioural adaptation often involves seeking refuge in an unfavourable microclimate, particularly during critical periods of the day. For example, to cope with heat stress conditions, the zebu cattle native to tropical climates possess various behavioural, physiological, morphological, and genetic adaptive traits (Singh et al. 2018). Hence, animal adaptation assumes significance in comprehending observable animal responses, which is depicted in Figure 24.2.

Thermal adaptation predominantly emphasises maintaining an elevated and consistent body temperature. Heat transfer can vary with the area of contact in the short term by behavioural means (basking, burrowing, wallowing, huddling, erecting or concealing appendages, etc.), while thermal conductivity, which is the function of external surface (hair, coat, skin, etc.), is often the most important long-term adaptive element in determining conductive heat gain or loss. Some ecological rules, in particular to domestic animals, explain morphological adaptations, which are the basis for their climatic plasticity and breeding tract -

1. Bergmann's Rule (1847): Correlates body size with climate, indicating that smaller animals with larger surface area per unit weight are better suited to warmer climates, whereas larger breeds thrive in cooler temperate areas.
2. Allen's Rule (1877): Links extremity length to climate, observing that animals in cold regions tend to have shorter extremities compared to those in warmer climates within the same species.

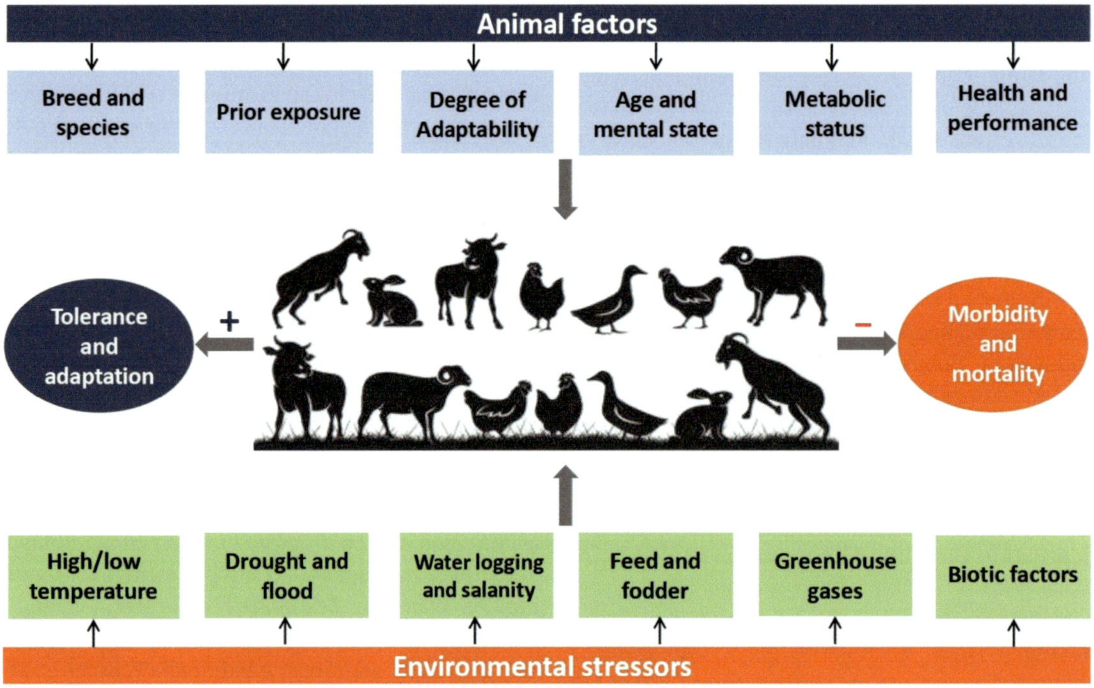

FIGURE 24.1 Bimodal interplay of animal factors and environmental stressors.

TABLE 24.1
Key Adaptive Mechanisms of Various Species of Animals Thriving in Different Environmental Conditions

S. No.	Environmental conditions	Adaptive mechanism	Species
1	High temperature	High surface area to body weight ratio, skin folds, long external appendages, loose coarse wool, hair shedding in summer	Cattle, sheep, goat, and donkeys
2	High humidity	Dark pigmentation, smooth and scanty hairs	Buffalo, cattle
3	Solar radiation	Long limbs, short reflective coat	Camel
4	Low temperature	Long hair intermixed with fine hair, minimum exposed extremity, large body size, thick heavy coat	Temperate cattle, sheep, yak
5	Polar region	Layers of fur and long hair, layer of fat under skin, wide paws, strong sense of smell	Polar beer, reindeer, fox, musk oxen
6	Desert	Hard tissue and thick skin around mouth, better recycling rate of urea, efficient digestive system, low metabolic requirement	Camel, sheep, goat
7	High altitude	Polycythemia, high O_2 carrying capacity in blood, high rate and depth of respiration, high efficiency in extracting nutrients from feed	Llama, alpaca

3. Wilson's Rule (1854): Pertains to insulating cover and climate, highlighting that breeds in cold climates possess dense, heavy coats, while those in warm climates exhibit shorter, glossy, and thinner hair.
4. Gloger's Rule (1833): Relates skin pigmentation to climate, noting that animals in cooler or hot, dry climates tend to have lighter skin colour, which darkens with increased temperature and humidity.

24.4 BIOLOGICAL RHYTHMS

Biological rhythms in animals refer to recurring and predictable patterns of biological activities or behaviours that follow a specific time frame. These rhythms are inherent in various physiological processes, including sleep-wake cycles, hormonal secretion, feeding behaviours, body temperature, metabolism, reproduction, migration, and other essential functions. Circadian rhythm represents a fundamental biological rhythm observed in domestic animals and refers to a roughly 24-hour cycle that regulates various physiological and behavioural processes. The animals active during the daytime have their minimum temperature in the early morning and maximum temperature in the late afternoon, while the reverse is found in animals that are active during the night. Biological rhythms can be influenced by internal factors such as internal body clocks or external cues like light-dark cycles, temperature changes,

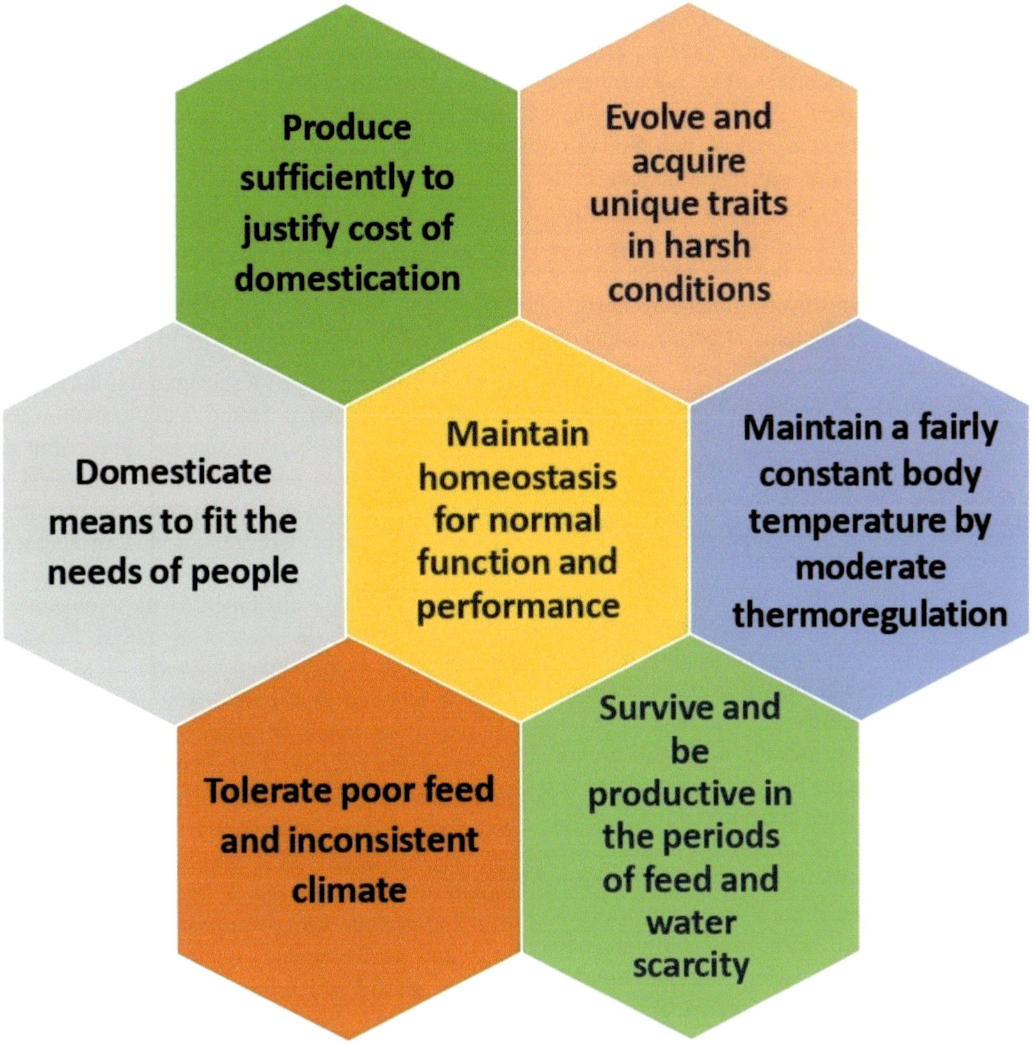

FIGURE 24.2 Implications of adaptation process in the process of selection.

or social interactions. In domestic animals, such as dogs, cats, cattle, horses, and others, the diurnal rhythms synchronise their activities with environmental changes and play a crucial role in governing their daily activities, feeding schedules, rest periods, and overall behaviour. In addition, changes in daylight duration play a significant role in regulating the breeding seasons of different animal species. Short-day breeders are animals whose reproductive activities are stimulated or influenced by shorter day lengths, often in response to decreasing daylight. Examples include sheep, goats, deer, and many small mammals like hamsters. On the other hand, long-day breeders are species whose reproductive activities are triggered or influenced by longer day lengths. Birds like chickens and turkeys, as well as some species of horses, fall into this category. In vertebrates, the pineal gland serves as a neuroendocrine organ that translates information about the duration of environmental light exposure into a biochemical messenger called melatonin. The magnitude and duration of the production of melatonin are dependent on the length of the dark phase of the photoperiodic cycle. The suprachiasmatic nucleus (SCN) receives signals regarding light and darkness through the retinohypothalamic pathway which regulates the secretion of melatonin by pineal cells (Aggarwal and Upadhyay 2013).

24.5 ANIMAL THERMOTOLERANCE

The thermal environment is the most important ecological factor as normal body functions require relatively constant temperature and any deviation on either side alters the chemical processes and efficient functioning of the body. Based on the stability of the body temperature, animals are classified as homeotherms and poikilotherms. The homeotherms, also known as warm-blooded animals, can control their body temperature despite varying external temperatures. These groups of animals are called temperature regulators, e.g., mammals and avian species. Contrarily, poikilotherms, or cold-blooded animals, cannot maintain a constant body temperature as their body temperature varies with the temperature of the environment. These groups of animals are called temperature conformers, e.g., reptiles, amphibians, fishes, and invertebrates. The homeotherms behave as endotherms and they generate their own heat through metabolic heat production and do not depend on the environment for their heat thus can maintain their body temperature above the ambient temperature. In addition, they employ substantial physiological and biochemical adjustments like changes in blood flow or respiratory rate, an increase in shivering or nonshivering thermogenesis, or the production of heat-shock proteins (HSP), aquaporins, and antifreeze molecules.

The capacity to manage body temperature is an evolutionary adaptation seen in warm-blooded animals, enabling the use of temperature as a signal to regulate internal processes. Thermosensitive receptors in the body, both in peripheral areas and within the central nervous system, detect temperature changes and trigger appropriate responses. The peripheral and core receptors transmit nerve impulses to the specific centres in the hypothalamus involved in feed intake, enzymatic activities, hormone synthesis and heat loss or production. For livestock, maintaining a stable body temperature relies on balancing heat generation and dissipation, primarily achieved through three key pathways: adjusting the body's temperature to the surrounding environment through behavioural thermoregulation; dispersing heat through processes like vasodilation, conduction, convection, radiation, and evaporation; and by making alterations in metabolic rate. Accordingly, these pathways of heat exchange are categorised into non-evaporative or sensible heat exchange which includes conduction, convection and radiation, and evaporative or insensible heat exchange (loss) as evaporation of water from the skin and respiratory passages. The various factors which affect this heat exchange include physical and optical properties of hair coat, length and density of hair follicles, hair colour, properties of the skin, etc. Similarly, when environmental temperature decreases, animals prevent a decline in their body temperature by employing physical regulatory mechanisms to minimise heat loss. If these physical regulatory methods become insufficient in maintaining the body temperature, animals resort to increasing heat production through chemical regulatory processes as a defensive measure. In domestic animals, the reduction of heat loss is taken care of by behavioural response, circulatory adjustment, piloerection and increased fur insulation.

At an animal level, we primarily consider microenvironments, or microhabitats, or microclimates because it define the actual experience by the animal on an appropriate spatial and temporal scale. The range of temperatures as depicted in Figure 24.3, within which the animal uses no additional energy to maintain its body temperature is called the zone of thermal comfort, a subset within the thermoneutral zone. This zone represents the temperature range where the physiological costs of animals are minimal and productivity is maximum. Within this zone, fluctuations in environmental temperature do not impact metabolic heat production, and circulatory adjustments effectively sustain a consistent body temperature. Beyond this thermoneutral range, circulatory adaptations are insufficient to maintain heat equilibrium, and this range is influenced by factors such as age, body mass, insulation, acclimatisation, evaporative cooling capacity, feed intake and activity of the body.

24.6 IMPACT OF CHANGING ENVIRONMENT ON BODY PHYSIOLOGY, REPRODUCTION, AND PRODUCTION

The environment exerts a substantial influence on the physiology, reproductive potential, and production performance of animals across various species. The rising demand for livestock products presents market prospects and financial gains for small-scale, marginal, and landless farmers. However, the global livestock production sector encounters mounting challenges due to adverse environmental consequences, primarily attributed to the prevalence of erratic climatic patterns (Soumya et al. 2022). Sudden spikes in temperature or prolonged periods of high temperatures lead to reduced feed intake, altered metabolic processes, decreased reproductive efficiency, and indirectly through water scarcity affect hydration, feed quality, digestion, and overall health of animals. Likewise, acute cold stress reduces glucose availability for normal development and growth in multiple species and is accompanied by enhanced thermogenesis at the expense of tissue accretion and growth. Moreover, losses in muscle mass, growth rate, and feed efficiency occur due to the partitioning of energy toward shivering and non-shivering thermogenesis, increased maintenance costs, heat production, altered lipid metabolism, and other digestive inefficiency (Young 1981; Steffen 1988). Predictably, many of the metabolic and physiological effects of heat and cold stress are biologically contrasting; however, the ultimate consequence of heat and cold stress is similar as both are homeorhetically orchestrated to prioritise survival at the cost of production (Figure 24.4).

24.7 MANIPULATION AND MODULATION APPROACHES

Reports indicate that adaptation often comes with a tradeoff in performance, where animals displaying lower performance levels tend to have better survivability. This advantage stems from their minimal body requirements and higher tolerance, significantly impacting the economic aspects for associated animal farmers. The adoption of genetic selection for stress-tolerant animals, along with environmental modifications, improved nutrition, and crossbreeding for particular phenotypes, can mitigate the impacts of environmental stress (Habimana et al. 2023). Traits like tolerance to heat, cold, seasonal changes, parasite infestation, poor quality feed, and metabolic disorders

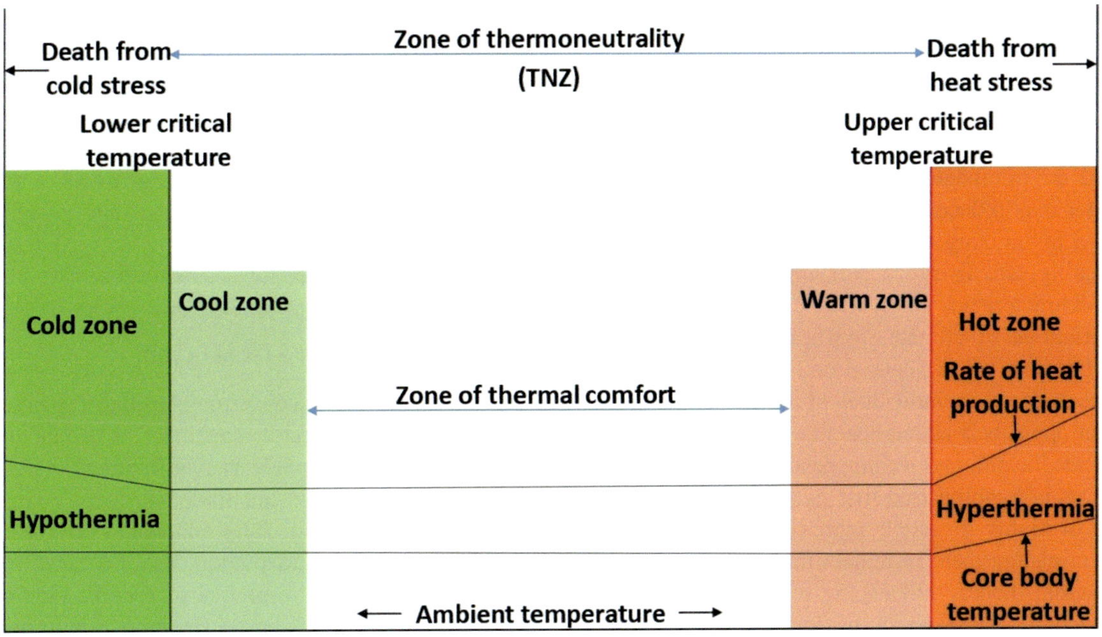

FIGURE 24.3 Thermoneutral zone.

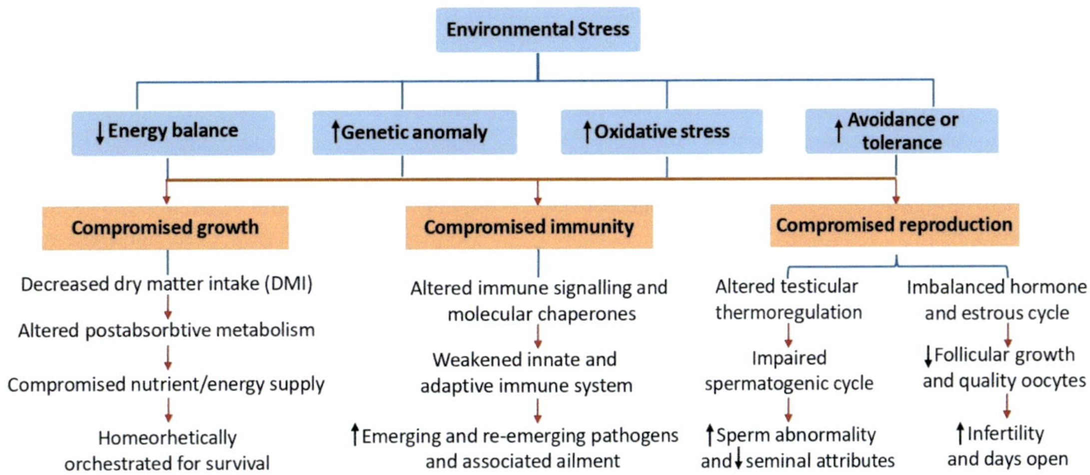

FIGURE 24.4 Impact of environmental stress on overall growth and physiology of animal.

are worthwhile in tropical climates. To achieve optimal production under changing climatic scenarios, a holistic approach is needed, tailored to environmental conditions and available resources employing vital strategies like genetic improvement, breeding management, grazing management, nutritional management, water management, shelter management, disease management, utilisation of unconventional feed resources, antioxidant supplementation, and hormonal therapy.

24.7.1 Physical Modification of Environment and Shelter Management

The common strategies for improving the comfort of animals and sustaining their growth and production performance during inclement environmental conditions are mainly executed by physical modification of the production environment. Modification of the environment involves heat abatement and cold abatement technologies that alter the environmental conditions to prevent undesirable stress. In addition, modification of the environment to provide physical and behavioural comfort is important both for welfare and production reasons (Mohankumar et al. 2012; Sejian et al. 2015). Modifications in response to heat stress include the management of shelter, such as proper ventilation, floor space, watering facilities, and other strategies that enhance heat exchange between the animal and the environment. The common strategies employed nowadays include the application of fans, misters, cooling pads, tunnels, and sprinklers that can enhance passive ventilation, improve body heat loss, decrease body temperature, and enhance dry matter intake (DMI).

24.7.2 Nutritional and Feeding Strategies

Prolonged environmental stress with periods of limited or no water may reduce the quality and quantity of forage production for grazing and hay making. Modelling studies evaluating safe livestock carrying capacity based on resource characteristics and climate data suggest that even slight variations in climate significantly impact pasture growth. Feeding adjustments aimed at augmenting feed intake are tailored to individual production systems and contingent upon their added benefits in enhancing animal performance, health, and welfare relative to the associated expenses. It is hypothesised that nutrition, especially the inclusion of highly digestible ingredients, plays a crucial role in diminishing heat increment through efficient nutrient utilisation, thereby influencing overall animal growth and production. The provision of decreasing the forage to concentrate ratio by providing degradable starch such as cereal grains (e.g., wheat grain and corn grain) reduces the amount of heat produced and also increases the absorption of ration. During periods of scarcity, strategic utilisation of unconventional feed resources can help in fulfilling an animal's nutritional needs by reducing the impact of antinutritional factors (ANF). Further exploration into the application of anti-stress minerals, vitamins, and herbal supplements, along with increased grain and supplemental fat, is warranted to mitigate the adverse consequences of drought on livestock health and productivity.

24.7.3 Genomic Responses During Acclimation and Genetic Selection

Genetic variations at protein and DNA levels show increased diversity when exposed to environmental and genomic stress, a well-documented phenomenon by numerous researchers. Environmental diversity and stress contribute to the development of genetic variations, especially in environments with dynamic fluctuations, which can generate intricate cycles and behaviours reminiscent of chaos (Nevo 2001). Apart from the selection of phenotypes, as determined by genes and their constituent DNA, permanent changes in protein structure produce subtle changes in various biochemical and signalling activity which have effects on the expression of diverse proteins serving as raw material for adaptation. For instance, at the biochemical level, there may be increased thermal tolerance of an enzyme, or expression of a more pH-stable allozyme; physiologically, there might be altered signalling regimes, secondary messenger activity, or Na-K pump, all this, in turn, may lead to permanent heritable changes in the development of an organism. Estimation of genetic parameters against complex multiple stressors is crucial for designing optimal breeding schemes and for predicting selection response. For example, the genes found to be associated with heat tolerance can be used for molecular detection of thermotolerance and as candidate genes to explore their use in crossbreeding schemes to transfer their adaptation abilities and rusticity. Data from gene expression or genome-wide association studies (GWAS) can be applied to further enhance the accuracy of selection, genome editing technology and crossbreeding. Identifying and exploring signatures of selection, genomic diversity, and candidate genes and modelling the effect of a breed as a function of time and environment is an important progression toward better breeding strategy.

24.8 ADAPTATION STRATEGIES

The primary emphasis concerning animal response should centre on adaptation and management. When an animal is not acclimated or adapted, its physiological, behavioural, and metabolic reactions are likely to differ compared to when it is acclimated or fully adapted. Therefore, within discussions on animal adaptation, it is crucial to acknowledge that an animal's responses to specific stressors can evolve over time as it adjusts. However, it's important to note that performance may not revert to pre-stress levels despite these adjustments (Bagath et al. 2019). So, it is necessary to build or improve the adaptive capacity of the animal through the implementation of certain adaptation strategies and policies that aim at:

- Gaining a deeper understanding of how animals interact with their macro and micro climate.
- Understanding the interaction of climatic alterations with behavioural changes like pain, frustration, aggression, malaise, etc. and their responses on production traits.
- Understanding the mechanisms of climatic adaptation and stress tolerance of dairy cattle breeds reared in tropical countries.
- Identification of agro-ecological zone specific climate-resilient thermotolerant animals.
- Advancing the use of sequence data from phenomics, genomics, and transcriptomics technologies and the application of marker-assisted selection for traits related to adaptation.
- Integrating indigenous knowledge on cattle genetic diversity for locally adapted breeds via selective breeding.
- Enhancing feeding, housing, and selection criteria for improved animal breeding practices.
- Disseminating new insights to farmers for effective herd management against climatic stress.
- Implementing short and long-term climate adaptation policies for sustainable livestock development and resilience to climate change.

24.9 CONCLUSION AND WAY FORWARD

Given the ongoing emergence of environmental challenges, the exploration of environmental physiology is poised to gain escalating significance in scientific inquiry. The global influence of climate change is evident, leading to prolonged disruptions in typical weather patterns to which our livestock are accustomed. Animals thriving in various regions have developed multiple strategies to adapt to their environments and the changes, essential for their survival, thereby giving rise to diverse genotypes through distinct mechanisms. Still, even slight alterations in an animal's normal body physiology significantly affect production, activating specific mechanisms that divert energy from productive purposes to maintenance needs. One of the primary means by which the environment influences animals is through the exchange of energy which is the function of heat gain from the environment, metabolic heat production and heat loss from the body. The presence of vegetation produces very stable microclimates as it regulates the amount of radiation reaching the ground and humidity level. Moreover, presence of an animal in a small enclosed space also modifies the microclimate for other animals, as it emits both heat and water vapour, and also produces liquid and solid wastes which equilibrate with the surrounding air. So, it becomes essential for animals to maintain a poised situation and remain within their thermoneutral zone, where there is no need for additional energy expenditure to regulate cellular physiology. Within dairy cattle systems, employing zebu breeds in hot climates proves more advantageous than temperate breeds as zebu cattle consistently exhibit enhanced adaptation to heat, diseases, and harsh conditions owing to their evolutionary roots. A holistic, integrative analysis of the organism-environment interaction can form the basis of understanding about evolutionary processes of adaptation and speciation, linking molecular and organismal organisational levels under complex biotic and abiotic stressors. Lastly, the study of environmental physiology may offer insights into the tolerance and resource usage efficiency of diverse animals which can aid in formulating selection criteria and breeding policies to ensure sustained ecological balance amidst anticipated climate change in the forthcoming decades.

BIBLIOGRAPHY

Aggarwal, A., and R. Upadhyay. 2013. *Heat stress and animal productivity*, 188, Springer, New Delhi, India.

Bagath, M., G. Krishnan, C. Devaraj, et al. 2019. The impact of heat stress on the immune system in dairy cattle: A review. *Research in Veterinary Science* 126: 94–102.

Bernabucci U., N. Lacetera, L. H. Baumgard, et al. 2010. Metabolic and hormonal acclimation to heat stress in domesticated ruminants. *Animal* 4: 1167e83.

Collier, R. J., and K. G. Gebremedhin. 2015. Thermal biology of domestic animals. *Annual Review of Animal Biosciences* 3: 513–532.

Dawson I. K., S. Carsan, S Franzel, et al. 2014. *Agroforestry, livestock, fodder production and climate change adaptation and mitigation in East Africa: Issues and options. ICRAF-178*. World Agroforestry Centre, Nairobi.

Gaughan, J. B. 2012. Basic principles involved in adaption of livestock to climate change. In: *Environmental stress and amelioration in livestock production*, edited by V. Sejian, S. M. K. Naqvi, T. Ezeji, et al. Springer-Verlag GMbH Publisher, Berlin, Germany.

Habimana, V., A. S. Nguluma, Z. C. Nziku, et al. 2023. Heat stress effects on milk yield traits and metabolites and mitigation strategies for dairy cattle breeds reared in tropical and subtropical countries. *Frontiers in Veterinary Science* 10: 1–21.

Hatfield, J., K. Boote, P. Fay, et al. 2008. Agriculture. In: *The effects of climate change on agriculture, land resources, water resources, and biodiversity in the United States. A Report by the U.S*, 362. Climate Change Science Program and the Subcommittee on Global Change Research, Washington, DC.

Kadzere, C. T., M. R. Murphy, N. Silanikove, et al. 2001. Heat stress in lactating dairy cows: A review. *Livestock Production Science* 77: 59–91.

Mohankumar, S. M. J, P. Balasubramanian, M. Dharmaraj, et al. 2012. Neuroendocrine regulation of adaptive mechanisms in livestock. In: *Environmental stress and amelioration in livestock production*, edited by V. Sejian, S. M. K. Naqvi, T. Ezeji, et al. 263–298. Springer-Verlag GMbH Publisher, Berlin, Germany.

Nevo, E. 2001. Evolution of genome–phenome diversity under environmental stress. *Proceedings of the National Academy of Sciences* 98: 6233–6240.

Sejian, V., R. Bhatta, J. Gaughan, et al. 2015. *Climate change impact on livestock: Adaptation and mitigation*, 491. Springer.

Sejian, V., V. P. Maurya, and S. M. K. Naqvi. 2010. Adaptive capability as indicated by endocrine and biochemical responses of Malpura ewes subjected to combined stresses (thermal and nutritional) under semi-arid tropical environment. *International Journal of Biometeorology* 54: 653–661.

Singh, S. V., S. Soren, C. G. Shashank, et al. 2018. Global warming: Impact, adaptation and amelioration strategies for bovine under tropical climatic conditions. *Indian Journal of Animal Sci*ence 88: 5–20.

Soumya, N. P., R. Banerjee, M. Banerjee, et al. 2022. Climate change impact on livestock production. In *Emerging issues in climate smart livestock production* S. Mondal and R. L. Singh (eds.), pp. 109–148. Academic Press, Amsterdam.

Steffen, J. M. 1988. Glucose, glycogen, and insulin responses in the hypothermic rat. *Cryobiology* 25: 94–101.

Upadhyay, V. R. 2019. *Effect of late gestation heat stress on colostrum quality of Sahiwal cows in relation to the immune system and growth performance of neonate calves* (Doctoral dissertation, NDRI, Karnal).

Willmer, P., G. Stone, and I. Johnston. 2009. *Environmental physiology of animals*. John Wiley & Sons.

Young, B. A. 1981. Cold stress as it affects animal production. *Journal of* Animal Science 52: 154–163.

25 Physiology of Aging

Jyotsana Shakkarpude[1], Deepika Diana Jesse A[1], Pramod Kumar[2] and Rajesh Kumar[3]

[1]Department of Veterinary Physiology and Biochemistry, College of Veterinary Science and Animal Husbandry, Mhow, Indore, MP, India
[2]Department of Veterinary Physiology and Biochemistry,
[3]Department of Veterinary Gynaecology and Obstetrics, College of Veterinary Science and Animal Husbandry, Acharya Narendra Deva University of Agriculture and Technology, Ayodhya,UP, India

25.1 INTRODUCTION

Aging is a biological process that leads to an increasing probability of degenerative diseases. All animals have a definite life cycle which includes birth, a period of rapid growth, puberty, a gradual cessation of growth, senescence, and finally death. The pattern is rather rigidly fixed by the genetic background of the organism and can be varied only by rather drastic measures. Animals have very efficient repair mechanisms, which allow them to recover from almost any type of damage or disease and yet, in the absence of demonstrable malfunction, the organism degenerates and finally succumbs to one of the degenerative diseases. The impact that aging has on organisms is a complex interaction between the processes of aging at a cellular, organ, and integrated systems level, and the effects of environmental factors such as nutrition, infection and trauma.

Biological aging, termed senescence (the process of aging), is one of the most complex biological processes. Aging can be defined as the time-related deterioration of the physiological functions necessary for survival and fertility. Aging is defined as the progressive accumulation of damage over time, leading to disturbed function on the cellular, tissue, and organ levels and eventually to disease and death. It is a multifactorial process where genetic, endogenous, and environmental factors play a role. Aging has been associated with a loss of complexity in a wide range of physiological processes and anatomic structures. Rather, they would consider aging to be the default state occurring after the animal has fulfilled the requirements of natural selection. Lipsitz and Goldberger (1992) first proposed that the aging process may be characterized by a progressive loss of physiologic complexity. We contend that this loss of complexity results in the functional decline of the organism by diminishing the range of available, adaptive responses to the innumerable stressors of everyday life. After its offspring are born and raised, the animal can die. The maximum life span is a characteristic of the species. It is the maximum number of years a member of that species has been known to survive.

There are several levels at which we can study aging, including cellular, biochemical, and genetic studies. Reactive oxygen species (ROS) can damage cell membranes, inactivate proteins, and mutate DNA. Mutations that alter the ability to make or degrade ROS can change the lifespan of the mutants. Mitochondria may be a target for proteins that regulate aging. Aging is the time-related deterioration of the physiological functions necessary for survival and reproduction. The phenotypic changes of senescence (which affect all members of the species) are not to be confused with diseases of senescence, such as cancer and heart disease (which affect individuals) (Figure 25.1).

25.2 PHYSIOLOGICAL CHANGES

Physiological changes occur with aging in all organ systems. The cardiac output decreases, blood pressure increases, and arteriosclerosis develops. The lungs show impaired gas exchange, a decrease in vital capacity, and slower expiratory flow rates. The creatinine clearance decreases with age although the serum creatinine level remains relatively constant due to a proportionate age-related decrease in creatinine production. Functional changes, largely related to altered motility patterns, occur in the gastrointestinal system with senescence, and atrophic gastritis and altered hepatic drug metabolism are common in the elderly. Progressive elevation of blood glucose occurs with age on a multifactorial basis, and osteoporosis is frequently seen due to a linear decline in bone mass after the fourth decade. When placed in a cold environment, an old rat takes much longer to adjust its metabolism to the changed condition than does a young one. The epidermis of the skin atrophies with age, and due to changes in collagen and elastin, the skin loses its tone and elasticity. Lean body mass declines with age, primarily due to the loss and atrophy of muscle cells. Degenerative changes occur in many joints, and this, combined with the loss of muscle mass, inhibits elderly patients' locomotion. These changes with age have important practical implications for the clinical management of elderly patients: metabolism is altered, changes in response to commonly

DOI: 10.1201/9781003426851-25

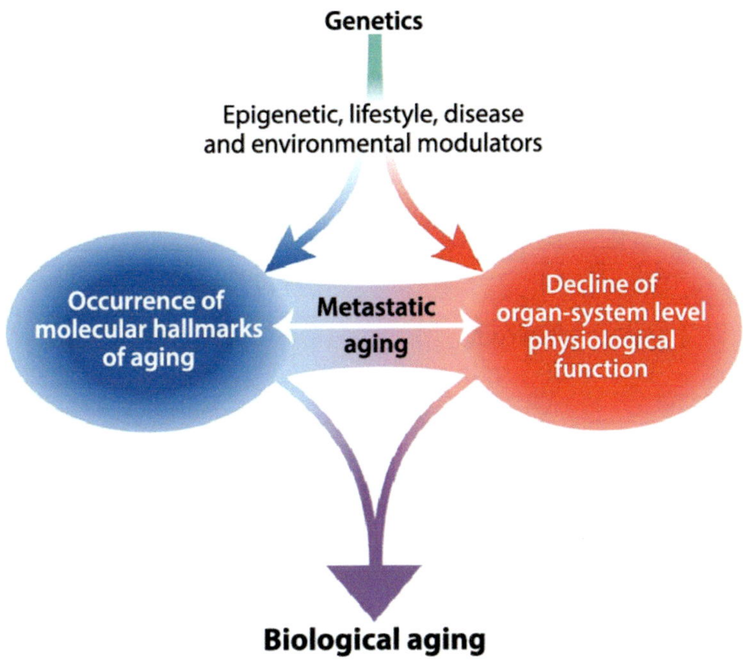

FIGURE 25.1 Biological aging.

used drugs necessitate different dosages, and there is a need for rational preventive programs of diet and exercise in an effort to delay or reverse some of these changes.

25.3 ANATOMICAL CHANGES

In most mammals, growth virtually stops soon after puberty and body weight remains relatively constant for the rest of the life span. Each animal has its preferred body size at any age and will attempt to attain this size if proper food is available. The common observations with regard to aging are

1. Decline in muscular performance:
   As the animal grows and increases the size and strength of the muscles, the size of the individual fibers increases but not their numbers. In senescence, the number of fibers decreases, the muscle mass tends to decrease with age, but the amount of decrease is quite variable.
2. Loss in central nervous system:
   It has been estimated that the average man loses about 10,000 brain cells per day out of about 10 billion.
3. Accumulation of Collagen:
   Once the collagen Once collagen is deposited, it is very insoluble and tends to remain in the tissues for years. Many of the connective tissue cells die and are replaced by collagen fibers. These fibers tend to shrink and join in the course of time, forming a dense network. This accounts for the wrinkled appearance of the skin of old persons. In addition, the dense network tends to impede normal function. Most of the other organs of the body tend to decrease in size as the individual ages. The liver and kidneys decrease by about 20% in man between 25 and 70 years of age, and the spleen by as much as 40%. Another tendency in both animals and humans is to accumulate fat as they grow older. This obesity varies greatly among different animals and even among individuals of the same species (Figure 25.2).

25.4 DEGENERATIVE CHANGES

The decline in organ function is very important in relation to aging, but the end result of aging, death, is related to this decline in only a very peripheral way if at all. In general, the cause of death of older domestic animals is degenerative diseases. These comprise the various forms of cancer and cardiovascular diseases that can be characterized as arising primarily from within the body itself. For some people, some of the cells in the body do not function properly, and this may cause the entire machinery of the body to malfunction.

25.4.1 Maximum Life Span and Life Expectancy

The maximum life span is a characteristic of the species. The maximum lifespan of a species is how long its longest observed member has lived. Life expectancy is the time at which approximately 50% of the members of a given population of a species still survive. It is the maximum number of years a member of that species has been known to survive. The maximum human lifespan is estimated to be 121 years (Arking, 1998). The life spans of tortoises and

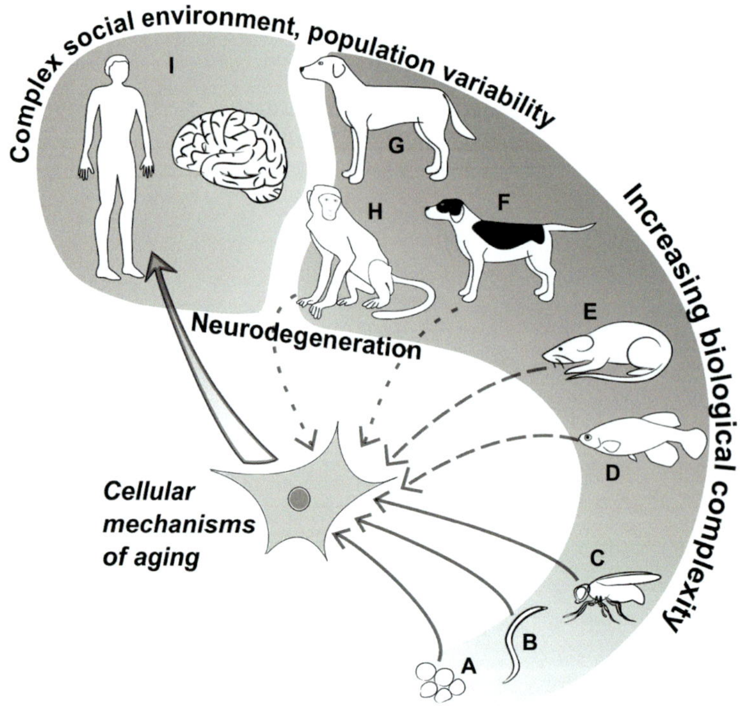

FIGURE 25.2 Model organisms of aging

lake trout are both unknown but are estimated to be more than 150 years. The maximum lifespan of a domestic dog is about 20 years, and that of a laboratory mouse is 4.5 years. Life expectancy, the amount of time a member of a species can expect to live, is not characteristic of species but of populations. It is usually defined as the age at which half the population still survives.

The phenomena of senescence and the diseases of aging are much more common today than they were a century ago. In 1900, people did not have the "luxury" of dying from heart attacks or cancers. These diseases generally occur in people over the age of 50. Rather, people died (as they are still dying in many parts of the world) from infectious diseases and parasites (Arking, 1998). Similarly, until recently, relatively few people exhibited the more general human senescent phenotype: graying hair, sagging and wrinkling skin, joint stiffness, osteoporosis (loss of bone calcium), loss of muscle fibers and muscular strength, memory loss, eyesight deterioration, and the slowing of sexual responsiveness. Life expectancy is the key metric for assessing population health. The ages of some wild animals can be determined by annual marks in scales or bones similar to treerings.

25.5 THEORIES OF AGING

Aging is a multi-factorial process. Most hypotheses on the underlying mechanisms of the aging process involve the deterioration of the maintenance of homeostatic metabolic, inflammatory, and/or redox processes in cells and tissues.

25.5.1 General Wear-and-Tear Theory

Dr. August Weismann, a German biologist, first introduced this theory in 1882. He believed that the body and its cells were damaged by overuse and abuse. The organs and skin are worn down by toxins in the diet and in the environment. Wear and tear are not confined to the organs; it also takes place at the cellular level. As one gets older, small traumas to the body build up. Point mutations increase in number, and the efficiencies of the enzymes encoded by our genes decrease. Moreover, if a mutation occurred in a part of the protein synthetic apparatus, the cell would make a large percentage of faulty proteins. If mutations arose in the DNA-synthesizing enzymes, the number of mutations would be expected to increase markedly. Faulty DNA polymerases in senescent cells have been evident. Likewise, DNA repair may be important in preventing senescence, and species whose members' cells have more efficient DNA repair enzymes live longer. Moreover, genetic defects in DNA repair enzymes can produce premature aging syndromes inhumans.

25.5.2 Neuroendocrine Theory

This theory was described in the year 1954 by Dr. Vladimir Dilman, a Russian scientist. The neuroendocrine theory of aging states that "the effectiveness of the body's homeostatic adjustments declines with aging – leading to the failure of adaptive mechanisms, aging and death." This theory has also been referred to as the aging clock theory. Consistent with this theory, the hypothalamic-pituitary-adrenal (HPA) axis, the main regulatory system controlling homeostasis in humans, loses efficiency with aging. The HPA system works by the interplay of various hormonal signals that initiate reactions in target tissues coupled with a negative feedback mechanism to allow for fine control of body functions, such as blood pressure, fluid and electrolyte levels, and body temperature. This theory proposes that aging is due to changes in neural and endocrine functions that are crucial for 1. Coordinating communication and responsiveness of all body systems with the external environment, 2. Programming physiological responses to environmental stimuli, and 3. Maintaining an optimal functional state for reproduction and survival while responding to environmental demands. Hormones are vital for repairing and regulating bodily functions, and when aging causes a drop in hormone production, it causes a decline in the body's ability to repair and regulate itself as well. Thus, hormone replacement therapy, a frequent component of any anti-aging treatment, helps to reset the body's hormonal clock and can reverse or delay the effects of aging and thus keeping young.

25.5.3 The Genetic Control Theory

This theory states that animals are born with a unique genetic code, a predetermined tendency to certain types of physical and mental functioning, and that the genetic inheritance has a great deal to say about how quickly one becomes aged and how long lives. Each person has a biological clock. When that clock goes off, it signals the bodies first to age and then to die. However, the timing on this genetic clock is subject to enormous variation, depending on what happens with the growth and on how one actually lives (quality of life, feeding, sanitation and health care practices). This theory should be restricted to cases where there is a specific control of the onset of aging by an identifiable metabolic process and a functional role for senescence can be demonstrated for that species, e.g. the rapid aging and death of the Pacific salmon after spawning (Figure 25.3).

25.5.4 Waste (Undegradable By-products of Metabolism) Accumulation Theory of Aging

In the course of the lifespan, cells produce more waste than they can properly eliminate. This waste can include various toxins (free radicals, histones, aldehydes, lipofuscins), which, when accumulated to a certain level, can interfere with normal cell function, ultimately killing the cell. The only mechanism by which these agents are diluted in the cells is cell division. This only applies to replicative cells. The challenge for multicellular organisms, such as the animal is that many cell types lose replicative capacity or divide slowly, even though they remain active throughout the lifespan. These cells, including cardiomyocytes and brain neurons, accumulate metabolic waste that eventually affects normal cell functioning. A common by-product of cellular metabolism seems to be lipofuscin (LF). Lipofuscin (age pigment) is a brown-yellow, electron-dense, autofluorescent material that accumulates progressively over time. LF is composed of highly oxidized cross-linked macromolecules (proteins, lipids, and sugars) with multiple metabolic origins. LF cannot be digested in the ubiquitin-proteasome system because of highly oxidized proteins and polymeric and highly cross-linked nature. LF cannot be cleared by exocytosis and accumulates within the lysosomes and cell cytoplasm of long-lived post-mitotic and senescent animal cells. LF inclusions from different animal and human tissues (myocardium, brain, liver, thyroid) are composed mainly of proteins and lipids, 30–70% and 20–50%, respectively. A small amount of carbohydrates (4.7%) and traces of metals are also found in LF granules. The accumulation of non-degradable material can occur in the intra- and extracellular environments. Among the extracellular deposits found in humans, cholesterol-containing plaques and their oxidized derivatives in blood vessels, as well as protein polymers, such as β-amyloid in the central nervous system.

In unstressed situations, protein homeostasis is balanced by folding and stabilization of proteins by chaperones of

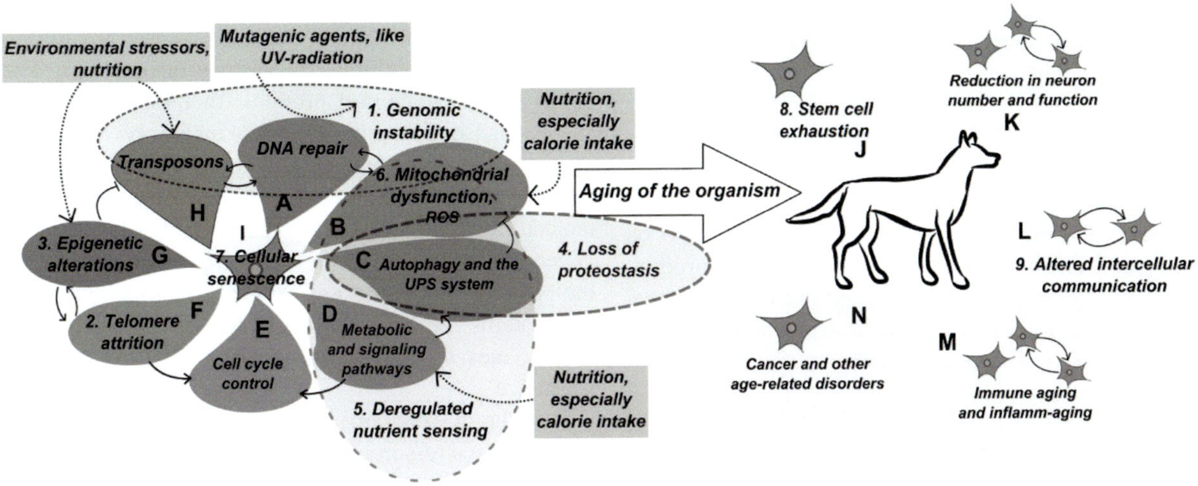

FIGURE 25.3 Main mechanisms of aging.

the HSP family and the controlled degradation of proteins by the proteasomal system. The proteasome exists in different forms, and its activity is modulated by multiple regulators. The 20S core proteasome contains the proteolytic activity and selectively degrades a multitude of oxidized proteins, as well as other substrates, in an ubiquitin- and ATP-independent manner. When the core 20S proteasome combines with two 19S regulators, the 26S proteasome is formed, which selectively removes polyubiquitinated proteins. Under stress conditions, and therefore most notably during aging, the balance between protein damage and clearance of damaged proteins is disturbed, leading to a malfunctioning of proteostasis and an accumulating mass of oxidized proteins, aggregate and aggresome formation, and finally to the accumulation of highly cross-linked materials such as lipofuscin, compromising cell viability. Accumulation of aggregates in post-mitotic cells seems to be especially dramatic since they are not able to dilute this material by cell division.

Consequences of oxidative stress on proteins, the proteasomal system, and transcription factors and formation of age pigment lipofuscin. When proteolytic capacity declines below a critical threshold of activity required to cope with oxidative stress, the final consequence is the accumulation of aggregated proteins which may, instead, cross-link with one another or form extensive hydrophobic bonds. This material might undergo further reactions and finally form the age pigment lipofuscin, which has toxic properties and accumulates in the lysosomal system. It is likely that spontaneous modification of proteins also plays a role in the formation of metabolic debris. Metabolic waste also includes, to a certain extent, spontaneously modified sugar-bound proteins, mainly glucose molecules. Glycation involves interaction between the amino groups of lysine and the aldehyde groups of glucose via a Schiff base reaction. It is followed by rearrangement of the double C=N bond, known as Amadori products, to yield a wide range of advanced glycation end-products such as glucosepane. The main consequence of spontaneous glycation is impaired elasticity, which is essential to blood vessels. In addition, spontaneous glycation affects protein functioning. This process well describes the concept of accumulation of metabolic waste that promotes aging.

25.5.5 Errors and Repairs Theory

In 1963, Dr. Leslie Orgel, a British Chemist, suggested that "an error in the machinery for making protein could be catastrophic." The production of proteins and the reproduction of DNA sometimes are not carried out with accuracy. The body's DNA is so vital that the natural repair processes kick in when an error is made. But the system is incapable of making perfect repairs on these molecules every time, and the accumulation of these flawed molecules can cause diseases and other age changes to occur.

25.5.6 Autoimmune Theory of Aging (Immunosenescence)

This theory was proposed by an American scientist, Dr. Roy Walford in the year 1969. Aging is associated with declines in adaptive and innate immunity established as immunosenescence. With age the system's ability to produce necessary antibodies that fight disease declines, as does its ability to distinguish between the antibodies and proteins. In a sense, the immune system becomes self-destructive and reacts against itself. Thus, elderly individuals usually present chronic low-level inflammation, higher infection rates and chronic diseases. These theories suggest that with age comes decreased ability to recognize "self" from "nonself". Antibody structure becomes modified with aging, causing them to acquire antigenic potential against normal body cells. Cells hidden during embryonic development appear later in life. Ab-Ag complexes from previous normal responses cause cumulative lesions characteristic of aging. Medications may form complexes with body proteins that are identified as foreign.

25.5.7 Cross-Linkage Theory of Aging

Developmental aging and cross-linking were first proposed in 1942 by Johan Bjorksten, US Chemist. He applied this theory to aging diseases such as sclerosis, a declining immune system and the loss of elasticity in the skin. The cross-linking hypothesis is based on the observation that with ageproteins, DNA, and other structural molecules develop inappropriate attachments or cross-links to one another. These unnecessary links or bonds decrease the mobility or elasticity of proteins and other molecules. Denatured proteins are irreversibly altered. Denaturation is caused by cross-links between peptide strands within a protein or between proteins. With age comes new cross-links, leading to irreversible protein structural changes and altered protein functioning.

It is thought that these cross-links begin to obstruct the passage of nutrients and waste between cells. Proteins that are damaged or are no longer needed are normally broken down by enzymes called proteases. However, the presence of cross-linkages inhibits the activity of proteases. These damaged and unneeded proteins, therefore, stick around and can cause problems. One of the main ways crosslinking occurs is through a process called glycation. Glucose molecules can stick to proteins, then transform into brownish molecules called advanced glycation end products (AGEs) When both of the sticky ends of AGEs adhere to neighboring proteins, they form permanent cross-links that disable the proteins' functions. Commonly affected body proteins a. Enzymes b. Collagen (fibrosis) c. Elastin d. Ground substance.

Cellular dysfunctions in aging or in age-related diseases by oxidative stress imbalance. 1. Cell metabolism generates reactive oxygen species (ROS) and reactive nitrogen species (RNS), which in turn causes oxidative/nitrosative damage. 2. Proteins are the most affected macromolecules by oxidative stress, undergoing several modifications that avoid their being correctly degraded and recycled by the proteasome, thus generating impaired protein function. 3. Oxidative stress also directly affects cytoskeletal proteins, causing structural damage and signaling alterations. 4. On affecting the mitochondria, oxidative stress alters energy production and 5. on affecting peroxisomes, oxidative stress alters correct metabolic functioning. 6. Oxidative stress also affects the cellular membrane. 7. Finally, all of the previously mentioned affections cause an alteration in the transcriptional activity of the cell, leading to an altered gene expression that in turn leads the cell to the aging process or to degenerative disease.

25.5.8 The Mitochondrial Free Radical Theory of Aging

With aging, mitochondria become highly susceptible to morphological changes. These changes result in reduced function due to oxygen radical damage, which eventually causes the aging of the organism (Fakouri et al., 2019). Sites of ROS production at the mitochondrial electron transport chain. The figure shows the four complexes of the respiratory chain (CxI–IV). ROS are mainly produced at CxI and CxIII. Rotenone (ROT) blocks electron transport from ubiquinone (Q) to complex I, thus avoiding complex I ROS production when succinate is used as substrate. AA, antimycin A. The Mitochondrial Free Radical Theory of Aging The mitochondrial respiratory chain, illustrating electron transfer from NADH and FADH2 to oxygen. The primary function of mitochondria is respiration, which promotes energy production. Mitochondria break down organic compounds into water and carbon dioxide to release energy in the form of adenosine triphosphate (ATP).Mitochondrial respiration generates a proton gradient across the inner membrane and a transmembrane potential through respiratory chain complexes (I–IV), enabling electron flow from the reduction equivalents NADH and FADH2 to oxygen. Simultaneously, the energy released in the oxidation of NADH and FADH2 is used to pump H+ ions out of the matrix into the space between the outer and inner membranes. During respiration, oxygen is reduced in several stages, producing a superoxide radical and hydrogen peroxide. Most commonly, these molecules, known as reactive oxygen species (ROS), remain bound to cytochrome c oxidase until the reduction of oxygen to water is completed. In contrast to the common sequence of oxygen reduction by cytochrome c oxidase, oxygen molecules can occasionally form superoxide species by reacting with the reduced components of the electron transport chain. This typically occurs at the level of complexes I and III in the respiratory chain (Figure 25.4).

25.6 OXIDATIVE DAMAGE

One major theory sees our metabolism as the cause of our aging. According to this theory, aging is a by-product of normal metabolism; no mutations are required. About 2–3% of the oxygen atoms taken up by the mitochondria are insufficiently reduced insufficiently to reactive oxygen species (ROS). These ROS include the superoxide ion, the hydroxyl radical, and hydrogen peroxide. ROS can oxidize and damage cell membranes, proteins, and nucleic acids. Evidence for this theory includes the observation that Drosophila that overexpress enzymes that destroy ROS (catalase, which degrades peroxide, and superoxide dismutase) live 30–40% longer than controls (Parkes et al., 1998). Moreover, flies with mutations in the methuselah gene live 35% longer than wild-type flies. The methuselah mutants have enhanced resistance to paraquat, a poison that works by generating ROS within cells (Lin et al., 1998).

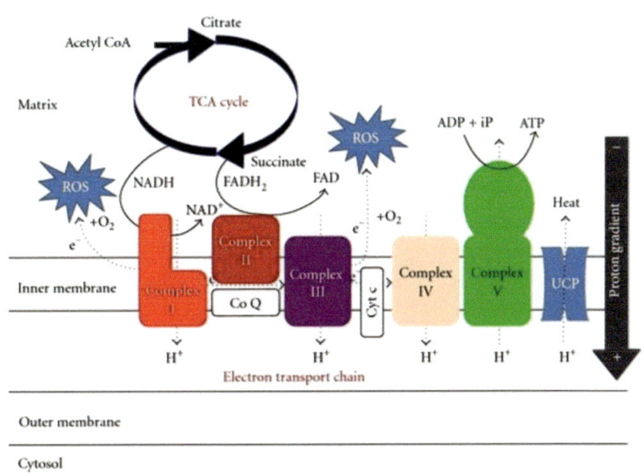

FIGURE 25.4 The electron transport chain responsible for ATP and ROS production in Mitochondria

These findings not only suggest that aging is under genetic control but also provide evidence for the role of ROS in the aging process. Individuals with mutations that increase the synthesis of ROS-degrading enzymes live much longer than wild-type. The evidence for ROS involvement in mammalian aging is not as clear. Mutations in mice that result in the lack of certain ROS-degrading enzymes do not cause premature aging. However, there may be more genetic redundancy in mammals than in invertebrates, and other genes may be up-regulated to produce related ROS-degrading enzymes. Migliaccio and colleagues (1999) have observed mutant mice that live one-third longer than their wild-type littermates. These mice lack a particular protein, p66. They develop normally, but the lack of p66 apparently gives them cellular resistance to ROS, and thus higher resistance to oxygen-induced stress on membranes and proteins. The p66 protein may be a component of a signal transduction pathway that leads to apoptosis upon oxygen stress, and it may be involved in mediating the life spans of mammals. Another type of evidence suggests that ROS may be important in mammalian aging: aging in mammals can be slowed by caloric restriction (Lee et al., 1999). However, caloric restriction can also have other effects, so it is not certain if it works by preventing ROS synthesis. Also, vitamins E and C are both ROS inhibitors, and vitamin E increases the longevity of flies and nematodes when it is added to their diet (Kakkar et al.,1996). However, results in mammals are not as easy to interpret, and there is no clear evidence that ROS inhibitors work as well as in invertebrates (Arking, 1998) (Figure 25.5).

25.7 CELLULAR SENESCENCE

Cellular senescence can be divided into two categories: replicative senescence and stress-induced premature senescence. Replicative senescence refers to the premature senescence that occurs after a limited number of divisions and a gradual shortening of telomeres at the ends of chromosomes. This results in cell proliferation stagnation and loss of differentiation ability. Stress-induced premature senescence refers to the premature senescence that occurs in response to pathological stimuli, such as DNA damage and oxidative stress. These two types of senescence share many regulatory molecules, and they both cause cell cycle arrest mainly through the p53/p21 and p16Ink4a/retinoblastoma protein signaling pathways. The number of senescent cells increases with age, obesity, and diabetes, and clearing senescent cells can alleviate many aging-related diseases and prolong the lifespan in mice. Senescent cells induce the formation of a complex, multicomponent SASP by secreting a range of cytokines, inflammatory factors, and adhesion factors. In the local microenvironment, the SASP alters the biological behavior of adjacent cells through autocrine and paracrine signaling. For instance, a systemic environment that shapes aging-related diseases has been reported to exist, as the infusion of young cerebrospinal fluid (CSF) improves the memory function of aged brains. This improvement is mainly mediated by serum response factor (SRF)-mediated oligodendrocyte progenitor cell (OPC) proliferation after exposure to young CSF. Elderly individuals often present with chronic low-grade inflammation, which is collectively referred to as immune aging. In the process of aging, the numbers of monocytes/macrophages, dendritic cells, and natural killer (NK) cells increase, possibly because of an increase in the number of aging cells, resulting in increases in the numbers of macrophages and NK cells, thereby eliminating aging cells. Immune cells may further induce the production of more pro-inflammatory cytokines and aggravate the progression of aging-related diseases.

Senescence may be one of the main features of immune senescence. Premature T-cell failure may accelerate aging in multiple organs and systems, with thymic degeneration, mitochondrial dysfunction, genetic and epigenetic alterations, and imbalance in protein homeostasis being the four main hallmarks of T-cell senescence. In mice with mitochondrial transcription factor A (TFAM) deletion, T cells with mitochondrial dysfunction induce a variety of aging-related phenotypes, such as metabolic disorders, cognitive impairment, and cardiovascular diseases, which ultimately lead to the premature death of mice.

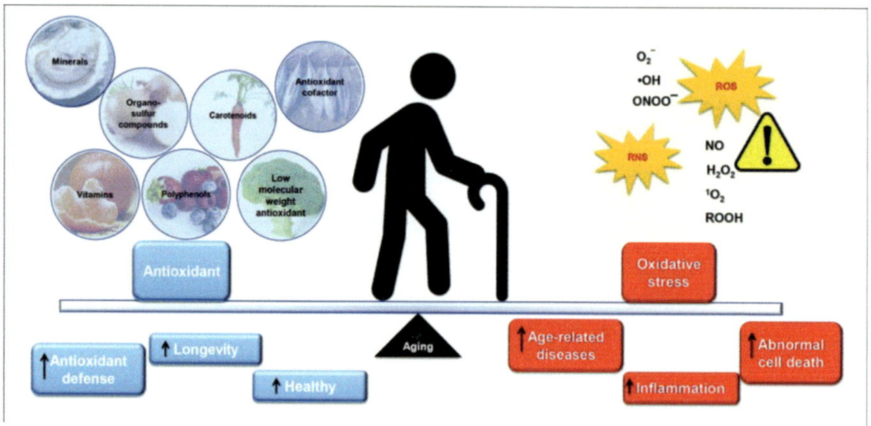

FIGURE 25.5 The balance of antioxidants and oxidative stress in aging

25.7.1 Mitochondrial Genome Damage

The mutation rate in mitochondria is 10–20 times faster than the nuclear DNA mutation rate. It is thought that mutations in mitochondria could 1. lead to defects in energy production, 2. lead to the production of ROS by faulty electron transport, and/or 3. induce apoptosis. Age-dependent declines in mitochondrial function are seen in many animals, including humans (Boffoli et al., 1994). A recent report (Michikawa et al., 1999) shows that there are "hot spots" for age-related mutations in the mitochondrial genome, and that mitochondria with these mutations have a higher replication frequency than wild-type mitochondria. Thus, the mutants are able to outcompete the wild-type mitochondria and eventually dominate the cell and its progeny. Moreover, the mutations may not only allow more ROS to be made but may make the mitochondrial DNA more susceptible to ROS-mediated damage.

25.7.2 Telomere Shortening

Telomeres are repeated DNA sequences at the ends of chromosomes. They are not replicated by DNA polymerase, and they will shorten at each cell division unless maintained by telomerase. Telomerase adds the telomere onto the chromosome at each cell division. Most mammalian somatic tissues lack telomerase, so it has been proposed that telomere shortening could be a "clock" that eventually prohibits the cells from dividing any more. When human fibroblasts are cultured, they can divide only a certain number of times, and their telomeres shorten. If these cells are made to express telomerase, they can continue dividing. However, there is no correlation between telomere length and the lifespan of an animal (humans have much shorter telomeres than mice), nor is there a correlation between human telomere length and a person's age. Telomerase-deficient mice do not show profound aging defects, which we would expect if telomerase were the major factor in determining the rate of aging (Rudolph et al., 1999). It has been suggested that telomere-dependent inhibition of cell division might serve primarily as a defense against cancer rather than as a kind of "aging clock."

25.7.3 Hayflick Limit Theory

This theory of aging suggests that the human cell is limited in the number of times it can divide. The human cell's ability to divide is limited to approximately 50 times after which they simply stop dividing and hence die. Nutrition has an effect on cells with overfed cells dividing much faster than underfed cells. This theory indicates the need slow down the rate of cell division if we want to leave long lives.

25.7.4 Death Hormone Theory (DECO)

According to this we are born with roughly 12 billion of nerve cells and over a lifetime about 10% perish. This theory states that as we age the pituitary begins to release decreasing oxygen consumption hormone (DECO)or death hormone which inhibits the ability of cells to use thyroxin, a hormone produced by thyroid governing basal metabolism, the rate at which cell convert food to energy. The metabolic rate brings on and accelerates the process of aging.

25.7.5 Thymicstimulating Theory

The thymus is the master gland of the immune system. The size of this gland reduces from 200 to 250 grams at birth to around 3 grams by age 60. Studies have shown that thymic factors are helpful in restoring the immune systems of children bone without them as well as rejuvenating the poorly functioning immune systems of elderly. Thymic hormones may also play a role in stimulating and controlling the production of neurotransmitters and brain and endocrine system hormones which means they may be the pacemakers of aging itself, as well as key regulators responsible for immunity.

25.7.6 Caloric Restriction Theory

Underfeeding particularly restriction of calorie intake increases the life span. The effect of the restricted energy intake was to slow down the growth rate and to delay the attainment of maturity. In other words, the restricted calorie intake delayed the onset of senescence by prolonging the juvenile period.

25.7.7 The Rate of Living Theory

This theory states that we are each born with a limited amount of energy. If we use this energy slowly then our rate of aging is slowed.If the energy is consumed quickly aging is hastened. Other rate of living theories focuses on limiting factors such as amount of oxygen inhaled or number of heartbeats spent.

25.7.8 Order to Disorder Theory

From the time of conception to sexual maturation, our body undergoes a system of orderliness. Most of our energy is directed to fulfill a genetically determined plan for the orderly production and arrangement of an enormous number and variety of molecules. After sexual maturation, however, this energy starts to diminish in efficiency. Disorder occurs in molecules in turn causing other molecules to produce errors and so on. These chaotic changes in our cells, tissues and organs causes aging. Disorderliness varies from individual and this may be the reason why our tissues and organs deteriorate at different rates.

25.7.9 Psychosociological Theory

In addition to theories of aging based on molecules and cells, there also exists a "psychosociological" theory of aging. As people grow older, their behavior changes, their social interactions change, and the activities in which they engage change. The psychosociological theory of aging can be divided roughly into four component theories: disengagement, activity, life-course, and continuity theories. Disengagement theory is based on hampered relationships between a person and other members of society. Activity theory emphasizes the importance of ongoing social activity and suggests that a person's self-concept (self-perspective) is related to the roles held by that person. Life-course theory is based on the developmental stages proposed by German-born American psychoanalyst Erik H. Erikson. According to Erikson's stages, maturity is a process that continues into old age, and in each stage the individual encounters new psychosocial demands. Continuity theory states that older adults try to preserve and maintain internal and external characteristics (e.g., values, personality, preferences, and behavior patterns) throughout life, despite changes in their health or life circumstances.

25.8 EXTERNAL ENVIRONMENTAL AGENTS OF AGING

25.8.1 Ionizing Radiation

The shortening of life caused by ionizing radiation (e.g., X-rays) has been determined for many species, including mice, rats, hamsters, guinea pigs, and dogs. The occurrence of some diseases, such as leukemia, may increase disproportionately after irradiation, with the degree of increase influenced by age and sex. The permanent nature of radiation damage is shown by the comparison of life spans of irradiated and control populations. An irradiated population dies out like a chronologically older unirradiated population. Members of a population given a single dose of X-rays or gamma rays in early adult life die of the same diseases that afflict the unirradiated control population, but they die months or even years earlier.

Continuous irradiation throughout life at low dose rates speeds the mortality process. Studies of animals and of cells grown in culture suggest that large doses of radiation kill by producing deleterious rearrangements of chromosomes in the proliferative cell population. Such aberrations also increase with age, but they seem to be less important in the natural aging process. At low radiation doses, chromosome aberrations become relatively less important than other effects, and the primary radiation damage in these conditions may bear a closer relation to the aging lesion. Natural radioactivity in the body, arising mostly from radioactive potassium and radium, and natural background irradiation, from Earth and from cosmic rays, are not major contributors to the aging process, even in the long-lived human species. They are responsible, however, for a small percentage of cancer incidence.

25.8.2 Temperature

Flour beetles, fruit flies, fishes, and other poikilothermic (temperature-variable) organisms live longer at the lower range of environmental temperature. An organism's life span is dependent on some critical substance that is exhausted more rapidly at higher temperature. An important factor that has not yet been adequately taken into account is the relation of metabolic efficiency to temperature. The energy cost of the biosynthetic processes studied has been discovered to be minimal at an intermediate temperature in the range to which the species is adapted and to increase at higher or lower temperatures. A related phenomenon holds for longevity; the number of calories expended by fruit flies per lifetime is maximal at an intermediate temperature, so the rate of aging per calorie is minimal at that temperature. Thermal denaturation of proteins is predominately a disruption of the folding of molecules, which requires the breaking of numbers of low-energy bonds. Mutations may arise to a significant degree from thermal denaturation.

Research has suggested that humans might live longer if their core body temperatures were lower, since in shorter-lived species there is a relationship between high metabolism, which increases core temperature, and short life span. In a study of mice engineered to have a lower-than-normal core body temperature, a reduction of about 0.5 °C (0.9 °F) was associated with a roughly 20 percent increase in life span.

25.8.3 Physical Wear of Nonrenewable Structures

One of an animal's most important assets is its chewing apparatus, including jaws and teeth. Adaptation to tooth rate of wear is especially important for animals that consume large quantities of grass and herbage. Such adaptations include higher tooth crowns (hypsodonty), larger grinding area, and longer tooth growth period. Tooth wear may be limiting for survival in adverse environments, but, on the whole, it is not an important life-limiting characteristic. The same can be said for other external organs subject to physical wear.

25.8.4 Infectious Disease and Nutrition

The populations in poor environments, characterized by high rates of infectious disease and poor nutrition, have higher death rates than populations in good environments at all ages, yet there is no positive evidence that disadvantaged populations experience a higher rate of aging. Rats kept on diets restricted in calories live longer and have lower cancer incidence than do rats that are allowed to eat at will. Maximum longevity, however, is achieved at a nutritional level that keeps the animal sexually immature and below normal weight.

25.9 INTERNAL ENVIRONMENT OF AGING: CONSEQUENCES OF METABOLISM

The metabolic activities of organisms produce highly reactive chemicals, including strong oxidizing agents. The internal structure of the cell, however, minimizes the harmful effects of such agents. The critical reactions take place within enclosed structures such as ribosomes, membranes, or mitochondria, and counteractive enzymes such as peroxidases are present in abundance. It is nevertheless likely that low concentrations of those reactive substances can reach vital molecules and contribute to the characteristic rate of aging injury. Experiments in which mice are fed low levels of antioxidants such as butylated hydroxytoluene (BHT) have been encouraging but are still somewhat equivocal.

Membranes are the site of much of the metabolic activity of cells; they provide the barriers that keep incompatible reactions separated. Membrane-bound structures known as lysosomes contain enzymes capable of digesting the cell if released. The stability of cells and organisms is therefore very much bound up with the stability of membranes. A number of drugs, including corticosteroids, salicylates, and antihistamines, act by stabilizing cell membranes against inflammatory stimuli. The mode of action of these drugs is connected to substances called prostaglandins, which can alter specific membrane characteristics.

25.10 ANTI-AGING AND LONGEVITY

The majority of anti-aging research has focused on understanding and finding ways to manipulate the metabolic pathways that are implicated in the progressive decline of biological function associated with senescence.

25.10.1 Antioxidants

One area of research into the process of aging concerns the generation of free radicals that cause oxidative stress. Reactions in which free radicals are released within cells in significant quantities can result in the oxidation of proteins and other cellular components, which can trigger programmed cell death (apoptosis). Although natural antioxidant molecules occur in cells and act to scavenge potentially harmful radicals, the development of antioxidant drugs to facilitate this process has been investigated extensively. Compounds such as retinol (vitamin A) have been found to combat skin aging by stimulating the growth of new collagen, which reduces skin roughness and wrinkling. Retinol can be incorporated into lotions, enabling its absorption directly into the skin. Several antioxidants, including selenium and resveratrol (a substance found primarily in grape skins), have been formulated into drugs for the treatment of cancer and obesity, respectively. There are a number of antioxidants sold over-the-counter; however, the dosing and safety of those agents, as well as whether or not they really have anti-aging benefits in humans, remain disputed.

25.10.2 Calorie Restriction and Longevity

The use of drugs designed to increase life span in humans is surrounded by ethical issues associated with the artificial prolongation of life. However, longevity researchers have identified certain dietary factors that influence the cellular and metabolic processes underlying age-related diseases in humans and other animals. Among humans, for example, research has shown that individuals who consume a primarily plant-based diet tend to live longer than persons who regularly consume red meat and other animal products. These discoveries are being used to understand aging in humans and to develop new approaches in the prevention and treatment of age-related diseases.

One area of anti-aging research that concerns longevity and that has revealed important information about diseases and aging is calorie restriction – the reduction of calorie intake to create a significant energy deficit while attempting to simultaneously maintain a balanced diet. Calorie restriction was first shown to increase life span in mammals in the 1930s. Subsequent research confirmed that reduction in calorie intake resulted in an increase in longevity in mice, rats, fruit flies, yeast, worms, and fish. In certain rodents, a diet reduced by 30–40% of normal calorie consumption was found to increase life span by as much as 40%. A study in rhesus monkeys demonstrated that, over the course of the animals' lifetime, reducing calorie intake by 30% translated to visible delays in aging and gains in longevity. Furthermore, monkeys on calorie-restricted diets had a significantly reduced incidence of cardiovascular disease relative to those animals raised on unrestricted diets. The metabolic and stress responses induced by calorie restriction in primates require more research before these findings can be used to accurately predict the impact of a low-calorie diet on human longevity.

25.10.3 Sirtuins

Calorie restriction can activate genes known as sirtuins (*Sir2* in yeast, *Sirt1* in mice and *SIRT1* in humans). In yeast *Sir2* regulates genes across large segments of chromosomes. Studies have shown that in organisms maintained on fewer calories than normal, *Sir2* suppresses the activity of those genes, in effect reducing the likelihood of the genes' acquisition of mutations that contribute to aging. Similar effects of sirtuin occur in mammals. The development of drugs aimed at mimicking the effects of calorie restriction on the sirtuin gene in humans has been pursued for the treatment of age-related diseases, including some cancers and diabetes mellitus.

25.10.4 Rapamycin

A compound called rapamycin (sirolimus) can increase the life span of adult mice by up to 14% and of young mice by 28 to 38%. Rapamycin is an immunosuppressant agent valuable in the prevention of transplant rejection. It has also been investigated for use as an anticancer agent, since it can inhibit the proliferation of certain types of cancer cells. Similar to drugs under development for sirtuin activation, rapamycin may prove useful in the prevention and treatment of age-related disease in some people.

25.10.5 Stem Cells

Stem cells have a longer life span than other cells and retain a capacity to proliferate and differentiate (mature into specific cell types, such as epithelial or muscle cells). They also have the ability to resist and repair changes in the genome, enabling them to defend against the shortening of telomeres, a process that normally determines cell life span, and to prevent the accumulation of mutations. These mechanisms of defense are central to self-renewal.

Adult stem cells play an important role in organ homeostasis and regeneration, and these functions can be impaired by aging. The aging of stem cells can lead to their transformation, rendering them carcinogenic (able to cause cancer). Aging of stem cells can be caused by intrinsic molecular alterations, such as oxidative damage that leads to decreased mitochondrial function, or by extrinsic changes in the stem cell microenvironment. There is some evidence from studies of parabiotic pairings (anatomical or physiological union) of aged and young mice that the aging of stem cells can be reverted by exposure to a young systemic environment. Research has also suggested the transplantation of embryonic stem cells may have beneficial effects in treating aging-associated conditions such as Parkinson's disease.

REFERENCES

Arking, R. 1998. *The Biology of Aging,* 2nd Ed. Sinauer Associates, Sunderland, MA.

Boffoli, D., Sacco, S.C., Vergari, R., Solarino, G., Santacroce, G., Papa, S. 1994. Decline with age of the respiratory chain activity in human skeletal muscle. *Biochimica et Biophysica Acta* 1226: 73–86.

Fakouri, N.B., et al. 2019. Toward understanding genomic instability, mitochondrial dysfunction and aging. *FEBS Journal* 286: 1058–1073.

Kakkar, R., Bains, J.S., Sharma, S.P. 1996. Effect of vitamin E on lifespan, malondialdehyde content, and antioxidant enzymes in aging *Zaprionus paravittiger. Gerontology* 42: 312–321.

Lee, C.K., Klopp, R.G., Weindruch, R., Prolla, T.A. 1999. Gene expression profile of aging and its retardation by caloric restriction. *Science* 285: 1390-1393.

Lin,Y.J., Seroude, L., Benzer, S. 1998. Extended life-span and stress resistance in the *Drosophila* mutant methuselah. *Science* 282: 943–946.

Lipsitz, L.A., Goldberger, A.L. 1992. Loss of 'complexity' and aging: Potential applications of fractals and chaos theory to senescence. *JAMA* 267(13): 1806–1809.

Michikawa, Y., Mazzucchelli, F., Bresolin, N., Scarlato, G., Attardi, G. 1999. Aging-dependent large accumulation of point mutations in the human mtDNA control region for replication. *Science* 286: 774–779.

Migliaccio, E. 1999. The $p66^{shc}$ adaptor protein controls oxidative stress response and life span in mammals. *Nature* 402: 309–313.

Parkes, T.L., Elia, A.J., Dickinson, D., Hilliker, A.J., Phillips, J.P., Boulianne, G.L. 1998. Extension of Drosophila lifespan by overexpression of human SOD1 in motoneurons. *Nature Genetics* 19: 171–174.

Rudolph, K.L., Chang, S., Lee H.W., Blasco, M., Gottlieb, G.J., Greider, C., DePinho, R.A. 1999. Longevity, stress response, and cancer in aging telomerase-deficient mice. *Cell* 96: 701–712.

26 Experimental Physiology

B. Deepika

Department of Veterinary Physiology and Biochemistry, Veterinary College and Research Institute, Theni, Tamil Nadu Veterinary and Animal Sciences University, Tamil Nadu, India

26.1 INTRODUCTION

Physiology is an experimental science. Physiological research contributes to our understanding of the intricate processes that govern and control the behaviour of living organisms. Through a thorough investigation of the various systems involved, we continue to learn more about basic functions, such as the regulation of heart rate or vision.

There are many biological similarities between humans and animals. Therefore, it is thought that studying animals gives us a platform to comprehend the phases of a typical biological process or the progression of a disease, enabling researchers and students to identify the underlying events in a systems-wise approach.

Since the turn of the century, the use of animals in physiology classes for undergraduate and graduate students in India has remained contentious, leading to a hostile climate brought on by polarised opinions. However, it is important to consider and evaluate whether the current educational model, which ostensibly does not expose students to real animal experiments, really gives Indian graduates in the basic and medical sciences the chance to fully comprehend the intricacies at all levels of physiological comprehension.

26.2 AN OVERVIEW OF THE CURRENT SCENARIO

According to a global review on the use of animals in biomedical science education, live animals are still used to teach animal physiology, pharmacology, and behaviour, along with computer simulations, models, and videos, all in accordance with a policy that upholds respect for all living things. It is recognised that this type of training helps biomedical scientists to:

(i) cultivate a respect for living things;
(ii) sharpen their ethical sense; and
(iii) Equips them to support the promotion of informed public opinion regarding living things and the environment.

Live animal dissections and experiments are taught at primary and secondary levels in the United States. "The appropriate and humane use of animals in the elementary and secondary classrooms can provide significant educational benefits to the students," according to the American Association for Laboratory Animal Science. "A positive interaction between students and animals in the classroom enhances not only scientific learning but also provides an additional avenue promoting the development and growth of the students' sense of responsibility and respect for all living things," the organisation states. Many different fields at universities, such as anatomy, anaesthesiology, biochemistry, pharmacology, physiology, zoology, surgery, neurosurgery, and urology, use animals as teaching aids.

In Europe, animals are used extensively in basic biological science, medicine, veterinary medicine, and dental research and development

However, in India, biology, life sciences, and medical sciences are taught in a completely different manner than in Western countries. The use of animals in undergraduate and postgraduate teaching of zoology and life sciences has ceased since the ban issued in 2014 by the University Grants Commission, a statutory body established by the Indian Union Government and charged with the coordination, determination, and maintenance of higher education standards in India.

The Medical Council of India (MCI), which oversees medical education in the majority of public and private medical institutes, banned animal dissection in undergraduate medical courses in 2014. In a gazette notification (No. MCI-34(41)/2013-Med./64022 clause A.1.9), MCI advised using computer-assisted modules to teach Physiology and Pharmacology in the undergraduate curriculum. However, in 2019, the MCI approved animal experiments in postgraduate physiology training as part of its competency-based medical education programme.

26.3 BACKGROUND AND IMPACT

The revolutionary discoveries of earlier centuries gave rise to the Age of Enlightenment. Vesalius, a Flemish physician and anatomist, is frequently recognised as the father of anatomy. Greek physician Galen (c. 129–216 AD) had already developed basic medical theories that were taken for granted for centuries. The understanding of physiology by Galen was greatly impacted by the works of Hippocrates (460-377 BC), Aristotle (384-322 BC), and Plato (427-347 BC). Galen believed that the balance of four humours (blood, yellow bile, black bile, and phlegm) was essential to human health.

DOI: 10.1201/9781003426851-26

Each humour had two elements from fire, earth, air, and water and had two primary qualities: hot, cold, wet, and dry. Galen believed that abnormalities in these humours might harm individual organs or the entire body. He provided precise medicines to restore the body's homeostasis. Galen's views ruled medicine until the mid-1600s.

The animals utilised in various studies and inquiries are linked to the evolution of humanity. Although there is evidence that Aristotle used animals to understand the human body in ancient Greece, the main breakthrough in animal models occurred in the eighteenth and nineteenth centuries with scientists such as Jean Baptiste Van Helmont, Francesco Redi, John Needham, Lazzaro Spallanzani, Lavoisier, and Pasteur who studied the origin of life using animal models. At the same time, human physiology, anatomy, disease, and pharmacology were explored using animal models.

Since therapeutic success and drug safety are the most crucial factors for a medication or medical device considered for use in the human model, the significance of animal models has increased significantly in recent decades due to impressive advancements in drug discovery, biomedicine, and pre-clinical trials. Since animals' physiologies and morphology are so similar to those of humans – especially mammals – animal models have been used scientifically for ages to study biology and develop new medications.

26.4 REGULATIONS OF ANIMAL EXPERIMENTATION

Animal experimentation in India is governed under the Prevention of Cruelty to Animals Act (PCA Act) of 1960, as well as the Rules enacted under the revised Acts of 1998 and 2001. This is carried out by the "Committee for the Purpose of Control & Supervision of Experiments on Animals (CPCSEA)." It is a legislative body established in 1964 under Section 15(1) of Chapter 4 of the PCA Act by the Ministry of Environment, Forestry, and Climate Change. The CPCSEA sets rules for conducting animal research and maintaining animal houses.

Animal houses must be registered with the CPCSEA and renewed every three years. Currently, 1723 facilities in India have registered with CPCSEA to undertake animal studies. In addition to the CPCSEA's norms and procedures, the Indian National Science Academy (INSA) and the Indian Council of Medical Research (ICMR) have developed recommendations for the care and use of animals in scientific research and medical colleges. In 2003, the Pharmacy Council of India (PCI) directed all Indian pharmacy colleges to replace classroom animal experiments with computer-aided learning software (Figure 26.1, Figure 26.2, Figure 26.3, Figure 26.4 & Figure 26.5).

Also in 2003, MCI's Executive Committee concluded: "As an alternative to these animal testing, JIPMER, Pondicherry, has developed EX-PHARM Blank CD. This CD has been carefully produced as a complete replacement for animals used in undergraduate courses in Medicine, Pharmacology, and Veterinary Science." More recently, in 2011, the University Grants Commission announced directions to end dissection and experimentation on live animals in zoology and life science courses. Since 2013, India has prohibited the testing of cosmetic items sold in the country, becoming the first country in South Asia to do so. Not only that, but India has prohibited the import of animal-tested cosmetics from other countries! India has also prohibited animal testing of soaps and detergents manufactured in the country.

In April 2016, Union Minister Maneka Gandhi and the Indian Ministry of Health and Family Welfare banned animal testing on domestic products manufactured in India.

26.5 SELECTION OF ANIMAL MODELS

The selection of models that best reflect the physiological and/or pathophysiological similarities between human and animal species is the most crucial criterion. A thorough assessment of the ability of specific medications, chemicals,

FIGURE 26.1 Frontier Mediville – Animal Facility is registered with CPCSEA (Registration number: 871/PO/ReRcBiBtS/ReRcL/2005/CPCSEA).

FIGURE 26.2 Large animal housing faciliy – Entrance I.

FIGURE 26.3 Large animal housing facility Exit II.

or technology to replicate a disease or pathology at a level comparable to human beings. The size and availability of the animal species in question. Long lifespan of the studied animal species. Many sub-species are easier to find in a model with a larger animal population. Numerous animal species, including Xenopus (frogs), Danio rerio, or zebrafish (fish), Drosophila (insects), and Caenorhabditis elegans (nematodes), as well as mammals, including mice, rabbits, rats, cats, dogs, pigs, and monkeys, have gained global acceptance due to their evolutionary similarity to humans. Choosing an acceptable animal model is sometimes a time-consuming task that is based on assumptions and the convenience of the study and researchers without regard for whether the model is appropriate or not. Irrational selection of an inadequate animal model for scientific investigations will result in inaccurate results as well as misallocation of resources and lives. Furthermore, it leads to erroneous, redundant, and inappropriate experiments. Recent advances in research have allowed scientists to create animal models that are highly specific to the study being conducted, which helps to reduce these issues. Through the use of retroviral vectors or direct injection of foreign DNA into the embryo, they created bespoke transgenic animal models. Researchers can examine how diseases affect the organism in a manner akin to that of humans by introducing human cells into the recipient animals. It is primarily dependent on the type of medication or medical device being studied when choosing animal models. Several animal models together may be able to represent a technique when one animal model is unable to adequately represent a human disease.

26.6 ANIMAL MODELS

Animals used as research models are classified into two groups:

The small animal category requires approval from the local animal ethical commission.

- Rats
- Mice
- Guinea pigs
- Rabbit (largest animal in this group)

FIGURE 26.4 Small animal testing facility.

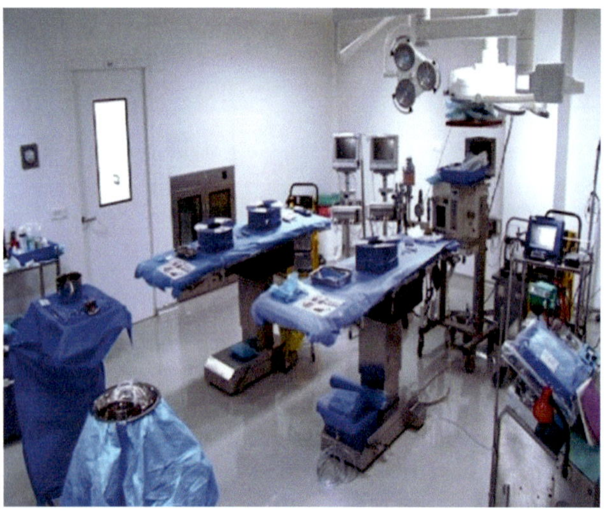

FIGURE 26.5 Operation suite.

Certain large animal categories require clearance from both the central and local animal **ethics committees**.

- Dogs
- Goats
- Primates

The use of animals in scientific study and treatment is a long-standing tradition that is frequently debated in our communities. The striking anatomical and physiological similarities between humans and animals, particularly mammals, have spurred researchers to examine a wide range of systems and test innovative therapies in animal models before translating their findings to humans. However, not all results acquired on animals can be immediately translated to people, and this insight is emphasised by those who deny the relevance of animal research. In addition, there is constant discussion over the role that animals play in our contemporary cultures, especially in relation to the permissible use of animals for human advantage when there is a chance that the animals would suffer as a result. Politicians and the general public are not helped to understand the issues by the frequent mixing of these two factors in convoluted debates. This was especially true during the assessment of the "Stop Vivisection" European Citizen Initiative (ECI) that was just submitted to the European Commission.

The animal model applies to non-human species because it can replicate the course of an illness, its diagnosis, and a human-like course of therapy. Finding a drug or component, apparatus, toxicological research, dosage, and adverse effects are all investigated in vivo before being used on humans in the future, taking ethical considerations into account. This highlights the significance of the animal model and its wide application in biological research.

Numerous aspects of animal models closely resemble human illness circumstances, including autoimmune disorders of the system, rheumatoid arthritis, epilepsy, Alzheimer's disease, cardiovascular diseases, atherosclerosis, diabetes, and many more. In addition, the model is extremely valuable for research on bone and cartilage regeneration, tissue engineering, medication development, medical device development, wound healing, and vascular and spinal disc regeneration surgery. Despite this, the use of animal models in research has allowed scientists to conduct studies that, given ethical constraints, would not have been feasible to complete on humans.

Mammals, including humans, are extremely complex animals with highly interconnected and controlled organ systems performing various physiological roles. A complex network of hormones, circulatory substances, cells, and cross-talk between cells in every compartment is involved in relationships. Biologists examine organisms on a variety of levels, including molecules, cells, organs, and physiological processes in both healthy and sick states. To fully describe and comprehend the mechanics, all levels of inquiry are necessary. Through the use of in vitro techniques (such as cell culture), the first two and occasionally the third levels of organisation can be investigated. These days, very advanced procedures are used to replicate the intricate and three-dimensional structures of tissues. They have supplanted the usage of animals and represent significant scientific advancements. However, investigating physiological processes and the relationships across organ systems necessitates an entire organism. This is true, for instance, of the majority of hormone controls, the spread of microbes during infectious disorders, the impact of gut microbes on immune response, and the development of brain functions. In many of these instances, research on humans and animals is still required because there is yet no in vitro model that can accurately replicate these interactions.

In vitro research can lead to the emergence of hypotheses and models, but these must then be verified and evaluated in a complete organism to avoid remaining theoretical. The functioning of a complex organism cannot be predicted by scientists based only on the study of individual cells, tissues, and organs. Therefore, in vitro approaches cannot yet totally replace animal studies, and it will be some time before they can, despite the claims made by proponents of the ECI.

A wide range of scientific issues, from fundamental research to the creation and evaluation of innovative vaccinations or treatments, have been addressed through the use of animal models. Animals are used for research not just because human diseases frequently affect other animal species but also because most mammals share a large number of biological similarities. In particular, this is true for the majority of viral diseases as well as for many common ailments like Type I diabetes, high blood pressure, allergies, cancer, epilepsy, myopathies, and so forth. Not only do these illnesses share same causes, but 90 percent of veterinary medications used to treat animals are the same or extremely similar to those used to treat humans due to these related mechanisms. Tests using animal models and observations have led to some significant advances in basic science and medical research. The majority of vaccinations have been successfully created using animal models, saving millions of lives each year, both human and animal. Insulin therapy for Type I diabetes was initially developed in dogs by Banting and McLeod, who were awarded the Nobel Prize in 1921. Animal tests have been conducted on engineered cellular therapies that use stem cells to regenerate tissue. Several surgical methods have been developed and refined in a variety of animal species prior to being used on humans. Animal models have been essential in many discoveries that have resulted in multiple Nobel Prizes.

It is apparent, therefore, that additional research on humans may not always corroborate the findings from studies conducted on animals. One can evoke several explanations. First, despite their many similarities, different animal species and humans have unique characteristics. For instance, more than 95% of the genes in humans and mice are similar, but there are some variations as well, such as in the members of gene families, in gene redundancies, and in the precise control of gene expression levels. These genetic variations result in physiological variations that are becoming more well defined and comprehended. While some, like the proponents of ECI, use these distinctions to argue against the usefulness of animal models, many others, including ourselves, fervently support expanding our understanding of these distinctions and incorporating them into the planning of experiments and the interpretation of data. Furthermore, these variations can present chances to identify new pathways and develop creative treatments.

Genetic and physiological differences within species or between closely related species account for the second cause. Inbred strains of laboratory mice with a highly uniform genetic makeup were established in order to improve experiment statistical power and reliability. "The mouse model of..." is a term frequently used in reports on animal models of human problems, however it actually refers to observations performed in a particular genetic context. If an alternative strain of mice is taken into consideration, the clinical presentation frequently differs. A startling illustration comes from a research team's November 2014 publication in Science, wherein they found that certain mice strains are completely immune to the Ebola virus, while others die from the disease without exhibiting any symptoms and yet others suffer from deadly haemorrhagic fever. These examples show why it is important to take into account animal models: no single animal model can accurately represent a particular human disease, which is polymorphic in and of itself among patients; however, the variations among strains or species offer unparalleled opportunities to comprehend the development of diseases and the varying host response, and ultimately to discover novel treatments.

Animal welfare and protection are the second concern when it comes to using animals for scientific research. The European Directive 2010/63/EU, which established the guidelines for all animal research, covers this territory. Three basic concepts have been acknowledged by scientists for many years as being crucial, and they have formed the core of the European Directive. First, where there are equally relevant and reliable non-animal-based experimental methods available, they should be employed instead of animals in experiments. Secondly, the quantity of animals employed needs to be reduced to the bare minimum required to draw a conclusion. Third, every precaution needs to be used during the process to reduce any damage done to the animals. These guidelines – also referred to as "the three Rs rules" – for replacement, reduction, and refinement have developed into the benchmark by which all projects involving the use of animals are assessed.

Animal research is carried out in accordance with legal regulations that address the licensing and inspection of animal facilities, the education and experience of project designers, the performance of animal procedures and animal care, and the requirement that all projects be approved by a responsible body following an ethical assessment by an Animal Ethics Committee. The evaluation criteria are founded on the 3Rs guidelines and a cost-benefit analysis to determine whether substantial advancements in our understanding of human or animal health outweigh the potential harm to animals, which must be minimised to the lowest feasible degree. Ethics committees are required by law to have members who care about animal welfare but are not engaged in animal experimentation. In a statement released on June 3, 2015 [8], the European Commission responded to the ECI by highlighting the continued importance of animal experimentation in advancing both human and animal health. In addition, it is dedicated to encouraging the creation and approval of non-animal methods and to making sure that everyone, even the scientific community, abides by the 3Rs. As a result, Europe has put in place one of the tightest legal structures to safeguard research animals.

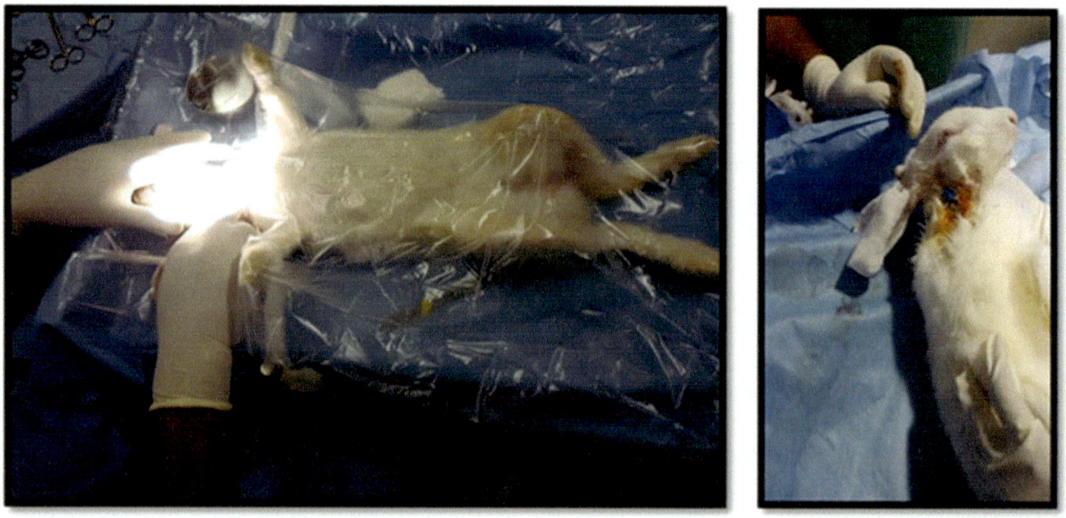

FIGURE 26.6 Rabbit jugular vein catheter implantation – Frontier Mediville.

26.7 EXPERIMENTS ON LABORATORY SMALL ANIMALS

The remarkable significance of rat and mouse models in biomedical research has been demonstrated. In addition, other small domestic animals, both mammalian and non-mammalian, such as hamsters, rabbits, ferrets, guinea pigs, amphibians, fish, flies, and worms, are equally important in terms of their anatomical and physiological similarities to humans.

Approximately 95% of laboratory animals are mice and rats, with mice being the most often utilised animal in biomedical research. For a number of reasons, including their small size (which makes them easier to house and maintain), short lifespan and reproductive cycle, generally mild-mannered and docile nature, abundance of information about their anatomy, genetics, biology, and physiology, and the potential to breed genetically altered mice and mice with spontaneous mutations, mice are frequently used as animal models.

Mice have been employed as study animals in a variety of fields, including engineering, psychology, and biology. In an effort to discover remedies or cures, they are employed to simulate human illnesses. Hypertension, diabetes, cataracts, obesity, seizures, respiratory issues, deafness, Parkinson's disease, Alzheimer's disease, several malignancies, heart disease, muscular dystrophy, and spinal cord injuries are a few of the illnesses they simulate. In addition, mice are employed in research on behaviour, senses, ageing, diet, and genetics. The domestic mouse is being used in study by biologists, geneticists, and other scientists at a rapid pace, therefore this list is by no means exhaustive.

A common choice for mammalian research models is the domestic mouse, Mus musculus and related subspecies, because to its small size, short lifespan, adaptability, docility, inexpensive husbandry expenses, fertility, and clearly characterised health and genetic backgrounds. Genetic modification, such as "humanization" through the use of human genes, tumours, immune cells, or microbiome components, is also comparatively simple. The use of mice as study subjects has been greatly expanded and enhanced by the advent of embryonic genome editing techniques, most notably the CRISPR-Cas9 system and next-generation prime editing, in the production of GEM (genetically modified model) mice. These developments have led to the development of innumerable mutant genotypes of mice, which range from minor immune function abnormalities to severe hereditary disorders that are almost exactly similar to those that affect humans and other mammals (Figure 26.6).

26.8 EXPERIMENT ON LABORATORY LARGE ANIMALS

Farm animals, or FAs, are commonly utilised in scientific research for investigations into animal husbandry, basic safety, veterinary and agricultural science, and comparative and translational medicine. Since many organs' anatomical features, sizes, and functions are essentially the same in FAs and humans, FAs make highly interesting models for studying human diseases. Examples of this include the use of sheep as a surgical model in orthopaedic studies and pigs in cardiovascular or transplantation research. Moreover, transgenic FA models are getting easier to obtain and are being employed more often. FAs are especially helpful comparison models since they can produce results with great biological relevance and, in some cases, are even dual-use because they promote improvements in both animal and human health (Figure 26.7 & Figure 26.8).

26.9 RULES OF EXPERIMENTS ON ANIMALS

Similar to humans, animals are capable of perceiving external stimuli, which means they can experience discomfort and perhaps suffer. It is difficult to measure an

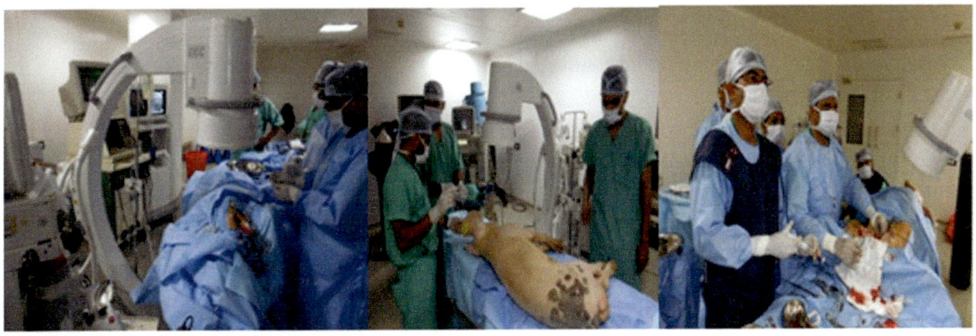

FIGURE 26.7 Coronary stent implantation in a Porcine model at Frontier Mediville.

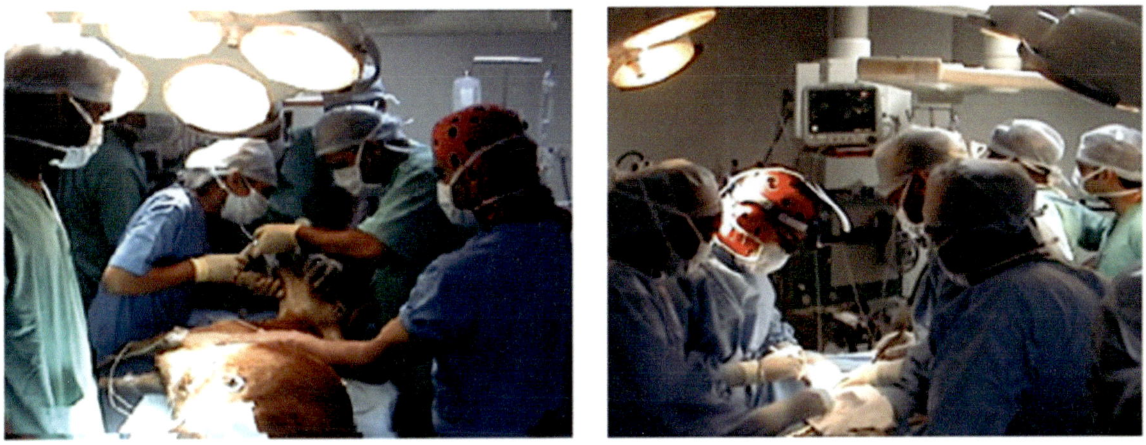

FIGURE 26.8 To assess the efficacy of an indigenously designed mini heart-lung machine in a sheep model at Frontier Mediville.

animal's level of suffering objectively, although it is generally acknowledged that vertebrates – especially mammals – struggle with some degree of suffering. Consequently, the experimenter has a fiduciary duty to prevent suffering from the subject of the experiment or, in the event that it cannot be avoided entirely, to minimise it. The body overseeing animal welfare has reviewed and authorised all of the practical's experiments.

The following considerations need to be made while organising and carrying out animal studies.

- Animals utilised in research should be fed correctly, housed in suitable conditions, and given the required social interactions. Veterinary treatment should be provided by professionals to sick animals.
- The academic background and experience of the researchers and educators supervising the experiments should be suitable, and they should be knowledgeable about the bioethical concerns surrounding animal experimentation.
- The duration of animal tests should be minimised;
- Muscle relaxants without general anaesthesia should only be used if pain elimination is also guaranteed;
- The experiment must have a strong scientific foundation.
- Only establishments authorised by the relevant authorities may conduct animal experiments,
- Adhere to the three Rs of humane animal experimentation:
  - **Reduce** the number of animals used in tests.
  - Attempts should be made to **replace** animal experiments with tissue cultures or microbiological preparations.
  - Experiments should be **refined** to reduce the pain and discomfort that an unanaesthetized animal must undergo without the ability to avoid it.
- Experiments may only utilise animals obtained from an official supplier or locally produced; however, captured wild animals may also be used in some cases. However, conducting studies on stray individuals of pet species (dog, cat) is strictly prohibited.

Authors of scientific journals must provide a declaration confirming compliance with animal experimentation laws and regulations. Manuscripts detailing experiments where these principles were ignored or violated should be rejected.

26.10 ANIMAL EXPERIMENTS

Studying the physiology of organs, organ systems, and the organism as a whole under both natural and experimental situations is the primary goal of the physiology labs. Research is consistently conducted on living subjects or isolated, functional organs. Sufficient physiological conditions ought to be supplied in both situations. Instrumental measurement and recording enhance and support visual observation. You should document your findings, observations, and conclusions in your notebooks.

26.10.1 Anaesthesia and Pain Relief

Several methods exist to prevent pain perception during an operation. Local anaesthesia refers to the reversible inhibition of synaptic transmission or nerve conduction in the operative region. This prevents pain signals from reaching higher centres in the central nervous system. General anaesthesia refers to the total inhibition of synaptic transmission in the nervous system. Adequate anaesthesia means the prevention of pain perception while not interfering with other physiological activities. These characteristics can be addressed by employing agents with a diverse range of actions. Current narcotics achieve anaesthesia levels in the brain, subcortical regions, cerebellum, medulla, and spinal cord, in that sequence (Figure 26.9).

Medication effects can be controlled by carefully choosing the dosage. Cumulative tables provide the anaesthetic dosages per body weight for the most common laboratory animals. The range of efficacy, or the difference between the median lethal dosage (LD50) and the median effective dose (ED50), is also displayed in the tables. (A dose known as the LD50 is one that is fatal for 50% of test participants, whereas the ED50 is a dose that acts in 50% of a population.) The use of the particular medicine is less dangerous the greater this range is. Both the tone of the muscles and the respiratory and circulatory systems are maintained under sufficient anaesthesia (a muscle relaxant is administered to facilitate the release of the latter). The majority of issues that arise during anaesthesia are breathing-related. It's also critical to understand that anaesthesia causes a drop in body temperature. Animals undergoing lengthy procedures should therefore be kept warm. The majority of anaesthetics are fat-soluble, meaning that they work by altering the structure of nerve cells' cell membranes.

General anaesthetics are classified as follows:

- volatile liquids (ether, chloroform, halothane)
- gases (nitrous oxide, ethylene, cyclopropane)
- intravenous aesthetic drugs, solutions (barbiturates, pregnanedion, ketamine and urethane)

Volatile or gaseous anaesthetics are commonly utilised in surgery nowadays. In animal research, however, solutions are preferred because they are simple. They can be delivered intraperitoneally (i.p.) or intravenously. The first method is easier, while the second is more effective and allows for more exact drug administration. Depending on the dose and duration of anaesthesia, distinct depths of anaesthesia are distinguished:

- **Analgesic time:** There is no pain perception and a lowered response threshold.
- **Exciting period:** Loss of cortical inhibition, absence of consciousness, unconscious escape reactions, speaking sounds, and heightened vegetative functions.
- **Period of Tolerance:** Surgical procedures are favoured. Subcortical motor centres are blocked, resulting in diminished tone and extinction of somatic reflexes, including the cornea reflex.

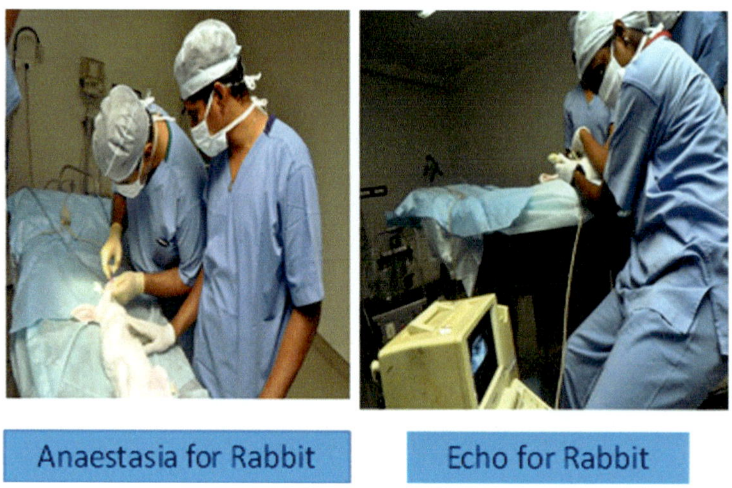

FIGURE 26.9 Myocardial patch for rabbit model at Frontier Mediville.

- **A period of respiratory arrest** with no medulla activity.

In physiological demonstrations, many anaesthetics are utilised. Short surgeries may be performed under ether anaesthesia (inhalation technique). For longer operations, ketamine with xylazine, nembutal (sodium pentobarbital), or urethane (ethyl carbamate) are employed.

Urethane is only used in operations that end with the animal being sacrificed. Urethane is a powerful aesthetic for long-term experiments because it influences a wide range of neurotransmissions. Chloralose, commonly in the form of a urethane-chloralose cocktail, is employed in various neurological investigations. It has no effect on reflex or cerebral electrical activity, but its administration is difficult and takes some skill.

Procaine or lidocaine local anaesthetics typically block nerve conduction.

26.10.2 Immobilisation of the Animal

In order to execute the procedures precisely and safely, the animals should be attached to the operating board after general anaesthesia or decapitation (frog). Depending on the species and the kind of procedure, there are many ways to fix their bodies. When it comes to frogs, puncturing the limbs to the waxy bottom of the dish can effectively immobilise them. On the other hand, while dealing with rats, the limbs must be secured to the operating board using thread. It is also recommended that the animal's head be held in place by the upper incisors. When operating with non-anaesthetized (alert) rats (for example, drawing blood samples from the tail vein), the animal is placed in a cone about 5 cm in diameter that only allows access to its tail. Alert rabbits can be investigated in a stock. Rabbits are the least sensitive to pain and the most tolerant to medical procedures of any experimental animal. Before operation under anaesthesia, the rabbits' forelimbs must be knotted crosswise beneath their backs and their hindlimbs extended apart. The position of the head should also be secured using a thread put between the upper incisors.

26.10.3 Injections

In physiological experiments, parenteral injection of medicines, fluids, and other substances may be required. In these circumstances, agents should be injected. Types of injections include:

26.10.3.1 Subcutaneous (S.C.)

Lift the skin slightly and pierce it in one step. Check for tissue injury and inject the fluid. If there is a big amount of fluid, draw the needle back slightly to prevent it from pooling in this location. Frogs have lymph sacs under their skin. The injected fluid is rapidly absorbed by them. The lymph sacs are divided by septa.

26.10.3.2 Intramuscular (I.M.) Injection

This mode of administration allows for faster and more equal medication absorption. Injections also cause less pain. Injections should be administered immediately and with a forceful movement.

26.10.3.3 Intraperitoneal (I.P.) Injection

For tiny laboratory animals, the abdominal cavity is the recommended place for injections because alternate routes of administration are difficult to implement. Injection involves penetrating the peritoneum. Avoid injuring the internal organs. Drugs delivered intraperitoneally are swiftly absorbed, but their efficacy is significantly influenced by the amount of chyme in the gastrointestinal tract. The doses stated in the table apply to animals that were hungry for 24 hours before to I.P. injection; medication effects cannot be determined in animals that were not starved prior to the experiment.

26.10.3.4 Intravenous (I.V.) Injection

It enables fast and precise administration. Its downside is the difficulty of finding an appropriate vein. Rats and mice use their tail veins. It can be somewhat dilated with warm water. Only a thin needle will yield an effective result. The rabbit's ear vein is easily localised and can be suitably dilated using a xylene-soaked cotton swab. During this procedure, the rabbit is placed in stocks. Injection of cats and dogs necessitates extensive understanding of the various procedures. It is also possible to execute it in an alert condition with tame animals (especially dogs). Otherwise, it is recommended to use a short-term pre-anaesthesia, such as ether. When administering an intravenous injection, be sure to put the needle parallel to the vein to avoid passing through the vessel wall. Compression of the proximal end of the vein causes a buildup of blood at the injection site, making it much easier to maintain the needle level. If surgery requires numerous intravenous injections, it is advisable to place a venous cannula in pre-anaesthesia.

26.10.4 Preperation of the Animal for Experiment

The animal's fur disrupts surgical labour since it contaminates the region of operation and, in chronic operations, it can become a source of infection. As a result, it should be shaved off before the surgery. Hairs in rats and mice can be plucked off by hand or with forceps (they regenerate within a few weeks following chronic surgeries). With larger animals (rabbit, cat, or dog), the fur can be shaved with scissors (Cooper's scissors), but be careful not to hurt the skin. Cut hairs can be removed using a moist cotton swab. A easier alternative is to use a small animal clipper with a cutting head designed to leave just extremely short hairs. Before performing chronic surgeries, the region to be operated on should be painted with iodine or alcohol. This is unnecessary for acute experiments.

26.10.5 Types of Operation

In physiological demonstrations, we usually use acute interventions, which involves killing the animals at the end of the trial. There are other experiments, however, when chronic operations are required. This means that the animals are taken out of anaesthesia and allowed to recover. The most important sorts of operations performed in physiological studies are listed below.

26.10.5.1 In Situ Preperation: Isolated Organ

The organ that will be studied is taken out of its original location and placed within a container. It performs admirably in the synthetic environment of the organ bath. "Straub's heart preparation" and "The Magnus-type isolated intestinal segment" are two examples of isolated organ preparations.

An organ is removed entirely or is destroyed in part. The effects of an organ's loss or injury on function are examined. Using this method, the functions of the endocrine glands and specific brain regions can be investigated.

26.10.5.2 Cannulation

It is possible to inject drugs or drain fluids from organ cavities by inserting cannulas. For instance, an arterial cannula can be used to measure changes in blood pressure, while the cannula placed into the trachea guarantees that the animals under anaesthesia can breathe without problem. Similar to this, gastric and intestinal cannulas can be used to take samples of digestive fluids, while venous cannulas can be used to inject fluids into the body. A ureter cannula can be used to collect urine samples.

26.10.5.3 Implantation of Electrodes

Electric stimulators (electrodes) are primarily applied to nerves. The vagus nerve is stimulated to investigate its functions in the heart, lung, and stomach, whereas the ischiadic nerve is stimulated to investigate sensory and motor processes. Electrodes are linked to an appropriate recording apparatus to make recordings.

Naturally, the basic types of operation can be used in combination. A sophisticated physiological experiment may necessitate cannulation, electrode insertion, and organ isolation.

26.10.5.4 Physiological Requirements

To examine animal physiology, appropriate physiological circumstances must be given during the investigations. Three requirements are particularly important. Administered drugs, organ bath, temperature, respiration, heart activity, and blood pressure of animals are all important factors to consider.

"Physiological saline" is the basis for some of the fluids that are injected into the body or used to perfuse isolated organs. Osmotic pressure, ion content, and pH value-wise, they are comparable to the bodily fluids they are meant to replace. Their true makeup may differ according on the experiment's objectives.

It is necessary to modify the temperature of the injected fluids in addition to their pH value. It is ideal to have a normal body temperature, especially while administering medication intravenously. Controlling the body temperature of animals under anaesthesia is also necessary. A hot water bottle or an electric pad can be used to keep the body temperature of smaller animals at a comfortable level (warming is especially important in the "Complex Physiological" and "Sensory Evoked Potential" tests). Warming up bodily cavities that have been surgically exposed is essential because interior organs are more susceptible to temperature drops. If this isn't possible, cover the exposed wound with a cotton swab that has been moistened in warm physiological saline.

One possible cause of respiratory failure is anaesthesia. Pre-operative tracheal cannulation can prevent difficulties such as saliva getting into the trachea of the animal lying on its back or the tongue bending back and obstructing the pharynx, which can exacerbate the condition. If this isn't possible for any reason, give atropine sulphate to lessen the amount of saliva produced.

26.11 GENERAL DESIGN OF EXPERIMENTAL SURGERY

The steps involved in preparation include choosing the animal, administering anaesthesia, securing the animal's position on the operating table, and cleaning the surgery environment.

26.11.1 Preperation of the Surgical Instruments

Surgical exposure involves cutting the skin using scissors or a scalpel. Generally, to prevent excessive bleeding, muscle bundles are divided by blunt dissection; however, in the case of the abdominal wall, the tendinous linea alba can be cut with scissors or a scalpel. Using bone forceps and a trephine, the skull is revealed.

Hemostasis, or stopping bleeding caused by cutting, can be achieved in a variety of ways depending on the kind of bleeding. For example:

- bleedings from capillaries can be mopped up;
- vessels can be temporarily constricted by artery forceps (such as Pean's, Kocher's, and mosquito forceps) to promote hemostasis;
- ligation, in which the only way to stop persistent bleeding is to pass a silk thread beneath the bleeding vessel, being cautious not to sever surrounding tissue;
- use of auxiliary materials, such as "fibrostan," a sponge containing fibrin-like substance to encourage coagulation;
- application of bone wax, an aseptic surgical wax consisting of boiled beeswax and paraffin to the bone's cut surface to halt bleeding;

- cauterisation, which is frequently employed in major operations to coagulate tissue proteins by high frequency alternate current flowing between the cautery's small surface and a large reference electrode.

However, the very goal of some research is to avoid coagulation. It is possible to stop blood from clotting in blood samples or blood that has entered the tonometer's cannula by applying citrate, which can bind Ca2 + ions, or heparin, a physiological anticoagulant.

26.11.2 Preparation

The anatomical region to be researched is made visible by removing or separating any surrounding tissues, if needed. The principal arteries of the organ that has to be removed – the heart, for example – is cut between two ligatures. Connective tissue should be removed from vessels chosen for cannulation and nerves to be used for stimulation or recording. If required, the prepared area should be secured in place with a thread for any additional manipulation (e.g., ligation, pulling up, etc.). Similarly, sutures are used to fix electrodes or cannulas. Make sure the organ is not damaged or twisted.

26.11.3 Postoperative Treatment

The animals are covered and brought to a warm chamber until they awaken following lengthy procedures. The administration of medications, including antibiotics, must be based on need. The animals should be fed light food on the first post-operative day before being gradually weaned back to regular diet. After the procedure is finished, collect and dispose of the debris. Following acute procedures, animals should be put to death with an aesthetic overdose and disposed of in accordance with the law. Washing and rearranging surgical tools, cannulas, dishes, and benches is recommended. The practical ends after all electrical equipment has been turned off, the workspace and surrounds have been cleaned.

26.12 CONCLUSION

Considering that the future of physiology education lies in the administration of strategically blended courses that use animal experiments in a controlled and restricted manner in conjunction with virtual, animal-based, and computer-assisted interactive learning at undergraduate and postgraduate levels, teachers have a specific responsibility to develop courses and contents that align with the aforementioned blending method of teaching-learning physiology while putting the new strategy into action. Furthermore, it is imperative that physiology students, both undergraduate and graduate, receive adequate exposure to the ethical and legal aspects of animal experimentation, as well as the technical, statistical, and administrative aspects of this practice. This is particularly important for students who may choose to pursue careers as medical or wildlife management researchers. Instructors must think carefully about how humane scientific teaching approaches might meet learning objectives while also involving students in the ethical issues surrounding the usage and dissection of animals. All stages of physiological science education must use an inquiry-based approach with the goal of fostering "a view of science as a way of knowing, rather than a body of knowledge," by allowing students to participate in the scientific process.

BIBLIOGRAPHY

Andersen ML, Winter LMF. Animal models in biological and biomedical research - experimental and ethical concerns. *An Acad Bras Ciênc.* 2017;91(Suppl 1):e20170238.

Barraza-Flores P, Fontelonga TM, Wuebbles RD, Hermann HJ, Nunes AM, Kornegay JN, et al. Laminin-111 protein therapy enhances muscle regeneration and repair in the GRMD dog model of Duchenne muscular dystrophy. *Hum Mol Genet.* 2019;28(16):2686–95.

Barré-Sinoussi F, Montagutelli X. Animal models are essential to biological research: issues and perspectives. *Future Sci OA.* 2015;1(4)

Bertoloni MD. Early modern experimentation on live animals. *J Hist Biol.* 2013; 46:199–226.

Carroll RG. Using animals in teaching: APS position statement and rationale. *Physiologist* 2005;48:206–8.

Cheng L, Liu Y, Zhao H, Zhang W, Guo Y-J, Nie L. Lentiviral-mediated transfer of CDNF promotes nerve regeneration and functional recovery after sciatic nerve injury in adult rats. *Biochem Biophys Res Commun.* 2013;440(2):330–5.

Cook J, Fox D, Malaviya P, Tomlinson J, Farr J, Kuroki K, et al. Evaluation of small intestinal submucosa grafts for meniscal regeneration in a clinically relevant posterior meniscectomy model in dogs. *J Knee Surg.* 2006;19(3):159–67.

Dam DV, Deyn PPD. Animal models in the drug discovery pipeline for Alzheimer's disease. *Br J Pharmacol.* 2011;164(4):1285–300.

Ernst W. Humanized mice in infectious diseases. *Comp Immunol Microbiol Infect Dis.* 2016;49:29–38.

Farah Z, Fan H, Liu Z, He J-Q. A concise review of common animal models for the study of limb regeneration. *Organogenesis.* 2016;12(3):109–18.

Fitzpatrick N, Smith TJ, Pendegrass CJ, Yeadon R, Ring M, Goodship AE, et al. Intraosseous transcutaneous amputation prosthesis (ITAP) for limb salvage in 4 Dogs. *Vet Surg.* 2011;40(8):909–25.

Gad SC. Animal models in toxicology. In: Wexler P, editor. *Encyclopedia of toxicology.* Boca Raton: CRC/Taylor & Francis; 2005. p. 138–40.

Gutpell KM, Hrinivich WT, Hofman LM. Skeletal muscle fbrosis in mouse validates its suitability as a murine model of Duchenne muscular dystrophy. *PLoS One.* 2015;10(1):e0117306.

Hansen LA, Boss GR. Use of live animals in the curricula of U.S. Medical schools: Survey results from 2001. *Acad Med.* 2002;77:1147–9.

Heber-Katz E, Leferovich JM, Bedelbaeva K, Gourevitch D. Spallanzani's mouse: a model of restoration and regeneration. In: Heber-Katz E, editor. *Regeneration: stem cells and beyond.* Berlin: Springer; 2004. p. 165–89.

Leppik LP, Froemel D, Slavici A, Ovadia ZN, Hudak L, Henrich D, et al. Efects of electrical stimulation on rat limb regeneration, a new look at an old model. *Sci Rep*. 2015;5(1):18353.

Lewis DI. *The pedagogical benefits and pitfalls of virtual tools for teaching and learning laboratory practices in the biological sciences*. Heslington and New York, UK: The Higher Education Academy, STEM; 2014.

Michael J. Where's the evidence that active learning works? *Adv Physiol Educ*. 2006;30:159–67.

Modell HI. Can technology replace live preparations in student laboratories? *Am J Physiol*. 1989;256:S18–20.

Monamy V. *Animal experimentation: a guide to the issues*. Cambridge University Press; 2009.

Moran CJ, Ramesh A, Brama PAJ, O'Byrne JM, O'Brien FJ, Levingstone TJ. The benefts and limitations of animal models for translational research in cartilage repair. *J Exp Orthop*. 2016;3(1):1.

Oakley J. Science teachers and the dissection debate: Perspectives on animal dissection and alternatives. *Inter J Environ Sci Educ*. 2012;7:253–67.

Oliveira KMC, Barker JH, Berezikov E, Pindur L, Kynigopoulos S, Eischen, Loges M, et al. Electrical stimulation shifts healing/scarring towards regeneration in a rat limb amputation model. *Sci Rep*. 2019;9(1):11433.

Oparin AI. *The origin of life on the earth*. 3rd ed. New York: Academic Press Inc.; 1957.

Pehlivanovic B, Dina F, Emina A, Ziga Smajic N, Fahir B. Animal models in modern biomedical research. *Eur J Pharm Med Res*. 2019;6(7):35–8.

Quiroga MD, Choate JK. A virtual experiment improved students' understanding of physiological experimental processes ahead of a live inquiry-based practical class. *Adv Physiol Educ*. 2019;43:495–503.

Ra'anan AW. The evolving role of animal laboratories in physiology instruction. *Adv Physiol Educ*. 2005;29:144–50.

Raske M, McClaran JK, Mariano A. Short-term wound complications and predictive variables for complication after limb amputation in dogs and cats. *J Small Anim Pract*. 2015;56(4):247–52

Simmons D. The use of animal models in studying genetic disease: transgenesis and induced mutation. *Nat Educ*. 2008;1(1):70.

Simon F, Oberhuber A, Schelzig H. Advantages and disadvantages of different animal models for studying ischemia/reperfusion injury of the spinal cord. *Eur J Vasc Endovasc Surg*. 2015;49(6):744.

Thaman RG, Saggar S, Kaur H, Dhillon S, Gupta S, Kaur P, et al. Neuromuscular physiology experiments: Computer assisted learning versus traditional lecture demonstrations. *Pak J Physiol*. 2012;8:28–32

Udovic D, Morris D, Dickman A, Postlethwait J, Wetherwax P. Workshop biology: Demonstrating the effectiveness of active learning in an introductory biology course. *Bioscience* 2002;52:272–81.

Zaccagnini G, Gaetano C, Della Pietra L, Nanni S, Grasselli A, Mangoni A, et al. Telomerase mediates vascular endothelial growth factor-dependent responsiveness in a rat model of hind limb ischemia. *J Biol Chem*. 2005;280(15):14790–8.

Zaccagnini G, Palmisano A, Canu T, Maimone B, Russo FML, Ambrogi F, et al. Magnetic resonance imaging allows the evaluation of tissue damage and regeneration in a mouse model of critical limb ischemia. *PLoS One*. 2015;10(11). doi: 10.1371/journal.pone.0142111

Index

F

I

J

N

O

P

W

X

Y

Z